Controlling International Technology Transfer

The Technology Transfer Trilogy

INTERNATIONAL TECHNOLOGY TRANSFER: Guidelines, Codes and a Muffled Quadrilogue
by Perlmutter and Sagafi-nejad

CONTROLLING INTERNATIONAL TECHNOLOGY TRANSFER: Issues, Perspectives and Policy Implications
edited by Sagafi-nejad, Moxon and Perlmutter

TRANSNATIONAL CORPORATIONS, TECHNOLOGY TRANSFER AND DEVELOPMENT: A Bibliographic Sourcebook
by Sagafi-nejad and Belfield

Pergamon Titles of Related Interest

Bhalla TOWARDS GLOBAL ACTION FOR APPROPRIATE TECHNOLOGY
Davis MANAGING AND ORGANIZING MULTINATIONAL CORPORATIONS
Feld MULTINATIONAL CORPORATIONS AND UN POLITICS
Hill/Utterbach TECHNOLOGICAL INNOVATION FOR A DYNAMIC ECONOMY
Kidd MANPOWER POLICIES FOR THE USE OF SCIENCE AND TECHNOLOGY FOR DEVELOPMENT
Negandhi FUNCTIONING OF THE MULTINATIONAL CORPORATION
Sigurdson TECHNOLOGY AND SCIENCE IN THE PEOPLES REPUBLIC OF CHINA
Standke/Anandakrishnan SCIENCE, TECHNOLOGY AND SOCIETY

Related Journals*

BULLETIN OF SCIENCE, TECHNOLOGY & SOCIETY
ENVIRONMENT INTERNATIONAL
FUTURICS
HABITAT INTERNATIONAL
LONG RANGE PLANNING
RECLAMATION REVIEW
SOCIO-ECONOMIC PLANNING SCIENCES
UNDERGROUND SPACE
WORLD DEVELOPMENT

*Free specimen copies available upon request.

PERGAMON POLICY STUDIES ON INTERNATIONAL DEVELOPMENT

Controlling International Technology Transfer

Issues, Perspectives, and Policy Implications

Edited by
Tagi Sagafi-nejad
Richard W. Moxon
Howard V. Perlmutter

Pergamon Press
NEW YORK • OXFORD • TORONTO • SYDNEY • PARIS • FRANKFURT

Pergamon Press Offices:

U.S.A.	Pergamon Press Inc., Maxwell House, Fairview Park, Elmsford, New York 10523, U.S.A.
U.K.	Pergamon Press Ltd., Headington Hill Hall, Oxford OX3 0BW, England
CANADA	Pergamon Press Canada Ltd., Suite 104, 150 Consumers Road, Willowdale, Ontario M2J 1P9, Canada
AUSTRALIA	Pergamon Press (Aust.) Pty. Ltd., P.O. Box 544, Potts Point, NSW 2011, Australia
FRANCE	Pergamon Press SARL, 24 rue des Ecoles, 75240 Paris, Cedex 05, France
FEDERAL REPUBLIC OF GERMANY	Pergamon Press GmbH, Hammerweg 6, Postfach 1305, 6242 Kronberg/Taunus, Federal Republic of Germany

Library of Congress Cataloging in Publication Data

Main entry under title:

Controlling international technology transfer.

(Pergamon policy studies on international development)
1. Technology transfer--Addresses, essays, lectures. I. Sagafi-nejad, Tagi, 1941- II. Moxon, Richard W., 1941- III. Perlmutter, Howard V. IV. Series.
T174.3.C63 1980 338.91 80-28329
ISBN 0-08-027108-4

Printed in the United States of America

To our families
and
to the contributors
with gratitude

Contents

Preface

This book is the culmination of an extraordinary amount of effort and teamwork.

The idea of studying the important issue of technology transfer control instruments was first initiated by the first editor over three years ago, in 1977. The institutional support which the idea received, thanks to the efforts of the other two editors, provided the necessary momentum to turn it into a research project.

Initial financial support for the project came from the Pacific Rim Fund at the University of Washington Graduate School of Business and from the Worldwide Institutions Research Center at the Wharton School, the University of Pennsylvania. Subsequent funds were provided by the Anspach Institute at the University of Pennsylvania and the Battelle Memorial Research Institute. The greatest financial support came from the National Science Foundation, through grant number PRA-78-15528. This was supplemented by an additional generous grant from the U.S. Department of State, via the National Science Foundation.

This multiplicity of sources of financial support made it possible, in turn, to assemble an impressive group of scholars and policymakers from around the world. A systematic proposal and screening process resulted in the selection of scholars to prepare original articles on given topics prior to the conference. Discussants were also given manuscripts prior to the conference, and intense and informative discussion ensured during the conference, which took place in two phases, in Philadelphia, February 8-10, 1979, and in Seattle, April 5-7, 1979. Due to space constraints, this book contains only the revised versions of most of the papers and comments. The Introduction and Conclusion have been added by the editors.

To the extent that we have succeeded in this effort, it is thanks to the generous support of the institutions that provided the funds and to the participants who provided the stuff of which this book is made.

We would like to express, therefore, our deepest gratitude to the National Science Foundation and the State Department, as well as the

Anspach Institute and the Battelle Memorial Research Institute, for funding the project. We also thank the University of Washington Graduate School of Business and the Wharton School's Worldwide Institutions Research Center at the University of Pennsylvania for cosponsoring the project. In addition, we would like to express our deep appreciation to the participants for contributing to a very constructive and stimulating learning experience regarding a pioneering topic.

Needless to say, none of the sponsoring or contributing institutions are responsible for the contents of this book. Views expressed in each article represent the personal views of the authors.

The present volume is the second in the Technology Transfer Triology series. We hope it, together with the other two volumes, will contribute to a better understanding of this complex issue.

Introduction

The international transfer of technology (TT) and attendant issues have come under increasing scrutiny from various perspectives, primarily as a result of a heightened awareness of the importance of technology to transnational corporations (TNCs)(1) and to nation-states. The host, home, and international environments within which technology transfer occurs, as well as the modes and methods of transfers and the actors involved in this process, underwent fundamental changes in the 1970s. One major consequence of this transformation was the emergence of myriad control systems, policies, and instruments devised or discussed at the corporate, national, regional, and international levels to influence the quality and quantity of technology flows in the desired direction. These include national laws, self-imposed corporate codes of behavior, and international codes and guidelines. In view of the widespread recognition of the strategic and pivotal importance of technology at the national as well as the corporate level, the escalation of interest in the subject is understandable.

As interest in the international flow of technology across boundaries has increased, so has concern with imposing on this flow some form of regularity and control. A plethora of control systems and regulatory mechanisms, presumably aimed at increasing the effectiveness of the technology transfer process, is now in existence, and more are being deliberated under various auspices. The 1970s, in particular, witnessed an accelerated awareness of the importance of technology, its generation, transfer, and use. The phenomenal growth in the technology transfer literature is one symptom of this acceleration.(2) Myriad international conferences and meetings on the subject are another. In the last few years, efforts to create a more structured environment for the international transfer of technology have intensified. Selected recent events with bearings on alternative approaches to control systems include the following:

- In December 1970 the Andean Common Market members set up stringent rules and regulations on the common treatment of foreign capital, trademarks, patents, licensing agreements, and royalties (Decision 24). They have refined, modified, and somewhat liberalized their common policies in subsequent decisions.
- In 1970 the Pacific Basin countries included provisions for regulating foreign investment and technology transfer in their Draft Charter on International Investment.
- In December 1972 the Mexican government adopted a law on registration of the transfer of technology and the use of working patents, trade names, and trademarks. A National Registry for the Transfer of Technology was also established to implement this law.
- The World Intellectual Property Organization (WIPO) has been revising patent laws and regulations to reflect the new interest in the transfer of technology and attendant patent and trademark issues.
- In the past few years, the interest of the transnational enterprises in drafting statements on company codes of conduct, standards of behavior, or corporate responsibility has increased substantially. Since technology is a major asset and concern to them, these statements have invariably included a stand on international transfer of technology.
- The United Nations Commission and the Centre on Transnational Corporations are engaged in studying alternative means of regulating these corporations and related issues, including technology transfer.
- In the Seventh Special Session of the United Nations (September 1975) the "Development and International Cooperation" resolution called for a code of conduct for the transfer of technology "prior to the end of 1977."
- The Group of 77 (the developing nations bloc within UNCTAD) issued the Manila Declaration in February 1976, seeking a code of conduct for the transfer of technology.
- UNCTAD IV (May 1976) passed a resolution appointing a group of experts to draw up a draft code for UN adopted by the end of 1977. Under the auspices of UNCTAD discussions and negotiations are continuing, with a view to formulating a draft code of conduct for TT, to be presented to the U.N.

Among the more recent and current international forums where technology has been a major agenda item were UNCTAD IV in Nairobi and UNCTAD V (1979) in Manila, the UN Conference on Technical Cooperation among the Developing Countries (TCDC), the United Nations Conference on Science and Technology for Development (UNCSTED) in Vienna (1979), UNIDO III (1980) in New Delhi, the Brandt Commission (1978-80) and its 1980 report,(3) the UN Conference on an International Code of Conduct on the Transfer of Technology (1978-80), the adoption of voluntary guidelines on restrictive business practices under UNCTAD auspices in April 1980, and the 1980 Special Session of the U.N. General Assembly on a New International Development Strategy.

As we enter the 1980s, debate and dialogue at every level about technology and its flow across national boundaries are assuming an increasingly eminent position. There is, furthermore, a shift in the overall thrust of much of that debate. From its originally negative and confrontational posture, the debate is gradually moving toward concrete proposals and some action, due for the most part to some fundamental changes that took place in the last decade within the international technology market.

In reaction to the gradual transformations taking place with regard to technology transfer, many forms of control over the flow of technology are emerging across the globe. The International Conference on Technology Transfer Control Systems for which the papers collected in this book were written was planned at this juncture to assess issues and perspectives and to examine alternative policy imperatives from all perspectives. Articles and discussion papers presented at the two phases of that conference are presented here. To put them in perspective, we will now undertake a brief review of the actors, issues, and alternatives in this debate.(4)

In the international transfer of technology, there are a number of primary issues that separate the actors in the arenas of debate and negotiation. First, it is important to identify the actors, that is, those who have a stake in the system. In a study aimed at identifying the issues of consensus and dissensus regarding TT, Perlmutter and Sagafi-nejad(5) identified four primary groups of actors that they referred to as the "stakeholders." These are (a) the firm supplying (selling, renting, licensing, or otherwise making available) the technology; (b) the home country government, under whose national jurisdiction the supplying firm is established and by whose laws it is bound – at least at the home market; (c) the recipient firm, that is, the firm that receives (purchases, rents, licenses, or otherwise obtains) the technology; and (d) the recipient (or host) government.(6)

The major issues of debate between the above groups and in relevant international fora are grouped into the following categories:

1. Contributions of technology to development and industrialization goals, policies, and strategies. This constitutes the heart of the concept of "developmental impact." And there is now increasing emphasis by the less industrialized countries on the relationship between imported technology on the one hand and developmental needs, capabilities, and policies on the other.

2. Who controls the technology after it has been transferred, and the impact of different levels of control over technology upon the user's ability to utilize it as well as upon the loss of competitiveness of the supplier. Many transnational corporations argue that technology transfer is most effective for the host country if it takes place within the firm in the form of a wholly owned subsidiary. This way, they say, the TNC is assured of control and can transfer more and better technology unhindered by anxieties over the loss of confidentiality and control associated with modes that leave the supplier with less than total control over the technology.(7) Others argue the opposite: that supplier

control leaves little opportunity for internalization by the recipient firm or for dissemination and assimilation in the host country.

3. Restrictive transfer practices that supplier firms are accused of. These have been extensively examined by UNCTAD (1971), the OECD (1973), and UNIDO (1975) and include grant-back provisions, market (territorial) restrictions, field-of-use restrictions, abuse of quality-control measures, tying clauses, package licensing, price-fixing, and manipulation of deletion allowances. Views differ, as might be expected, on the definition of "restrictive business practices" in general as well as on more specific restrictive transfer practices. There is also disagreement about how widespread these practices are.

4. Price paid for a purchased, leased, or rented technology. This has also become a major point of controversy. Depending on how one views the international technology market, one can argue that price is a product of market forces or a result of the tyranny of the licensing executive. Major disagreements, as well as substantial ambiguities, surround this issue.

5. The proper role of the government. This, too, continues to be a bone of contention among the actors. Often firms wish less intervention. Supplier governments in advanced market economies concur. But the ideas of host country governments preoccupied with macroeconomic, societal, and welfare considerations run in another direction. The result is lack of consensus on the proper role of the government, and hence the ascension of this issue to an agenda item.

6. Protection of the supplier firms' proprietary rights. This has always been a very sensitive issue, since the process of entrepreneurship rests on the provision of a certain amount of monopoly rights and the protection of this right for a limited time – thus the patent system. But today's conditions in the less industrialized countries are far different from the circumstances that warranted this provision in the West; and a policy of "abuse of monopolistic advantage" does not suit the majority of those countries today. But intensified pressure for documentation and justificatory data threatens to erode this position, and suppliers feel that the protection of their proprietary rights is being jeopardized. No easy solution is on the horizon; case-by-case handling of this sensitive issue appears to be the most effective method thus far.

7. Increasing pressure on technology suppliers to provide assessment data in support of a price tag. There is, as well, pressure to unbundle the package and to provide convincing evidence to justify both the price and the extent of "packagedness." Recipients are not at all convinced that they need, or can use, all the elements and components of what is put together by the supplier and offered to them as an unbreakable package. More assessment and justificatory data is demanded by both host governments and recipient firms.

8. And, finally, questions of conflict resolution and dispute settlement. Their occurrence is, understandably, unavoidable and constitutes a major debating item. The less industrialized countries who tend to jealously guard their sovereignty resist any perceived infringement upon it. On TT as on other jurisdictional matters, they want their own courts

and laws to be operative. Suspicious of the intent or the manner of implementation of these national laws and regulations, suppliers of technology would prefer international tribunals or their own home country's courts.

The above issues constitute major agenda items in the ongoing debates and discussions among TT actors.(8) The first phase of the Technology Transfer Control Systems Conference in Philadelphia (February 8-10, 1979) took a closer look at these issues, in order to explore major policy issues that must be dealt with in any TT control system. A working definition of technology transfer control systems would be "any instrument, policy measure, rule, regulation, law, or other action that influences the flow of technology between the supplier and the recipient." In the international flow of technology across boundaries, these instruments can be initiated or located at any point between the supplier and the recipient. Besides the origin of the control systems, several additional questions suggest themselves:

- What is the main objective?
- What is the scope of coverage?
- How explicit and detailed is the instrument?
- What is the intended target? (Whom is it applicable toward?)
- What about implementation?
- What is its legal status?

Thus the authors and discussants were asked to address questions such as: How should this issue be handled in any proposed control system? What are its major facets? How amenable is it to control? And what do we know about these various dimensions? Policy implications were also examined. Besides these questions, additional issues were raised under each specific topic. These included:

1. Patent systems. How would a control system bear upon the existing national and international patent laws, conventions, and regulations? How would the present patent system be altered in light of the alternative control or regulatory measures?
2. Pricing of technology. What are the alternative methods of pricing a technology package? What methods exist or are being considered for controlling the price of technology? Which method is most effective or has the best prospect of success? What are the attendant problems in assessing control systems aimed at technology-pricing policies?
3. Taxation. How do the present systems of national taxation policies with respect to R&D expenditures, royalty revenues, intercompany transactions, and so on lend themselves to various forms of control? What are the problems in the present tax systems and how would they be altered to accommodate various control alternatives?
4. Restrictive business practices. What are the points of agreement and disagreement among parties regarding the definition of RBPs? How are they treated in various national, regional, and international

frameworks? How facilitating or restraining are these vis-a-vis other alternative control systems for TT?
5. Transfer channels and modes. How do various channels, modes, and methods of technology transfer lend themselves to control? Is there a necessary and inevitable correlation between different channels and their amenability to various control systems?
6. Nature of technology. To what extent are control instruments affected by the nature of technology being transferred? What are the product-specific, industry-specific, or technology-specific issues that any TT control system must take into consideration?
7. Conflict resolution. What issues are more likely to be the root causes of conflict in TT arrangements? What alternative approaches are available for the resolution of such conflicts? How do these approaches compliment or contradict one another? How would various new control systems deal with conflict resolution?

The second part of the conference (Seattle, April 5-7, 1979), explored alternative approaches to TT control systems and their implications. Each author presented an objective evaluation of one of the perspectives and alternatives on controlling the TT process, assessing how each party views the prospects of each regulatory system and how each particular perspective or approach would relate to the major issues discussed in the first part of the conference.

The major questions addressed by all participants in this phase of the conference included the following: Having explored the issues and the perspectives on TT control systems, what is the most desirable, feasible, and probable outcome of the current debates? What alternative control system is most desirable and feasible from the particular vantage point to which each particular paper is addressing itself? What seems to be the most probable or likely control system? And how is this likely to be received from this vantage point? Among the policy perspectives covered by papers and discussions included were:

1. Technology-supplying firms. Papers in this category dealt with the perspectives of the technology-supplying firms on control. Different types of supplying firms' views on self-control (corporate codes) as well as other forms of control were considered. How do various transnational enterprises (suppliers) view the TT control issues?
2. Technology-receiving firms. Here the views and perspectives of different types of recipient firms (wholly owned subsidiaries, joint ventures, and independent recipients) were examined. How are the recipient firms likely to perceive control measures?
3. Technology-importing national perspectives. National views of technology recipients are a major element in considering any control mechanism. Since views of various governments do not necessarily coincide, several papers examined different perspectives.
4. Technology-exporting national perspectives. Since exporting countries have diversities as well, a variety of perspectives were also dealt with here, in different papers and in discussions.

5. Regional perspectives on TT control. Regional entities, notably the Andean Group, have attempted to arrive at a regionally coordinated TT policy. An assessment of the past history and performance of such efforts, as well as their future prospects, sheds light on their effectiveness and advisability.
6. International perspectives. Various international regulatory and control mechanisms are being deliberated upon. The two major forums are the United Nations Conference on Trade and Development and the World Intellectual Property Organization, as well as the UNCTC. International attempts were therefore examined as well.
7. Contrasting perspectives. What are the critical factors that promote or hinder the legitimacy and viability of various TT control systems? What lessons can be learned from contrasting the views of different groups?

Parts I and II of the present volume contain most of the articles and discussion papers presented at the TTCS Conference. Essays in Part 1, "The Underlying Issues," pertain to the first phase, and those of Part II, Policy Perspectives and Implications," belong to the second phase. In Part III we present our conclusions based on the enormous amount of informed discussion that was generated during the two conferences.

NOTES

(1) Here as well as elsewhere in this and the other two volumes of the TT Trilogy, the terms "transnational corporations" (TNCs) and "transnational enterprises" (TNEs) are used interchangeably. So are the terms "less developed countries" (LDCs) and "less industrialized countries" (LICs. We are, however, cognizant of the subtle differences.

(2) A comprehensive bibliography on the subject in 1976 identified over 1,200 references on TT and related issues. See Taghi Saghafi-nejad and Robert Belfield, Transnational Corporations, Technology Transfer, and Development: A Bibliography (Philadelphia: Worldwide Institutions Group, Wharton School, 1976). By the second edition in 1980 the number had swelled to over 1,500. See Tagi Sagafi-nejad and Robert Belfield, Transnational Corporations, Technology Transfer, and Development: A Bibliographic Sourcebook, Book #3 in the Technology Transfer Trilogy (New York: Pergamon Press, 1980).

(3) Willy Brandt, North-South: A Programme for Survival, Report of the Independent Commission on International Development Issues (Cambridge: MIT Press, 1980).

(4) The reader is also referred to Book 1 in the Technology Transfer Trilogy. See Howard V. Perlmutter and Tagi Sagafi-nejad, International Technology Transfer: Codes, Guidelines, and a Muffled Quadrilogue, Book 1 in the Technology Transfer Trilogy (New York: Pergamon Press,

1981), especially chaps. 1, 2, and 10. See also Richard W. Moxon, "The Costs, Conditions and Adaptation of MNC Technology in Developing Countries," in Economic Policy and the Multinational Enterprise, ed. Robert Hawkins (New York: JAI Press, 1979), and to Farok Contractor and Tagi Sagafi-nejad, "International Technology Transfer: A Review of Major Issues, Impacts, and Policy," (1981), mimeo.

(5) Perlmutter and Sagafi-nejad, International Technology Transfer.

(6) There is a fifth group of stakeholders identified in that study: the international organizations, which have an increasingly salient influence on technology transfer and the related matters of TNCs, the new international economic order, and development. However, since they derive their power and legitimacy from nation-states, they are not considered primary stakeholders.

(7) See Jack Behrman and Harry Wallender III, Transfers of Manufacturing Technology Within Multinational Enterprises (Cambridge, Mass.: Ballinger, 1976).

(8) Perlmutter and Sagafi-nejad, International Technology Transfer, chap. 2.

I

The Underlying Issues

1 Alternative Channels and Modes of International Resource Transmission

John H. Dunning

One aspect of the debate over the international transmission of technology and other resources(1) that is receiving increasing attention concerns the mode and institutional form of that transmission. Experiences of the last two decades or so suggest that the way in which foreign technology, organization, marketing and financial skills, and management expertise is imported by host countries may be as important, if not more so, in influencing the consequences of the transfer as the resources themselves (United Nations, 1978). For channels and modes may influence the extent and kind of the resources transmitted – how, if at all, they are packaged (supplied jointly); and what, if any, control is exercised on their use and the terms of the transmission. Governments seeking to advance their development programs have been faced with the dilemma that to obtain the resources needed to implement their plans, they may have to import other resources that they may not need, or accept control over the use of the resources that they do not want. In the 1970s the position has changed to some extent as the options open to governments have widened. But the basic dilemma still remains.

This chapter first describes the main channels of international resource transmission and constructs a simple taxonomic apparatus to identify their main characteristics. Using this framework, it traces the history and examines the present-day structure of resource transmission in terms of the three main determinants: (1) the international, economic, and technological environment; (2) the policies of home and host governments towards foreign direct investment; and (3) the strategies of the main resource-transmitting economic agents, the transnational corporations. It then compares some of the effects of these channels from the viewpoint of country and firm objectives. A final section of the paper briefly speculates on the factors likely to influence future modes of transmission and some of the features of such modes.

FEATURES OF TECHNOLOGY TRANSMISSION AND THE MAIN MODES

In the present context the most important distinction of types of technology is between proprietary and nonproprietary technology. The former is specific to a particular institution, in that it can only be acquired by another institution with the permission of the first. Nonproprietary technology consists of general knowledge and information, which, although it usually costs something to acquire, is not the exclusive property of another institution. Such public technology is available in capital or consumer goods, in the migration of labor, and through trade, professional, and technical literature, conventions, conferences, and seminars, and word of mouth, and from education and training at foreign educational establishments. The word "proprietary" means "holding property or the (exclusive) right to property." Proprietary property includes both tangible and intangible assets, and patented and unpatented technology, as long as it is institution-specific. Although a country may attempt to control the outward or inward flow of public technology – for example, by disallowing the export or import of certain types of goods, machinery, literature, blueprints, or formulae, or by controlling immigration, emigration, foreign visits, or participation in conferences and so on – this chapter will focus attention on the nature and consequences of firm-specific technology, recognizing at the same time that governments may sometimes "disproprietarise" such technology by, for instance, not accepting international patent conventions.

Basically, there are three mechanisms by which resources are transferred and disseminated across national boundaries. The first is the market, the second is decision making within a firm, and the third is government fiat or intergovernment agreement. In the first, technology, whatever its form, is transmitted between independent buyers and sellers at market ("arm's length") prices. Depending on the conditions of the market, these prices will range from monopsonistic though perfectly competitive to monopolistic prices. The second mechanism, unlike the first, involves no change in the ownership of resources; they are transmitted within the same firm at internally determined prices, subject to any constraints imposed by governments of the recipient or sending countries. There is no a priori reason why such transfer prices should be higher or lower than "arm's length" prices, although it is not difficult to theorize when, and in which direction, they will diverge (Lall, 1973). The point about this mode of transfer is that de jure, at least, direction over the use of the resources remains with the sending firm; in other words, it is jointly supplied with the resources. In practice, the boundary between an external and internal transmittal of resources is often blurred. A transmittal of resources between a parent company of a transnational corporation and its 100-percent-owned affiliate is obviously an internal one; de jure, a transmittal between a TNC and an affiliate in which it has a 51 percent or greater equity stake is also internal. Where a company has less than a 50 percent participation, the matter is less straightforward. On the one hand, part of the payment for resources transmitted takes the form of profits; on the other, although there is no de jure right to control how the resources are used once transmitted, de facto such control is often exercised. Moreover, some kinds of contractual agreement may provide the transmitting firm as much direction over the resources transferred as if they were transferred internally.

The third mechanism of transmittal is by government fiat or agreement. Governments, of course, may engage in the first two mechanisms in their own right. Thus state-owned transnational corporations may internalize resource flows, while governments may use the facilities of international capital markets. Governments may also transmit resources to each other on terms negotiated among themselves; they may help through educational programs, they may sponsor technological transfers at below-market prices. Most government efforts related to aid programs are, in effect, resource transmittals by fiat. Nevertheless, the more important role of governments is their influence over the conditions under which the first two mechanisms of resource transfer are undertaken.

Resources may be transmitted separately or as part of a package. Usually intrafirm transmissions take the form of a joint supply of capital, technology, and management skills; indeed, rarely is it possible to identify the precise contribution of each component. Moreover, as we have already said, this kind of transmission is usually accompanied by control over decision making. Control, or direction, is not usually considered as a separate resource in the literature (but see Houston, 1977), although it is implicit in the concept of entrepreneurship. Yet the fact of who decides how resources are allocated – as opposed to the efficiency of resource allocation, which is the task of management – may vitally effect the outcome of such allocation. This is especially relevant when conflicts of interest arise among the possible decision makers.

There are two main reasons why a parent company may wish to control decisions over its affiliates' operations. The first is that it may perceive its own management to be more efficient in its decision making than that of its affiliates. The second is that the goals of the parent company may diverge from those of its individual affiliates, or may be given different priority; in this case, direction is exercised for reasons that have nothing to do with managerial shortcomings, but rather because decisions taken by affiliates will not always be in the interests of the transnational entity of which they are part. This is particularly likely in the case of TNCs that pursue a global production and marketing strategy.

In the case of resources transmitted between independent parties, there is less reason to suppose these are part of a package. The supply of portfolio or finance capital is a case in point. On closer inspection of many contracts, however, the joint supply of resources is more common than may appear. Partly this depends on how narrowly one defines a resource, but rarely, for example, is production technology supplied without some accompanying managerial or organizational knowledge – even if it is only a question of information about the running and maintenance of a machine. When whole systems of technology, management training programs, organizational set-ups, and so on, are transmitted, then inevitably there are elements of joint supply. Moreover, the contract in question may embody clauses relating to managing and/or controlling the resources once they are transmitted. Though these may be entered into on a time-limited basis, the effect on the resource allocation within the recipient firm might be as far-reaching as that of any intrafirm transmission. I will return to this point later.

SOME THEORETICAL CONSIDERATIONS

Now let us examine some principles underlying the modes of resource transmission, concentrating on two of the three main channels described earlier, namely the market and corporate fiat. We will view situations under which each of these two routes will be preferred, in terms first of the supplying firm, and second, the recipient country.

The firm will choose the market route for exporting technology or other resources whenever its objectives appear more likely to be advanced than through its own control procedures.(2) The benefits of the market route embrace not only the price paid for the resources but the broader effects on the rest of the organization that are expected to result from the exchange. They also include avoidance of the costs of internalized transactions. The price of choosing the market route is essentially that of negotiating and contractual costs and the consequences of market failure – knowing whether one has charged the right price, uncertainty over future market conditions, not being able to fully capture the economies of interdependent activities, lack of control over supplies or markets, protection of property rights, and so on.

Now, it is possible to identify certain variables whose value will determine which of these routes are chosen. These are set out in table 1.1. The crux of the matter is the extent to which markets operate efficiently. If they are perfect, then the firm cannot do better than use this mode of transfer (which, incidentally, will in that case also be the best route for the acquiring firm or government). If they are not, then the selling firm may find it preferable to internalize the appropriate transactions. According to some writers (Coase, 1937; Penrose, 1959; Hymer, 1960), much of the growth of large firms can be traced to the market imperfections (sometimes aided and abetted by governments) that encourage firms to internalize activities that would otherwise have been performed by separate institutions. The growth of upstream vertical integration by manufacturing companies, to resources or intermediate goods, is explained by the uncertainty over the amount, quality, or regularity of supplies or prices of inputs. Downstream integration is generally designed to fully capture the benefits of the production or primary products, or to protect or exploit property rights over technological innovation. Another motive for forward integration concerns the selling of end products. Where the market fails to compensate producers for the possibly inferior quality of tertiary services and/or after-sales facilities, the selling firm may decide to undertake these activities itself. Horizontal and lateral integration – a feature of many conglomerate enterprises – may be encouraged in order to take advantage of such practices as predatory pricing and cross-subsidization. Integration brings with it the economies of joint products or production processes, which yield benefits over and above those that arise from the control over a single product or production process.

Table 1.1. Illustrations of Market Failure, With Some Internal and External Solutions

Problem	Example	Market Solution	Internal Solution
Contractual costs			
Inadequate property rights	Lack of patent protection on technology	Enforce right of exclusion by resort to secrecy	Diffuse technology freely within the firm
	Unenforceable forward contracts (for intermediate products in capital-intensive multistage processes)	Contract forward in terms of another good whose price is stable in terms of the good concerned	Instigate corporate long-term planning of interdependent activities
Costs of negotiation	Time costs of continually renegotiating short-term contracts with the same individual in a fluctuating market	Use long-term contracts	Replace decentralized bargaining by centralized authority
	Costs of sanctions enforced as part of bargaining strategy in bilateral monopoly	As above	As above
Policing for misrepresentation	Quantity or quality not as specified in the contract	Both buyer and seller measure quantity and check quality	No incentive for misrepresentation, hence duplication of effort unnecessary

Table 1.1. (Cont.)

Problem	Example	Market Solution	Internal Solution
Enforcement against defaulter	Cost of bringing a defaulter to court or of preventing a debtor from going bankrupt	Tie in other party using imperfections in capital markets as a basis for sanctions (e.g., make subcontractor take trade credit)	Potential defaulter will be employee, who can be tied in on a similar principle
Intervention costs			
Quotas	Capital controls prevent movement of funds across currency areas	None	Transfer pricing
	Income controls prevent attribution of all taxable income to sources in minimum-tax fiscal area	None	Transfer pricing
Tariffs	Ad valorem tariffs	None	Transfer pricing
Price Controls	Statutory regulation of the price of an intermediate product	None	National internal prices
Buyer uncertainty	Sale of a secret technology (no patent protection)	Seller guarantees commercial value of technology to buyer using contingent contract	Secrecy unnecessary within the organization
Resale problem	Pricing discrimination enforced in an intermediate product produced with economies of scale	Supply a differentiated intermediate product to each segment of the market	Charge each division of the firm at marginal cost and impute the loss to overheads
	Buyer of a secret prevented from reselling it to others	Sell to all buyers simultaneously	Secrecy unnecessary within the organization

Source: Derived from M.C. Casson, Alternatives to the Multinational Enterprise (London: Macmillan, 1979), chap. 2.

It is not difficult to classify types of transactions that favor one form of allocative mechanism over another. The sale of some kinds of intermediary products through the market generally offers less rewards to firms than sale through administrative fiat. Take technology as an example. At the time of its production, its possessor is the innovator who wishes to exploit it most profitably; it is costly and takes time to produce, but there is no futures market in it; it is difficult for a potential buyer to value, as its usefulness can only be determined after it has been bought. Moreover, in most cases technology needs backup resources. These qualities apply especially to the kind of knowledge that cannot easily be patented, for example financial systems, information sourcing organizational skills, marketing know-how, managerial judgment, and so on. All of these characteristics encourage a firm to internalize the transmission of resources rather than sell them to independent firms. The market fails to allow the supplying firm to fully capture the benefits from the sale of its knowledge.

Another example is the internalizing of the supplies of essential raw materials. In this case there are advantages to the buyer from internalizing transactions. Here the imperfections of the market lie in its inability to avoid uncertainty over future supplies and price and to ensure control over timing and quality. This prompts firms seeking to buy raw materials from abroad to engage in foreign direct investment and, by so doing, to provide intermediary inputs and access to the host economy.

These two examples suggest that the propensity to internalize transactions varies directly with the imperfections of the market. But firms may also wish to exploit the advantages of market failure. Such imperfections may be either structural or market-oriented; in international markets, they may also reflect differential policies of governments (which can only exist if governments are able to prevent resources from moving internationally). Examples of market imperfections include the ability of transnational corporations to engage in discriminatory pricing, to restrict the kind of technology that is transferred, and to control the export markets of their affiliates. Examples of market failures due to differential government policies include the practice of transfer pricing to avoid taxation and minimize exchange risks, engaging in leads and lags and the adjustment of the distribution of short-term assets between different currency areas. Not all of these practices are specific to internalizing decisions, but most involve some kind of conflict of interest between buyers and sellers.

The final advantage of internalization is based on capturing the externalities or spillover effects of external transactions. These benefits accrue to the company undertaking the integration, rather than any affiliate of the company; they are also dependent on the affiliate being centrally managed or controlled. Such integration arises from the international specialization of the activities of transnational corporations. This may be horizontal (involving product specialization of the Ford, Philips, or International Harvester type, especially as practiced in Europe and Latin America) or vertical (involving process specialization, as practiced by U.S. companies engaging in labor-intensive processes of production in low labor cost areas). Equally – indeed, perhaps, more important – TNCs may engage in functional specialization. It is estimated, for example, that over 90 percent of the research and development undertaken by U.S. and Swedish transnationals is located in their home countries. Head or regional offices of companies may also house central financial, purchasing or marketing divisions, which may oversee the activities of affiliates in these areas. It is this hierarchial division of functions, with the key areas of decision making reserved headquarters of TNCs, which is the core of the center-periphery debate that has been well articulated by Hymer (1970), Sunkel (1970), Murray (1972), and others.

In summary, resources may be transmitted across national boundaries for a variety of reasons. Sometimes the motive may be to ensure future supplies of natural resources, particularly oil and hard minerals; sometimes to allow a local affiliate to service a market previously supplied from exports; sometimes to advance the regional or global strategy of a transnational corporation. In some cases resources may be singly provided; in others, as a package, over the use of which the supplying firm may wish to continue to exercise direction. These technical requirements of firms must be set against other variables. In some cases market forces work well and the firm may not be able to improve on them; in others, they do not and the firm finds it preferable to internalize transactions. Generally speaking, the more a firm supplies or acquires resources jointly, and the more it seeks to integrate its global strategy, the more it is likely to avoid the market route of international resource transfer.

To recipient countries, the most appropriate mode of resource transmission is viewed from a different perspective. Their goals are both more numerous and more complex than those of firms; moreover, they may be very different from each other. But most countries seek to acquire resources in a way that will both maximize their level and rate of growth of real national product and allow them full discretion to use resources as they wish. These twin goals are rarely completely attainable; nor are they necessarily mutually consistent. Whenever this best solution is not possible, recipient countries face the dilemma of how to get the resources needed to advance development goals without surrendering control over their use.

One basic difficulty faced by most countries, and particularly by developing countries, is that they often lack information on the contribution that different modes of resource transmission can make towards their objectives. In the great majority of cases countries have only a very hazy idea of the relative costs and benefits of obtaining technology through equity investment and contractual agreements. Judgment and policies, then, tend to be based on fragmentary and imperfect data, case studies, what is known of the experience of other countries, general economic and political principles, and hunches, impressions, and prejudices. The current vogue among most developing countries is to assume that the equity investment route is a more costly way of obtaining resources (in terms of payments of royalties, profits, and so on), and one giving more control of decision making to the investing company, compared to contractual arrangements. Purchasing a bundle of resources from the same source is thought to be more expensive than buying the ingredients of the package from different sources. An affiliate closely integrated with the rest of the organization of which it is a part is considered less desirable than an affiliate that acts independently of the parent company and indentifies itself completely with the local economy.(3) Rarely are these judgments based upon a careful documentation or appraisal of the facts, nor do they take account of the total effects of resource transfer. Rarely, too, are the consequences of alternative modes systematically evaluated; the belief that control is inextricably linked with a foreign presence is strongly entrenched.

Conceptually it is not difficult to compare the net benefits to recipient countries of different forms of resource transmittal. The data constraints are, however, formidable, and it is doubtful that any useful generalizations can be made about which route is preferable. However, it is possible to identify situations and types of resource transfer that favor one mode or the other. Clearly there are situations in which only internalized transfer of technology can make an effective contribution; an obvious example is where there are insufficient local resources to provide an efficient backup to the transferred resources if they are separately supplied. At the other extreme, a specific resource (or right to use a resource) may be needed that requires no backup resources. Unfortunately, it is often the case that the countries that most want to be self-sufficient and autonomous in decision making are those that need the totality and continuity of resources that only an internalized transaction can provide. This does not always mean that such transactions have to be integrated with the rest of the organization of which the affiliate is a part, but in the last analysis, decisions on what products to produce, where to locate production units, and which markets to serve must depend on the overall strategy of the parent company.

Other factors that may influence a host country in choosing the mode of resource acquisition are supply and demand conditions and the bargaining power of the parties to the transaction. Assuming that benefits of the transmission are the same, then the modality that offers the most favorable terms to the host nation (that is, can be obtained at least cost) will be the one most preferred. In practice, particularly for the smaller nations, this means that the more perfect the supply conditions, the more the external route is likely to be favored. On the other hand, if a country could improve its bargaining strength, its choice of modes might be widened, particularly where there are greater benefits from internalization to be captured from the supplying institution.

The possession of a combination of complementary resources and a strong bargaining position both better enables a country to opt for contractual arrangements and enhances its power to gain a larger share of the particular benefits generated by internalized and integrated transactions of transnational corporations.

THE DEVELOPMENT OF ALTERNATIVE MODES OF RESOURCE TRANSMISSION

In the nineteenth century most international transmissions of resources were undertaken separately and independently of each other and between buyers and sellers at "arms length" prices (Nurkse, 1954). Almost all finance capital was channeled through the international capital market, while technology and human capital migrated separately, or were exported (or in some cases smuggled out) in the form of plant and machinery, drawings, blueprints, formulae, and so on. Firms were small and unincorporated, and because of transport and communication constraints they rarely found it profitable to supply markets beyond their regional or national boundaries. Technology was simple, easily transferred, and quickly assimilated; the production process took only a short time; information was readily available and speedily disseminated; there was no international patent system. Where there was equity investment, this usually took the form of backward vertical integration into raw materials, minerals, and agricultural commodities, to ensure price, quality, and certainty of supplies. Most of the early overseas ventures of British firms were of this kind, an internalization that in the main reflected the absence of a satisfactory futures market. Countries that were prospective hosts to resource transmission were rarely in a position to judge the efficacy of different vehicles of transfer. In any case, the involvement of governments in the management of economic affairs was strictly limited. For the first half of the nineteenth century a free-trade philosophy dominated the pattern of international commerce; in the latter half, increasingly protectionist measures were introduced, which, together with the emergence of managerial capitalism, stimulated direct investment by European manufacturing firms in the United States (Franko, 1976).

In this economic and technological environment, with home and host countries adopting a largely laissez faire policy toward foreign investment and with the organization and strategy of firms kept simple, resource transfer took place through the market mechanism. Even in 1914, foreign direct investment accounted for only 15 percent of the total British capital stake overseas. (But for a rather different interpretation of the balance between direct and portfolio investment in 1914, see Svedberg, 1978).

In the twentieth century up until the mid-1960s, circumstances increasingly favored the internal transfer of reosurces by TNCs. The revolution of transport and communications technologies, culminating in supersonic air travel, the microcomputer, and telecommunication in the 1970s, has drastically reduced the problems of organizing a large firm over distance. The technology of product innovation and production has become enormously complex, and increasingly technology must be supplied as a package or system. Institutional changes, most notably the widespread adoption of the patent system, have encouraged technology-producing firms to internalize their use of the technology, to capture both monopoly rents and the economies external to the technology but internal to the firm. These latter economies are not capturable to the same extent through licensing, in part because the buyer, uncertain of the benefits of the technology, will not be prepared to pay the price required by the seller to compensate for the technology's disinternalization (Buckley and Davies, 1979). Finally – and this applies particularly to the post-World War II period – while the general atmosphere to trade has been favorable, firms have increasingly serviced foreign markets through local production facilities rather than exports.

At the same time that the international environment has favored the transfer of resources within the same firm, government policies toward the mode of transmission have had an identifical effect. The most obvious example is the gradual movement toward import restrictions by most host governments, including the advanced industrialized countries. On the other hand, in the 1950s and early 1960s, governments unreservedly welcomed foreign direct investment for the contribution it was thought to make to economic development. The mode of resource transmission was not at issue. For one thing, licensing was often not a practical alternative because of the lack of backup resources; for another, the international capital markets were in disarray after the war; for a third, the bundle of resources transmitted internally by foreign companies appeared particularly attractive, simply because it was believed to be the quickest and most effective way of acquiring the ingredients. Each of these ingredients was wanted for its own sake, but because countries did not know whether or how they might be separately obtained, the opportunity to obtain them in one package from one source saved the host country considerable search and contractual costs. To this extent, the interests of firms and countries to avoid the market mechanism or resource transmission coincided.

In the 1950s and early 1960s there was also a great deal of ignorance about the costs and benefits of foreign direct investment, and, indeed, of its characteristics. In the case of developing countries, few development plans had been formulated, and political objectives and economic aspirations were still unclear. In the main, foreign companies were seen as channels for providing much-needed resources for growth; questions relating to what kind of resources were being transferred, at what price, and for what purpose came later. Initially the only concern was that the foreign company might use any monopolistic power it might have to raise the price of supplying the resources above the level acceptable to the host country. Certainly at that time little attention was paid to the extent to which foreign companies allocated resources against the host country's best interests or locked the host country into an interntional economic order it believed was not in its best interests. Partly this was because, outside the resource-based sectors, foreign companies in both developed and developing countries were mainly engaged in import-substituting activities, and their international activities replicated, albeit on a smaller scale, those of their domestic operations. There was little attempt to integrate international activities or to practice process or product specialization; intragroup trade was generally limited to imports of capital equipment and some inputs that were not available locally. Few foreign companies were genuinely multinational either in their geographical diversity or their philosophy.

Between the early 1960s and early 1970s the growth of transational corporations considerably outpaced that of world output and trade, and was most marked in those sectors in which the advantages of internalized resource transfer were most prominent, namely capital-intensive resource-based industries and high-technology manufacturing and service industries (Buckley and Casson, 1976). There was increasing pressure on the larger developed industrial countries to search outside their national boundaries for raw materials to meet their needs. Unlike the industrial revolution of the nineteenth century, that of the twentieth rested on resources that were not indigenous to the industrialized countries – copper, bauxite, oil, and some of the rarer metals. And firms within the manufacturing sector, anxious to protect their market positions, were reluctant to surrender control over their foreign affiliates. Indeed, as markets expanded they were encouraged to rationalize their product processes and markets on a regional or global basis.

As a result of these forces, the strategy and organizational structure of transnational corporations was changed in such a way that the internalization of foreign operations, and particularly of those having to do with resource acquisitions and technology transfer, became more rather than less desirable.(4) Events concurrent in the international economic climate, the uneasy move to variable exchange rates, the growth of Eurocurrency and analogous markets, and different government policies, such as with respect to taxation and treatment of earnings of foreign companies, added to firms' incentives not to choose the market route for resource transmission.

Yet while companies were becoming more multinational in their operations and strategies, most host governments were moving in the opposite direction. As they sought to increase control over their political and economic destinies, they became aware of the differences between the way that foreign and domestic companies contributed toward that goal, in terms of performance and also in their repsonse to domestic policies and the international economic environment. Because of their multinationality, TNCs tended to be more response-elastic to the conditions that affected the relative profitability and liquidity of their various operations than were their indigenous competitiors (Dunning and Gilman, 1976). Moreover, because so many transactions were internalized, transnational entities were better able to adapt their operations to changing circumstances in a way that best suited the company as a whole, rather than that of individual affiliates. This fairly decisive step from viewing a TNC as primarily a provider of resources to perceiving it organizing its foreign operations to advance its international strategy fundamentally altered the attitude of host countries toward foreign direct investment. As I have pointed out elsewhere (Dunning, 1977), it was at this point that governments switched their attention from the phenomena of foreign direct investment to that of the institutions making the investment. It was at this point, too, that attention began to be seriously focused on obtained technology and other resources by routes that did not involve continued outside direction over the use of such resources once they were transferred. The appropriate mechanism for resource transmission became – and still is – one of the focal debating points on the role of transnational corporations in economic development.

Thus from the early ninetheenth century up to the First World War, resource transmission mainly took the form of external transactions between independent parties. From 1918 on, and at an increasing pace through to the mid-1960s, resource transmission began to be internalized within companies engaging in foreign direct investment – companies that became known as multinational entities in the late 1960s and transnational corporations in the mid-1970s. At first this trend was acceptable to companies and countries alike. The main features of such internalization were that control over resource allocation remained with the enterprise owning the resources and that resources were transmitted as a package. As the internalizing of resource transmission assumed greater prominence, and the results of it became better understood, and as governments pursued a more interventionist policy in both demand management and development planning, governments' reactions to foreign direct investment became more critical.(5) Such criticisms, mainly directed at the large transnationals, peaked in the early and mid-1970s, aided and abetted – and sometimes sidetracked – by various political issues.

The debate was further fueled by the not unrelated concern about the international econommic order then operating, one that, in spite of the upheaval of the early 1970s, was still very much a product of decisions made at Bretton Woods and Havana in the late 1940s. To many developing countries this order seemed to perpetuate an international division of labor and distribution of income favoring the industrialized countries. The TNCs were regarded as part and parcel of that order, and, because of their domination of high-technology sectors, the main inhibitor of indigenous technology-creating activities and of the advancement of the dynamic comparative advantage of developing countries. Through this inhibition of the competitive advance of the developing countries, industrialized countries (not firms) were, in effect internalizing their comparative advantages, a process that – so it was argued by the developing nations – was a poor substitute for efficient market transactions or (host) government fiat. Hence the desire to frame a new international order and the wish to modify the mechanisms for the international transmission of resources go hand in hand.

THE CURRENT SITUATION

Partly as a result of actions taken by governments, ranging from outright expropriation of transnational assets to a host of persuasive devices to influence their behavior; partly because of the growth of non-U.S., and particularly Japanese, TNCs with rather different attitudes toward foreign direct investment; partly because of the opening of new sources of resources (notably technology from the Soviet Union and finance capital from the international capital market and Arab countries); partly because of the improving technological and absorptive capacity of many host countries, particularly of complementary resources to hardware technology (engineering maintenance, repair, service facilities, and so on) and partly due to the response of TNCs to government attempts, particularly in developing countries, to make internalized transactions less desirable, the mid- and later 1970s have seen the emergence of many new modes of international resource transmission, all of which involve some disinternalization of decision making (compare the equity investment route). The literature on these alternative forms is still very fragmentary, and there has been little serious analysis of their comparative merits from either a company of a country viewpoint. Most monographs and journal articles have concentrated on one or other type of contractual arrangement; typical of these is Gabriel's classic work on management contracts (Gabriel, 1967), Hayden's analysis of East-West technology transfer (Hayden, 1976), Franko and Tomlinson's studies on joint ventures (Franko, 1971; Tomlinson, 1970) and Grub's analysis of franchising (Grub, 1972).(6) Neither has there been any substantive work on the effects of government policies that have attempted to disinternalize resource acquisition, for example by the expropriation of foreign enterprises. How have countries managed to acquire the foreign technology and management skills that the expropriated firms previously provided? What have been the resulting cost and benefits?

Looking at the current trends in the North-South resource transmission, one might well be forgiven for speculating the the peak of the TNC qua TNC is now past. This is not so for intra-OECD transfers, nor for West-East technology flows, where the opportunities for foreign direct investment are increasing. In terms of the analysis developed earlier, host government constraints and the strategy of TNCs operating in developing countries are such as to make externalized transactions (or no transactions at all) a more likely mode of resource transmission than that which has taken place over the past two decades.

We have said that host governments' policies are mainly directed to acquiring the foreign resources they need, while minimizing the control over these resources by the supplying institutions. It is being increasingly recognized that poorer countries, at least, cannot achieve these goals overnight, and that a choice has to be made about priorities. The nature of the trade-off is very much influenced by the country's need for the resources and the extent to which it wishes to avoid control.

Countries may pursue a variety of policies. To maximize the net benefits of resource acquisitions where the equity investment route is preferred, their attention is directed to (1) the screening of sectors in which equity investment is permitted, or permitted only to a certain extent, (2) restricting the percentage of foreign particiation in indigenous firms, (3) insisting on certain criteria for the behavior of foreign affiliates, particularly as regards restrictions on the use of technology, employment of nationals, and subcontracting, and (4) limiting payments extracted by investing companies. In each of these areas, controls have become more widespread over the last two decades, with the result that the 100 percent owned foreign affiliate, with its wide maneuverability over resource transmission and the conditions attached to it, is a much less common phenomenon.

There is a fairly well-documented literature about the ways host governments influence the transmittal of resources by TNCs.(7) Although this is not the main concern here, it is germane in that it is an alternative policy to that of seeking new transmitting outlets. It is also worth exploring the extent to which joint ventures, which are a halfway house between the 100 percent equity investment and a purely contractual venture, affect the balance between control over decision making and resource transfer.

Joint ventures are generally favored by host countries, as they are believed to provide most of the benefits of foreign direct investment while decision making is shared. The presumption is that even a minority share held locally may influence decision making in favor of local interests, while a majority local holder will ensure that this is done. Joint ventures may also be desired either to capture part of the economic rents of foreign direct investment or, by obtaining access to more information about costs and pricing policies of the investing firm, reduce the opportunities for these rents to be earned.

When are joint ventures likely to be successful? Basically, when they are in the interests of the TNC itself (hence when local negotiators have more leverage), and when the local partner has a unique managerial, marketing, or technical contribution to make, particularly in areas complementary to the needs of the TNC. This would suggest that import-substituting joint ventures, in which knowledge of local production and marketing conditions is required, are likely to be among the most successful. Technology transmission to joint ventures will normally take the form of new products, production processes, and research and development (mainly development) programs. However, since the TNC may have less incentive to integrate its affiliates into its network of work markets, the host country may lose some of the benefits of the international division of labor. Nonetheless, where only local markets are served, the TNC with a less than 100 percent equity stake might not have the same incentive to develop this market, relative to others in its network where conditions are more profitable.

Where the host country is in a weak bargaining position, the transition from a fully owned affiliate to a joint venture may make little difference to the amount of control exerted by the investing company. Since the profits have to be shared with the local partner, the foreign partner may feel reluctant to relinquish control that might enable him to recoup the lost profits by other means. Alternatively, depending on the marginal cost of supplying the technology and the use made of it in markets that could be better served from more profitable affiliates, the TNC might choose to reduce the flow of technology, or fail to respond in the way the host country would like to demands for a more appropriate technology. This is particularly likely in the case of production destined for export markets.

For these and other reasons (U.N., 1978) host countries have become disenchanted with joint ventures, although they may still be preferred to 100-percent-owned foreign affiliates. Attention has instead been focused on encouraging various forms of nonequity arrangements.

In the 1950s and early 1960s there were three main forms of time-limited contracts: the concession agreement, which was particularly popular in resource-based industries, particularly mining and oil exploration; the licensing and marketing agreement, dealing with the sale or licensing of patents, trademarks, or franchises for a royalty and/or fee; and management contracts, covering the sale of various forms of management (including administrative) services. These agreements have multiplied over the years and, as has been pointed out, all tend to involve the exchange of resources or the right to use resources, both singly and jointly, usually with some control implicitly or explicity exercised by the source firm.

In some cases the contracts have been with TNCs as part and parcel of direct foreign investment – this is the case with most concession agreements. But in others agreements may be made with firms that have no equity participation, for example, management contracts.

Though it is generally not well publicized, there has been growing dissatisfaction by many host countries, particularly developing countries, with these forms of resource transference, on two basic counts. The first is that, because of their stronger bargaining power, TNCs are able to tilt the terms of any contract in their favor – so much so, in some cases, that the result of the resource transfer is a net outflow of income from the recipient country. The second is that control over decision making either explicitly or implicitly exercised, allows the secondary or downstream portion of resource-based activities to be undertaken in the contractor's country rather than the contractee's country. This means that the technology of downstream operations is retained in the home country, even though it may be in the interests of the host country to develop that capability.(8) Whether this was due to companies' inability to make any profit in the host country or not enough profit varied from case to case, but there can be little doubt that increase bargaining power on the part of some nations, or groups of nations, has had some success in forcing TNCs to relocate their secondary processing operations. This suggests that there may well have been some surplus profits available for rechanneling, or at least that companies would prefer to engage in local processing rather than lose the benefits of the original investment altogether.

In the 1970s the terms and conditions of contracts of the traditional kind concluded between TNCs and host countries have generally swung in the latter's favor (UN,1978), while new forms have emerged by which the recipient country buys resources it lacks or in which it has a comparative disadvantage in supplying. Hence, particularly for developing countries, management contracts became popular in the 1960s and still remain among the most acceptable forms of resource imports. Usually, however, technology accompanies the management, and fixed-term technical-service agreements have come to incorporate both knowledge and the use made of that knowledge. Such agreements might also embrace a training element. Countries, particularly those short of marketable products, favor the acquisition of trademarks or franchises, so that they can offer for sale consumer goods with well-known brand names and/or be host to hotels of international repute.

Another group of contractual arrangements followed the movement toward regional economic integration. Coproduction and complementation agreements were especially widespread in areas and industries where there were benefits to be derived from product or process specialization. While TNCs are often involved in this specialization, the decisions on what is produced in which country is taken by the agreement of countries concerned, rather than by the TNCs themselves. The experience of the Latin American Free Trade Area is perhaps the best illustration of the value and limitations of this form of contractual arrangement.

The experience of recipient countries with the various contractual forms has been mixed. Success appears to be very much dependent on their ability to absorb and use the resources being imported in the most efficient manner. This requires that complementary and backup resources be locally available. I have suggested that countries tend to overestimate the extent to which they possess these capabilities. Similarly, resources may need to be jointly supplied. It may be cheaper to acquire a package of resources because of the spreading of joint overheads; it may be more efficient because of the related skills possessed by the technology producers, managers, technicians, chemists, and engineers whose knowledge is part of the system. In general, governments tend to underestimate the benefits of joint supply. Finally, the control over decision making often associated with the resources might itself enhance the benefits the benefits of the resources; this could result, for example, from the integration of the local affiliate (or contractee) with the TNC (or contractor); or it could result from the global sourcing, access to markets, personnel deployment, and so on that international productions make possible. But governments usually look more at the negative aspects of control than at the benefits it may bring.

The extent to which direction is an integral part of any transmission of resources depends as much on characteristics specific to the country, industry, and firm involved as on the mode through which it is transferred. This is another lesson that governments find difficult to learn. The usual presumption is that control is a unique feature of the internalizing activities of TNCs that can only be, and usually is, exercised where they hold a 51 percent or greater equity stake in the foreign affiliate. But in practice control over key areas of decision making – for instance the location of research and development activities, what sort of technology is transferred, who appoints managers, who decides which markets are to be served and where inputs are to be sourced from – are not unique to foreign direct investment. They may be explicitly written into a contractual agreement, or they may be part of the operating strategy of a contractor that the contractee, due to lack of knowledge about the options open to him or lack of bargaining leverage, can do little about. The experience of countries moving toward the contractual route of importing resources confirms that, de facto, the control they seek to avoid is not embodied in the equity investment per se but in the resources these countries still need to import.

It is not difficult to cite examples where control and resource transmission go hand in hand independently of the mode of transfer.

The first of these is when the recipient firm is catering to an international market. Then the product or service has to be of a particular quality, and unless the supplier is assured that the host country is able to provide this quality, the supplier will not be prepared to make technology available unless it can keep control over the way it is used. Examples are goods with internationally acknowledged trademarks – cigarettes, chocolate, rubber tires, detergents, razor blades – where a company's name is involved and goodwill is important. Outside of manufacturing, the hotel industry provides what is perhaps the best illustration of control being exercised over resources that have been transferred. One has only to read the brochures of Hilton International and Intercontinental Hotels to realize that, although they have not invested any equity capital, these chains exercise enormous control over day-to-day resource allocation through management contracts. Indeed, the host country is importing not only management but control.

Second, and this goes with the first, is where there is a lack of technological and organizational capability in the recipient country. Where there are few or no backup facilities to the resources being transferred, where a host country doesn't know what to do with the resources, then, inevitably, the sending enterprise can insist upon some control over the resources before they are made available. If the sending enterprise does not, then the receiving country may find that it has bought resources but cannot deploy them to best advantage. This seems to be the fundamental dilemma facing the poorest of the developing nations.

Third, insofar as the contractor wants to exert some control over the technology once it is transferred, his ability to so so will be greatest when his bargaining position is strongest vis-a-vis the host country. This, in turn, will depend on the need of the country for the resource and the alternative sources of supply for it, and on the alternative opportunities open to the TNC and how profitable the particular transfer is likely to be.

Whereas most of the emphasis of host governments in the mid-1960s and early 1970s was directed to maximizing the net benefits of foreign direct investment and/or encouraging the acquisition of technology through alternative routes, in the later 1970s much more attention was given to the contents and terms of contracts concluded between foreign firms and host governments (or host firms), independent of the form of the acquisition. This, in turn, has forced countries to search for new ways of bundling the resources, or some of the resources, currently provided by foreign direct investment. The mode of the contractual route, in terms of the likely control associated with it, is also being reappraised. Concession agreements are being replaced by technical service agreements; turnkey operations are being supplemented with some kind of continuing contractual relationship. Countries are realizing that the indigenisation of control cannot be accomplished overnight, but must be a gradual process; that, initially at least, the costs of self-sufficiency may be too great; that without some external direction of resource usuage, resources either will not be provided or will be inefficiently deployed. It is therefore necessary to build into any contractual relationship sufficient flexibility to allow both parties to reconsider their resource-providing and resource-control roles, subject to a reasonable period of time over which such relationships remain stable. This period must be long enough to allow the resource-providing firms to undertake the necessary investment or to make the contract worthwhile, but short enough to allow either party to renegotiate the terms as soon as it is reasonably possible, when justified by changing circumstances.

SOME THOUGHTS ON FUTURE MODALITIES

From the published literature, and the attitudes and stances of governments, particularly those expressed in UNCTAD and by the U.N. Commission of Transnational Corporations, one might suppose that the mechanisms international resource transmision will command the attention of both exporting and importing countries over the next two decades. My own opinion is that this is not likely to be so. Like the whole debate about transnational corporations, which has gone through fairly identifiable and predictable phases, discussions about the costs and benefits of various modes of resource transmission will, I suspect, go through similar phases, and gradually the issue will narrow down to questions about market structure and the process of moving from one stage of foreign involvement to another.

Viewed from the perspective of world development, I believe that the question about new forms of resource transmission is much more interesting and worthy of attention in international fora than the present debate on the merits and demerits of TNCs.(9) The crucial question is how to harness the resource capabilities of these corporations, and of other suppliers of capital and technology, in a way that will help create an economic infrastructure that can satisfy basic needs of the poorer countries. Here is where more cooperation between governments and private enterprise may be necessary to both create the stimulus for and bring about the implementation of schemes that may involve many billions of dollars. Irrigation ventures, exploration of the sea bed, and electrification schemes are the type of activities in which consuming governments are likely to be entrepreneurs, with financing obtained perhaps from international agencies or the capital market, technology from both Western firms and East European governments, and management and marketing expertise from a consortia of TNCs or other firms. Some of the these issues have been touched on in international fora, and the role of TNCs in them has been discussed, but because of the current climate of opinion about transnationals little progress has been made.

Yet forerunners do exist of what may well be the organizational pattern of resource transmission for the future, not only in grandiose schemes such as those just mentioned but in others involving resources that might best be supplied from different sources. To quote a speculation by Howard Perlmutter: "A Japanese and U.S. exploration company would use U.S. nuclear devices to help find Siberian oil in conjunction with a Soviet firm; the oil would then be distributed in Western Europe by European oil companies." A precursor to Perlmutter's vision already exists in today's multinational banking consortia and the agreement between U.S. General Electric's and France's SNECNA to make low-pollution aircraft engines. In the case of the building of the F-16 fighter plane, more than thirty-five European companies are involved in a production program that is being undertaken on both sides of the Atlantic simultaneously.

On a more traditional but multilateral front, we are likely to see a further development of complementation and coproduction schemes supported by governments. Take for example the agreement worked out by IBM with Argentina, Brazil, Chile, and Uruguay. Computer components are to be manufactured in Brazil and assembled in Argentina in IBM affiliates; a Chilean firm is to be licensed to produce punchcards, and the entire operation is headquartered in Uruguay. In the motor vehicles industry, the Ford Motor Company is reputed to be planning to spend $1,000 million by 1980 in integrating its plants in Southern Asia. Other illustrations of recently concluded agreements are cited by Jack Baranson (1978). It is worth noting that, in these arrangements, decisions on the allocation of resources are often taken with the permission of or directly between governments. To this extent, individual host governments have voluntarily surrendered control over resource allocation to a regional association. They have not gained self-sufficiency, except insofar as they are part of a more self-sufficient region. This reinforces a point made earlier: that the question of the extent of dependence is less the real issue than on whom one is dependent (Dunning and Gilman, 1976).

There can be little doubt that governments, both home and host to resource transmissions, are likely to play a more important role in determining both the vehicle of transmission and the terms and conditions attached to it. In a number of resource-based and high-technology sectors, governments either directly (through specialized agencies) or indirectly (through state-owned corporations) act as entrepreneurs or contractors. In addition, through their control on pricing, markets, and location of production, they may strongly influence the behavior of foreign firms. With their greater knowledge of and experience with the technology of choice, TNCs and other firms may play the role of contractees. On the other hand, because of the growth of indigenous capacity, more and more TNCs may choose the subcontracting route of international involvement, rather than licensing local firms or setting up affiliates. Much will depend on market conditions for the alternative modes of involvement – which may themselves be influenced by both national and international institutions and policies.

In summary, then, my contention is that the past twenty years have seen an acceleration of the evolution of the form and content of international resource transmission. The nineteenth century was the first state in this process, in which resources (or the right to their use) were transferred separately, at "arm's length" prices, with control over their use being transferred with the resources. For most of the twentieth century, up to the mid-1960s, resources were increasingly transmitted within the supplying institutions, with control maintained by the transferer and the package of resources becoming both more complex and difficult to disentangle. The last decade has witnessed two main developments: first the beginning of a return to externalized transmissions, as firms have been forced to disinternalize their activities, while the markets for resources have become less imperfect; second, increasing government intervention, in effect a replacement of private administrative fiat by public administrative fiat in resource allocation. At the same time, the sources of capital, knowledge, and human skills, the forms of their packaging, and the manner of their marketing have widened, so much so that it is no longer appropriate to focus on a particular institutional form of resource. Indeed, in the 1980s economists should shift their attention away from institutional forms to the structure of markets in which resources are transferred. These are now becoming as numerous as product markets, and as yet relatively little attention has been paid to them (but see Casson, 1979). This is a challenging area for researchers and one that is of no less interest to policy formation, for both resource-sending and resource-receiving countries.

NOTES

(1) A word on terminology is appropriate here. First, I use the word "transmission" rather than "transfer." The Oxford English Dictionary defines "transfer" as "to make over the possession of some asset or right" – implying that the transferee's gain is the transferor's loss – while "diffusion" is defined as "spreading out," implying no loss on the diffuser's part. The word "transmit" seems to embrace both "transfer" and "diffusion," for the emphasis of the word is on "the communication of passing on or ideas, assets, etc." By "resources" I mean either a stock or productive assets or the right to use such assets, wherever these are mobile across national boundaries.

(2) For a discussion of the incentives for TNCs to internalize their activities, see especially Buckley and Casson (1976), Dunning (1977), and MaGee (1977). For a discussion of the relevance of internalizing activities to the theory of the firm, see Arrow (1969), Alchian and Demsetz (1972), and Williamson (1975).

(3) For an excellent account of the extent to which TNCs do integrate themselves with the economies or regions of which they are a part, see Vaitsos, 1977.

(4) It is interesting to note that up to the mid-1960s the extent to which foreign companies controlled decision making in their affiliates was closely related to the age of the affiliate and its size relative to that of the parent company; see, for example, Dunning, 1958. Subsequently, a reverse trend set in as the larger and well-established affiliates began to integrate their activities, e.g., in the EEC and Latin American Free Trade Area, as the result of which control returned to the parent company.

(5) There is an exact parallel here to the substitution of a managed currency for the gold standard in the interwar period and the protectionist trade policies that followed.

(6) A brief survey of this literature is contained in a paper written for the U.N. Centre on TNCs, Dunning and Pearce, 1977.

(7) For a summary of the current situation, see U.N., 1978.

(8) It must be admitted that many other considerations affect the economic viability of downstream operations, not least the lack of available marketing opportunities and technological and social infrastructure. The change in this situation is most strikingly shown in the oil-producing countries.

(9) One possible exception is the debate about the role of new types of TNCs, e.g., from Japan and some of the developing countries. See, e.g., Ozawa (1975), Wells (1977).

REFERENCES

Alchian, A., and Demsetz, H. 1972. Production, information costs and economic organization, American Economic Review 62 (December).

Arrow, K. 1969. The organization of economic activity: Issues pertinent to the choice of market and non-market considerations. In The analysis and evaluation of public expenditure: The PPB System. U.S. Congress, Joint Economic Committee. Washington: U.S. Government Printing Office.

Baranson, J. 1978. Technology transfer to developing countries. New York: Praeger.

Buckley, P.J., and Casson, M.C. 1976. The Future of the Multinational Enterprise. London: Macmillan.

Buckley, P.J., and Davies, H. 1979. The place of licensing in the theory and practice of foreign operations. Unpublished paper.

Casson, M.C. 1979. Alternatives to the multinational enterprise: An exercise in the pure theory of proprietary information. London: Macmillan.

Coase, R.H. 1937. The nature of the firm. Economica 4 (November).

Dunning, J.H. 1958. American investment in the British manufacturing industry. London: Allen and Unwin.

Dunning, J.H. 1977. Trade, location of economic activity and the MNE: A search for an eclectic approach. In The International Allocation of Economic Activity, ed. B. Ohlin, P. Hesselborn, P.M. Mijkman. London: Macmillan.

Dunning, J.H. 1978. Multinational business and the challence of the 1980s. Multinational Business, no. 1.

Dunning, J.H., and Gilman, M. Alternative policy prescriptions and the multinational enterprise. In The multinational enterprise in a hostile world, ed. G. Curzon and V. Curzon. London: Macmillan.

Dunning, J.H., and Pearce, R.D. 1977. Alternative forms of international resource transmission. Paper prepared for Centre on Transnational Corporations, November 1977.

Franko, L. 1971. Joint venture survival in multinational corporations. New York: Praeger.

Franko, L. 1976. The European multinationals. Stamford: Greylock.

Gabriel, P. 1967. The international transfer of corporate skills. Cambridge: Harvard University School of Business Administration.

Grub, P. 1972. Multinational franchising: A new trend in global expansion. Journal of International Law and Economics (June).

Hayden, E.W. 1976. Technology transfer to East Europe: U.S. corporate experience. New York: Praeger.

Houston, T. 1977. A simple theory of the consequences for the metropolitan industry. Paper presented to the European Institute of International Business, Uppsala, December 1977.

Hymer, S. 1960. The international operations of national firms: A study of direct foreign investment. Ph.D. diss., MIT.

Hymer, S. 1970. The multinational corporation and the law of uneven development. In Economics and world order, ed. J.H. Bhagwati. New York: World Law Fund.

Lall, S. 1973. Transfer pricing by multinational manufacturing firms. Oxford Bulletin of Economics and Statistics 35 (August).

Magee, S.P. 1977. Technology and the appropriability theory of the multinational corporation. In The new international economic order, ed. J.H. Bhagwati. Cambridge: MIT Press.

Murray, R. 1972. Underdevelopment, international firms and the international division of labour. In Towards a new world economy, ed. J. Tinbergen. Rotterdam: Rotterdam University Press.

Nurkse, R. 1954. The problem of international investment today in the light of nineteenth century experience. Economic Journal 64 (September).

Ozawa, T. 1975. Peculiarities of Japan's multinationalism: Facts and theories. Banca Nationale del Lavoro Quarterly Review (December).

Penrose, E. 1959. The theory of the growth of the firm. New York: Wiley.

Plasschaert, S. 1979. Transfer pricing and multinational corporations. Farnborough: Saxon House.

Sunkel, O. 1972. Big business and 'dependencia.' Foreign Affairs 50.

Svedberg, P. 1978. The Portfolio-direct composition of private foreign investment in 1914 revisited. Economic Journal 88, no. 352 (December): 763-77.

Tomlinson, J.W.C. 1970. The joint venture process in international business: India and Pakistan. Cambridge: MIT Press.

United Nations. 1978. Transnational corporations in world development: A reexamination, Sales No. E78 II A.5. New York: U.N. Economic and Social Council.

Vaitsos, C. 1977. Regional integration cum/versus corporate integration. Paper presented for the Center on Transnational Corporations.

Wells, L.T. 1977. The Internationalization of firms from developing countries. In Multinationals from small countries, ed. T. Agmon and C.P. Kindleberger. Cambridge: MIT Press.

Williamson, O.E. 1975. The economics of internal organization: Exit and voice in relation to market and hierarchies. American Economic Review: Papers and Proceedings 66 (May).

Comments

R. Hal Mason

It has been a great pleasure to read Professor Dunning's paper on technology transfer, because so many of the thoughts expressed there are so agreeable to my own.* However, this fact alone makes the job of discussant and critic doubly difficult. My job would be ever so much easier if he had embraced the conventional wisdom as it has traditionally been applied to the multinational firm. Then I would be able to suggest that the phenomenon of the multinational firm does not fit nicely into traditional theory and go on from there to make several of the good points he has already made. I am placed in a dilemma similar to that suggested by the old saying, "Don't just stand there, do something!" As discussant, I do have to do something. Accordingly, I have decided to examine some points not included in Professor Dunning's paper – points that I consider compatible with the models of resource transmission presented by him. Also, I would like to raise a few questions with respect to some of Professor Dunning's points. I think there are alternative approaches and interpretations in several instances. My mode of presentation will be to raise questions about specific aspects of the Dunning paper, followed by some points that might well have been included in the paper but were not.

A PROBLEM OF DEFINITIONS

I do have some difficulties with Professor Dunning's treatment of some aspects of technology transfer. In particular, these are: (1) the distinction between proprietary and nonproprietary technologies, and (2) the

*Many of the thoughts underlying this discussion were developed during 1977-78 while I was an IBM Postdoctoral Fellow. I wish to express my appreciation to IBM for its support.

distinction between transfers via markets and those taking place within firms.

In my opinion, the distinction between proprietary and nonproprietary technology needs to be clarified. Whether a technology is proprietary or not often may depend upon how the technology is to be applied. For example, a machine may be sold in the open market and anyone can purchase one. From this point of view it is nonproprietary. However, if it must be used as a part of a mechanical system and only transnational corporations can supply and operate such a system, it makes little difference whether the machine itself is nonproprietary. Most modern technological systems can be fabricated using nonproprietary technologies or standardized equipment and machinery; indeed, most plants the world round are so constructed. It is either the products and services they produce or the peculiar expertise required in the production of these outputs that is proprietary. Occasionally there are proprietary processes in some product fields, such as drugs and chemicals, but most manufacturing uses standardized processes available to all. The point is that it is not so much the technology per se that is proprietary, but the ability to operate the technology and/or the end product of the technology. Moreover, the product itself may be and probably is protected by patents, trademarks, or brand names. It is largely the end product of technology that is desired, and it is the ability to produce it effectively that allows TNCs to appropriate economic rents in their international operations.

Therefore I am unsure how helpful it is to dichotomize transactions across national boundaries into those that take place through the market and those which take place within an enterprise. There are subtleties that are obscured by such a classification. One can buy through the marketplace the basic technology to produce many products. One can even buy the rights to use patents and trademarks using licensing agreements and other contractual arrangements. And all of these can be opened to bidding so as to identify the lowest-cost source of supply. However, I am dubious that the market can effectively be used to reduce the supply price of the technology and its final output when compared to within-firm transactions. Unfortunately, the market is almost never pure, because certain types of information cannot be transferred readily. Large enterprises represent a stock of physical capital, human skills, and accumulated learning. And somewhere along the line, in complex industrial processes, there are critical junctures where the proprietary technology cannot be readily disentangled and sold as a separate package. Perhaps it can, given enough time and resources. However, it is generally less costly to bring the owner of the proprietary technology into a long-term transactional relationship through some form of implicit or explicit partnership arrangement. This may involve one or more methods, including technical service and management contracts, supply contracts, complex licensing agreements, and perhaps even an ownership role.

Under such conditions, between-firm transfers of technology may be little different from within-firm transfers. Moreover, the TNC may be in a position to appropriate its economic rents just as it would if all transactions were internalized. Indeed, I believe that ownership of know-how, rather than of physical technology, is the key to the generation of most of the appropriable economic rents attributed to technology. Such know-how is created by the firm, and the cost of duplicating this know-how is what dissuades competition and provides a continuing flow of economic rents. It is these rents that make the acquisition of technology appear to be expensive when it comes in the form of foreign wholly owned subsidiaries. However, from my own studies of American, European and Japanese firms I have come to the conclusion that the rents are collected even when there is little or no ownership position and the transactions take place through the market mechanism. This suggests that there is an imperfect distribution of knowledge across countries, and certainly there is. If there were not, there would be little reason for technology transfers or for the existence of TNC.

A MODEL FOR ANALYZING TECHNOLOGY TRANSFERS

My own concern has been to examine the question of why firms transfer technology and why they use the methods they do. A simplistic answer is that they are attempting to maximize profits. This is hardly revealing. It does not explain why enterprises, and even entire industries, follow the practices they do. If they are maximizing profits, they certainly have found a variety of ways to do so. For example, why has the Japanese textile industry aggressively invested abroad, while the American textile industry has not? Also, are so many Japanese firms so much more willing to engage in joint ownership ventures than are their American or European counterparts? There must be some fundamental, underlying reasons for the differences we observe.

The Japanese, at least superficially, appear to be much more willing, partially or totally, to unpackage their technologies, through greater use of joint ventures, licensing, personal service contracts, supply contracts, and the like. They rely less on the wholly owned subsidiary than do American and European firms. However, it should also be noted that data seem to indicate that European firms are more flexible than are American firms in terms of the extent to which a variety of transfer instruments is used. In light of these observations I have developed my own working hypothesis. It draws on Hymer's thesis and Vernon's "product life cycle" hypothesis.(1) Moreover, it is very much akin to the framework implicit in Professor Dunning's paper. It might also be interpreted to be a part of the "market failures" genre beginning with Coase and extended by Arrow, Alchian and Demsetz, and Williamson.(2)

Simply stated, the hypothesis is that enterprises have unique capabilities that are created over time and that result in competitive advantages. These advantages include both the creation of technology and the ability to internalize resources or decision-making capabilities in such a way as to either become more efficient than the marketplace or be able to regulate and control access to these advantages, or both. In this sense, the enterprise substitutes for the market mechanism and appropriates economic rents by controlling access to its own unique capabilities.(3) The extent to which the enterprise can control access to these proprietary advantages will largely depend on how difficult or costly it is to duplicate these advantages, which include technology, know-how, reputation, and identity. The more costly it is to replicate these advantages, the higher will be the barriers to entry by other firms(4) and the greater will be the owning firm's incentive to limit or control access to the advantages or technology.(5) At the other extreme, the easier it is to replicate the technology, the greater will be the incentive for the owning firm to use the market mechanism and sell the technology outright. Perhaps this explains why a variety of devices are used to apply technology. Moreover, these devices differ in the extent to which they allow the owners of technology to maintain control over or limit access to the technology.

To restate the above proposition: if the technology is easily replicated and/or the economic rents generated by internalization are small, the technology will be sold freely in the marketplace, but if the technology is costly to replicate and/or the economic rents from internalization are large, then the owning firm will attempt to control access and appropriate economic rent. Under these circumstances it will seek to compromise the market mechanism even to the point of internalizing all aspects of the technology's application by owning all of the facilities used to apply the technology, that is, it will engage in direct foreign investment and may insist upon 100 percent ownership.

As with any economic decision, the degree to which a firm will attempt to control access to the technology and the method of control chosen will depend not only on how great are the appropriable economic rents but also on the costs of control and the extent to which various control devices can be used to capture the rents. Schematically, the methods of transferring technology in terms of their degree of control to technology owners might be envisioned as in figure 1.1. By using the schema shown in figure 1.1, I do not mean to imply some direct functional relationship. Moreover, it is quite common for two or more of these devices to be used in combination. Perhaps it is feasible to obtain just as much effective control by using a minority joint venture in combination with a license, a management contract, and a technical service contract as would be obtained by using a wholly owned subsidiary.

There is another element to the argument, and that is that the degree to which one or another instrument provides effective control also depends on the nature of the technology being offered. Stated differently, the enterprise appropriating economic rents certainly is not going to transfer those aspects of the technology that yield the rents. If the total technology must be protected in order to capture the rents, then the firm will choose only the wholly owned subsidiary. IBM and Texas Instruments are good examples of this strategy. For example,

IBM absolutely will not establish a joint venture that allows access by the partner to the technical side of the business. It is only willing to joint venture in such a way that the partner owns the marketing arm and IBM owns the production arm of the total enterprise. When the Indian government found this approach unsatisfactory, IBM withdrew; it was unwilling to share its technology.

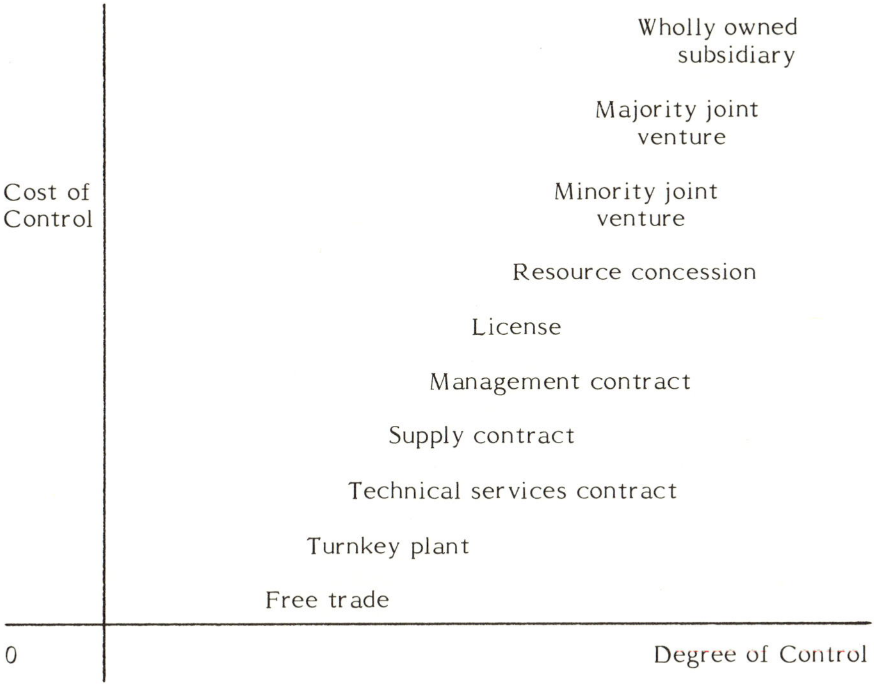

Fig. 1.1. Methods of technology transfer and degree of control.

There are vast differences among firms as to how far they must go in order to protect their technology. Unlike IBM, the Coca-Cola Company needs not own and operate facilities abroad in order to protect the technology that yields it economic rents. Indeed, the Coca-Cola Company's policy is one of avoiding direct foreign investment wherever possible. It does not need to invest abroad or operate foreign bottling plants. A simple franchise and supply contract is sufficient, because Coca-Cola does not allow others access to the Coca-Cola syrup formula – the Coca-Cola Company supplies the syrup. Through the pricing of the syrup and the franchising fee charged for the use of Coca-Cola's trademark, the company captures the economic rent.

Thus firms need not invest abroad in order to appropriate economic rents. Whether they need to or not depends on the technological chain, the end product of the technology, and the extent to which one or another transfer instrument can be used to protect those aspects of the technology that yield the economic rents.

Host countries have concluded, in many instances, that they can reduce the economic rents appropriated by TNCs in their technology transfers. Perhaps they can, but I am dubious. It seems apparent that, through regulation, they have elevated the cost of control to the TNCs. But at the same time they may have reduced the flow of technology. There appears to be some recognition of this. At least some countries have backed away from highly stringent controls on direct foreign investment, ownership structure (local equity and joint ventures), ceiling rates and controls on licensing agreements, local content requirements, and the like. My examination of the ASEAN group indicated to me that these countries are becoming more sophisticated regarding technology acquisition. They recognize that various industries differ with respect to the value of the technology they have to offer. This is partly a recognition of the fact that locally owned enterprises can handle many of the technologies and that the technologies are broadly available under competitive conditions, that is, that there is little need for foreign involvement. It also is partly a recognition of the fact that owners of some technologies must be given preferential treatment if the technology is to be acquired at all.(6) High-technology industries are generally accorded greater latitude and offered more incentives than are low-technology industries. Also, industries that involve highly differentiated products or provide access to foreign markets are dealt with more leniently than those facing more competitive conditions, reflecting the recognition that some types of know-how are not available locally. However, such policies raise questions about how successful countries have been in reducing or even sharing in the economic rents generated by foreign technology. Foreign enterprise is largely excluded from industries where economic rents are low, but is allowed to operate under incentive in those industries where economic rents are high.

AN APPLICATION OF THE MODEL TO SUPPLYING COUNTRIES

I have suggested elsewhere that Vernon's model, when coupled with the model outlined above, can be extended to explain, at least in part, why Japanese and European firms are more willing than American firms to engage in joint ventures.(7) As Vernon has pointed out,(8) countries differ in their technological capabilities, with the advanced countries being the pioneers of new technologies and the developing countries the users of standardized technologies. There is a trickle-down effect whereby the pioneer has a monopoly for some time during which it is a net exporter of the outputs created by the technology. But gradually the technology moves to other advanced countries, and ultimately, after it is standardized and innovation has come to an end or nearly so, the technology moves to developing countries. At this stage, the Ricardian model of comparative costs prevails. Since there are no intercountry differences in labor productivity, the low-wage-cost country has the comparative advantage. However, this model does not mean that one country will always be the technological leader in every industry. One can ask "why it was that so few American firms in the textiles, wearing apparel, steel, clay, glass, and other low-technology industries invested abroad." One answer relys on Hymer's reasoning.(9) In highly competitive industries there is no advantage to investing abroad; locally owned firms are as capable of applying the technology as are foreign-owned firms. Or stated differently, there is little in the way of economic rents to be appropriated and therefore the incentive to internalize these activities is too low, that is, the costs of control are not offset by the returns to control. Then why is it that Japanese firms in these same industries have invested abroad? Here we must come back to Vernon's model. Perhaps the technological capability is highest in the United States, followed by Europe and finally Japan, in descending order. This does not mean that Japan has no firms that are just as technologically capable as some firms in the United States and Europe. It merely means that on average Japanese enterprises are lower down on the technology curve. Based on Hymer's reasoning, firms investing abroad would tend to have internalized certain capabilities and these would tend to spring from the comparative advantages offered by the resources of the home country. Any capturable economic rents would spring from these advantages. There must be something unique within the firm that allows it to make a greater profit than locally owned firms in the same or similar industries. Otherwise it has no incentive to invest abroad.

There must be some explanation of why Japanese technology transfers and investments, even in manufacturing, have gone mainly to developing countries in low-technology industries, whereas American and European investments have gone more heavily to advanced countries in higher-technology industries.

In the low-technology industries, Japanese firms do compete with local firms. This goes against Hymer's reasoning, if we assume that local firms are equally proficient in the low-technology industries when compared with the Japanese, but it is still consistent with Vernon's hypothesis. Perhaps the Japanese firms in these industries have developed superior technology or some other advantage that firms from the United States or Europe have lost. This would be in line with the product cycle hypothesis, if indeed the United States, followed by Europe is higher up on the product cycle curve. There is no inconsistency between this hypothesis and a country's loss of its comparative advantage to another in certain industries. Firms across industries must allocate resources into innovation and technological development if they are to maintain comparative advantage over time. However, they must also compete with one another across industries for the service of scientists, engineers, and technicians. Thus in a reasonably competitive economy the high-technology, high-profit industries will attract the innovative capability away from the low-technology industries. But what is a high-technology and low-technology industry will differ across countries. Perhaps in Japan the textiles industry, along with several other industries that might be considered low-technology in the United States, is higher up on the product cycle curve than it is in the United States. Thus in Japan the textiles industry may be gaining (or at least not losing) innovative capability, while in the United States it is. The consequence would be that the United States, followed by Europe, would lose its ability to compete in textiles and other low-technology industries, and it would not be surprising to find few overseas U.S. and European investments in these fields.

Perhaps the reason so many Japanese firms are willing to enter joint ventures rests on the underlying technological capabilities. If the Japanese technology, on average, is less sophisticated, there would be fewer economic rents to be appropriated and hence less to be gained by protecting the technology. However, it is notable that the Japanese firms I studied in high-technology industries seemed just as anxious to have 100-percent-owned subsidiaries as their American counterparts. This is quite consistent with the analysis presented here.

SOME CONCLUDING COMMENTS

I find Professor Dunning's thinking quite compatible with my own. Perhaps we are operating with the same paradigm but a somewhat different terminology. I think we might differ as to what is considered embodied as opposed to disembodied technology. And perhaps we are even defining the term "technology" somewhat differently. Nevertheless, I think we arrive at the same destination, although we take somewhat different routes to get there.

I do have some minor disagreements that should be mentioned in passing. These are not meant to detract from his paper, which I consider to be a fine one indeed.

I feel that in his discussion of modalities he has not placed enough emphasis on the costs of controlling access to technology. There are costs involved in capturing economic rents, and perhaps the cost of complete control, in many instances, is just too high to warrant overseas investment. There needs to be recognition that the market is likely to be imperfect where the potential economic rents are high – that is, it pays to internalize. This is only feasible where the costs of duplication are high. Also, creating technology is costly. And, although Professor Dunning recognizes this, not enough is made of it. Internalization and protection is likely to be in evidence where costs of creation are high because this is the only way the enterprise can assure recoupment of its costs.

It also would be helpful to make some distinction as to the types of industries involved, the complexity of the technology being transferred, and the need for access to international markets. All of these tend to affect the economic rents and the willingness to transfer using various instruments or modalities.

Some points are overly generalized in the discussion of technology – financial systems and other services, for example. Most such know-how can be purchased competitively from accounting firms, consulting firms, and the like, and like any other know-how or technology, much of this can be purchased by competitors or potential competitors. That they do not purchase it suggests that the market is operating and that firms that have already developed and internalized such know-how can provide the services internally at less cost than other firms that must buy it in a competitive market.

Timing is extremely important to the decisions of large transnational firms. So, too, is the time value of money. This may be a key reason why TNCs internalize decisions and limit access to their own know-how. When opportunities arise, the enterprise that can move most quickly often has the advantage. Thus it is profitable to internalize and collect the economic rents rather than allow someone else to purchase the technology and reap the economic rents. Because of this, and because of the international extension of the patent system, Dunning may be correct in his conclusion that, over time, the barriers to outright acquisition of technology have increased. International registration of patents probably does limit the number of potential suppliers. However, it should be noted that the developing countries are the major critics of the system; yet it is in the developing countries that inventive capability is low. I do not see, under these circumstances, how there would be any great burst of invention and emergence of competitive sources of supply, even if reciprocal patenting were to be eliminated.

I disagree somewhat with Dunning's appraisal of developing countries' abilities to assess technology and to locate competing sources of supply. I agree that they are shorthanded but my own exposure to the problem suggests that their capabilities are not so starkly meager as he suggests. There is considerable exchange of information among boards of investment and similar agencies across countries. They have considerable expertience and are backed by able young technocrats who examine alternative sources of technology. Governments have become much more knowledgeable regarding the value of various technologies and the available methods of transferring them. They recognize that they do not have to accept 100 percent ownership in order to obtain lower-level technologies. As I have noted, knowledge requirements differ among industries and among the technologies being applied. Disinternalization has taken place in the low-technology and primary producing industries. This is largely a result of the learning curve effect. Countries now have many of the complementary resources required in the way of skills. This in turn has allowed them to drive a harder bargain and acquire certain technologies on more competitive terms. Yet it would be misleading to conclude that the TNCs have lost their ability to collect economic rents. For example, the major international oil companies are using technical aid agreements and management contracts to achieve many of the same ends they achieved through 100 percent ownership.

A final comment: It would have been helpful if Dunning, in his mention of LAFTA's experience with complementation programs, had given a few examples to illustrate, as he says, "the value and limitations of these forms of contractual arrangements" (meaning between LAFTA countries and TNCs).

NOTES

(1) Stephen Hymer, The International Operations of National Firms: A Study of Direct Foreign Investment (Cambridge: MIT Press, 1976). R. Vernon, "International Investment and International Trade in the Product Cycle," Quarterly Journal of Economics 80 (May 1966): 190-207.

(2) R.H. Coase, "The Nature of the Firm," Economica (n.s.) 4 (1937): 386-405. K.J. Arrow, "The Organization of Economic Activity," The Analysis and Evaluation of Public Expenditures: The PPB System (U.S. Congress Joint Economic Committee. Washington: 1969). A. Alchian and H. Demsetz, "Production, Information Costs and Economic Organization," American Economic Review 62 (December 1972): 777-95. O.E. Williamson, Markets and Hierarchies: Analysis and Antitrust Implications (New York: Free Press, 1975).

(3) R. Hal Mason, "Technology Transfers: A Comparison of American and Japanese Practices in Developing Countries," Working Paper, Japan Society Workshop, UCLA Graduate School of Management, June 1978.

(4) For an examination of entry barriers, see Joe S. Bain, Barriers to New Competition (Cambridge: Harvard University Press, 1956).

(5) To simplify the terminology I define "technology" to include all of the mentioned advantages.

(6) For an examination of this situation, see R. Hal Mason, "Technology Acquisition in the Pacific Basin: Direct Foreign Investment versus Un-Packaged Technology," in International Business in the Pacific Basin, ed. R. Hal Mason (Lexington: D.C. Heath, 1978), pp. 117-29. See also Mason, "Technology Transfers," pp. 20-26.

(7) Mason, "Technology Transfers."

(8) Vernon, "International Investment."

(9) Hymer, International Operations.

Comments

Igor E. Artemiev

The diffusion of technology is one of the main elements of technological change in any economy of the world. Technology can be diffused by action of the owner of the technology (transfer of technology) or by action of other parties through parallel development of technology, competitive imitation, and other modalities.

Socioeconomic environment is an important factor in technology diffusion, one that determines to a large degree the channels and modes thereof. For example, the main modes of technology diffusion within the market-economy countries are commercial technology transfer and competitive imitation, while within the nonmarket-economy countries the transfer of technology is planned and competition among enterprises does not exist, minimizing duplication of research and development.

International technology transfer, its channels and modes, are also influenced by the socioeconomic environment. The paper by Professor Dunning is an important contribution to the study of two alternative channels of technology transfer used by the countries with market economies: the interfirm channel, where technology is traded on the market as goods; and the intrafirm channel, where technology crosses national borders but stays within the ownership of transnational corporations. Dunning identifies governments as the third channel of technology transfer, one that plays an important role in influencing the conditions under which the first two types of mechanisms of technology transfer are undertaken.

The channels of international technology transfer used by the countries with nonmarket economies seem to be different from the above-mentioned systems. Here government-to-government agreements have the priority, and within this structure different modes of commercial and noncommercial technology transfer are developed.

Different types of technology transfers result in different scales of technology diffusion in the recipient economy.(1) The scale of diffusion is a function of a number of factors, including efforts of the transferrer to provide his counterpart with all the elements needed to successfully apply the technology (completeness of the technology transfer, reasonable pricing of technology, restraint from restrictive business practices); and absorptive capabilities of the transferee (stage of technological development, availability of complimentary resources, intrastructure). The pattern of application and scale of diffusion of technology in the host country's economy is likely to depend to a certain extent upon the channel of transfer chosen by the counterparts.

The interfirm transfer of technology seems often not to fully realize the principal goals of the private corporations of the market-economy countries – maximization of profits and corporate growth. Dunning explains this by the concept of the imperfect market, emphasizing that corporate practices may lead to further imperfections in the market because firms exploit or capture the advantages thereof.

In the twentieth century technology has increasingly been transferred with foreign direct investment in packaged form. The intrafirm channel of technology transfer met more sufficiently the demands of the private corporations, often allowing them to extract monopoly profits due to their exclusive access to technology, sources of raw materials, or markets. This pattern of technology transfer began to cause concern among both developed and developing host countries as far back as the early 1960s. But not until the 1970s was the issue raised in the international arena of the technological dependence of the developing countries on the TNCs based in the market-economy countries. It was raised in connection with the demands of the developing countries aimed at achieving economic decolonization and more rapid rates of development.

Many developing countries have resorted to government regulation of technology acquisition, to try to reduce the economic and other costs of technology imports. The regulation of technology transfer in coordination with national goals is not an easy task. It is difficult to identify all costs and benefits and evaluate different modalities of technology acquisition. The attitude of many countries toward the equity investment route – that it is an expensive way of technology importation – is based not only on a priori assumptions, imperfect data, and prejudices, but on the experience of many years, including the long period of colonialism as well as the decades that have elapsed since the liberation of these countries.

The experience of these countries showed that in many cases the importation of technology through intrafirm channels gives rise to a limited technology diffusion in their economies, due to the fact that the affiliates orient their activities to their parent companies' interests and have weak links with the host countries' economies. Many countries have chosen to encourage contractual forms of technology transfer and joint ventures, hoping to gain access to the foreign technology without losing control of some industries in their economy. But if the bargaining power between the parties is in favor of the transferrer, foreign control could come, as Dunning put it, "through the back door" with contractual arrangements. In the 1970s many buyers of technology paid more attention to the terms and content of the contracts, favoring in some cases agreements with shorter-term obligations and lump-sum payments instead of running royalties.

Lack of coordination in supply of resources, as Dunning shows in his paper, is a major problem with the interfirm acquisition of technology by developing countries. The buyer often has to acquire capital, technology, training, and so on from different suppliers once he chooses not to take the direct investment route. Dunning leaves the impression that there is no alternative to the intrafirm transfer of technology if the buyer needs the resources to be supplied not separately, but in an integrated way. Nevertheless, the third channel of technology transfer – intergovernmental agreements – could be of importance in this context.

The identification of the government channel as a separate mechanism of technology transfer is one of the very strong points of Dunning's paper. This seems to be a clear alternative to the direct investment route without many of its inherent costs of technology transfer.

The experience of the USSR and other nonmarket-economy countries with intergovernmental agreements on economic and technical cooperation shows that these agreements could become a framework for integrated transfer of resources to the developing nations. The agreements provide the recipient economies with complementary resources and services without which they would have difficulty utilizing the technology. These complementary resources and services include loans, credits, feasibility and economic studies, assistance in organization and management, training of the labor force, supply of machinery, materials, and spare parts, geological prospecting work, and construction. The integrated transfer of technology is not viewed by the nonmarket-economy countries as an unbreakable package. The developing countries can and do acquire only those resources they need for development, without tying-in conditions.

Within the framework of intergovernmental agreements the following modes of technology transfer are being developed: construction of industrial, agricultural, and regional complexes; turnkey operations and plants; delivery of complete plants or production lines with payment in goods produced with the assistance of this equipment; and subcontracting and manufacture under license. These modes require major efforts on the part of the transferrer in supplying technology and supplementary resources. Some of these modes foresee long-term involvement of the parties to the cooperation. Intergovernmental communications guarantee mutual benefit and the equality of the partners involved in the commercial modes of technology transfer, excluding possibility of excessive payments and restrictive practices. In many cases the public sectors of developing countries are the counterparts of the nonmarket-economy transferrers.

Machinery, equipment, and other resources transferred to the developing countries by nonmarket-economy organizations usually pass into the ownership of the developing nations without outflow of dividends from the recipient countries.

A substantial amount of technology is being transferred by nonmarket-economy enterprises to the developing nations at practically no cost except for the expenses incurred in the preparation of the documentation. This usually occurs in the process of construction of complete plants, production lines, or regional complexes. If certain adaptation of technology to the conditions of importing nations is needed, the buyers are not usually charged for technological modifications made by the transferrer. If a substantial volume of extra work or a search for new solutions is required, the buyer reimburses only the costs of redesigning processes for appropriate adaptation.

The transfer and diffusion of technology is considered important by the nonmarket-economy countries, but only as one stage of a process of technological change that is to be accompanied by the growth of the national scientific and technological capabilities of the developing countries. The intergovernmental agreements between nonmarket-economy countries and developing nations encourage a number of programs to enhance the technological potential of the latter, such as the training of the labor force and of qualified personnel in the process of construction of plants and regional complexes; assistance in the development of national educational systems; the training of personnel from the developing countries in the nonmarket-economy countries; and assistance in the foundation of R&D laboratories, design units, scientific and technological data banks, and so forth. The multilateral cooperation between the Council for Mutual Economic Assistance on the one hand and developing countries on the other is a growing contribution to bilateral efforts in the sphere of utilization of science and technology for development.

In summary, the nonmarket-economy countries consider intergovernmental agreements a framework within which they transfer technology in an integrated way using a wide range of modalities. In our view this channel has stability and maximizes the interests of both parties, something that is not always possible to achieve, as Dunning implies, using inter- and intrafirm channels of technology transfer.

"There can be little doubt that government, both home and host to resource transmission, are likely to play a more important role in determining both the vehicle of transmission and the terms and conditions attached to it," writes Dunning. Governments could encourage the development of new forms of cooperation, from bilateral coproduction schemes up to multinational consortia for the solution of scientific, technological, and world economic problems. To illustrate his point, Dunning quotes a speculation of Perlmutter's: "A Japanese and U.S. exploration company would use U.S. nuclear devices to help find Siberian oil in conjunction with a Soviet firm; the oil would then be distributed in Western Europe by European oil companies."

But unfortunately there are a number of impediments to East-West economic and technological cooperation that could prevent the realization of the large-scale multilateral scheme envisioned by Perlmutter, which would probably receive support from a majority of people both in the USSR and in other countries involved in the proposed project.

First, the U.S. firm participating in the project would be unable to find government-source financing, because the USSR is effectively deprived of credits and guarantees from the Export-Import Bank of the United States. U.S. law specifically limits financing to natural resource exploration and development projects within the USSR.

Second, export licenses must be obtained by firms that want to use the sort of equipment that Perlmutter refers to in the USSR. These licenses are issued by the U.S. government and approved by all members of the COCOM. The issue of licenses is based on single, occasional, and unpredictable extension of permission, and, furthermore, issue of a license does not guarantee that the license will not be revoked, even if the contracts are signed and the parties are engaged in the fulfillment thereof.

Third, if this project were to include trade not only in raw materials but in manufactured goods as well, Soviet exports to the United States would run up against discriminatory tariffs, because the USSR together with a number of other socialist countries is deprived of "most favored nation" status in trade, a status provided by the United States to about 140 other countries.

The socialist countries consider the principle of nondiscrimination in trade and technology exchange to be an important component of restructuring international economic relations on the basis of peaceful coexistence of different socioeconomic systems.

Dunning comes to the conclusion that the debates about costs and benefits of the main channels of technology transfer will narrow down in the 1980s to questions about the market structure of resource transmission. This forecast is based on the growing imperfections of markets and the increasing intervention by governments into the process of transfer of technology and other resources.

We certainly have to know more about the "markets for resource transmission" and how they differ from the product and commodity markets. The problem of choosing alternative channels of technology transfer was raised because of countries' efforts to make more effective use of technology for development. And while the channels and modes of technology transfer (in spite of their transformations in the 1970s) differ, in the degree of technology diffusion, distribution of costs and benefits of the application of technology and so on, this issue will, it appears, remain on the agendas of international forums.

NOTE

(1) Only technology and management skills can be transferred and diffused (or transmitted, to use Dunning's term). Other resources (money capital, labor) can be applied only at one place at one time, and hence can be transferred but can in no way be diffused (or transmitted) without loss to the transferrer. Some resources (land, climate, and undeveloped natural resources, for instance) cannot even be transferred, because they are immobile.

2 Transfer and Style: A Historical Account

Thomas P. Hughes

Some years ago "ugly American" was a catchphrase suggesting that Americans abroad, especially those in less developed countries, were not sensitive to local material conditions and needs, and forcibly imposed the American style of technology in inappropriate situations. Americans also seemed too often to assume that technology throughout the world should be measured against the American model; if it differed, then by definition it was the foreign technology that was falling short. Foreign peoples and their societies were expected, by implication, to adjust to an international, essentially American, style of technology.

Numerous articles and books published more recently clearly demonstrate that the "ugly American" is giving way to one who more carefully considers local conditions, short- and long-term, before recommending technological solutions. Now it is conventional to point out, for example, that many nations should have labor-intensive technology rather than American labor-saving technology. Others recommend technology that is not large-scale and capital-intensive for less developed regions.

The willingness to consider technology in context suggests an awareness of technological style. The term "style" is appropriate because it is broad and conveys a commonsense meaning that can be applied by analogy to technology. "Style" denotes a set of characteristics giving a person or thing an identifiable quality. Persons who consistently display dress characteristics have a style of dress; there are also styles of music and architecture. Increased understanding of technology leads to an explicit or implicit realization that likewise there are styles of technology. Capital-intensive, labor-saving, large-scale, mass-producing technology is an American style of technology. Like fur coats and long underwear, it is not appropriate everywhere.

Detailed information about the qualities of a technology, as distinguished from its essential function, usually tells us something about the geography, economics, industry, politics, and history of the region in which the technology originated. For example, simply providing the dimensions of a bridge says little about its style; a description of its material, its form, and its structure reveals much about its location in time and place. The material often reflects the availability of particular natural resources, its form might say much about aesthetics of the region, and its structure indicates the level of scientific knowledge and industrial development. This can be summarized by stating that technological style reflects cultural context, defining "culture" as a sum of economic, political, social, and psychological, as well as historical, factors.

This chapter argues that highly industrialized nations or regions have well-defined technological styles that emerged in interaction with a cultural context. Furthermore, highly industrialized regions – similar to less developed ones – tend to reject unadapted technology imported, or transferred, from other regions having well-defined cultures of different characteristics. They reject the technology, not just its style. This behavior is not unexpected. Far more surprising – and often overlooked by students of technology transfer – is the tendency of highly developed regions to reject indigenous technological inventions deviating in their implications from the developed region's own prevailing style.

American inventions are today finding receptive soil for development aborad rather than at home. Managers of multinational corporations can cite cases of this outflow. The historians can support this hypothesis with a longer, richer body of evidence of the flow, or transfer, of inventions. Before continuing, however, a distinction should be made between inventions and innovations. Inventions are undeveloped innovations, or devices and processes that have not been practically utilized. It is common for innovations originating in heavily industrialized regions to be transferred to – and adopted by – less developed regions. This chapter argues, by contrast, that a number of inventions, instead of being developed and used in the originating, industrialized region, are transferred, adapted, developed, and used in less developed ones.

To sustain the hypotheseis, several case histories will be presented. They are selected from the history of electric light and power systems in England, the first of the industrialized regions. During the period 1880-1920, from which the histories are selected, England was the region with the longest history of industrialization and the one with the most pronounced, highly industrialized technological style. Today the United States has the most pronounced industrialized style, and technology transfer may be taking on the unusual character observed in England earlier.

THE CASE OF THE ALTERNATING-CURRENT TRANSFORMER

The transfer history of the alternating-current transformer is complex. Although the principle of the transformer had been known since Michael

Faraday's remarkable discoveries in 1831, various inventors claimed to have experimented with transformers before Lucien Gaulard and his partner, John Gibbs, displayed their transformer system in London in 1883. Most historians, however, give Gaulard and Gibbs credit for attracting the electrical world's attention to the technological and commercial potential of the transformer.(1) Their contribution falls somewhere between an invention and an innovation. Subsequent to the display of their system in London, the history of the transfer of this technology followed a circuitous route through Turin, Italy, Budapest, Hungary, and East Pittsburgh in the United States. Further development and commercial innovation occurred along the way, and in other hands.

Technical literature, exhibitions, full-scale demonstrations, and patents provided modes of transfer for the alternating-current transformer. On September 18, 1882, Gaulard and Gibbs were issued a British patent "for a new system of distributing electricity for the production of light and power." Technical journals described their inventions, but this information seems to have been taken far less seriously by potential users than working displays and full-scale demonstrations. Functioning artifacts impressed practical engineers and manufacturers more than journal articles that were in any case little more than thinly veiled advertising.(2) Gaulard and Gibbs demonstrated their system in 1883 by lighting four of London's Metropolitan Railway Stations with incandescent and arc lamps using transformers on a sixteen-mile circuit. In September 1884 they connected their transformers in a fifty-mile circuit at an exhibition in Turin, Italy. The jurors awarded them the grand price of 10,000 francs for the display. They mounted a display, again in 1885, this time at the Inventions Exhibition in South Kensington, London.(3)

English legislative controls forced the outward flow of Gaulard and Gibbs' invention. The Gaulard and Gibbs system bore the mark of the Electric Lighting Act of 1882. If this law is seen as a factor in the environment, then the Gaulard and Gibbs transformer specifications can be likened to acquired characteristics. Clause 18 of the 1882 law stipulated, for example, that electric lighting systems could not impose the use of any one manufacturer's lamp on the consumer. Since the various lamps on the market operated at different voltages, the law could be intepreted as requiring that the consumer be supplied current at the voltage desired. This complex law shaped the characteristics of the Gulard and Gibbs transformer.

The transformer, a Rube-Goldberg-style device, had no less than sixteen secondary bobbins, or coils. These could be connected in series, parallel, or series-parallel to obtain a variety of voltages. In addition, an iron core could be cranked in and out of the transformer to generate more subtle variations. Historians trying to explain the Gaulard-Gibbs device without taking into account the legislative environment have been flaggergasted by its complexity.

The Gaulard and Gibbs transformer might be described as county-specific. When foreign engineers and manufacturers saw it at exhibitions, they quickly realized that it could be greatly simplified and improved for other, less politically structured environments. Hungarians and Americans took advantage of the situation and developed the first practical transformers. In addition, an ingenious British engineer redesigned the Gaulard and Gibbs transformer and used it in a circumscribed local situation in which the law of 1882 was not applicable.

The Hungarian engineers worked for the well-known firm of Ganz and Company. According to Lucien Gaulard, Otto Bláthy, favorably impressed by him and his system at Turin, said he would recommend to Ganz the purchase of rights for Hungary. Bláthy later acknowledged that he and another Ganz engineer, Charles Zipernowski, had seen the system at Turin, but learned only how a transformer should not be constructed. Nevertheless, in May 1885 Ganz and Company displayed a transformer at the Hungarian National Exhibition in Budapest and soon afterwards began widespread installation. Hungarian historians insist that the Ganz engineers invented the practical transformer. Considering the controversy from the point of view of technology transfer and on the basis of evidence of transformer design, my conclusion is that the Gaulard and Gibbs system spurred the Hungarians to develop a more practical and simpler device unfettered by regulatory legislation. Ganz's failure to take out a license from Gaulard and Gibbs made them quite bitter.

George Westinghouse also learned of the Gaulard and Gibbs transformer from an exhibition, the Inventors' Exhibition in South Kensington. He saw an analogy between the transformer system of transmission and distribution and the system he had developed of distributing gas under high pressure from central gasworks. Westinghouse ordered several Gaulard and Gibbs transformers, paid a young Annapolis-trained engineer, William Stanley, to improve on the design, and, unlike Ganz and Company, took out a license under the Gaulard and Gibbs patents. Reginald Belfield, an English electrician who had been in charge of the Gaulard and Gibbs London exhibition, also entered the employ of Westinghouse and improved the transformer. Westinghouse introduced alternating current in America on a large scale after 1886. In its simplicity, his system was similar to that of Ganz. And like Ganz, Westinghouse was also free of legislative encumbrance.

The design characteristic indicative of the Westinghouse contribution to the development of a transformer system was the provision for machine production, which permitted Westinghouse to organize mass production, with its attendant economies and market advantages. A noted Westinghouse electrical engineer and patent lawyer, Franklin L. Pope, perceived the essence of the Westinghouse contribution when he wrote in 1887, "I saw that there were great possibilities in the [Gaulard and Gibbs] system although it was not worked out practically at all – not worked out as we would work it out in this country by any means – and I was so well satisfied . . . of the possible results which we could get when we put our young men on it in this country."(4) Another writer lamented at about the same time that "it is not a very complimentary reflection for European electricians and capitalists that although all ideas and experimental work need have come from Europe . . . it should be reserved for an American firm to take up the system and make it the commercial and practical success which the Westinghouse Company is now doing.(5) Westinghouse manufacturing and marketing of the system was impressive: by September 1887 the company had in service or under construction central stations with capacity for 134,000 sixteen-candle-power incandescent lamps; by November 1890 the number had risen to 700,000.(6) Ganz and Company had connected 100,000 lamps by 1900.

For Gaulard and Gibbs the history of the transformer took a melancholy turn. In London the Grosvenor Gallery electric lighting compnay used their transformer in a system that avoided the Electric Lighting Act of 1882 by stringing distribution lines across rooftops instead of using the public streets. In 1886, however, the Grosvenor Gallery engineer, S. Z. de Ferranti, modified the system by connecting the transformers in parallel, as the Hungarian engineers had earlier. His compnay then refused to pay Gaulard and Gibbs for rights to their system. Heaped upon this difficulty were the time and expense of defending their patents in court – litigation that was finally decided against them. In 1888 young Gaulard died in Sainte Anne's madhouse in Paris, "brokenhearted, a victim of science and his own greatness."(7)

PARSON'S TURBINE ABROAD

After the alternating-current transformer, and after the generalized, or universal, polyphase system of power appeared at Niagara Falls,(8) the introduction of the steam turbine was the next major step in the history of electric light and power systems. Contemporaries and historians have explained its significance. "Few revolutions in the mechanical world have been more rapid and sweeping," wrote T. Commerford Martin in 1907 about the introduction of the steam turbine. Five years later Martin, the expert on electrical matters for the U.S. Census Bureau, added, "It is very doubtful if any more large reciprocating units will be installed for power-station purposes."(9) A historian of the electrical supply industry concluded in a similar vein that the steam turbine "proved to be the most potent factor in the history of electrical engineering throughout the world over the past sixty years."(10)

The steam turbine offered many advantages over the reciprocating steam engine. A much smaller and lighter turbine could equal a reciprocating engine's output. In one early installation, for example, a 5,000-kilowatt turbogenerator weighed only one-fifth as much as the comparable reciprocating unit. An early turbogenerator occupied one-half to one-third the floor space taken by reciprocating units of the same capacity. Later turbines had an even greater size advantage. First costs and operating turbines were less. Savings on lubrication oil were substantial, and control and regulation were easier. Also, one man who could care for only one of the giant reciprocating engines could tend two to three turbines.(11)

Turning to the history of the turbine, we find the English inventor Charles Parsons, who while a student at Cambridge University concentrated on the theory and design of rotary-cylinder steam engines. Subsequently employed by an engineering firm near Newcastle-Upon-Tyne, he focused on the turbine, whose essential principle had been long and widely known, but whose practicality had been doubted. Applying a mixture of theoretical sophistication and mechanical ingenuity, young Parsons took out his first turbine patent in 1884. Soon he designed high-speed generators suitable for turbine drive and installed a number of turbogenerators for shipboard, where space was at a premium. Dissatisfied with his employer's conservative development of the engine, Parsons left to establish his own company at Newcastle-Upon-Tyne, and in the mid-1890s offered turboalternators for central stations.(12) The commissioning in January 1890 of two Parsons' 75-kilowatt turbines at Forth Banks Power Station near Newcastle-Upon-Tyne attracted considerable attention as the first in a public central station. Parsons, it should be observed, was a board member of the utility making this installation.(13)

Parsons' patents were a mode of transfer. In 1900 the Swiss firm of Brown Boveri acquired Parsons' rights. Because the Swiss firm could expect small demand for steam turbines in Switzerland, where water power was abundant and coal scarce, this is an extremely interesting decision. The firm's excellent machine (or engineering) work helps explain the decision, for turbine design was extremely complex and only the highest-quality work could ensure reasonably reliable operation. C.E. L. Brown strengthened his firm's position in 1901 by inventing and patenting a rotating field for generators driven by extremely high-speed turbines. The combination of new rotor and turbine built according to the Parsons patents gave the Swiss firm an advantage, and it thrived on exports despite tariff barriers.(14) To this day the Swiss firm is an interesting example of a multinational corporation able, for example, to see turbogenerators to utilities in the United States and throughout the world.

In America, both General Electric and Westinghouse showed interest in the Parsons patents. General Electric, however, advised by Charles Steinmetz, its senior research engineer, refused the opportunity to buy rights because Steinmetz believed the Parsons design to be still too complex. Steinmetz advised G. E. to begin its own experimentation immediately. On the other hand, Westinghouse purchased the American patent rights in 1895 and completed its first turbogenerator in 1896. In 1901 Westinghouse made its first central station installation at Hartford, Connecticut, with a 1,500-kilowatt unit, which, however, performed erratically at best.(15)

General Electric, aware of the commercial potential of turbines, provided an American inventor, C. G. Curtis, with staff and facilities for experimentation and development of turbine design. Not until five years later, in 1901, however, did General Electric have an operational turbine, first a 500-kilowatt unit and then a 1,500-kilowatt unit.(16) Meanwhile, in 1900 Parsons's English firm constructed two 1,000-kilowatt units for the Elberfeld power station in Germany, a pioneer installation that became famous throughout engineering circles.(17)

In a roundabout way the German installation of Parsons turbines led to the first installation of a General Electric turbine in a major American utility. This resulted more from the utility's imagination and risk-taking than the manufacturer's. Also, the consulting engineer's role in technology transfer emerges in this episode. Samuel Insull, the innovative head of Commonwealth Edison of Chicago, a utility believed by many experts to be the country's best-managed during his tenure, saw the Elberfeld installation.(18) Enthusiastic about its advantages, Insull sent Louis Ferguson, his chief engineer, and Fred Sargent, the head of Commonwealth Edison's consulting firm Sargent and Lundy, to inspect turbine technology in Europe in the summer of 1901. The two engineers visited the works of Brown, Boveri in Baden where they saw turbines under construction for the Edison utility in Milan and where they learned of an order for an unprecedentedly large 4,000-kilowatt turbine for Frankfurt, Germany.(19)

Insull turned to General Electric, then developing the modified Francis turbine. He may have done so because he believed that American design, manufacure, and installation suited American conditions. "European conditions," he remarked, "individualized design of each machine, the careful coddling of new devices and the infinite pains and complication which is of the nature of European engineering gave an utterly different environment from American conditions."(20) Insull, British by birth, realized that regional styles of technology differed, but throughout his career as a utility manager and entrepreneur he often turned to England and the Continent, seeking opportunities for technology transfer and adaptation.(21) Like the English writer quoted earlier, Insull expected Amerians to take up the system and make it "practical and commercial."

Because General Electric experienced serious turbine problems and knew that many were yet unsolved, it drew back from giving performance guarantees for units larger than 1,500-kilowatts. Insull, however, determined to "bore with a big auger" rather than "living from hand to mouth,"(22) ordered a trend-setting 5,000-kilowatt unit from General Electric in 1902 and agreed to assume all responsibility for performance – or nonperformance – and attendant costs. Years later Insull confided that only the fourth turbine installed was "really efficient."(23)

The role in technology transfer of Sargent and Lundy and other consulting engineering firms throughout the world needs study. Samuel Insull not only employed Sargent and Lundy to appraise foreign developments in electrical engineering, but also established close contacts with the leading British consulting firm of Merz and McLellan. Charles Merz, the eminent electrical engineer who head the Newcastle-Upon-Tyne firm, and Insull became close personal friends as well as professional allies. By letter and face to face, the two men exchanged information about technological developments during the period of dramatic growth in electrical light and power from 1900 to 1930. Merz sought Insull's advice when attempting to introduce a single, coherent electrical supply system to fragmented London during the decade before World

War I. Merz had Insull testify before London County Council committees considering reform of the electrical system. In turn, Merz's firm prepared a number of reports for Commonwealth Edison evaluating equipment and system advances contemplated by the Chicago utility. In this way Merz and McLellan brought their intimate knowledge of British and continental technology to the American scene. Their accumulated experience, derived from study of American designs and installation, was in turn transferred elsewhere in the world.(24)

DIFFUSION OF GRID SYSTEMS

The widespread introduction of the steam turbine and also of the water turbine into electrical power stations after 1900 helped bring about another phase in the history of electric light and power. Generally, the efficiency of turbines increased with the increase in their capacity. By 1920, therefore, units ranging up to 50,000-kilowatts were installed. Central stations with units of this magnitude, however, had to reach out with very high voltage transmission to find the load to absorb their capacity. This was especially true for large hydraulic stations located in relatively thinly populated and sparsely industrialized areas rich in water power sites. Coal-fired stations located near mines or even at mine mouths also had to transmit over a large region in order to obtain sufficiently diverse loads to produce economical load curves. By the 1920s, therefore, engineers were designing regional power networks, systems, or grids in various parts of the industrial world.

The regional systems, whether resulting from connecting formerly independent systems or springing forth newly created from drawing boards, required some new technology. For instance, transmission lines and towers crossing open country needed lightning arresters and special insulation as protection against lightning. Frequency indicators and controls had to be used to interconnect systems of slightly different characteristics. Engineers developed boilers able to withstand higher pressures and temperatures and auxiliary equipment, such as heat exchangers, to exploit the savings possible when outputs were extremely large. Inventors and engineers also designed control or dispatching centers to monitor and operate the large integrated systems. By 1930 analog computers were analyzing and solving network problems, superseding the laborious algorithms of paper-and-pencil calculation.

Despite the need for some invention and development, the point-to-point transfer of technology common during early years seems to have been relatively unimportant. The historian is able to trace the transfer from place to place of alternating current transformers and steam turbines, but tracking the transfer of regional power-systems technology is virtually impossible. Instead of point-to-point transfer, the historian finds engineers and managers in the industrially advanced nations drawing from and contributing to a great pool of technology equally accessible to all. Gaulard and Gibbs, Ganz, Westinghouse,

Parson, and Brown, Boveri come to mind in the earlier phases of technology transfer, but a host of prominent men and firms emerge during the development of regional systems technology.

Consulting engineers, a professional group involved earlier in the transfer of turbine technology, had unusually prominent roles in the history of the diffusion of regional power systems. This professional group should be singled out as instrumental in the new technology transfer associated with the "pool of technology" concept. These engineers, many with expert knowlege of management, had easy access to knowledge and apparatus because they were unconstrained by parochial attitudes or by sharply defined attachments to the product line of a particular manufacturer or the technological style of a particular region.

There are innumerable examples of consulting engineers who designed and monitored the construction of regional power systems during the decades between the world wars. In England Merz and McLelland and the firm of Kennedy and Donkin were retained as consulting engineers for the planning and design of the national grid. In Germany immediately after World War I the consulting firm of Oskar von Miller designed a unified system of transmission for all of Bavaria. In the United States, Stone and Webster took a leading role in the development of regional systems. Hugh Cooper, an experienced American designer and builder of hydroelectric power systems throughout the world, won the Order of Lenin from the Soviet Union for his contribution as a consultant to the great dam and related power development on the Dnieper river in the Ukraine. (Cooper's work is an extremely interesting chapter in the history of technology transfer.)

These engineers applied the essential principles of the regional system and sought the economies of scale and mass consumption, the advantages of applying location theory to power plant siting, the flexibility of high-voltage transmission, the optimization of the varied characteristics of interconnected power plants, the increase in reliability and reduction of reserve capacity, and the economies of load diversification, forecasting, and management that regional systems allowed. These principles, constituting the generalized state of the art, were adapted to local regional conditions by detailed specifications. Usually the client, a private utility or a government, chose whichever manufacturer who could supply the specified equipment.

Consulting engineers had to take into special account legislative or political controls. Once again, England provides an excellent sample of the effect of legislation – first constraining, then supportive – in the case of regional systems. Until 1926, English legislation restrained technological change. The electric lighting acts of 1882 and 1888 limited the geographical extent of electric utilities, whether government or privately owned, to long-standing politial boundaries such as parishes, boroughs, municipalities, counties, and other local authorities. This became the governing constraint after 1900, when turbines and other technology made regional supply possible. Under pressure from private interests, Parliament authorized power companies after 1898, but few large area suppliers flourished because existing urban utilities, many owned by local authorities, would not interconnect with power companies to create a large, diverse, and economical load.

The break came after World War I. During the war, the increased production that the rational concepts of large area supply brought were amply demonstrated throughout the warring industrial world. Aware of this and also acutely sensitive to the relative decline of the nation as an industrial power, the British Parliament responded to a national problem by finally deciding against parochial interests. The parliamentary legislation of 1926, based on the report of the Weir Committee, authorized a nationally interconnected series of regional transmission systems linking the more economical power stations with a widely diversified load. Until the Tennessee Valley Authority was established in 1933, the British grid was politically, if not technologically, the boldest of the world's growing number of regional systems.(25)

CONCLUSION

Modification and adaptation of general knowledge resulted in regional systems with distinct styles. The British grid had relatively low transmission voltages because the distances between industrial and urban centers were small as compared with, say, California, where the water power sites of the Sierras were at least a hundred miles from the load centers. In Bavaria, Oskar von Miller's regional system reached northward from the water power of the Bavarian Alps to connect with the coal-fired plants of lower Bavaria. The extensive system of the Rheinisch-Westfaelisches Elekrizitaetswerk in the Ruhr included giant cooling towers for power plants located at the lignite mines where cooling water was scarce. And other distinctive regional characteristics could be cited.

A minor theme, as well as a major hypothesis, emerges from this survey, of the history of electric light and power systems. The minor theme concerns various modes of transfer; the major hypothesis, transfer and style.

When the alternating current transformer spread, publicity in the technical literature and exhibitions were particlarly effective modes of transfer. Patents diffused general information, but engineers needed to expand on the information contained in them. The licensing of patents served to transfer Gaulard and Gibbs's technology to Westinghouse, but Ganz in Hungary avoided acquisition of rights. On the other hand, Parsons transferred technology through the sale of rights abroad. He was probably able to protect these because manufacturers needed much detailed and refined know-how to supplement the basic information. Finally, the consulting engineer played a major transfer role, a role not fully appreciated even today.

This survey of transfer modes holds no great surprises. On the other hand, the pattern of outflow from England suggests a major hypothesis deserving further study and testing. The Gaulard and Gibbs transformer was an invention that only found fertile soil for full development aborad. Parsons installed turbines in England, but the largest and most numerous installations were also abroad, where regional systems took full advantage of the technological implications of the turbine.(26)

A definitive explanation for the outflow and retarded development of electrical inventions in England must be long and complex. This short survey, however, suggests a testable model. By the close of the last century English technological style bore the mark of English politics. The British Parliament, in a series of legislative acts beginning with the Electric Lighting Act of 1882, exercised effective control over the shape of electrical supply technology. The British political system, reflecting especially the power of local government, the trend toward municipal socialism, the abhorrence of monopoly, and the power of vested technological interests, such as gaslight, reacted sensitively to impending technological change.(27) By contrast, the United States Congress seldom reponded to technological change then and on the local level politicians often compliantly provided for it. Gaulard and Gibbs's transformer, for instance, embodied a reaction against industrial monopolies. The circumscribed development of turbine-based regional power systems manifested the power of local authorities, many of whom were committee to socialization of public utilities.

The general model suggested involves the gradual development in industrializing nations of more effective political control of technological change. The technological style therefore carries an increasing political component. The style tends to reject technology that does not conform to the prevailing emphasis, be it parochialism or environmentalism. As a result, inventions often flow outward, to developing countries where the prevailing emphasis on economic development and production proves less restrictive. Needless to say, the implications of such a model for technology transfer policy are manifold.

NOTES

(1) Controversy about the introduction of the transformer usually centers on the question of the practicality of the Gaulard and Gibbs transformer as compared to those introduced by Max Deri, Otto Bláthy, and Charles Zipernowski in Hungary, S. Z. de Ferranti in England, and William Stanely and George Westinghouse in the United States.

(2) For example, Edison asked one of his foreign representatives how the publicity being given a rival system in a journal could be countered. He as told that the reporters could be influenced by consulting fees.

(3) On the Gaulard and Gibbs episode, see P. Strange, "Transformer Patent Litigation: I. 1880-90," Papers Presented at the Weekend Meeting on the History of Electrical Engineering, July 13-15, 1973 (Institute of Electrical Engineers), pp. 8/1-8/8.

(4) Telegraph Journal and Electrical Review 20 (April 15, 1887): 349.

(5) Telegraph Journal and Electrical Review 21 (November 4, 1887): 446.

(6) Harold Passer, The Electrical Manufacturers, 1875-1900 (Cambridge: Harvard University Press, 1953). p. 149.

(7) Letter from E. H. Cadiot, November 27, 1888, in Telegraph Journal and Electrical Review 23 (1888): 620. Months before his death he appeared at the Elysée demanding to see the president, declaring, "I am God and God does not wait."

(8) On the transfer of technology to and from Niagara Falls, see Robert Belfield, "The Niagara System: The Evolution of an Electric Power Complex at Niagara Falls, 1883-1896," I.E.E.E. Proceedings, 64 (September 1976): 1344-50.

(9) Department of Commerce, Bureau of the Census, Central Electrical Light and Power Stations, 1907 (Washington, D.C.: GPO, 1980) p. 97 and Central Light and Power Stations, 1912 (Washington, D.C.: GPO, 1915), p. 114.

(10) Percy Dunsheath, A History of Electrical Power Engineering (Cambridge: MIT, 1962), p. 196.

(11) Passer, Electrical Manufacturers, p. 312, and C. Matschoss, E. Schultz, and A. Gross, 50 Jahre Berliner Elektrizitaestswerke, 1884-1934. (Berlin: VDI Verlag, 1934), pp. 129-32.

(12) Conrad Matschoss, Grosse Ingenieure (Munich: Lehmanns, 1954), pp. 294-99.

(13) R. H. Parsons, The Early Days of the Power Station Industry (Cambridge, England: Cambridge University Press, 1940), pp. 171-72.

(14) 75 Years Brown Boveri (Baden: Brown, Boveri & Co., 1966), pp. 36-38.

(15) Glenn Weaver, The Hartford Electric Light Company (Hartford, Conn.: Hartford Electric Light Co., 1969), pp. 87-89.

(16) Ludwig Spennemann, "Die Entwicklung der Dampfkraftwerke," in Das Zeitalter der Elektrizitaet (Frankfurt am Main: V.D.E.W., 1967), p. 21.

(17) George Siemens, History of the House of Siemens, trans. A. F. Rodger (Munich: Alber, 1957), I: 225.

(18) Insull reported that he saw a Parsons installation in Germany before the summer of 1901; outside of Elberfeld there were no major German central-station turbine installations.

(19) Commonwealth Edison Company, Edison Round Table, November 1928, pp. 2-3.

(20) Ibid., p. 3.

(21) On Insull, see Forrest McDonald, Insull (Chicago: University of Chicago Press, 1962).

(22) Ibid., p. 11.

(23) Ibid., p. 12.

(24) Numerous reports prepared for Commonwealth Edison and other utilities are in the Merz and McLellan Archives in Newcastle-Upon-Tyne. Some of these reports on new turbine designs and other matters are dated 1906, 1911, 1915, 1919, 1920, 1923, 1926, and 1929, showing how the long-standing relationship and transfer continued while Insull headed Commonwealth Edison.

(25) For the history of the grid and of British electric supply see Leslie Hannah, Electricity Before Nationalization (Baltimore: Johns Hopkins, 1979).

(26) I have discussed the major exception to the English rule in my essay on Charles Merz's early regional system centered in Newcastle-Upon-Tyne. See "Managing Change: Regional Power Systems, 1910-1930," in Business and Economic History, ed. Paul Uselding (Urbana: University of Illinois, 1977), series two, 6: 52-68.

(27) Leslie Hannah thoroughly discusses the relevant legislation in his Electricity Before Nationalization. I have discussed the ramifications of the electric lighting act of 1882 and 1888 in "British Electrical Industry Lag: 1882-1888," Technology and Culture 3 (1962): 27-44.

Comments

Gustav Ranis

Professor Hughes's paper contains insightful historical observations that, unfortunately, are not fully explored for the purposes of this conference. For example, Professor Hughes introduces but does not explain the meaning of technological "style." He sees England in the nineteenth century and the United States today as having the "most pronounced" technological style. Is this equivalent to technological leadership, to who moves out the technological frontier in the pure engineering sense? If not, the concept is too absolutist, since "styles" are clearly relevant to culture and endowment. Moreover, the heart of the issue before us is how "styles" are transmitted and altered with respect to the private and social objectives of less developed countries, hopefully with a minimal gap between the two and, if need be, with the help of public policy.

Second, Professor Hughes emphasizes the well-accepted distinction between inventions and innovations and subtly raises the question of whether it matters if innovations are transferred from developed to less developed countries and adapted, or if instead inventions are transferred and adapted en route to becoming innovations. But having stimulated us in this way to think about the issues of induced innovation – the location as well as the direction of R & D underlying the production of innovations, including the possible relevance of induced institutional innovations – he never really returns to the issue.

Let me be more specific. Since we are supposed to be mainly concerned with the overall environment affecting the flow of technology across countries, it would have been of considerable interest to explore the question of the impact of differences in resource endowment situations, or in the expectation of future such situations, on the strength, and especially the bias of innovations. Moreover, pushing into the area of control systems or policy packages, we could then examine how these can be altered to accommodate greater support of entrepreneurial and/or societal objectives from the side of technology change.

It would, moreover, have been very helpful if the role of technology transfer itself, either through the transfer of inventions closer to the science end of the spectrum or of innovations at the technology end, were differentiated from the role of indigenous experimentation and/or adaptation processes. In brief, the role of in-country modification, as contrasted to what is transferred from abroad, is not focused on sufficiently.

In researching matters of this kind, we have three laboratories available to us: cross-sectional evidence, historical evidence in the North-South axis since World War II, and historical evidence from the Western transition to modern growth. Professor Hughes has made a commendable effort to bring to bear evidence from the last of these. It is a shame that he has placed too little emphasis on a broader interpretation of his findings.

Let us examine some of the episodes presented in his paper. The evidence he has marshalled is no doubt interesting, but its interpretation in terms of a broader set of generalizations is unfortunately left almost entirely to the reader. The story of the alternating-current transformer, for example, focuses on an English system that was never fully successful at home but was transferred to both Hungary and the United States. The relative success of the licensed U.S. transplant seems to underscore the importance of a large market for mass production accompanied by scale economies, on the one hand, and the existence of a superior legal system providing incentives for innovation, on the other. It is, of course, generally accepted that Great Britain was the world's technological leader in the eighteenth and early nineteenth centuries, followed by France and Germany on the Continent and by the United States. There is also substantial agreement on the factors that gave Great Britain its original position of prominence, including higher and better-distributed income, relatively better endowment in natural resources, especially coal, plus, most important of all, greater progress in throwing off internal feudalistic and mercantilist interferences. Yet in spite of all efforts to keep such advances "bottled up" – including through the prohibition of exports of skilled workers and machinery – by the middle of the nineteenth century Germany had taken the lead in the chemical, pharmaceutical, and electrical engineering industries, with the United States moving ahead in mechanical engineering. By the time of World War I Britain had become an importer of skilled labor and technology. The why of this change in leadership position has never been fully resolved, but seems to me relevant to the episode explored by Professor Hughes. Legislation seems to have limited the scale of electric systems in England, which was apparently responsible for the commercial failure of the Gaulard and Gibbs alternating-current transformer at home and its success abroad.

Professor Hughes's second episode deals with the steam turbine. In this instance the patents were transferred to Switzerland where, utilizing that country's comparative advantage in precision machining, the necessary scale was achieved in the export markets. Patents were also transferred to the United States, where General Electric benefited once again from the large size of the U.S. internal market, in contrast to the higher cost and individualized design characteristic of the small internal markets in Europe. The technology originally invented in England and commercially applied on a massive scale in the United States was then transferred back to England once it became possible to have one coherent system for London, a system no longer controlled by balkanized regulations.

One somewhat more elusive characteristic of what Professor Hughes calls "regional system technology" as part of a general pool available for selection is that it is "equally accessible to all." Does that mean nonappropriability – at least once the patents have been exhausted – or the absence of private monopoly power? It is this second dimension of the environment making for or against "appropriateness" in technology transfer mechanism that seems to be given short shrift here. The presence of undue private market power may have as great an effect on selection, diffusion, and adaptation processes as any set of government policies – although, admittedly, the former may indeed be facilitated by the latter.

The strength of local authorities and the fear that monopoly power will restrict markets is indeed referred to as a possible obstacle to economies of scale-related technology change. But what seems to be missing is the recognition that any type of satisficing or rules of thumb behavior on the part of entrepreneurs receiving large windfall profits is likely to be highly related to the reduced social appropriateness of the technology transfer mechanism. What Professor Hughes calls "parochial attitudes or attachments to product lines" can be translated into indulging one's preferences for prestige goods and techniques, a preference much more likely to be sacrificed in the presence of greater competitive market pressures.

Professor Hughes, finally, tries his hand at some generalizing via the introduction of phases in technology transfer, from "general acquaintance" (trade fairs, journals, exhibits) to "diffusion" (patents and licensing) and "adaptation." But the connective tissue between it and the historical episodes presented is not in evidence. That, in essence, is the basic weakness of his paper. Forays through the historical laboratory don't automatically yield a better understanding of today's problems of technology transfer to the less developed countries. The one very general generalization suggested in the paper is that government regulations are likely to inhibit the smooth functioning of the technology transfer mechanism, but even here Professor Hughes's comments that the typical contemporary less developed country's environment is less restrictive in this respect than was England's indicates perhaps some lack of familiarity with the central problems at issue here.

These problems are by no means simple. But their solution demands that we search for some superior (if not ideal) mechanism to effect technology transfer from rich to poor countries and to ask how policies, at the national, international, and corporate levels, can be marshaled to improve these mechanisms. Some further comments seem appropriate here.(1)

First, the relation between invention and innovation is a peculiarly difficult area since, contrary to the simple-minded view, there exists a real possibility of a reversal of the common causal ordering running from invention to innovation. New inventions and even basic scientific progress may be stimulated by new data, new tools, and new "puzzles" that emerge in the course of the application and modification of technology. Thus the original smelting of iron may have proceeded without full understanding of the chemistry of oxidation or reduction. Yet the fact that the Bessemer process initially worked in England but not on the Continent (where the iron ore had a higher phosphorous content) led to new scientific inquiries into basic metallurgy and, in turn, to the improved Thomas-Gilchrist steelmaking process. Similarly, the difficulties encountered in the transplanting of improved seeds from one country's agriculture to another's have led to substantial new breakthroughs in agricultural chemistry.

It may therefore be useful to think of inventions and innovations globally as more of a closed mutually reinforcing and mutually dependent circle, and to view both scientific and technological advances as moving points on a spectrum, some indicating major cataclysmic or epochal "jumps," others less spectacular advances in understanding and productivity.

What history may have to tell us here is that the success of the United States in catching up with England in the last half of the nineteenth century was due not only to its large market, which permitted new production with interchangeable parts, but also to its freedom to experiment, tinker with and improve largely imported technology. Rosenberg sees the Americans borrowing "freely and extensively from Europe," with very little "genuinely inventive activity" in evidence during the colonial period. There was little government support – nor much government regulation. Beginning around 1850, nevertheless, the United States began to innovate meaningfully in the area of production engineering and the application of improved mechanical skills; the McCormack reaper, the Colt .45, the cotton gin, and the typewriter were among the products that revolutionized in a fundamental way factory production methods generally. While putting out and handicraft production persisted in Europe, the United States quickly became the undisputed leader in industries that lent themselves to the introduction of labor-saving machinery for the mass production of a standardized product. Before World War I Singer Sewing established a subsidiary outside Glasgow said to be the "most advanced" in Britain, if not in Europe – shades of Professor Hughes's steam turbine example.

A second important historical lesson might well have been an inquiry into why Britain lost its engineering-intensive industrial leadership to the United States but lost its more science-intensive industrial leadership to the Continent, especially to Germany. The explanation here undoubtedly has much to do with differential innovation inducement mechanisms at work in the two places. Germany acutely felt its relative disadvantage in coal, iron, and geography, and reacted to its higher fuel and iron ore costs. As Landes put it, "Continental ironmasters were making more of their resources than their competitors across the Channel, and since fuel economy was the key to efficiency in almost every stage of manufacture, the tentative advances of the 1830's and 1840's were the starting point of a scientific metallurgy that was to pay off in major improvements a generation later." There can be little doubt that Germany's spectacular success in the science-intensive chemical industry was very much related to a strong nationalistic drive aimed at finding substitutes for deficiencies in natural resources at home and colonies abroad.

With respect to the United States, that country could increasingly take advantage of its relatively much more abundant wood supply to manufacture light textile and other machinery. Such machinery was first considered an amusing oddity, but later was generally recognized as technically superior. The relative abundance of the United States' natural resources base also gave it the continuing advantage of a cheaper supply of fuel, first based on steam, then built on the scientific advances made elsewhere in the field of induction, on electric power. Moreover, the United States' relative labor shortage removed most institutional as well as economic obstacles to a thoroughgoing exploitation of labor-saving technological opportunities. From textiles to metallurgy and to the many later applications of machine-making in routinized mass production industries, the response to changes in the environment was usually rapid. As the comparative U.S. advantage in cheap wood dwindled, we find iron replacing wood and coal and coke replacing charcoal as the primary source of fuel. The rapid overall pace of industrialization was also accompanied by increasing capital intensity and associated scales. Increasing pressure for labor-saving technology in industry, together with the existence of a large, dependable and expanding domestic market, propelled by the expansion of the railroads and the tinkering of the railroad workshops, provided the cornerstone for the "American System" of mass production.

This reminds one very much of the innovation-inducement mechanism proposed by Binswanger, Hayami, and Ruttan(2) and applied historically mainly to the contrasting types of agricultural advances in Mexico and Japan. The point is that the technological patterns that evolved in Britain, the Continent, and the United States over time were also due to government policies that acted either to facilitate or to obstruct the system's accommodations to its changing relative endowments and capacities. With respect to Britain, for example, a policy of substantial laissez faire vis-a-vis the more mercantilist and still somewhat feudalistic countries of Europe was interventionist relative to the United States.

Germany, on the other hand, responding to its raw material shortage, resorted to substantial government intervention – at the more science-intensive end of the spectrum – either via protective tariffs, as in the case of the rise of the important beet sugar industry, or via support of scientific research labs and scientific education, as in the case of the von Humboldt reforms. Prussia went so far as to set up costly state enterprises and issue invitations to moneyed private parties to establish factories; but mostly, in contrast to the heavy interventionism of the French, the German effort was a more indirect one, using expositions, awards, subsidies, and technical advice, as well as the establishment of a whole network of technical and scientific institutions at various levels to provide formal training, from engineers and mechanics to manual arts and design. This reminds one of Professor Hughes's first, "general acquaintance" phase of technology transfer.

The role of government in the United States, while clearly more limited than in Continental Europe, also facilitated development in a similar but not identical direction. Rosenberg, commenting on the growth of the multifaceted U.S. machinery industry, speaks of a surprising volume of public/private collaboration: "visiting one another's plants, sharing new technological knowledge and even occasionally borrowing one another's workmen." Other observers place heavy emphasis on the role of widespread general education, which provided for a measure of technical literacy at lower skill levels – and for substantial empirical problem-solving capacity at higher levels. Even if most technology was borrowed and even if there was no first-rate scientific establishment in evidence, Rosenberg observes that the United States was "highly discriminating in borrowing patterns and highly selective in the uses to which imported technologies were put." Clearly the mechanical skills and ingenuity required for this task were considerable. And while the United States produced little in the way of contributions to frontier science until much later, the diffusion of labor-saving technology change and adaptations, from firearms to clocks, from watches to harvesters and typewriters, all part of the "American System," required engineers who had at least a grounding in science and its use, even if they were not active contributors to it.

Little wonder that the nineteenth-century United States attitude toward science and technology has often been called extremely pragmatic. While higher-risk basic science was neglected, technology was borrowed and improved upon. The continuing shortage of labor resulted in continuing labor-saving technology bias. Only agriculture was, to some extent, an exception; with private risks larger, so was the role of government. The unique institutional framework that focused on the land-grant college system was able to generate substantial technology change tied to progress in the chemistry-related agricultural sciences and diffused widely after the turn of the century.

The nineteenth-century United States may thus be characterized as a frontier society presiding over what seemed to be unlimited natural resources, including fuel, and therefore, unlike Germany, not much inclined to invest heavily in basic science. Nevertheless, innovative activity, based largely on imported technology and assisted by public-sector action, especially in education and agricultural research and extension, proceeded at a very rapid pace, associated with rapid increases in per capita income. This divergence of the historical paths taken by the two early followers of Britain, Germany, and the United States, tells us that a combination of differential endowments and policies may lead one country to participate in growth via basic science and science-intensive industries, another by borrowing technology and using a broadly based scientific literacy to improve upon and diffuse such technology.

What, then, is the relevance of such historical insights for contemporary policies regarding improved technology transfer mechanisms?

First, today's less developed countries, just like the United States and Japan in the nineteenth century, don't need to be pioneers in invention. But they do need to develop a definite indigenous capacity to select wisely and adapt imported technology. This ability does not just happen. It is related to the educational system and to the national ethos, as well as to the types of interventions, direct or indirect, practiced by governments. An educational system that imparts a modicum of scientific understanding to a substantial portion of the population, a pragmatic "catchup" philosophy which accompanies "latecomer" status, and national governments' willingness to move away from dirigist mercantilist interventions can provide basic building blocks for this type of capacity at a relatively early stage of a country's development. As the experience of both Japan and (especially) the United States also illustrates, the same country may later, in its modern growth phase, acquire the capacity to advance the international frontiers of science and invention.

Second, while no government policies can be expected to legislate away the basic initial endowment, they can provide an important assist to the transition effort of a developing economy as its endowment changes with time. Analogously, if a country is dominated by narrow vested interests or is lacking in historical perspective, such policies can attempt to draw a veil over the endowment and lead the system into expensive scientific-technological dead ends and economic stagnation. While there is no rigid unidirectional sequence of phases that every country must somehow traverse on the path to mature growth, some attention to the changing roles of science and technology in terms of a changing resource endowment and, especially, changing human capacities is essential in all but the most unusual cases.

At the micro and institution-building level, the appropriate role of government in the mixed economy context is, of course, not unrelated to the appropriability or nonappropriability of the new knowledge acquired. Investment in inventions carries a high risk, in part because of its at best indirect and long-term relationship with technology and growth, but partly also because an invention is generally an international good appropriable not even by a country, let alone by any private party within the country. As we move from basic international science to changes in technology, risks are reduced and private appropriability becomes much more important. As the extent of appropriability rises, so, normally, does the level of private R & D expenditures.

Appropriability, of course, depends not only on how basic the research effort is but also on the overall state of competitiveness or noncompetitiveness of the system. This is partly a function of the overall policy environment; for example, during periods of intense import-substituting industrial protection and large unearned profits, there would seem to be less interest on the part of industrialists to search for the best technology; instead, satisficing behavior and the use of inappropriate (often prestige) technology seems to frequently displace maximizing behavior – Professor Hughes's "parochial attitudes." But for any given industry or sector the state of competitiveness also depends on conditions peculiar to the particular market, with respect to goods, information, or technology. Agriculture, for example, is typically the most competitive field, and therefore exhibits the least private appropriability possibilities and the least willingness (or capacity) by individual farmers to incur R & D expenditures. Consequently, not only basic scientific agricultural research but also the search for appropriate adaptive technology and even its dissemination to individual farmers usually represents activities (and costs) that fall to the public sector. Ditto for health – except perhaps even more so. This is fairly well recognized. But what is perhaps less well understood is the fact that there exist other industries – again on a continuum moving through agricultural processing and input industries to light consumer goods, some services, and beyond – where similar characteristics abound: a competitive market structure, the relative absence of scale advantages, and thus the need for possible government involvement in R & D, education, and extension. If, in the absence of pronounced market imperfections, new technology can be selectively borrowed from abroad, the burden of high-cost domestic R & D is reduced and a minimum of government support can lead to rapid diffusion of technology change. This certainly was the case for Japan. Here the profound technology change associated with the switch from mule to ring cotton spindling in the late nineteenth century was diffused as rapidly as the agricultural practices of the "veteran farmers."

Whether or not, in mature market economies, competitive or noncompetitive industrial configurations yield relatively more private R & D activity remains an as yet unresolved empirical issue. Competitive industries have more incentive but less capacity. With respect to the less developed countries, it seems to me, any viable science and technology policy must begin with an examination of the extent that overall competitive pressures are felt by individual decision makers with respect to economic versus engineering choices. It must include sensitivity to differences in the market structures of specific industries, and consideration of selective government action in creating appropriate social overheads. Such government interventions may be restricted to ensuring that existing technological alternatives are known to all sizes of firms, or to helping expand the range of alternatives via the support of university or R & D institute activity. Activities supported within these fields must carry built-in devices to ensure that the criteria of ultimate contributions to social and economic development and not any exclusively internal criteria of the "invisible college" are addressed. One such device, frequently referred to, is government subsidy of R & D institutes set on a long-term declining basis, with private sector contracts forced to fill the widening gap. Another is to concentrate scarce attention on the more competitive nonappropriable sectors and, in fact, to ensure that access to information as well as to the required complementary inputs is relatively equal across firms.

In short, we now recognize that there exists a substantially larger number of borrowing options by country and by type than we used to believe, but many remain obscured by a lack of information and other institutional impediments, some related to the public and private capital transfer mechanism. The options that exist with respect to indigenous or adaptive technology change, new or derived, are more numerous still. Governments can do much at the aggregative level to ensure that the veil between relative prices and endowments that must sometimes be drawn is neither excessively thick nor kept excessively long; governments can help ensure that sufficient workable competition exists so that entrepreneurs are interested in finding the most appropriate technology in the first place, rather than being in a position to indulge their preference for prestige and the "quiet life." And, perhaps most important, a complement to these aggregative measures is the possible intervention of the public sector in institutional areas, in ensuring a freer flow of information on market, quality, and technology options and in providing support to technical education and R & D, especially of the unspectacular adaptive nonappropriable type.

We would expect, in short, the developed countries to continue to focus their control systems on the uncovering of new inventions at the frontier. They can, moreover, help with the establishment or redirection of informational and R & D networking systems within and among countries at different levels of development. Less developed countries should focus their attention on improving their indigenous capacity to make better innovation import choices, diffusing existing information on technology options within the country, and improving related R & D capacity to pursue indigenous adaptive technology changes. LDC's might benefit, in this respect, from the recognition that R & D that is meaningful for these purposes often takes place in such unconventional locations as repair shops and factory floors, and that its encouragement – whether through institutional reforms or the restructuring of decentralized commercial banking functions – may be one of the best ways to enhance a country's technological problem-solving capacities.

NOTES

(1) Drawn in part from my "Science, Technology and Development: A Retrospective View," in Science, Technology and Development, ed. William Beranek and Gustav Ranis (New York: Praeger, 1978).

(2) For a full statement see Binswanger, Hayami and Ruttan, Induced Innovation: Technology, Institutions and Development (Baltimore: Johns Hopkins Press, 1978).

3 The Role of the International Patent System in the International Transfer and Control of Technology

Sigmund Timberg

PATENTS, KNOW-HOW AND THE OTHER COMPONENTS OF THE INNOVATION AND DEVELOPMENT PROCESS

The international patent system may be defined as the aggregate of national laws and international treaties and conventions (starting with the Paris Union of 1883 and coming up to the Common European Patent Treaty) that provide for the granting of patents, and the body of national commercial law and business practices that provide for the licensing and assignment of the patents thus granted.

There can be no doubt that the international patent system plays an important role in the international transfer of technology, particularly in those situations where a technologically sophisticated multinational corporation in an industrialized country licenses or assigns its foreign patent rights to an equally sophisticated multinational located in another country. But even here, in industries where the basic patents have expired the emphasis shifts to the transfer of the unpatented technology, the so-called know-how, of the licensor.

On the other hand, when technology is transferred to inexperienced licensees or to the developing countries, the technological transfer may consist preponderantly (and often exclusively) of know-how, or unpatented technology. In the case of the developing countries, their primary need is to develop low-technology industries to improve the basic living standards of their people. For the developing countries, licensing of technology alone is of limited economic effectiveness, unless it is accompanied by the making available of other vital factors of the production and distribution process – physical plant and facilities, capital resources, marketing technology, management techniques, and the training of management, technical, and labor personnel. The value and impact of the technology transfer, patented or unpatented, cannot be realistically appraised on its own merits, but must be viewed as part of an overall economic package.

Inventiveness, and a national patent system that fosters and promotes inventiveness, are important foundation-stones for the building of an industrial economy, but will not by themselves create such an economy. A great deal more is needed before patents can be commercialized and inventive concepts and technical ingenuity can be converted into a functioning industry. That tertium quid is the complex of business talents, practices, and procedures that falls under the heading of "innovation."

Even in the United States there is concern that the domestic rate of growth of innovation is declining perceptibly, as contrasted with those of the German and Japanese industry.(1) For the developing countries, whose primary concern is with the introduction of industries and of innovation rather than their rate of growth, innovation is an even more vital concern. Any assessment of the impact of the international patent system on the international transfer of technology is valueless unless it can be assumed that the other factors involved in industrial innovation are present or can be made available.

In defining what is meant by "innovation," reference may be had to Professor MacLaurin's brilliant historical analysis of the technology of the radio industry,(2) where he distinguishes between science, invention, and innovation. The work of the scientist, although basic to all inventions and all technology, is not patentable. A basic, or even an improvement, invention may result in the grant of a patent, but that grant means only that the patented product or process is physically operable, nor that it is a commercially viable product or process. (3) The ramified network of research and development and business practices and arrangements that is required to convert a patentable but merely physically operable invention into a commercial reality is what is known as innovation. Of the inventors described in MacLaurin's book, only Marconi can be classified as a business innovator, and he was aided by the resources of the British Navy, considerable financial backing, and the brains of the gifted lawyer-politician who later became lord chancellor of Great Britain and viceroy of India.(4)

Innovation has been defined as "the total complex process by which an invention, or an idea, is brought to commerical reality for the first time."(5) John Copp, formerly solicitor for one of the world's largest multinational corporations, which has licensed its technology broadly throughout the industrial and developing world, has defined innovation in the chemical industry as taking the following five forms:(6)

1. The invention of a new product;
2. The invention of a new and better process for making an existing product;
3. New uses for existing products;
4. Minor improvements to manufacturing processes;
5. Minor improvements to processes which use the manufacturer's products.

The most immediate and obvious method for accomplishing innovation is that of research and development, which is especially significant in the pharmaceutical industry. Professor Dunning points out that

> the pharmaceutical industry is par excellence a science-based industry. It has spent, in proportion to its turnover, more on research and development than any other industry, except the aircraft and electronics industries, and no other industry employs a higher proportion of qualified staff.(7)

Hoffman-LaRoche has reported that 11 percent of its total operating revenues are spent by its research division directly, and that, when hidden R & D costs in other Hoffman-LaRoche departments are taken into consideration, this figure could be raised to 16 percent.(8) A great deal of this research cost precedes the production of the patentable invention. Statistics of pharmaceutical research show that only one commercially salable compound emerges out of 6,000 to 10,000 chemical syntheses.(9)

The high costs of R & D are an important factor in the international transfer of technology. As Copp pointed out, "it is always more profitable to manufacturer yourself than to take a royalty for teaching someone else to manufacture."(10) Firms with high R & D expenses that are located in the industrialized countries and are asked to license their patents and know-how to the developing countries do so in the expectation that these costs will be recouped by royalties, and that these royalties will be commensurate with the royalties that they receive from their licensees in the industrial countries. The developing countries, with their low national incomes and balance-of-trade problems, frequently regard these royalty expectations as excessive. This is one of the causes of tension between the technology-transferring and technology-receiving countries.

R & D may not be enough. In many cases the new product or process covered by the patent is so revolutionary as to require the construction, start-up, and operation of expensive new plant or facilities. The technology involved in the construction and the subsequent operation of such facilities will consist mainly of designs, blueprints, manuals, formulas, and personal technical assistance, but additional patent rights may also have to be licensed.

The new plant construction will also require capital financing and engineering resources. Typically, outside engineering firms will be called on for the engineering resources needed to construct the new facility. However, the capital required for the construction of the facility will, in the case of a developing country, usually have to be obtained from the transferor of the technology. The transferor of the patented and nonpatented technology needed to construct the plant will have to continue supplying it for a period of time (five years is not uncommon) after the plant has started commercial operations. Inability to obtain necessary capital engineering resources and competent technical advice and administrative management will therefore constitute a barrier to the international transfer of technology, even where the host country is in a position to meet the royalty and other patent license requirements of the transferor of the technology.

Steele has thus summarized the major barriers to international technology transfer:

1. The investment barrier, arising out of lack of capital resources for building the needed plant or facility;
2. The adaptation barrier, which may involve rescaling or modifying the manufacturing process to meet local market needs or the local scale of production; and
3. The technical capacity barrier, caused by the low rudimentary level of local techical capacity.(11)

Obviously the international patent system, either in its present form or modified as some have proposed, can do nothing to remove these nontechnological barriers to technological transfer. Unless these barriers are overcome the international patent system, by itself, is powerless to promote the industrial innovation and national industrial development desired by the host country.

For countries anxious to embark on full-scale industrialization, improved disribution patterns must be instituted. Thus before the new (or more efficiently and cheaply produced) product can be marketed by the recipient of the technology, the potential of the market must be explored, which in some cases may involve advance testing of the market. Moreover, to bring the product to the attention of industrial users and individual consumers and to ensure effective marketing, advertising, and promotion techniques may have to be utilized, involving the use of trademarks, copyrights, logotypes, and other forms of intellectual property. Typically these distributive and promotional techniques have been developed by the transferor of the patent rights and the industrial technology involved, and are not in the possession of the transferee.

It is only after the necessary R & D, capital outlays, engineering resources, and local adaptation, advertising, and promotional efforts have been invested in patented inventions that the user of the inventions has satisfied the prerequisites of innovation, that is, has produced a commercially viable product. The original owner of the patent rights and technology, who has innovated and brought the patented product to the market, usually controls not only the invention but also the other resources and factors needed to make the invention a commercial reality. Where this is the case, when he is asked to transfer the foreign rights over his technology he expects to be compensated not only for his specific R & D expenditures in procuring the issuance of the patent but also for the technological, financial, and other costs involved in innovating that product and bringing it onto the market. This expectation also raises prolems for the developing countries, with their low per capita income, thin markets, and perennial deficits in international trade balances.

THE LEGAL FOUNDATIONS OF THE INTERNATIONAL PATENT SYSTEM: THE PARIS UNION CONVENTION OF 1883

The first part of this paper discussed the economic role of patented and unpatented technology in the overall process of industrial innovation and national economic development. This section will sketch the legal foundations of the international patent system, as it is currently operating, and how it facilitates the international dissemination of technology.

The core of the international patent system is the national patent systems of the individual countries. With the exception of the system for the issuance of regional patents covering the nine European Common Market countries,(12) patents are issued by the patent offices of individual countries, are enforced according to the national legal system of the patent-issuing state, and are restricted in legal effect to the territorial boundaries of the issuing state.(13)

Taken by themselves, national patent systems contribute to the industrial development of their respective countries, but they do not promote or facilitate the international diffusion and exchange of technology. The linchpin of the system for the international transfer of technology is the Paris Union Convention of 1883, as amended, Article 2 of which requires member states to accord "national treatment" to foreign inventors. Under this principle of "national treatment," nationals of foreign countries, or others who are domiciled or have effective industrial or commercial establishments in a country, are guaranteed equality of treatment with the nationals of the country granting the patent.(14) The Paris Union is adhered to by some eighty countries, but the principle of "national treatment" is so important to the international diffusion of technology that it is followed by some countries that are not members of the Paris Union.

Several countries qualify the principle of "national treatment" by the "principle of reciprocity"; this also is true both of countries that are and are not members of the Paris Union.(15) Some of the countries that accord "national treatment" to foreign inventors require a foreign applicant for a patent to designate an agent for service of process or to designate a resident of the country to represent him in all matters pertaining to the patent application and in subsequent legal proceedings relating to the patent. This procedure is authorized by Article 2(3) of the Paris Union.

The practical workings of the principle of national treatment are facilitated by another basic provision of the Paris Union (Article 4), which provides for the so-called right of priority. Under this right, an applicant for a patent that has filed a patent application in any one country may, for a period of twelve months after such filing, file a patent application in any other country recognizing the "right of priority," and such subsequent filing will be dated back to the date of the first filing.(16) Several countries that are not members of the Paris Union also recognize the right of priority.

Another provision of the Paris Union that facilitates the transfer of patent rights from one country to another is Article 4 bis, which establishes the principle of the independence of patents. Under this principle, the cancellation or expiration of a patent in one country of the Paris Union does not lead to the cancellation or expiration of a patent for the same invention in another member state of the Paris Union.(17)

The major substantive limitations imposed by the Paris Union Convention on the national patent systems of its member states are to be found in Article 5, which purports to limit the sanctions that national patent legislation imposes where the patent has not been worked at all, or has been insufficiently worked, in the country. But even this provision indicates the extent to which national legal systems control both the issuance of patents and the use of such patents after they have been issued. Under the national patent laws of most countries of the world with patent systems (with the exception of the United States), provision is made for the compulsory licensing or revocation of patents in the event that there has been no, or an inadequate, use of the patent within the country.(18) Article 5 of the Paris Union attempts to standardize, and render more lenient to the patentee, these national compulsory licensing and revocation procedures. It provides that patent revocation will be resorted to only if the compulsory license that has been granted does not suffice to prevent abuses resulting from the nonuse of the patent. It also provides that a nonused or inadequately exploited patent is not subject to compulsory licensing until after three years from the date of its issuance, or until after four years from the date of filing of the patent application (if the patent was issued within twelve months from filing), and that no patent should be subject to revocation until two years after the issuance of a compulsory license to an applicant. Important as these rules are considered to be, the national patent laws of some member countries of the Paris Union do not conform to the standard laid down in Article 5.(19)

Article 5 of the Paris Union Convention also prohibits the forfeiture of a patent on the ground of the importation into a country of patentable articles produced in other countries that are members of the Paris Union.(20) This provision safeguards the rights of patentees against xenophobic national legislation designed to prevent products made abroad under foreign patents from entering local markets.

To conclude this discussion of the current legal framework of international patent system, it should be noted that there is general agreement in the international community on four factors as justifying the grant of the patent. One is the concept of patents as confirming the private property of the inventor in his invention.(21) Another is the concept of the patent as a special grant of monopoly by the state to encourage inventions and national industrial growth and development.(22)

It is also recognized that the patent is an incentive to inventors not only to invent, but also to disclose the nature of their inventions and thereby stimulate the further growth and application of technology.(23) Finally, the patent grant is assumed to be a prime stimulus for expanded investment in the industry covered by the patent.(24)

From the standpoint of international law, the Paris Union Convention and the cluster of rights and principles associated with it implement two of the seven standards of classical international economic law, as laid down by Professor Schwartzenberger: the standard of national treatment and (for those countries that have adopted the principle of reciprocity) the standard of identical treatment.(25) Both of these standards differ widely from what is known as the minimum standard of international law that has been applied to the expropriation of foreign investments and has evoked so much criticism from the developing countries (particularly those in Latin America), because it discriminates in favor of foreign and against domestic nationals.(26)

In order to complete the theoretical international law framework of this chapter, it should be pointed out that the current efforts of the developing countries to change the international patent system involve recourse to still a fourth standard, which authorizes preferential treatment for the nationals of the technology-receiving countries and discrimination against the nationals of the technology-exporting countries. Schwartzenberger calls this the "standard of equitable treatment,"(27) and justifies it by saying that it "may involve formal inequity and discrimination to attain substantive equality and justice on the level of community ethics."(28)

The developing countries relate their proposals for changing the current international patent system to the demands of what they call the "new economic order."(29) Regardless of how Professor Schwartzenberger or others would label these proposals for change, they are clearly contrary to the position that has been held for a century by the industrialized countries and firms in those countries that are the primary exporters of technology. The conflict between these two viewpoints constitutes a barrier to an accommodation between the developing and industrialized countries, and it is much more than a linguistic one.

NATIONAL AND INTERNATIONAL TECHNOLOGY POLICY

Achieving an international consensus, embracing both the industrialized and the developing countries, on the proper role of the international patent system in the international transfer and control of technology involves the adjustment and accommodation of the conflicting views and policy objectives of the group of technology-exporting and the group of technology-importing countries. (The first of those can roughly be referred to as the OECD group and the second as the expanded Group of 77.)

The preceding two sections of this chapter have noted two of these basic conflicts: their different economic expectations as to the appropriate terms of the patent license bargain, and their opposed philosophies of international law. Both conflicts arise out of the different economic and political circumstances of the two groups.

This and the following two sections of this paper will discuss the extent to which the respective national policies in the fields of technology, competition, and international economic development promote or frustrate the achievement of a consensus on the appropriate future role of the Paris Union Convention and national patent systems in the international transfer and control of technology.

The preceding section of this paper has shown the Paris Union Convention of 1883 to be the leading international law instrumentality for the international diffusion and exchange of patented technology. But the basic legal mechanism through which this interchange is achieved is the national patent systems of the technology-transferring and technology-receiving countries, each of which embodies national technology policy with respect to inventions.

There is considerable theoretical agreement on the broad features of a national technological policy. Patents are issued by national patent offices for "inventions," which are technological developments that meet the tests of "novelty," industrial applicability," and "nonobviousness" (sometimes referred to as "inventive step" or "technical progress").(30) The grant of a patent confers upon the recipient a patent monopoly, limited territorially to the issuing country (31) and temporally to a fixed period of years, which is also a property right. The patent grant is conditioned on the disclosure to the public of the claimed invention.(32) The grant of patent monopoly is generally considered to be an incentive to invent, to innovate, and to invest, thereby promoting the public interest and the development of the national economy.

As long as the issued patent remains under the control of the original grantee, national technology policy has only two major further concerns. The first, already discussed, is to implement, where complaints and other appropriate circumstances exist, national legislation requiring that the patent be worked and exploited within the country. This concern may be of greater importance to the developing than to the industrialized countries and does not even exist under U.S. patent law.(33)

The second, which is fundamental to the entire concept of the patent, is to provide for the enforcement of the patent against infringers. This is accomplished through the judicial system of the country, and imposes on the national judiciary the difficult task of determining the validity and scope of the patent, the fact of infringement, and the damages arising out of the infringement.

While there is a large measure of agreement as to the ideal theoretical components of a national patent system, the practical realities fall far short of the ideal. Only among the industrial countries (and even there not among all) do there exist systems of search and examination that would enable the patent office to search for the relevant "prior art" or technology and determine whether a specific invention meets the necessary requirements of novelty, utility, and nonobviousness.(34) Even countries with advanced examination procedures, elaborate physical facilities, and large numbers of patent office personnel equipped to carry out those procedures – the United States, the United Kingdom, Germany, The Netherlands and Japan –

suffer long time delays and high costs in the issuance of patents. Even after the patent is issued, its enforcement in the courts may involve costly and time-consuming litigation.(35) Where the country has a weak or no examination procedure, the issuance of a patent tends to become a formality; no presumption of validity attaches to the action of the patent office in granting the patent. In many developing countries the issuance of a patent amounts in practice to rubberstamping the patent application.

In recent years, various international instrumentalities have been developed in an effort to alleviate the administrative burdens and cure the deficiencies of the national patent systems. One of the most important of these is the Patent Cooperation Treaty, first signed in Washington in June 1970, which provides for the filing of an "international application," which has the same effect as if separate application had been filed in each of the countries in which patent protection is desired; a search to discover the "prior art" relevant to the international application; and, if requested, a preliminary examination to find out whether the invention meets the statutory requirements of novelty, industrial applicability and nonobviousness. After the search and/or preliminary examination report is received, the individual national patent offices process the applications separately, and grant or reject the application in accordance with their national legal systems.(36)

Perhaps the most dramatic example of a truly international approach to the overall patent is the regional convention for the establishment of a European Patent Office, authorized to issue a single unitary patent for the countries belonging to the European Economic Community. Patents thus issued by the European Patent Office, which started functioning in Munich on January 1, 1978, "will constitute the first example of rights governed by the process of an international convention rather than of domestic legislation."(37)

For the developing countries, with their lack of trained personnel, physical facilities, and financial resources, the indicated course is to establish a regional patent office that would serve the needs of the entire region. The oldest example of such a regional approach is the Industrial Property Office (covering trademarks and copyrights as well as patents) established by the African and Malagasy Organization for Economic Cooperation under the so-called Libreville Agreement of 1962. This provides for uniforn national patent legislation; a system whereby only a single patent application is filed; and the operation of a joint centralized administrative procedure.(38) Involving less centralized administration, but directed to common planning of the rules to govern patent and other intellectual property in a geographic region, are the polices established by the so-called Andean Group of countries pursuant to the Cartagena Agreement.(39) There are other examples of interntional conventions and cooperation with respect to such matters as patent classification and the storage and retrieval of information.(40) However, the only cases where it can be said that national formulated technological policy is being replaced by a multinationally formulated technological policy are the regional agreements governing the European Common Market, the African-Malagasy Union, and the Andean Pact countries.

Before leaving the subject of national technology policy, a word should be said about its socioeconomic-political origins. The international patent system embodied in the Paris Union Convention of 1883 was originated by countries that were trying to accelerate their industrialization. They believed that their patent systems, strengthened by international agreement, would continue to promote both national industrialization and the international diffusion and exchange of technology. These countries had a restricted view of state participation in economic development and industrialization. Accordingly, patents and the patent system were conceived of as legal privileges and as mechanisms that would liberate the creative energies of inventors and of business innovators, whose technological and entrepreneurial capacities would be the chief factor in the furtherance of national development and the national economy.

This pervasive faith in the advantages of leaving inventions and the innovation process within the largely uncontrolled private sector of the economy is obviously not shared by countries within the Group of 77, where invention and innovation are in varying ways controlled, regulated, and guided by the state. This disparity in basic approach does not mean that economic coexistence cannot be worked out between the technology-exporting countries, where government pursues a laissez faire policy, and the developing countries, where government interference is the rule, but it represents a gap in viewpoint that must be bridged before the specific details of such economic coexistence can be worked out.

Having dealt with the formulation of technology policy at the national and multinational levels, let us now turn to two other policies, those relating to competition and to national economic development, that are closely intertwined with technology policy. What links these three policies is that the national patent systems of the world provide for the licensing and assignment of patents, which in turn makes it possible for the patentee to disseminate technology to third persons. In these patent licenses,(41) licensors and licensees prescribe the terms and conditions under which the licensed inventions will be used and the licensed products distributed. Where these arrangements are restrictive and give licensors or licensees market advantages not available to their competitors, or place them in a position to impose unreasonable conditions on purchasers, they are called restrictive business practices. When such practices unreasonably restrain commerce and have adverse effects on competition among licensors and licensees, or upon competitors and the general public, they may become the concern of national competition policy, which may or may not be incorporated in national antitrust legislation.

Only a few countries, such as the United States, Germany and Japan, have strong competition policies that are applied to patent license agreements. No other country has as broad a range as does the

United States of restrictive conditions and practices under patent license agreements that are prohibited, or considered doubtful, under national antitrust legislation.

However, it has been recognized that restrictive business practices in international license agreements tend to impede the free flow of international trade, and may have undesirable repercussions on the economies of the countries affected by such practices. It is therefore no accident that the European Economic Community, with its strong emphasis (directly derived from the Rome Treaty) on preventing "distortions of competition" from affecting trade among its member states, should have the second strongest antitrust enforcement policy in the world, although most of its member states have very weak or even no antitrust legislation. Similarly, there is every reason why both the OECD group, that is by and large committed to liberal international trade policies, and members of the Group of 77 should be interested in establishing some sort of control over patent license restrictions that produce effects at variance with national or international competition policy.

Many conditions in patent license agreements are considered by the developing countries to be unduly rstrictive of their industrial growth and development, and hence run counter to the economic development polices of those countries. In many of the industrialized states, national competition policy far outranks national development policy in importance, for the simple reason that those countries have attained satisfactory levels of development, and are of the view that further innovation and industrialization may be better achieved by promoting stronger competitive policies. In the developing countries, on the other hand, national economic development is the overriding objective, regardless of whether such development is implemented by monopolistic, restrictive, or competitive policies.

Despite the apparent diversity of purpose that characterizes technology, competition, and development policy, there are restrictive conditions in patent license agreements that may be contrary to the purpose of all three kinds of government policy. Dramatic illustrations of such restrictive conditions are so-called tie-in clauses in license agreements, which obligate the licensee to purchase unpatented raw materials, spare parts, intermediate products, capital goods, and/or competing technologies from the licensor or sources designated by the licenser.(42) Such tie-in restrictions have been held to violate the national technology policy of the British Commonwealth countries and the United States, because they are considered to fall outside the scope of the patent monopoly.(43) In the United States, Germany, and the European Common Market, such restrictions are, except in unusual circumstances, regarded as violations of antitrust laws.(44) And they are quite repugnant to the developing countries because they are considered to have detrimental consequences for national economic development policy and for the domestic economy. These detrimental effects are summarized as follows in the United Nations report on "The Role of the Patent System in the Transfer of Technology to Developing Countries":

194. Most of the goods that are currently produced or planned for production in the developing countries are available on the world market from several sources, and potential purchasers of these products in the developing countries can buy them at world market prices. But when contractual agreements tie part or all of the inputs to a single source of supply, developing countries are deprived of the possibility of exploiting market opportunities and are faced with a price structure determined by the unique supplier. Tied-purchase provisions thus result in a monopoly control of the supply of equipment and other inputs by foreign enterprises, leading to what has come to be known as "transfer pricing", "transfer accounting" or "uneconomic output".

195. By reason of his exclusive position, the supplier is able to charge higher prices than for comparable equipment and other inputs that could otherwise be obtained elsewhere. Overpricing of inputs in this way constitutes a "hidden cost" of the transfer of technology which is much the same as that of aid-tying.

196. Tied-purchase clauses connected with the transfer of technology not only affect production costs through the over-pricing of inputs but may have important indirect effects on the import substitution, export diversification and growth efforts of developing countries. When the source of supply is determined by the supplier, rather than by the receiver, of technology, a bias in favour of imports is only to be expected. Furthermore, since the imported technology itself originates in a developed country it is usually ill adapted to factor endowments and the availability of domestic resources in developing countries. Both these factors contribute to raising costs of production in developing countries and rendering the resulting product less competitive in world markets. The high cost of imported technology and inputs imposes a heavy burden on the balance of payments of developing countries. Together with reduced export possibilities, this affects adversely the rate of growth of the economy by preventing backward and forward linkages.(45)

Another group of restrictive provision in licenses that have adverse implications for technology, competition, and development policy are so-called grant-back provisions, under which the licensee agrees to cross-license or "grant-back" to the licensor any improvement the licensee has developed. However, where the grant-back clauses are nonexclusive in nature or relate to improvements that fall within the scope of the technology covered by the license, and where the licenser has entered into a similar undertaking to license any further improvements it may develop during the term of the agreement, such provisions are usually not considered illegal.(46)

Nevertheless, grant-backs (whether exclusive or nonexclusive) have been condemned on technology policy grounds as extending the patentee's original patent monopoly beyond its expiration date.(47) They may also violate competition policy where they tend to keep the licensor in a dominant or monopolistic positon, may substantially lessen competition, and may lead to a reduction in the innovative initiatives of both licensors and licensees.(48) That such grant-back restrictions may have adverse effects on development policy is indicated by the fact that they are considered susceptible of abuse and therefore subject to control by the Andean Group of countries, Argentina, Brazil, Mexico, and Spain, as well as under United States, European Economic Community, and West German law.(49)

There are some restrictive conditions that are rather unrelated to technology policy, may or may not (depending on the particular circumstances) have a bearing on competition policy, and are considered to have a very important bearing on national development policy. Perhaps the most important of these provisions are restrictions on exports, which have been said to be the most common limitations imposed on licensees and to take the following forms:(50)

> (a) total ban on exports; (b) prior approval by the licensor required before exports can take place; (c) prohibition of exports to certain countries; (d) exports allowed only to certain countries; and (e) requirements to channel exports through the licensor's agents.

Such restrictions on exports are forbidden under the laws of Argentina, Brazil, Mexico, Spain, and the Andean Group countries because they adversely affect the export potential of the technology-receiving countries.(51) They have a peculiar relationship to the national patent system and competition policy, in that they attempt to regulate international trade and are therefore outside the legal scope of the patent monopoly grant. However, where the patent rights in the countries to which the developing countries wish to export are held either by the licensor or persons licensed by him or his licensees, the prohibition against exports does not place the would-be exporter under any disability that he does not already have.

It is not feasible in a paper of this scope to review the entire list of restrictive conditions and practices under patent licenses that have varying implications for technology policy, competition policy, and development policy. Instead it may be helpful to present a summary of the restrictive business practices that the Council of the OECD, the diplomatic organ of industrialized countries, has indicated have a potential for harmful effects on national and international trade. In a recommendation adopted on January 22, 1974, the OECD Council:

> Recommends to the Governments of member countries:
>
> 1. That they should be particularly alert to harmful effects on national and international trade which may result from abusive practices in which patentees and their licensees may engage, and in particular, from the following:

(a) when negotiating or operating patent pools or cross-licensing agreements, unjustifiably imposing territorial, quantity or price restrictions or attempting to dominate an industry, market or new industrial process;

(b) by means of territorial restrictions in patent licenses affecting international trade, unjustifiably prohibiting exports of patented products or unjustifiably restricting trade in or exports of the patented products to specified areas;

(c) by means of clauses concerning tied sales, obliging the licensee to obtain goods from the licensor or his designated sources, when the tied sales are not justified, for instance, by technical reasons concerning the quality of the goods manufacturered under the license;

(d) by means of grant-back clauses, unjustifiably requiring the licensee to assign or grant back to the licensor exclusively all improvements discovered in working the patents when the effect of this practice is to reinforce the dominant position of the licensor or to stifle the licensee's incentive to invent;

(e) by means of clauses unjustifiably limiting competition, preventing one or more parties to the patent licensing contract from competing with others parties to the contract, or with third persons, in other industrial fields not covered by the licensed patent;

(f) arbitrary grouping and licensing all patents in a particular field and refusing to grant licenses for only some of the patents or using other forms of package licensing when these practices are coercive in character and when the selection of the patents is not negotiated for the convenience of the parties;

(g) contrary to national law, fixing the prices of patented products by means of patent licenses.(52)

AN INTERNATIONAL CODE OF CONDUCT: PRESENT STATUS AND FUTURE PROSPECTS

Negotiations for an International Code of Conduct to govern the international transfer and control of technology have been going on among the OECD group and the Group of 77 continuously since about the time the OECD Council issued the just-cited recommendation of January 22, 1974. These negotiations have from time to time included views and proposals emaning from the socialist countries, but for purposes of convenience this part of the paper will be limited to setting forth the major difference between the OECD group and the Group of 77, their causes, and the prospects for bridging them in the form of a mutually acceptable International Code of Conduct.

Joel Davidow, the chief negotiator of the U.S. team participating in these negotiations, has written that there are four vexing issues on which the OECD and the Group of 77 are still apart.(53) He summarized them as follows:

1. When is a restrictive business practice that is not per se illegal unacceptable? On this issue, the U.S. answer has been that a restrictive practice is unacceptable when it restricts competitive freedom under the standards, well known to U.S. antitrust practicioners, of the "rule of reason."(54) The response of the European Common Market is that such a practice is unacceptable when it has an adverse effect on the flow of international trade, an understandable position on the part of a supranational governmental entity committed under the terms of its governing treaty, the Rome Treaty, to the elimination of all barriers to trade among its member states.(55) The view of the developing countries, according to Davidow, is that restrictive practices should be designated illegal solely because of their adverse effect on the export or development goals of a country, even where the practice is not abusive from the standpoint of competition policy.

2. Should the International Code of Conduct apply to export prohibitions and purchasing requirements imposed by parent multinational corporations upon their foreign subsidiaries? The general approach of the United States and of the European Common Market, where most MNCs are headquartered, is that such business restrictions take place within what is, from an economic standpoint, a single "enterprise entity,"(56) and are therefore normally acceptable, except where they are employed by dominant firms to entrench themselves in their dominant position or to injure their competitive rivals.(57) the approach of the Group of 77, in which the subsidiaries of multinationals have their situs and carry on their operations, is necessarily keyed to the dictates of their national economic development policies.(58) Thus more than a dozen developing countries regularly screen and remove tie-in restrictions between the multinational parent and its local subsidiary.(59) The developing countries regard a subsidiary within their jurisdiction as an aggregate of physical and human resources, subject to their control, which carries on operations affecting their national economies and their positions in international trade, not as mere corporate alter egos of a multinational parent.(60) Also, it may be noted that there is precedent in U.S. antitrust law, under the so-called double trader doctrine, for treating a parent and its subsidiary as separate corporate entities, capable of conspiring and agreeing to adhere to restrictive business practices between themselves and therefore subject to antitrust prosecution.(61)

3. Should exceptions be permitted to the rules of conduct set forth in the International Code? The Group of 77 argues that its members' inferior status entitles them to preferential treatment under the International Code wherever such treatment is necessary to the support of their national economies and economic development. Included among such permissible exceptions, the developing countries urge, are per-

mision for them to form or join international cartels in primary commodities (for example, OPEC)(62) and the authority to exempt themselves from the application of any code provision they deem contrary to their national development policies and the national economic interests. The United States and the OECD countries, for whom competition policy is more vital than development policy, respond by saying that exceptions from a general code of conduct may be appropriate in the tariff and trade areas, but are not acceptable in the antitrust area.

4. Should the International Code of Conduct be legally binding on member states and on their multinational corporations, or should be regarded only as a voluntary guideline? The Group of 77 takes the view that the International Code of Conduct should be a legally binding document, under which all agreements not in conformity with the provisions of the code would become unenforceable. The United States and OECD disagree, saying that the international code should be regarded only as a set of guidelines, to be accepted on a voluntary basis by the signatory countries and their multinational corporations and other enterprises. Part of the justification for the "voluntary guidelines" position taken by the OECD countries is that, if the international code were to become binding, it would be a one-sided affair, discriminating against the industrialized countries and their enterprises. The OECD countries would be bound by the code to enforce its provisions against their own multinations, whereas the developing countries, by virtue of their position that the international code should give them preferences and allow them to make such exceptions to the code provisions as their economic development policy may dictate, would be free to disobey the international code in their own uninhibited discretion. The OECD countries have the futher legal objection to a binding code that, under their constitutional procedures and under established principles of international law, they cannot proceed against a parent multinational corporation because of business restrictions and activities taking place abroad that have no effect on their domestic economies.(63) The multinationals headquartered in the industrial countries are just as adamantly opposed to any international code that would compel the industrialized states to embark on such a course of extraterritorial antitrust enforcement.(64)

Davidow is of the view – which is the necessary standpoint of the experienced diplomatic negotiator – that the approaches of the OECD and of the Group of 77 are not far apart. In fact there has been considerable progress toward agreement on the purely verbal and drafting level since the negotiations began. I do not have Davidow's background in this area, but I have read the stream of papers that have been issued both at the United Nations and at UNCTAD, and I am somewhat skeptical. Both Davidow's excellent summary of the current differences and earlier sections of the prsent paper disclose fundamental differences in political, economic, and social philosophy and psychology that cannot easily be papered over. I suppose, however, that it is the function of this Conference to ventilate those differences and see whether they can be effectively reconciled.

NOTES

(1) Under presidential direction, a U.S. government task force has been working on this problem, as has a private-sector Advisory Committee on Industrial Innovation, convened by and reporting to the Secretary of Commerce. See nn. 24 and 35.

(2) Maclaurin, Invention and Innovation in the Radio Industry (Macmillan, 1949).

(3) Sigmund Timberg, "Equitable Relief under the Sherman Act," University of Illinois Law Forum, Winter, 1950, pp. 629, 647.

(4) See Hyde, Lord Reading (Farrar, Straus and Giroux, 1967).

(5) Brown, Invention and Innovation – What's Ahead, 1 Finnegan & Goldscheider, The Law and Business of Licensing, 445 (1977), hereinafter cited as F & G.

(6) Copp, International Licenses of Chemical Technology, 2 F & G. 734.21.

(7) Dunning, Science, Industry and the State, 15-34 (1965), cited in Von Oertzen, Licensing Opportunities in the International Pharamaceutical Company, 1 F & G 506.155 (1977).

(8) Von Oertzen, Licensing Opportunities.

(9) Ibid.

(10) Copp, International Licenses, n. 6.

(11) Steele, Barriers to Technology Transfer, 2 F & G 734.62 (1977).

(12) The European Patent Convention, which came into effect October 7, 1977, established a supranational entity, the European Patent Office, which is physically located in Munich and has the authority to issue a European Patent. However, the European Patent is effective only in those countries subscribing to the Convention that have been designated by the patentee at the time it filed its European Patent Application. Moreover, a European Patent is to be construed and enforced according to national law.

(13) Aluminum Co. of America v. Sperry Products, Inc., 285 F2d 911, 925, (6th Cir. 1960).

(14) See the report prepared jointly by the U.N. Department of Economic and Social Affairs, the UNCTAD Secretariat, and the International Bureau of WIPO, The Role of Patents in the Transfer of

Technology to Developing Countries (New York: United Nations, 1975), (hereinafter referred to as 1975 U.N. Report), page 47. The reader interested in a fuller report on the Paris Union may wish to refer to an earlier U.N. report, bearing the same title and published in 1964 by the U.N. Department of Economic and Social Affairs, for which this author wrote the legal chapters. There are also economic approaches in the earlier report that the reader may find worth exploring.

(15) 1975 U.N. Report, p. 9.

(16) Ibid., p. 44.

(17) Ibid.

(18) Ibid.

(19) Ibid., pp. 19 ff.

(20) Ibid., p. 44.

(21) Wigmore, The Patent "Monopoly," 25 J. Patent Office Society 687 (1943). The inventor's property right does not come into being upon his obtaining the patent, but exists (although in legally unenforceable form) prior to that that time, upon his reduction of an original invention to actual practice. Crown Die & Tool Co. v. Nye Tool & Machine Works, 261 U.S. 24 (1923); Six Wheel Corp. v. Sterling Motor Truck Co., 50 F.2d 568 (C.C.A. 9, 1931).

(22) Article 1, Clause 8 of the United States Constitution; Pennock v. Dialogue, 2 Pet. 1, 19 (1822); Kendall v. Winsor, 21 How. 322, 327-9 (1858).

(23) See Sinclair & Carroll Co. v. Inter-chemical Corp., 325 U.S. 327, 330-1 (1945); U.S. v. Dubilier Condenser Corp., 289 U.S. 178, 187 (1933). As is evident from the earlier discussion of the role of R & D and unpatented know-how, and has been noted in the 1975 U.N. Report, pp. 45-46, in many industries patent disclosures are inadequate for the purposes of industrial development.

(24) See U.S. Senate Judiciary Subcommittee on Patents, Trademarks and Copyrights, Study No. 1 (1956); Draft Report on Patent Policy (by Advisory Subcommittee on Patent and Information Policy), Dec. 20, 1978, pp. 3-4.

(25) Schwartzenberger, The Principles and Standards of International Economic Law, 117 Receuil de Cours (Hague Academy, 1966).

(26) Ibid., 78.

(27) Ibid.

(28) Ibid.

(29) See Resolution No. 3201 and 3202, United Nations General Assembly, Sixth Special Session, May 1, 1974 ("Declaration and Programme on the Establishement of a New Economic Order"), especially p. 8.

(30) 1975 U.N. Report, pp. 5-6.

(31) See n. 13.

(32) See n. 23.

(33) This statement does not cover situations where nonuse of patents is caused by antitrust violation or is clearly against the public interest.

(34) See 1975 U.N. Report, pp. 6-7.

(35) The expense of carrying on a patent infringement litigation frequently exceeds $250,000 per party; see Draft Report on Patent Policy, Advisory Subcommittee on Patent and Information Policy, Advisory Committee on Industrial Innovation, Proposal IV, December 20, 1978.

(36) 1975 U.N. Report, p. 15.

(37) Ibid., p. 17.

(38) Ibid., pp. 16-17.

(39) Common Regime on Treatment of Foreign Capital and Trademarks, Patents, Licenses and Royalties (Dec. No.24), 11 I.L.M. 126 (1972).

(40) See 1975 U.N. Report, pp. 18-19.

(41) International discussion and controversy has centered on patent licenses, but similar considerations are involved with respect to the outright assignment of patent rights and the conditions that might be imposed by the assigner in such assignments.

(42) 1975 U.N. Report, p. 24.

(43) Motion Pictures Patents Co. v. Universal Film Mfg. Co., 243 U.S. 502 (1917). For the British Commonwealth legislation, see Analysis of Governmental Measures Relating to Restrictive Business Practices, Economic and Social Council Official Records: Sixteenth Session, suppl. no. 11A, p. 27 (U.N. 1953), and the text of such measures in suppl. no. 11B, passim.

(44) International Salt Co. v. United States, 332 U.S. 392 (1947); International Business Machines Corp. v. United States, 298 U.S. 131 (1936).

(45) 1975 U.N. Report, pp. 25-26; footnotes omitted.

(46) See e.g., Binks Mfg. Co. v. Ransburg Electro-Coating Corp., 281 F2d 252 (7th Cir. 1960), cert. dismissed, 366 U.S. 211 (1961).

(47) This was the basis on which exclusive grant-backs were held to be per se exclusive violations of the antitrust laws by Circuit Judge Learned Hand in Stokes & Smith Co. v. Transparent-Wrap Machine Corp., 156 F.2d 198 (2d Cir. 1946). This ruling was reversed by a 5-to-4 vote of the Supreme Court, 329 U.S. 637 (1947).

(48) See United States Aluminum Co. of America, 91 F.Supp. 333, 410 (S.D.N.Y. 1950); United States v. General Electric Co., 82 F.Supp. 753, 815 (S.D.N.Y. 1949).

(49) 1975 U.N. Report, p. 25.

(50) Ibid., p. 24.

(51) Ibid.

(52) Ibid., pp. 28-29.

(53) Joel Davidow, International Antitrust Codes of Conduct: A Progress Report, Fordham Corporate Law Institute (International Antitrust), Nov. 14-15, 1978.

(54) Addyston Pipe & Steel Co. v. United States, 85 Fed. 271 (6th Cir. 1898), aff'd, 175 U.S. 211 (1899); Chicago Board of Trade v. United States, 246 U.S. 231 (1918); Continental T.V., Inc. v. GTE Sylvania, Inc., 433 U.S. 36 (1977).

(55) See, e.g., Consten & Grundig - Verkaufs - GmbH v. E.E.C., CCH Common Market Reports § 8046 (1966); Omega decision, 2 CCH Common Market Reports 9346, 9396 (1970); Sirena S.r.l. v. Eda GmbH, 2 CCH Common Market Reports § 8101 (1971); Deutsche Grammaphon v. Metro Sp. - Grossmarkte, 2 CCH Common Market Reports § 8106 (1971).

(56) U.S. Department of Justice, Antitrust Guide for International Operations, January 26, 1977, pp. 12-13; Centrafarm BV v. Sterling Drug, Inc. (1974) Common Market Law Reports 480; but cf. Re Kodak (1970) Common Market Law Reports, R.P.Supp. D19.

(57) See United States v. Yellow Cab Co., 332 U.S. 218 (1947); Perma Life Mufflers v. International Parts Corp., 392 U.S. 134 (1968).

(58) See Davidow, The United States, Developing Countries and the Issue of Intra-Enterprise Agreements, 7 Georgia Journal of International and Comparative Law 507, 512 (1977).

(59) Ibid.

(60) 1975 U.N. Report, pp. 35-36, 47-48, 56 ff.

(61) See n. 57.

(62) The exception desired by the developing countries for primary product price stabilization schemes has not surfaced in connection with the negotiation of a code on international transfer of technology, but in other international negotiations regarding restrictive business practices. However, it cannot fail to have an effect on the reception of the technology code by the industrialized countries.

(63) U.S. extraterritorial antitrust enforcement jurisdiction, which is the most far-reaching of all national antitrust laws, nevertheless extends only to activities that have a substantial effect on U.S. foreign or domestic commerce. See United States v. Aluminum Company of America, 148 F.2d 416 (2d Cir. 1945); and Section 18 of the American Law Institute Restatement of the Foreign Relations Law of the United States (Second). Thus the Justice Department's foreign enforcement program extends only to foreign acts substantially affecting the U.S. consumer, exporter, and investor; Justice Department, Antitrust Guide for International Operations, (1977) p. 6.

(64) See n. 58.

Comments

Alvin P. Adams

In his valuable and thoughtful analysis of the international patent system, Sigmund Timberg has drawn attention to an overarching question in the technology dialogue between developed and developing countries, namely, the disparity in basic approaches to the role of government and the extent of its intervention in technology transfer matters, particularly as it affects commercialized technologies.

The dialogue on this point takes place against a background of nearly universal macroeconomic difficulties (such as inflation, unemployment, and foreign exchange deficits) as well as rapid population growth and rising demands on scarce resources. In coping with these phenomena, governments are buffeted by a variety of pressures – sometimes conflicting – for dramatic action at the national and international level to ensure the achievement of their particular social, economic, or development objectives. It is not surprising, therefore, that there may be serious consideration in some quarters of notions such as "selective delinking" of national economies from the global economic system, or the establishment of bipolar or double-tier economic orders. At the same time, policymakers are often faced with demands for economic or social protectionism or "beggar thy neighbor" policies in investment, taxation, or trade.

It is clear, however, that these pressures raise profound questions about whether the present and evolving system of generally open international flows of trade, investment, and technology will be permitted to function efficiently and to the mutual advantage of all parties. Yet it is equally clear that this system – which rests heavily on self-sustaining enterprises – likely will be expected to provide the bulk of the resources needed to meet the challenge of development in future decades.

---

*The views expressed in this paper are strictly the personal views of the author.

In regard to technology transfer, the more precise issue we face is whether, in the economic and political turmoil of today's world, ways can be found to encourage the generation, disclosure, and international dissemination and effective application of technology on a basis that is commercially satisfying to technology suppliers and recipient parties, and does not entail dislocations to the economies of their respective countries. If ways cannot be found – and it is not necessarily the job of governments exclusively to find them – we run a serious risk that technology flows, particularly those between developed and developing countries, will become deficient and constricted. This is an outcome which, in the long run, is in no one's interest.

In the case of the United States, the policy of our government, as I understand it, is to maintain a posture of nonintervention in regard to private investment or technology transfer transactions, except where important security or overriding foreign-policy interests are affected. The government generally refrains from actions specifically directed at promoting or discouraging such transactions. At the same time, it offers assistance to enterprises, or imposes sanctions on them, designed primarily to support or improve the efficiency and competitiveness of the market. Examples are the Export-Import Bank, the Overseas Private Investment Corporation, the Small Business Administration, and the Justice Department's Antitrust Division.

As for action at the international level, the U.S. government is primarily concerned with multilateral and bilateral efforts to enhance the underlying commercial or legal climate for the flows of technology and investment. Examples are negotiations within a variety of fora, such as the OECD, UNCTAD, and the Transnational Enterprise Center in New York, to identify workable and balanced sets of expectations of conduct for enterprises and governments. Ongoing negotiations for revision of the Paris Convention on the Protection of Industrial Property provide another important example of official efforts to improve the international context in which transfers occur.

Within the context of a policy of nonintervention in technology markets and of a priority concern with basic human needs, the United States strongly supports efforts to strengthen the technological capabilities of developing countries. In addition to programs by the Agency for International Development to support technological and education infrastructure, the United States is engaged in a variety of other mechanisms to assist technology development. Among these are joint bilateral research projects by technical agencies, global programs in health and nutrition research, projects sponsored by Appropriate Technology Incorporated, a variety of information services provided by the National Technological Information Services through outlets in developing countries, and a proposed Foundation for International Technological Cooperation. Supplementing these programs are major contributions to the United Nations Development Program and to Multilateral Development Lending Institutions, as well as efforts to accord a higher priority within these institutions to technology development and the utilization of appropriate technology.

Against the above review of policy and programs, it might be useful for this conference to keep in mind the following general observations as it considers the specific question of control systems for technology transfers.

1. Lest we lose sight of the forest for its trees, it is important to recall that technology control systems have little rationale, and in some instances may be counterproductive, if they are not integrally linked with overall economic development policy and priorities, particularly with a concern for raising standards of living. In addition to their current emphasis on foreign exchange savings, control systems should therefore reflect a broader concern with employment, local revenue, training, and "technology rub-off" effects of specific technology transfers.

2. Technology transfer is basically an ongoing process of communicating knowledge, often confidential and proprietary, between individuals and effectively applying it to produce a desired good or service. Technology control systems should recognize the intimate, diverse, and long-term nature of effective transfer relationships. In this connection, the advantages of appropriate equity participation by technology suppliers should not be overlooked.

3. In addressing technology transfer and its control, it is important to focus on both supply and demand factors. An arguable case can be made, for example, that technology transfer is primarily constrained by such domestic factors as general absorptive capacity, inadequate management and technical skills, and deficiencies in the ability of technology users to identify, assess, and communicate their own technology needs to those who can help meet them. Such constraints suggest that the answers to technology and economic development may lie less in the realm of technology import controls than in concerted development efforts to improve domestic capabilities of technology users.

4. Given the importance of demand and general development considerations in technology transfer controls, it is essential that rigorous analysis be conducted in the potential or actual impact of regulations on the flow and utilization of technology. Few objective assessments, to my knowledge, have been carried out on this point. For example, any justification of legislation concerning restrictive business practices in technology transfers should examine not only the extent to which various clauses are contained in agreements, but the estimated or actual trade, investment, and employment and foreign exchange effects of their inclusion – or exclusion.

5. In light of the magnitude of the interests at stake in technology transfer controls, and of their effect on the welfare of people and enterprises, it is also vital that technology-using communities be more involved in national and international dialogues on these matters. These dialogues would benefit by greater attention to the actual needs of technology users, as well as to the substantive expertise they can provide. This is particularly appropriate because the volume and quality of technology flows depend as much on purely commercial factors (such as market size, effective demand, relative experience of transfer

partner, and cost and availability of needed materials) as they do on the kind of controls addressed by this conference.

6. As we proceed to a dispassionate examination of the specifics of technology transfer controls, we should not lose sight of the progress made at the national and international levels in the past decade in the field of technology transfer and development. In this period a number of developing countries have emerged as important sources of technology in their own right. At the same time developed countries have established a variety of national mechanisms and programs to assist in accelerated technology growth in developing countries. Private enterprises, similarly, are responding at various levels to the concerns of developing countries. Examples are the growth of corporate guidelines on conduct, instances of the "internationalization" of research and development facilities, and evidence that decisions by R & D managers may be taking into account, to a greater extent, local needs and conditions in developing or modifying specific products and services for different markets.

In conclusion, I look forward to a stimulating and informal exchange on the questions addressed at this conference. The fact that we are gathered in an international meeting to discuss matters largely involving national legislation only highlights a reality – actually a dilemma – in today's world: on the one hand the growth of an increasingly integrated and global economic system that has long since transcended national boundaries, and on the other, the persistence of an international political structure that divides the world into a large number of individual states, each a sovereign in its own domain. How to square these political and economic realities, is, in fact, the overarching issue at this conference, as it is in all other forums that address global questions.

4 The Impact of New Transnational Technology Transfer Control Systems on the International Patent System: A European Perspective

Stanislav J. Soltysinski

The aim of this chapter is to analyze the implications of selected universal and regional technology transfer control systems on the international patent system.(1) It will cover both the recent proposals discussed in The World Intellectual Property Organization (WIPO) and UNCTAD fora as well as the two main European regional schemes aimed at facilitating the flow of technology across national borders. More specifically, it will discuss the impact of the recent UNCTAD-and WIPO-sponsored reforms of the existing international patent system as embodied in the Paris Convention (1883) and in regional and national laws of its member countries.

There is as yet no universal technology transfer control system. Therefore the UNCTAD projects for an International Code of Conduct on Transfer of Technology deserve special attention. But it would be a mistake to limit the discussion to the global attempts to regulate the international flow of technology, however interesting and ambitious they are. Indeed, the ideas of a universal code of conduct and of the related reform of the Paris Convention stem from the recent efforts toward creating new systems of regulatory mechanisms by the governments of many developed and developing countries in this field. While the global solutions are still in statu nascendi, and there is no consensus on their feasibility and desirability, it is worth mentioning that several groups of countries have already succeeded in creating regional mechanisms of cooperation and control for the transnational transfer of technology.(2) Thus a study of the effectiveness of such systems and especially of their impact on respective regional and national patent systems may be helpful in evaluating the possible implications of global reforms in this area on the existing international patent system.

For those reasons, this paper will focus on both the U.N.-sponsored global reforms and selected East and West European regional technology transfer control schemes. However, since the discussion on various regional and national control systems belongs to the second stage of this conference, I will limit my observations to those aspects of selected legal institutions of the European Economic Community and Council for Mutual Economic Assistance that bear on their respective systems of protection of inventions and, directly or indirectly, on the international patent systems.

The following is a breakdown of the main issues covered by this paper. First, to get some perspective on the main objectives and mutual relationship between the two main global reforms, I will review the work of governmental experts on the UNCTAD Code of Conduct and the proposed Revision of the Paris Convention. In the course of the first part, I will also try to critically examine the premises of the main proposals submitted by the three groups of countries participating in the UNCTAD and WIPO work. In particular, I will try to answer the primary question: How would the already agreed-to proposals in Geneva bear upon the existing Paris Convention and on related regional and national patent systems? Of course, this question will be dealt with by addressing a list of specific subquestions. The aim of each of them will be to establish to what extent, if at all, the implementation of a given proposal would entail a change in the international patent system.

The second part of this paper is devoted to a study of the transnational technology transfer control systems existing in the two largest European economic communities, namely those of the EEC and CMEA countries. The majority of recent Western studies and publications addressing the issue of TT control systems concentrate their attention on the new regulatory mechanisms set up in the developing countries.(3) But it is also worthwhile to examine similar trends in the two main European markets. In fact, the two European economic integrations have nowadays developed a number of refined control systems aimed at achieving an improved flow of technological information across their borders. While analyzing the main subject of this paper from a European perspective, I will endeavor to prove that the "plethora of control systems and regulatory mechanisms" in this field cannot be characterized as a regional phenomenon restricted only to developing countries. Furthermore, I will try to examine the impact of economic and social background on the choice of methods applied in the EEC and CMEA to increase the effectiveness of the technology transfer processes.

In the concluding analysis, I will try to answer the following two related questions: How do these European regional TT control schemes bear upon their respective intracommunity systems for protection of inventions and the international patent system? And how would the existing system be altered if the UNCTAD Code of Conduct becomes a universally accepted set of guidelines or a legally binding convention?

THE CODE OF CONDUCT AND THE INTERNATIONAL PATENT SYSTEM

General Character of the Code

Pursuant to the resolutions of the Seventh Special Session of the United Nations of September 1975 and of the UNCTAD IV Conference of May 1976, the Intergovernmental Group of Experts on an International Code of Conduct on Transfer of Technology was convened at its first session in Geneva, November 8-19, 1976.(4) By the end of 1977 the group had held a total of four meetings.(5) Pursuant to its original mandate, the group was called on to elaborate a code of conduct "prior to the end of 1977." The experts did not complete this task by the end of June 1978, the next deadline proposed by the UNCTAD calendar. Several deadlines later, there is still no completely agreed upon code in early 1981. There is as yet no clear consensus among the experts on the legal character and the scope of the code, or even on its specific objectives.

At the close of the fifth session, held in Geneva in February 1978, there was not even a single chapter with a completely agreed-upon text. In the crucial chapters on guarantees of the parties to transfer of technology transactions and choice of law and forum the group of experts was unable to draw up composite texts.

Despite the differences of opinion among the experts of the three regional groups – namely, the developed free market economy countries (Group B) the developing countries (Group of 77), and the developed socialist countries of Europe and Mongolia (Group D) – it is encouraging that the participants have made significant progress in agreeing on an outline of the code and arriving at consolidated texts on principles and objectives, definitions and scope of application, national regulation of transfer-of-technology transactions, restrictive practices, and special treatment for developing countries. In addition, the three regional groups have agreed to elaborate three additional chapters dealing with guarantees of the parties to technology transfer transactions, applicable law and settlement of disputes, and final provisions.

It was inevitable that so ambitious a project should encounter deadlocks and delays, and it is evident that the sponsors of the first U.N. resolution of 1975(6) did not foresee all the practical difficulties connected with its implementation.

Despite the inherent complexity of the problem, the three regional groups have agreed to contribute to the elaboration of a draft on a universally applicable code governing a broad spectrum of TT transactions. It is interesting to note that the already agreed-on texts of the preamble and the principles show that the participating countries have accepted a number of basic objectives of the code, such as: (1) expanding the international flow of proprietary and nonproprietary technology to all countries, particularly to and from the developing countries; (2) establishing universally applicable rules for the international transfer of technology; (3) taking into account both the needs of developing countries and the legitimate interests of technology suppliers and technology recipients; and (4) eliminating restrictive business practices and all other forms of discrimination.

As it is now drafted, the code can be accepted by interested governments either as a set of voluntary guidelines or as a legally binding multilateral convention.(7) Apart from these two possibilities, there are still other solutions, providing for a binding character of only certain chapters of a final draft. Another solution might be agreement to periodically review the implementation of the code by interested governments and their enterprises (the parties within the meaning of the code), or agreement that the interested governments will postpone the issue of a binding code for a definite period of time (say five years) during which the rules will function as a set of voluntary guidelines. Such an agreement might provide that the initial period will serve as a basis for gathering much-needed experience. A subsequent diplomatic conference preceded by preparatory work might either transform the code into a binding convention or limit its tasks to the refinement of some of its rules while retaining the essentially voluntary nature of the document.

Because of the variety of "persuasion" mechanism(8) that can be embodied in a nonbinding agreement, the ultimate success of the UNCTAD code does not depend solely on its mandatory nature but rather on the merits of its final drafts and the willingness of a broad spectrum of countries to follow the guidelines it contains. This in turn will depend on reaching a genuine compromise, taking into account the legitimate interests of all groups of countries concerned and especially of the developing countries.

In light of these facts, it is worthwhile to examine possible implications of the code solutions for the existing international patent system that are already agreed on or still under consideration. The subsequent comments contain a review of those proposals presented in the UNCTAD forum that, if eventually adopted as either voluntary or binding rules, could have practical implications for the patent system. But in order to assess the significance of these postulates, we must first take cognizance of the main proposals for the revision of the Paris Convention that are being discussed both in UNCTAD and WIPO.

The Reform of the International Patent System: Main Lines for Revision of the Paris Convention

The focus of this paper is only on the revision of that part of the Paris Convention that deals with protecting inventions, that is, certificates of authorships, patents of importations, and so on; no attempt is made to cover matters such as industrial design, trademarks, and general administrative aspects. To begin with, I will examine the economic background for the revision of the convention.

For almost a century the advocates of the existing international patent system have been able to defend the view that the basic provisions and philosophy of the Paris Convention of 1883 are just and remain universally practicable.(9) While accepting the idea of "certain adjustments," the defenders of the present system are against the revision of the "basic principles" of industrial property, which include the granting of a monopoly right to a patent holder, the guarantees against its forfeiture or compulsory licensing (Article 5); the ban on importation of products manufactured by a patented process without the patentee's permission (Article 5), the rule of "national treatment" (Article 2); and the right of priority (Article 4). Until very recently, the successive revisions of the convention have mainly served to strengthen the position of the patent holder. This was consistent with the view that "the protection of inventions by the grant of the patent is apt to promote technical and economic progress, especially of developing countries."(10) But in light of many recent economic and legal studies, this optimistic appraisal no longer stands unchallenged. Eminent economists question the validity of the patent system even within the framework of a free-market economy. The antitrust "camps" in the United States and in the EEC have launched a remarkable offensive against their respective patent systems limiting certain traditional prerogatives of the patent holder. But the most critical evaluation of the system comes from developing countries. Armed with the arguments gathered in the latest UNCTAD and WIPO studies on this subject, the developing countries demand a meaningful revision of the existing international patent system. Their main economic argument is based on statistical data showing that nationals of the developing countries own only about 1 percent of all patents, whereas foreigners hold six times more grants in the developing countries than the nationals of these countries.(11) Furthermore, the same source indicates that 90 to 95 percent of the patents owned by the nationals of the developed countries are never worked there. Thus, according to the proponents of the reform, it is clear that in the developing countries the patent system does not serve its basic function of encouraging technology to be effectively disclosed and worked there. Because the Paris Convention forbids each member country to discriminate in favor of its nationals, the developing countries must give their stronger partners "unlimited freedom to utilize [their] power at the expense of the others."(12) Consequently the developing countries are demanding a sweeping reform of the convention. Their main proposals concerning patents may be briefly summarized as follows:

1. Each member state should be free to adopt any legislative measure deemed necessary to prevent the nonworking of patents, including forfeiture and granting of exclusive licenses. Recourse to any of these measures should not be constrained by the convention.(13)

2. The monopoly aspect of the grant should be limited by eliminating parts of Articles 5 of the convention. These permit the owner of the patent for a manufacturing process to prohibit the importation of products made abroad with the same process.(14) The developing countries argue that control over process is not enough to justify the import monopoly in the national market of the host country.

3. The revised convention should explicitly recognize the principle of special treatment for developing countries. Preferential treatment should be granted to nationals of the developing countries with regard to the choice of form of protection,(15) the length of the convention priority term, patent fees, and so forth.

The arguments of the opposition to these sweeping reforms may be summarized as follows:

1. The reformers have not proven their cause. The fact that 90 to 95 percent of the inventions patented in the developing countries belong to foreigners does not necessarily mean that all of them are exercised as import monopolies. Available data show that only 25 to 30 percent of foreign-held patents are actually used for this purpose. The remainder are not exploited at all – a universal phenomenon showing the risk accompanying patent investments in all countries.(16)

2. While certain modifications of the traditional concept of the patent monopoly are possible, such adjustments should not interfere with the basic foundations of the international patent system. For instance, the proposed extension of the twelve-month term of priority would lead to perturbations and uncertainties; the reduction of fees should take account of the applicant's resources and not his nationality; since it is impossible to work the patent in all countries of the union, the convention should permit the owner to justify his inaction for economic reasons rather than confer upon the member countries the right to grant exclusive licenses or forfeit the patent.(17)

3. The weakening of the patent system would reduce the present inducement to invest abroad, especially in the developing countries.

The Code and the International Patent System

Although the links between the code and the preparatory work to revise the Paris Convention for the Protection of Industrial Property are often emphasized, code provisions already agreed upon contain relatively few direct references to the main issues being discussed in WIPO today. This is partly because the subject-matter of the Paris Convention and the future code of conduct are different. Whereas the code deals with the transfer of technology primarily on a contractual basis, the convention regulates the processes of acquisition of patents and other industrial property rights by the citizens of the contracting countries and their protection in the member host countries. It is worth noting that the convention contains no single reference to licensing arrangements stemming from voluntary agreements. On the other hand, the often conflicting views on the role of the international patent system between the two largest regional groups have neutralized each other. There are examples showing that while the developed free-market-economy countries have been trying to submit proposals aimed at strengthening the patent system, the developing countries have been presenting ideas consistent with the position they have taken in the WIPO forum. As a result, their opposing views often remain "in brackets" or are finally withdrawn because they lack support from the other regional groups. One example of such a tendency is the Group B proposal concerning the necessity of "respect by parties and governments for industrial property protection."(18)

This clearly propatent declaration has not been included in the tentative composite text of the code. Likewise, the proposal contained in Provision 6.2 (VI) of the composite text requiring governments of developed countries to "establish preferential measures so that [a patent] holder in technology supplying countries should not be used . . . to restrict imports of products from developing countries" remains in brackets, indicating that it is solely a Group of 77 formulation.(19) Finally, it should be mentioned that as yet the agreed-on composite text of the same chapter does not include the Group D proposals aimed at preventing enterprises of developed countries from using patents as a means of blocking the development of industrial and research activities in developing countries.(20)

In short, it is doubtful that the code will contain a "principle" or "objective" that will either stress the virtues or emphasize the sins of the existing international patent system on a more general level. It is therefore to be expected that neither the pro- nor antipatent countries will be able to win the majority needed for the introduction of their respective declarations of principles.

This is not to say, however, that the UNCTAD code, if finally drafted and adopted, will remain neutral vis-a-vis the patent system. Such a result would be a de facto victory for the "traditionalists" and "conservatives," who either oppose all change of the Paris Convention or regard proposals for reform with distrust, favoring the preservation of the existing mechanisms of transfer of technology. It seems obvious, but it must be emphasized, that the proper evaluation of possible implications of the code for the international patent system can be ascertained only after a careful review of its all provisions. Unfortunately at this stage, when only approximately 50 percent of the code is elaborated in the form of composite texts containing first compromises or simply juxtaposed formulations of the three groups, it is almost impossible to ascertain what overall effect this document might have. In light of this fact, I will limit my observations to those parts of the code that were elaborated in the form of tentative composite texts. These include the preamble, the definitions and scope of applications, the principles and objectives, and the section on restrictive practices.(21)

As I have emphasized, there is no direct, explicit reference to the main issues discussed in WIPO in the tentatively agreed-on texts of the UNCTAD code. But the tentative composite draft of the preamble contains formulations that, despite the generality of their language, can ber interpreted as principles supporting the reformers' views at the WIPO forum. For instance, the document draws attention "to the need of having an unrestricted (improved) flow of technological information and in particular of information on the availability of alternative technologies . . . suited to the specific needs of the developing countries."(22) The reference to an "unrestricted" flow of technological information is supported by the Group of 77.(23) The two other regional groups have hitherto been reluctant to accept this formulation, possibly because of the apparent breadth of the term "unrestricted," which could support abolition of all restrictions in this field, including limitations stemming from industrial property rights.

A similar formulation is found in Item 7 of the preamble, which expressly recognizes "the need for the developed countries to grant special treatment to the developing countries in their technology transfer transactions."(24) Again, this unchallenged and just postulate is consistent with the main line of revision of the Paris Convention.

On the other hand, the preamble contains proposals that offer substantial help to the cause of the owners of patents. For instance, the document emphasizes that technology transfer transactions should be made on "mutually advantageous terms" and that the code should respect "the legitimate interests of all parties to the transfer."

The "battle of principles" continues throughout the next chapter, "Principles and Objectives," which further develops the ideas found in the preamble. Again, the principles of special treatment for developing countries and of mutual benefit for supplying and acquiring parties to TT transactions are present. The only new and interesting idea is found in Item V, which provides that one of the code's objectives should be "to facilitate the formulation, adoption and implementation of national policies, laws and regulations on the subject of transfer of technology by setting forth international norms."

This important provision, accepted by the experts of the three regional groups, clearly shows that we are slowly approaching a phase of negotiation between the suppliers and recipients of technology in which the tendency for multilaterally agreed-to international standards aimed at protecting the legitimate interests of all parties prevails over laissez-faire-ism and uncoordinated unilateral regulation.(25)

It is almost universally agreed that the heart of the UNCTAD code is its chapters on restrictive practices and contractual guarantees. Rules governing both restrictive business practices and guarantees are aimed not only at limiting the concept of freedom but also at imposing certain constraints on the patent holder-transferrer, so it is therefore worthwhile to examine their potential impact on the international patent system. There can be no doubt that the current tendency to condemn certain standard license practices "impinges most directly on the transfer process"(26) and the permissible scope of exercise of patent rights.

By the end of its fourth session the UNCTAD-sponsored group of experts had managed to elaborate eighteen composite draft texts dealing with various restrictive business practices, but only twelve of them were accepted, at least in principle, by all three regional groups. These include rules on restrictions after expiration of arrangement, grant-back provisions, challenges to validity of industrial property rights, restrictions on research and use of personnel, price fixing, exclusive sales or representation agreements, restrictions on adaptations, tying arrangements, export restrictions, cartel patent pool or cross-licensing agreements, and restrictions on publicity. The six other composite texts agreed to in principle by the Group of 77 and Group D cover limitations on volume, scope, and so on, use of quality controls, obligation to use trademarks, requirements to provide equity or participate in management, unlimited or unduly long duration of technology transfer arrangements, and limitations on the use of technology already imported.(27)

The Working Group II of the Intergovernmental Group of Experts also discussed two issues of a more general nature, namely the concept of exceptions granted by competent national authorities and the scope of application of the doctrine of intracorporate conspiracy in the context of technology transfer transactions. The following composite text on the issue of exceptions is a good illustration of the inherent complexity of the problem:

> [Notwithstanding]$^{D+77}$ [the provisions]^{D} [Section]^{77} [of this chapter on transfer of technology transactions or practices and arrangements contained therein shall be deemed]$^{77+D}$[non-objectionable]^{D} [VALID]^{77} [if]$^{77+D}$[Based upon exceptional circumstances,]^{77} [the competent national authorities of the]$^{77+D}$[acquiring party's] [technology acquiring]^{77} [country decide that it is in its public interest]$^{D+77}$ and that on balance the effect on its national economy will be not adverse]^{77D}[and it has no substantial adverse effects in other countries].

Significantly, both the Group of 77 and Group D seem to be more procompetitive and antitrust than Group B. It is worth noting, for instance, that the experts from the market-economy countries were unable to agree to ban unreasonable "field-of-use restrictions," the practice outlawed in the United States since 1938!(28) Another illustration of this tendency is Group B's inflexible position on the necessity of resorting to a broadly stated rule of reason limiting each rule of this chapter.(29) There is no doubt that, for a variety of reasons, the scope of application of the rule of reason in a universal code of conduct should be much wider than in domestic legislation or court practices. It is understandable that because of the lack of a common legal tradition and the variety of commercial situations, a universal code should offer flexible and concrete rules rather than rigid or indefinite commands. However, the experience of the United States and the EEC with their antitrust laws shows that there are a number of restrictive practices involving technology transfer transactions that are practically speaking without "redeeming social value."(30) If, however, there are justified reasons to refine some per se rules developed in the "cradle" of antitrust for the purpose of the UNCTAD code – for instance, in order to draw a clear line between an attempt to exact royalties after expiration of a patent and a delayed payment for the use of valid industrial property rights – then it would be better to draft specific exceptions, rather than qualify such rules by an indefinite "rule of reason."

It is understandable that the Group of 77 and the socialist countries of Group D expect the market-economy countries to agree to introduce into the UNCTAD code the same standards of competitive behavior that were developed and successfully tested in their own legal systems. Otherwise there will be an increasing likelihood of polarization and mistrust between the supplying and receiving countries. Indeed, it is difficult to explain why many well-established antitrust rules that are often applied to transnational arrangements when they produce negative effects within market-economy countries cannot be accepted as a basis for universal rules of conduct if they produce similar consequences in developing countries or elsewhere. The proper course seems to be to punish such practices on a global scale and not to let the wrongdoers escape unchastened in the latter case.

Let us now turn to the composite draft texts of the proposed code of conduct that imply the existence of a potential conflict with the well-established principles of the international patent system. Are there such rules that are likely to limit either the scope or exercise of patent monopolies in the future?

After examining the whole chapter, I found no rule that, in my view, could be interpreted as a direct attack on the institutions of patents and confidential know-how (trade secrets). However, there are several rules, especially those supported by the Group of 77 and to a lesser extent also by Group D, that are aimed at limiting the patentee's freedom to exercise his monopoly to the detriment of a recipient party. Here is the most interesting example of this:

(1) Export restrictions

> [Unreasonable]^{B} Restrictions which prevent or [substantially]$^{B+D}$ hinder export by means of territorial or quantitative limitations or prior aproval for export or export prices of products or increased rates of payments for exportable products resulting from the technology supplied [,unless justified]$^{B+D}$ [,for instance,]^{B} [to prevent export of such products to countries where they are protected by the supplying party's industrial property rights^{B+D}][or where relevant know-how has retained its confidential character]^{B} [, or where the supplying party has granted]$^{B+D}$ [an exclusive right]^{D} [a license]^{B} [to use the relevant technology]$^{B+D}$.(31)

Export restrictions are among the most effective deterrents of international trade. Indeed, they are inconsistent with the basic principle of free trade. They are very often used by transnational corporations and other transferrers to eliminate or limit the competition of their transferees. However, an outright condemnation of such restrictions would be difficult both for suppliers and recipients of technology. Many recipients in developing countries insist on obtaining exclusive licenses in their domestic and adjacent markets in order to limit the risk involved in entering technology transfer transactions. It is worth noting in this context that the developing countries favor a drastic revision of Article 5 A (4) of the Paris Convention, demanding that the proper national authority should be permitted to grant exclusive compulsory licenses as a remedy for nonworking of a patented invention. Although such change may be justified in order to overcome obstacles to commencing the exploitation of a nonworked invention in a developing country, it would be inconsistent to outlaw all export restrictions imposed on third parties by a transferrer to protect the territory of an exclusive licensee in a country that either tolerates or encourages such arrangements for the same reasons that constitute the background of the proposed revision of Article 5 A (4) of the Paris Convention. There is no doubt that there are cases where a potential licensee may require the same degree of territorial protection as would a would-be applicant for a compulsory license. In such cases, for instance, reasonable export restrictions imposed on the transferrer and his subsequent licensees have valid justification. Therefore, one of the specific exceptions to the ban on export restrictions suggested by the Group B text just cited relates to this very situation.

Another perplexing question is whether the transferrer should be permitted to forbid his transferee to export licensed products to countries where they are protected by the supplying party's patents. It is clear from the text presented above that only the answers of Groups B and D are in the affirmative. The different position of the Group of 77 is consistent with the developing countries' views on the role of the import monopoly aspect of the Paris Convention. It should be noted that a contractual obligation preventing a licensee from exporting to a country where his licensor owns a valid process patent strengthens his control over the market in the patent-granting country. This is due to the fact that remedies for the breach of technology transfer contracts that provide for resort to arbitration are usually more effective than actions for patent infringement brought before competent national authorities. The latter processes are both time-consuming and risky, because of the danger of a counterclaim alleging that the patent in question is simply invalid. Thus it is possible to argue that even in light of Article 5 of the Paris Convention, export restrictions are clearly outside the scope of the patent monopoly, and that they strengthen the import monopolies in developing countries.

But there are also valid reasons in favor of export restrictions to countries where the transferrer owns valid patents. To require the supplying party to license the protected technology without such restrictions would probably result in much higher royalties for the first recipient in order for the licensor to recover expenditures for developing the technology and for the worldwide patent protection of his invention. Thus although export restrictions should be characterized as an almost per se illegal practices, a proper prohibition should be subject to a few carefully drawn exceptions.

In view of the link between the Group of 77 proposals to outlaw Article 5 of the Paris Convention and to eliminate all export restrictions, the question arises whether the proponents of these measures consider them to be two alternate means of sufficiently protecting the developing countries' interests. In my view, an inflexible prohibition of all export restrictions coupled with the successive abolishment of Article 5 of the Paris Convention would seriously diminish the practical value of patent rights in those countries where there exists so-called weak patent protection for chemical, pharmaceutical, and other manufacturing processes. This in turn may reduce the existing incentives to license or may force suppliers to increase their prices. Besides, foreign-based technology owners would stop filing patent applications in such developing countries, leading not only to a reduction of patent fees and annuities (a fact of minor significance), but also perhaps to the frustration of the current proposals to make effective use of existing and new remedies aimed at improving the working or protected inventions. The objectives of the proposed revision of Articles 5 A(2) to 5 A(4) of the convention would be seriously frustrated if the overall effect of the current reforms under discussion in WIPO and UNCTAD would substantially limit the number of patent applications in developing countries.

The foregoing suggests that the prospects for abolishing the patentee's right to stop the importation of products covered by his patented process would be much better if the developing countries would agree to mitigate the owner's loss by "legitimizing" the proposed list of exceptions to a general rule prohibiting export restrictions in the UNCTAD Code of Conduct. Such a rule could be further refined, providing, for instance, that every export restriction be limited in scope and duration.(32)

The draft composite texts of the same chapter contain two related rules aimed at limitation or elimination of restrictions on the use of the technology transferred:

> (a) Restrictions after expiration of arrangement.
>
> Restriction on the use of the technology after the expiration or termination of the arrangement, [unless the technology is still legally protected, or has not entered the public domain] B+D [or after the know-how has lost its secret character] 77 independently of the acquiring party. . .
>
> (t) [Limitations upon use of technology already imported] [Limitations upon the diffusion and/or further use of technology already imported] 77

The first of the two proposed rules, (a), prohibits restrictions on the use of the technology transferred, whereas the second, (t), suggests that the technology, once imported, may be freely "diffused" and used. The latter proposal was supported only by the Group of 77. It is clear that the two rules partly coincide. Besides, the purpose and scope of item (t) is unclear. Does this rule prohibit field of use restrictions and practices restricting the recipient's volume of production? If so, then it involves duplication of principles already embodied in item (o). Does it suggest that the first recipient should be permitted to sublicense every potential customer within or outside his territory without the transferrer's permission? This would not be an advisable solution.

A considerable compromise was already achieved in item (a). All three groups agreed that, in principle, the recipient of technology should be free to use the acquired technology after the expiration or termination of the arrangement. However, the experts of the Group of 77 suggest that this rule cannot be qualified by the two exceptions introduced by their colleagues from the other regional groups. The formulation proposed by the developing countries would permit the recipient to use the technology transferred after the expiration or termination of the arrangement (say after five years) even if the supplier would still possess valid patents in the licensed territory. This principle, if adopted, would drastically curtail the content of the patent right. Each act of disposition over a patent monopoly would amount to a de facto exhaustion of the grant. If suppliers could be coerced to live with such an inflexible rule, coupled with those contained in items (t) and (l) (export restrictions), they would hardly want to license small or medium-sized firms and would try to reap maximum of profit by entering into long-term contracts only with licensees who would be willing to pay high royalties in exchange for the prospects of sharing the world market in the licensed product.

On the other hand, it is understandable that the recipient wants to be assured that he will not be required to stop the use of the licensed technology after the expiration of the agreement if it is still commercially useful. True, this aspect should be resolved by proper planning of the terms of the contract.(33) However, the code of conduct and national legislation should provide some guarantees of the acquiring party's ability to continue the use of licensed technology when its commercial life exceeds the term of the contract. A limitation of the transferrer's right to rescind or terminate a technology transfer arrangement is a good example of the kind of assurances that might be helpful to meet the justifiable expectations of the recipients of technology, especially those domiciled in developing countries.(34)

REGULATORY MECHANISMS IN THE EEC AND CMEA AND THEIR IMPACT ON THE EXISTING PATENT SYSTEM

The Council for Mutual Economic Assistance

The traditional system of inventor's certificates adopted by the majority of the socialist countries after the Second World War is based upon two basic principles. First, the employee-inventor is granted a certificate of authorship and the state acquires the exclusive and sometimes perpetual monopoly on his idea.(35) Accordingly, the state itself or a proper unit of the socialized economy is responsible for adequate utilization of the invention, thereby assuring maximum remuneration for the inventor, whose royalties depend on the total value of the economic and other benefits resulting from the utilization of his innovation. Second, the invention protected by a certificate of authorship may be utilized free of charge and without any authorization by all socialized economic units, that is, state-owned enterprises and cooperatives. The rationale behind this rule is that the principle of free use assures the fullest utilization and diffusion of the new technology and the highest remuneration to the innovator.

These models of the protection of inventions and dissemination of technology correspond to the requirements of a highly centralized system of economy where all decisions relating to R & D projects and utilization of state-owned inventions are decided by the proper governmental agencies.

Significantly, during the second session of the CMEA countries in 1949, the member states adopted a resolution providing for free transfer of technology in intracommunity relations.(36) It is obvious that in those circumstances patents played a very insignificant role, even in those countries that retained this type of protection, namely, Poland and the German Democratic Republic.

With the advent of the economic reforms of the mid-1960s, the system of free dissemination of technology became the target of severe criticism. Critics stressed its inconsistency with the principle of economic accounting. They also demonstrated that the principle of free use has a negative impact on the economic results of R & D institutions that are required to pursue the policy of economic accounting. Furthermore, the traditional system provided insufficient technical and economic guarantees to the potential user of the freely available technology; the original investor could not expect reasonable return and profit from its capital outlays and technical assistance.

As a result the system of free use was abolished in Yugoslavia, Hungary, and Poland.(37) At the same time the role of patents was strengthened both in purely domestic and in transnational economic relations.(38) The principle of free use was considerably revised in two other countries, Bulgaria and the German Democratic Republic. In Bulgaria, which retains the institution of a certificate of authorship, a subsequent user of a protected invention may be required to pay the original investor a partial or full equivalent of the cost of development of the innovation. In the GDR an enterprise that develops and uses a new invention may demand full compensation for the technical assistance and know-how it offers to a subsequent user.

In the same period, the principle of free use was limited in intra-CMEA economic relations. Since 1967 the member countries of the socialist community are free to choose between the traditional system of exchange of technical information and universally accepted licensing practices. The resolutions adopted by the CMEA in 1967 and 1973 provided that joint research projects and specialization agreements carried out by scientific institutions and enterprises located in different member countries should be formed on the basis of commercial agreements regulating the division of profits and outlays, protection of industrial property rights, licensing policies, and so on.

It is worth noting that, unlike in the EEC, there is no supranational body within the CMEA performing the task of a regulatory mechanism in the field of technology transfer. The resolutions concerning joint planning and cooperation are made by plenipotentiaries, usually authorized cabinet members of the countries involved. There is also no supranational judicial authority similar to the Luxembourg Court of Justice. The legal nature of resolutions adopted by various systems of the CMEA members are regulated by the domestic laws of each member country. In Poland, for instance, to become legally binding on juridical and natural persons such resolutions must usually be ratified and published. Thus, the introduction into practice of the "Complex Program of Further Extension and Improvement of Cooperation and Development of Socialist Economic Integration Among the Countries of CMEA" in the field of inventive activity and patents required an extensive legal collaboration among the member states, their enterprises and R & D institutions.(39) The program sets forth economic objectives and specific measures to be taken to ensure its realization for a period of fifteen to twenty years. To avoid duplication of effort and improve the exchange of new technological information, the program makes recommendations for joint R & D projects and specialization contracts that should be carried out in accordance with agreements made by competent governmental bodies, enterprises, scientific institutions, and so on. Such forms of specialization and cooperation raise questions of ownership of jointly developed innovations and their utilization within the community. The recommendations of the program in this field are very flexible. It should be emphasized that the parties to such agreements should enjoy the right to use the jointly developed innovations. The agreement should also regulate the principles of licensing policy, division of outlays and profits, and patent protection issues. The program of 1971 reiterates the already accepted principle that the exchange of technological information within the community – for instance, between the participants of a given specialization agreement and enterprises located in member countries that have not participated in the joint venture – may be carried out either free of charge or on the basis of reasonable compensation. Present practice shows that the latter method of transfer prevails.

The recommendations of the program have been further elaborated in the 1973 "Agreement on the Protection of Inventions, Industrial Designs and Trademarks Developed in Connection with Economic, Scientific and Technical Cooperation Among the CMEA Member States."(40) Again, it is interesting to note that the overwhelming majority of the agreement's rules can be classified as "yielding norms"; that is to say, the parties to a given specialization agreement or the owners of a joint venture are free to modify the model solutions envisaged in this convention. For instance, one of the articles of the agreement provides that the profits resulting from licensing a jointly owned patent shall be divided between the parties in proportion to their expenditures in R & D. However, the parties are free to decide otherwise. Another interesting point is that up until now there has been no law aimed at the elimination of restrictive business practices in this field. The absence of antitrust legislation within the socialist community permits the parties to technology transfer transactions to shape their mutual obligations freely. Because of the differences in the economic environment between the market-economy countries and the socialist countries, the legislators of the latter group are not coerced to pursue vigorous antitrust policy. Indeed, because of the economic control schemes connected with the mechanisms of central planning and the monopoly of foreign trade, many business operations that are considered per se or prima facie illegal under the antitrust laws of the United States and the EEC are valid in intra-CMEA relations.(41) However, both the program of 1971 and the Moscow Agreement of 1973 provide that the parties to cooperation agreements and technology transfer transactions shall take into account the common interest of all member countries of the socialist community.

Mention must be made here that all foreign trade transactions involving the transfer of technology are supervised by proper state agencies, such as branch and foreign trade ministries, which provide efficient although not always perfect safeguards against the abuse of economic power. Again, it is worth emphasizing that although there is no supranational supervisory institution similar to the European Commission, and although foreign trade enterprises of each CMEA country have considerable discretion in negotiating contractual clauses developed by international licensing practice, all technology transfer transactions are subject to administrative approval.

It is interesting to note that the new types of industrial and scientific cooperation between the CMEA countries and the contractual forms of dissemination of technological advances described above have strengthened the economic role of patents and certificates of authorship in both intra-CMEA and external relations. With the advent of the era of licensing practice within the CMEA, patents and other forms of protection are considered valuable corporate assets. Moreover, participants in cooperative agreements are interested in securing effective industrial property protection for the fruits of their R & D investments, both within the CMEA and abroad.

Therefore the CMEA program of 1971 stresses the need to improve cooperation among the member countries in the field of industrial property. At present, one of the most urgent problems in this area is that of bringing about a gradual unification of the legal regulations in the field of inventive activities. The goal is twofold: first, to assure effective and simple ways to protect valuable inventions in all member states, preferably with a single filing, and second, to find a legal solution assuring the rapid utilization of these inventions throughout the socialist community.

These objectives have been already partially achieved due to the adoption of the 1973 Moscow Agreement and the subsequent "Convention on the Mutual Recognition of Inventor's Certificates and Other

Forms of Protection for Inventions" of 1976.(42) In principle the 1976 agreement permits the applicant who has secured protection in one convention country to apply either for a certificate of authorship or for a patent. Although the process of recognition is simpler than a regular filing under the priority rules of the Paris Convention, it provides for the possibility of a full examination of the invention submitted because the respective definitions of a patentable subject-matter found in the domestic invention laws of the CMEA countries reveal certain differences.

The growing role of patents, certificates of authorship, and other forms of protection as well as the revival of licensing practices has not resulted in a "rebirth" of monopolistic practices in technology transfer transactions. The guidelines of the CMEA program of 1971 make it clear that the results of R & D cooperation agreements should be available on reasonable terms, including royalty-free transfer, to all interested parties in the socialist community. Similarly, the Polish Law of Inventive Activity of 1972 not only abolished the system of free use of inventions, but also introduced the unique and comprehensive system of licensing contracts between economic units.(43) The law declares that, as a rule, a socialist enterprise to which a patent has been granted cannot refuse to license another socialist economic unit. Thus in the socialist sector of economy the patent is no longer a classic monopoly, but a protective title ensuring the original investor the right to obtain a reasonable remuneration from subsequent users (licensees).

The European Economic Community

The basic purposes and structure of the EEC institutions are very well known. Therefore I will limit my comments to the following issues: Which legal institutions existing in EEC law can be characterized as technology transfer control schemes? To what extent, if at all, do current regulatory measures bear on EEC regional and national patent laws?

At the community level, there seem to be no other TT control mechanisms except the pertinent rules on competition that are supervised, enforced, and interpreted by the proper EEC organs, especially the Commission and the European Court.(44) There seems to be no doubt that the codified substantive antitrust rules and precedents established by the EEC authorities plus the institution of individual and group exemptions for certain restrictive practices form a powerful controlling mechanism in the field of technology transfer. The system of notifications of restrictive TT transactions, combined with the Commission's powers to issue requests for information, conduct inquiries in any sector of the economy of member states, and, finally, to impose sanctions on undertakings infringing rules of competition, has no equivalent in other regional economic integrations.

The Treaty of Rome makes an attempt to describe the relationship between industrial property rights and the community's competition rules. In accordance with Article 36, the condemnation of quantitive restrictions on importation and other equivalent measures between member states "shall not be an obstacle to prohibitions or restrictions in respect of importation, exportation or transit which are justified on grounds of . . . industrial . . . property. Such prohibitions or restrictions shall not, however, constitute either a means of arbitrary discrimination or disguised restriction on trade between member states."

The experience of the application of the community's antitrust laws shows that the treaty offers no clear delineation between the two branches of law. While propatent forces stress the assurances offered to the industrial property owners in the first sentence of Article 36 and in Article 222, the antitrust camp, which seems to occupy a dominant position within the Commission, points to the limitations provided for in Article 36 in fine.

As far as patents are concerned, the European Court has accepted the Commission's view that so-called parallel patents obtained in two or more member countries cannot be used as instruments of the division of national markets. Thus a person who enjoys a patent right under the legislation of a member state may not prohibit the importation of a produce protected by the patent if the goods in question have been marketed by the patentee or with his consent in another EEC country.(45) Thus to the extent that parallel patents are at stake, the patentee and his assignees, licensees, and so on have been deprived of the right to exercise their ius prohibendi (granted under domestic laws of each member state) in relation to patented products once they have been put into circulation by a "related" person.

But not all consequences of the doctrine of "exhaustion"(46) of patent rights are clear. Parallel developments in the field of trademarks suggest that the key concepts of "parallelism" and of a "related person" can be interpreted only on a case-by-case basis. For instance, are patents covering the same process owned by a French company in France and Poland "parallel" within the meaning of the Centrafarm v. Sterling Drug case? Does the act of marketing patented products undertaken by a licensee in the United Kingdom in violation of the contract made with its Dutch licenser, which owns parallel patents for the same invention in all EEC countries, deprive the patentee of his right to stop the products' importation to France or Italy? What about the products marketed by someone who acquired the right to make and sell by way of a compulsory license or confiscation of one of a cluster of parallel patents?

In the past decade, the Commission announced quite a few decisions concerning the legality of many clauses hitherto treated as unobjectionable in licensing practice. In general, the Commission's view corresponds to the established principles of U.S. antitrust laws. Thus, for instance, the following provisions found in notified transactions were held to be generally not exempt: postexpiration obligations,(47) no-challenge and no-competition clauses,(48) and agreements providing that the authorization of the licensor to grant subsequent licenses to third parties depends upon the consent of earlier licensees.(49) In some areas the Commission displayed an even more vigorous antitrust approach than do U.S. enforcement authorities. For instance, in its landmark decision in Re AOIP/Beyrard, it questioned the validity of exclusive licenses.(50)

It is clear from the above that the process of enforcement of the EEC rules has entailed considerable curtailment of the patent rights granted under the national laws of the member states. The blessing offered by the European Court to the exhaustion doctrine resulted in the suspension, if not curtailment, of the right to stop importation in case of parallel patents. Furthermore, the hitherto judicially untested rulings of the EEC Commission have substantially reduced the patentee's freedom to exercise his rights whenever it conflicts with free circulation of goods within the European Community.

These developments have met with the growing opposition of industrial circles, especially British industry, who complain about the uncertainty and red tape resulting from the Commission's regulatory policies.(51)

It should be stressed that the concept of exhaustion of patent rights has been formally adopted within the framework of Articles 32 and 81 of the EEC patent convention (the so-called Luxembourg Convention), and it will apply both to community-wide patents and to grants conferred by national authorities in the contracting states.(52) The definition of the exhaustion of patent rights shows that the propatent camp has managed to use formulations aimed at limiting the apparent breadth of the principle developed by the EEC authorities. For instance, Article 32 indicates that the proprietor of an EEC patent is barred from exercising his right to stop circulation of patented products only if they have been put on the market "with his express consent." Likewise, the definition of a "related person" is narrower than that suggested by the past precedents of the European Court in industrial property cases. Finally, the EEC patent convention expressly provides for the possibility of granting exclusive licenses.

At the present time, however, it is impossible to foresee all future ramifications of the exhaustion principle. Anyway, one has to remember that the Luxembourg Convention will be enforced and interpreted by the same EEC institutions that hitherto have been giving priority to the principle of free flow of goods within the community. Furthermore, unlike in other legal systems, the inherent weakness of the European Parliament limits the chances of a legislative branch of the community balancing the judicial activism of the European Court.

Another example of the regulatory policies of the Commission in this area is the idea of a group exemption on licensing agreements. The proposed regulation will cover only patent and certain know-how transactions. The published draft regulations distinguish between clauses that will be unconditionally exempted (the so-called white area), clauses that may never be exempted (the so-called black area), and clauses that may be exempted in some circumstances (the gray area).

This classification is based on the past precedents laid down by the EEC Commission and the European Court. Since the court has never tested the majority of the Commission's rulings condemning a number of notorious licensing practices, EEC industry raises serious objections to the draft.(53) This is especially true with regard to the treatment of postexpiration nonuse requirements and exclusive and territorial licenses.

The regulation, if adopted, would introduce two interesting innovations. One, which is of a substantive nature, would be the adoption of the notion of per se illegality. The clauses belonging to the "black" category (no-challenge clauses, no-competition clauses, requirements of payments beyond the expiration of the patent, and so on) could not be exempted under any circumstances and therefore would be considered as violating Article 85 (1). The second of the two basic modifications would introduce a change in the procedure for reviewing "gray area" clauses. Under the draft, agreements belonging to the latter category are exempted if the Commission does not raise an objection within six months from the modification.

CONCLUSION

All the technology control systems examined above, whether regional or global, whether of planned or market economies, directly or indirectly bear on the existing national and international patent laws.

While it is almost universally agreed that patents stimulate competitive P & D activities, their monopoly aspect has become the subject of severe criticism. Perhaps the most controversial feature of the patent is the power it confers to prevent importation of the products covered by the grant. It is obvious then that this element of the patent privilege has been attacked not only by the developed countries but also by EEC authorities. The purposes of the "exhaustion of patents" theory and of the proposal to abolish Article 5 of the Paris Convention have much in common: both are aimed at depriving the patentee of the right to isolate a national market and charge "exorbitant" prices.

There is also a universal tendency to impose certain restraints on the patentee's competence to freely exercise his monopoly. Interestingly, the limitations imposed by antitrust laws in market-economy countries and by the laws of planned economy states point in one direction: patents should gradually be transformed into legal titles ensuring the owner a right to obtain reasonable compensation for his investment. This process has been almost accomplished in CMEA countries, but it is also underway in the EEC countries and elsewhere. The curtailment of monopolistic prerogatives takes such forms as the elimination of the right to stop importation, compulsory licensing, limitation of exclusive licenses, and control over the patentee's pricing practices.

At this point it would be rather hazardous to predict whether the "proliferation" of the process of the transformation of patent monopoly into a less powerful legal title will be accomplished during the planned WIPO revision conference. Likewise, it is difficult to foresee whether the realization of the main proposals of the developing countries regarding the UNCTAD code of conduct and the revision of the Paris Convention will substantially improve their technological capabilities. Despite the strong opposition of business circles in market-economy countries and their pessimistic predictions, it is rather doubtful that adoption of a universal code of conduct and of reasonable limitations of the patentee's prerogatives in the Paris Convention would lead to less licensing. Indeed, as one commentator observed, many of the objectives sought by the Group of 77 are the same as, or similar to, those already implemented by the antitrust and patent laws of the developing countries.(54) Since technology suppliers have managed to live with the antitrust laws of the United States and the EEC, they should also be able to survive once analogous principles are codified in a universal code of conduct. Likewise, the prospective revision of the Paris Convention and the possible limitation of the monopoly element in the patent privilege is unlikely to entail the undesired consequences as long as the patentee retains the opportunity to obtain a fair compensation on the basis of freely negotiated contracts.

Finally, the foregoing analysis shows that regional TT control systems lead to the creation of regional patents that, surprisingly, seem to be less restrictive than national grants. However, even if a future EEC patent may prove to be cheaper than the total costs of protection of nine separate national patents, a small or medium-sized company, especially from a developing country, may be unable to afford it. And even assuming that such firms are required to pay lower fees, they may confront the problem of nonsufficient working within the European Community.

Paradoxically, a sudden proliferation of large-scale regional patents, if coupled with the elimination of all export restrictions and exclusive licenses limited to national markets, would seriously hurt small technology recipients and suppliers.

NOTES

(1) The existing international patent system is governed by the 1883 Paris Convention for the Protection of the Industrial Property Rights, and the regional as well as national patent laws of the member countries.

(2) For instance, the Andean Common Market and the CMEA and EEC regulatory mechanisms.

(3) For a comprehensive bibliography on the subject and related issues, see Tagi Sagafi-nejad and R. Belfield, Transnational Corporations, Technology Transfer, and Development: A Bibliography (Philadelphia: Wharton School, 1976).

(4) See TD/AC 1/4.

(5) The fifth session took place at the Palais des Nations, Geneva, February 13-24, 1978. This paper contains an account of the group's deliberations during the first four sessions. For the official reports of the group on its first four sessions see, respectively, TD/AC 1/4 (1976), TD/AC 1/7 (1977), TD/AC 1/9 (1977), and TD/AC 1/11 (1977). The sixth session, which took place in Geneva in July 1978, and the subsequent Diplomatic Conference on a Universal Code of Conduct for Technology Transfer (Geneva, November 1978) have not changed the content of the already agreed-upon texts of those code provisions that are discussed here. Although the negotiating regional groups have not solved their basic differences, they have agreed to continue negotiations in the future.

(6) According to the tentative plans proposed by the UNCTAD Secretariat in March 1977, the U.N. conference for the purpose of negotiating a code of conduct was scheduled for April-May 1978. See TD/AC 1/7, p. 2. While a draft code was prepared by mid-1978, many unresolved issues were put in brackets, and negotiations continued into 1981.

(7) While the Group of 77 strongly favors the idea of a binding instrument, Group B wishes the code to be a set of guidelines for the international transfer of technology. See the report of the group on its fourth session held in Geneva from October 31-November 11, 1977, TD/AC 1/11, Annex III/p.8. The tentative texts of the code proposed on behalf of Group D leave this question open, to be solved by a diplomatic conference. See also A. Finnegan, "A Code of Conduct Regulating International Technology Transfer: Panacea or Pitfall?" Patent Law Review, 1977, p. 559 ff.

(8) The issue of the possible alternatives to the binding character of the code is discussed at length in an unpublished paper entitled, "The U.S. and the Issue of the Binding or Voluntary Nature of International Codes of Conduct Regarding Restrictive Business Practices," by J. Davidow and L. Chiles.

(9) J.E. O'Farrell, "Industrial Property Rights and their Economic Aspects," AIPPI 1972/II, p. 7.

(10) Ibid.

(11) See the UNCTAD publication The International Patent System: The Revision of the Paris Convention for the Protection of Industrial Property Right, TD/B/C 6/AC 3/2 (1977), p. 25.

(12) Ibid.

(13) Ibid. at p. 33.

(14) Article 5 has special significance in those countries of the Union which, like the majority of developing countries, do not permit a patent new chemical substances, foodstuffs, etc. The definition of Art. 5 quater was recommended by experts from Latin American countries; see Industrial Property, 1976, p. 204.

(15) In some countries there is more than one form of protection for inventions – certificates of authorship, patents of importation, and so forth.

(16) See, for instance, AIPPI 1977 (I), pp. 92-98.

(17) That is to say, the principles of international treatment, convention priority, etc. Ibid.

(18) The text of this proposal, submitted on behalf of Group B at the fourth session, reads as follows: "The Code of Conduct is based on the following principles: . . . (V) Respect by parties and governments for industrial property protection is necessary in order to provide incentives for research, invention, development, disclosure and transfer of technology." See TD/AC 1/1, Annex III, p. 6.

(19) See the chapter on special treatment for developing countries prepared by Working Group I at the fourth session of the Intergovernmental Group of Experts, as continued in TD/AC 1/11 (1977), p. 15.

(20) See Chapter VI, 1/5 of the revised outline of the code as suggested by Group D; TD/AC 1/11 (1977), p. 14.

(21) See The Role of the Patent System in the Transfer of Technology to Developing Countries, 1964, U.N. publication, Sales No. E.65.II.B.1 para 321.

(22) The International Patent System: The Revision of the Paris Convention for the Protection of Industrial Property, Report by the UNCTAD Secretariat, TD/B/C,6/AC 3/2 p. 26.

(23) TD/AC 1/11,2.

(24) Ibid.

(25) The three-phase typology on mutual relations between receiving countries and transnational enterprises, embracing "the Laissez-Faire, the Restrictive-Unilateral and Multilateral-Regulatory periods," is developed by H.V. Perlmutter in his "The Perplexing Routes to Legitimacy," published in Codes of Conduct for the Transfer of Technology: A Critique, ed. S. Holland (New York: Fund for Multinational Management Education, 1976), pp. 1-34.

(26) See the thorough review of the first versions of draft Codes by M. Finnegan, "The Economic Effects of the Proposed Codes of Conduct Governing the Transfer of Technology: Their Impact on the Licensing and Protection of Industrial Property Rights and International Industrial Cooperation," in Conference on Significance of Protection of Industrial Property in International Industrial Cooperation (Budapest, 1977), pp. 29, 35.

(27) See Report, TD/AC 1/11, Annex I (1977), pp. 8-13.

(28) General Talking Pictures Corp. v. Western Electric Co., 305 U.S. 124, aff'g, 304 U.S. 175 (1938).

(29) The proposal justifying the inclusion of an all-embracing "rule of reason" into a preamble of this chapter was made by Group B during the fifth session.

(30) That is, the requirements to pay royalties after the patent has expired – Brullote v. Thys Co., 379 U.S. 29; licensee clauses – Lear v. Adkins, 395 U.S. 653; price fixing arrangements – United States v. Line Material Co., 333 U.S. 287 (1948). It should be stressed that the same practices are de facto illegal in the EEC.

(31) TD/AC 1/11 (1977).

(32) See, for instance, Dunlop Co. v. Kelsey-Hayes Co., 434 F 2d 407 (6th Ci. 1973); cert. denied, 415 U.S. 917 (1974).

(33) See Finnegan, "Economic Effects of the Proposed Codes," pp. 48-49.

(34) The prospects of legislative and judicial limitations upon the licensor's right to refuse to renew or terminate a licensing contract are discussed in an in-depth study entitled "Termination of License Agreements: Indemnities, Royalties, Etc." by A.N. Wise (unpublished).

(35) The certificate of authorship gives a state a perpetual, exclusive right to exploit the invention under the laws of the USSR and Czechoslovakia.

(36) During the years 1948 to 1968 the Soviet Union transferred freely over 24,000 sets of technical documentation and received about 14,000 sets on the same basis. M. Gorodisski, Protection of Industrial Property and Development of Relations Among Socialist Countries, 1972 II AIPPI, p. 105.

(37) See S. Soltysinski, "Patent and Know-How Licenses in Polish Law: The Domestic and International Perspective," International Review of Industrial Property and Copyright Law, 1977, pp. 228-30. For a compre-

hensive review of CMEA laws in the field of inventive activity, see W. Skripko, Ohrana praw izobretatelei i racionalizatorow w ewropejskich socialistischeskih srtanach (1975).

(38) See, e.g., S. Simanovsky, "Patent and Licensing Aspects of East-West Economic Cooperation," in Conference on Significance of Protection. Simanovsky stresses that the future EEC patent and CMEA protective document "could mutually complement each other, thus contributing a further development of industrial cooperation in the European Continent."

(39) The program was adopted in 1971 at the twenty-fifth session of the CMEA. See also M. Gorodisskij, Protection of Industrial Property and Development of Relations Among Socialist Countries, 2 AIPPI (1972), p. 105 ff; M. Boguslawskij and M. Cerviakow, Mezdunarodnoje sotrudnischevstvo w oblasti promyslennoi sobstviennosti (1972), p. 133 ff.

(40) Woprosy Izobretatelstva, no. 7, 1973, p. 38.

(41) Compare a Western view on this subject: W. Seeger, "Beschraenkunge der Vertragsfreiheit beim Abschluss von Kooperationsvertragen," in Conference on Significance of Protection, p. 472.

(42) See 1977 Industrial Property, Laws and Treaties, Multilateral Treaties, Text 2-002, p. 4. The rules under the agreement were published in 1977 Industrial Property, Text 2-003, p. 1.

(43) 1973 Industrial Property, p. 296 (English translation).

(44) See, generally C. Bellamy and G. Child, Common Market Law of Competition (1973); V. Korah, "Patents and Competition Law: Recent Decisions of the European Commission," European Law Review, 1976, p. 185.

(45) Centrafarm v. Sterling Drug, 1974 E.C.R. pp. 1147-68; see also Parke Davis v. Centrafarm, 1967-1970 cMR pp. 7810-826.

(46) The theory of exhaustion of patent rights is attributed to J. Kohler, a nineteenth-century German scholar, who argued that if a patented product (a manufacture embodying a patented invention) is put on the market, the offeree wants to acquire it free from any encumbrances, and the seller must be ready to consent to the use of the sold product by the buyer. In 1965, two officials of the EEC Commission published an article arguing that the exhaustion of patent rights relating to products marketed in an EEC country should be effective throughout the Community. See U. Shatz, The Exhaustion of Patent Rights in the Common Market, 1 IIC (1971), pp. 1-7.

(47) Decision of the Commission of December 2, 1975 in Re AOIP/Beyrard, 7 IIC 286 (1976).

(48) Ibid; Decision of the Commission of June 9, 1972 in Re Davidson/Rubber, 3 IIC 528 (1972).

(49) Decision of the Commission of July 25, 1975, in Re Bronbemaling OJEC L 249, 27 (1975).

(50) The Commission's rulings have also had a considerable economic impact upon rationalization cartels, joint ventures, and patent pools. See the EEC Commission decision of December 15, 1975 in Re Bayer/Gist-Brocades, OJEC L 30, 13 (1976).

(51) The Commission's recent draft regulation aimed at exempting certain categories of patent licenses from the impact of EEC competition rules is considered "highly unsatisfactory" by the European Industry. The project regulation listed seventeen clauses "frequently appearing in patent licensing agreements which it considered to be illegal under any circumstances." As a result, the British government has withdrawn its support for the project. See A. Herman, "Disagreement in the EEC over Patent Licensing Rules," Financial Times, January 5, 1978.

(52) See W. Albrechtskirchinger, The Impact of the Luxembourg Conference for the Establishment of a Community Patent on the Law of License Agreements, 7 IIC 447, 460 (1976).

(53) See A. Deringer, "Competition Policy" in Current Legal Aspects of Doing Business in the European Economic Community, ed. M. Joelson, R. Berry, P. Schliesser, and R. Schwartz (Chicago: American Bar Association, 1978), p. 186.

(54) Finnegan, "Economic Effects of the Proposed Codes," p. 34. Not surprisingly, "the UNCTAD talks . . . indicate that the Group of 77 would be unlikely to accept more restrictive licensing agreements than are admitted under the EEC Treaty in business between member states": Albrechtskirchinger, Impact of the Luxembourg Conference.

Comments

Louis B. Schwartz

Professor Soltysinski has given us a profound and sophisticated review of the international patent system under siege. The Paris Convention, that nineteenth-century pillar of industrial property, faces challenges on every side. There are revisionist proposals within the convention establishment itself. The underlying individualist private property assumptions of the convention have been called into question by the appearance in socialist states of an important nonpatent alternative for encouraging technological advance: the certificate of authorship, entitling the investor to compensation for his contribution, but not to a transferable monopoly and virtually unlimited power to prevent use of the invention. In the United States, the European Community, and elsewhere, antitrust laws have institutionalized a confrontation between the opposing policies of unrestricted competition and monopoly for invention. But the great modern challenge comes from the gradual political coalescence for some purposes of the so-called Third World or developing countries. On the patent front, as on other fronts, they demand specially favorable treatment, "affirmative action," by nations enjoying greater natural or historic advantages. The demand is heard in the debates of the WIPO and even more insistently in the congenial atmosphere of U.N.-sponsored discussions of an International Code of Conduct on Transfer of Technology.

Professor Soltysinski shows us that the reform movement focuses on four main demands: (1) discriminatory treatment in favor of the developing countries; (2) limitation of restrictive licensing, especially territorial and export restraints; (3) freedom to use licensed technology after the termination of the license contract; and (4) insistence on the power of national authorities to grant exclusive compulsory licenses, to bring about the working of patents within Third World countries where the invention is patented. The author demonstrates that these demands are not radical in principle, but rather express tendencies already manifest in the domestic law of the developed countries.

But while the principles are relatively noncontroversial, almost all sense of common purpose or tangible achievement is lost when the discussion turns to specific applications or enforcement mechanisms. The UNCTAD experts can tentatively agree on a preamble and principles, but even here critical issues remain unresolved. Is the goal to promote "unrestricted" or only "improved" flow of technical information? How is the principle of "affirmative action" favoring the Group of 77 to be reconciled with the parallel propositions that technology shall be transmitted in "mutually advantageous terms" that respect "legitimate interests of all parties to the transfer?" Irreconcilable tensions are exposed in the proposal relating to restrictions on exports: Are only "unreasonable" restraints to be condemned? Are restraints to be condemned if they "hinder" exports or must the hindrance be "substantial?" And which of such generally disapproved restraints may be "justified?" It is a lesson in realpolitik to reflect on the agreement between the developed market-economy countries and the developed socialist countries that export restrictions may be justified to protect the supplying countries' industrial property rights.

Readers of Professor Soltysinski's revealing analysis must wonder, despite his guarded optimism, whether the prospect of multilateral control of restrictive trade practices, within or without the patent system, has improved in the thirty years since the abortive Havana Charter. The sense of world solidarity is surely no greater now. Compassion for developing nations is confused when a Group of 77 speaks for both desperately impoverished pseudostates and stupefyingly rich exploiters of cartelized oil. It does not augur well for the UNCTAD endeavor that progress toward a unified law on patents and trade restraints within the EEC has been so agonizingly slow, despite the fact that EEC member countries share a common culture and have collective judicial, executive, and legislative institutions already in being.

This reader was also stimulated by Dr. Soltysinski's account to ask why the Group of 77 was in some respects so modest in its demands. Why was there no demand for access to patented technology on reasonable royalty? Why, indeed, does the discussion proceed on the assumption that a patent system is to exist at all in underdeveloped countries? Patenting is not part of the cosmos or of natural law. Some Western nations adopted the system quite late. Many have excluded specified fields from the general patent system, such as drugs and new animal and plant breeds. In matters deemed to be of the greatest urgency, such as those related to defense applications of nuclear energy, the United States barred patenting, made disclosure of inventions compulsory, and provided a system of compensatory payments to inventors in lieu of private patent monopolies.

If I were counseling most members of the Group of 77, I would have recommended against adopting any patent system, and in favor of their taking a free ride on the patent files and published technological literature of the advanced countries. The reasons for that position would be as follows:

1. The putative incentive effects of a patent system cannot work in a country that does not already have a high level of technological education and practice.
2. A highly developed bureaucracy and judicial system is required to administer a patent system; the United States itself has difficulty keeping the system rational, excluding trivia, mediating conflicting claims, and preventing fraud on the patent office.
3. Inevitably, the chief beneficiaries of a Third World patent system are the multinational corporations of high-technology nations; in any event, the royalty drain would usually more than offset the benefits of any technological advance stimulated by local patenting.
4. A more effective and selective stimulus to local technological development would be to subsidize research and development in areas where a particular Third World country has a comparative advantage.
5. The most promising general program for technology transfer to developing countries would be a multilateral institution whose business would be amassing and selecting the most useful technology freely available in the advanced countries, adapting it to the needs of particular developing countries, training potential technologists from those countries, and possibly promoting developmental use of that technology by assistance in negotiating appropriate development contracts, whether or not related to patented technology or trade secrets.

But I have come a long way from transfer technology – I am talking about education and developmental aid. "Transfer of technology" sounds as if it refers to a problem of communication. There is reason to believe that much of the discussion relates to something quite different, namely, a power struggle between the haves and the have-nots as to how much the haves are entitled to collect from the have-nots. It is not the least paradoxical and anguishing feature of this power struggle that any diminution of what the haves can collect or protect in the name of fairness in technological transfer they can generally make up in another way. As Professor Soltysinski points out, denying the supplying state or firm any protection against recipient-state exports may mean only that technology will not be transferred. Mandating free use of technology after the expiration of a license may only mean a compensatorily higher royalty rate for the license period. The refusal of a developing country to enact a patent law would possibly be treated by Big Brother as a hostility warranting denial of aid, of most-favored-nation treatment, or even of eligibility to join in a treaty of friendship and commerce. No wonder the gap between rich and poor countries continues to widen.

Comments

Herbert S. Levine

Professor Soltysinski has presented us with an excellent, thoughtful, well-organized paper. He writes not only from a background of legal study, but also from his own experience in dealing with some of the thorny issues he discusses. His paper adds considerably to the depth and substance of this conference.

My brief comments are those of an economist, not a lawyer. And though my specialization has been the study of planned economies, my work has primarily been on the Soviet Union rather than Poland or Eastern Europe. My comments will, therefore, draw heavily on Soviet practice; thus, in those cases where East European practice and experience differs from that of the Soviet Union, my comments would have to be modified in order to be generally relevant to the entire set of socialist countries.

The first point I would like to comment on is Professor Soltysinski's call for the introduction by market economies of codes of competitive behavior to cover the transfer of technology between the market economies and the socialist economies (of Group D). The problem here derives from the one-sidedness, the asymmetry in the competitiveness. For technology transfer between market and socialist economies involves economic relations between competitive firms in the market economy and the state trading organization in the socialist economy. It is frequently discussed in the United States whether or not some of the laws that enforce competition within the United States (antitrust laws, for example) should be relaxed in dealings between U.S. firms and socialist state trading organizations in order to protect U.S. firms from the unfair use of market power possessed in such relationships by the state trading organizations. Thus the point raised by Professor Soltysinski is a complex one, which requires and deserves deeper and further discussion.

In general, the paper deals primarily with West-West technology transfer. I would like to see Professor Soltysinski direct more of his attention, especially given his background and experience, to the special difficulties of economic relations between East and West. There are, of course, many such economic difficulties. One that is particularly relevant to the transfer of technology from West to East concerns the legal restrictions on property rights that can be granted to Western firms in socialist countries. This is of particular importance in regard to the Soviet Union. As a result of these limitations, the usual channel of transfer to the Soviet Union is licensing rather than the channels of fully owned subsidiaries or joint ventures. The latter two, involving equity ownership, are preferred by Western firms, especially large ones (the available data indicate that in the West licensing is preferred more by small firms than large ones). However, in dealing with the Soviet Union, the precontract costs and risks are substantial, and cannot easily be borne by small firms. Thus for this and other reasons it is mostly large firms that deal with the Soviet Union and are forced to use the licensing channel. The situation is further exacerbated by the reluctance of the Soviets to use running royalties as the mode of payment. They prefer a fixed sum. This, actually, is not beneficial to either party, for it reduces the interest of the transferrer of technology in the ultimate success of the transfer.

These systemic difficulties are hard to cope with, although several East European countries, and recently China, have proved to be more flexible than the Soviet Union in regard to the granting of property rights. In dealing with the Soviet Union, however, certain aspects of the advantages of property rights can be acquired by a Western firm through "creative contracting," such as profit sharing, and managerial powers in regard to quality control.

Finally, let me turn to a somewhat technical point, one, however, that an economist finds of some importance. In speaking of the CMEA system of protecting an inventor through a certificate of authorship rather than a patent, Professor Soltysinski argues that since the state is responsible for the adequate utilization of the invention, the inventor is assured of maximum remuneration. This does not necessarily follow. In socialist economies, and in particular the Soviet Union, weak managerial incentives and bureaucratic barriers to innovation substantially limit the introduction and diffusion of inventions, and thus reduce the rewards to inventors. Indeed, this is the basic reason why the Soviet Union and other industrialized socialist countries have, in the last decade, turned to the Western industrialized market economies for substantial transfers of advanced technology.

5 The Pricing of International Technology Transfers via Non-Affiliate Licensing Arrangements

Franklin R. Root

This chapter explores the "pricing" of international technology transfers that occur through licensing arrangements between U.S. firms and nonaffiliated (independent) foreign firms. Employing an "inside" rather than an "outside" perspective, it is directed primarily to an explanation of how U.S. firms view the trade-off between direct foreign investment and nonaffiliate licensing, and how, within this strategy context, they determine an acceptable level of compensation for their technology transfers to nonaffiliate foreign licensees. The chapter closes with a brief discussion of the implications of the technology pricing process for the less developed countries.

THE TRADE-OFF BETWEEN DIRECT FOREIGN INVESTMENT AND NONAFFILIATE LICENSING: MONOPOLISTIC ADVANTAGE THEORY

The monopolistic advantage theory of direct foreign investment postulates that the investing firm possesses monopolistic advantages that allow it to operate subsidiaries abroad more profitably than local competing firms. These advantages are specific to the firm; they are available to other firms only if the investing firm is able and willing to transfer them. In brief, they are proprietary.

Many, if not all, of the investing firm's monopolistic advantages derive from its "superior knowledge" or technology. Broadly defined, the firm's technology assets include all the <u>intangible</u> skills owned by the firm that afford a competitive advantage wherever it undertakes operations: industrial technology (patents and know-how), management and organization skills, financial skills, marketing skills (including trademarks), and so on.

Its possession of technology assets enables the firm to offer products in the marketplace that are <u>differentiated</u> either physically or psychologically from rival products. The firm thereby gains a significant degree of control over the prices and sales volumes of its products, which is the source of the economic rent on its technology assets.

The ownership of technology is a necessary but not a sufficient condition for horizontal direct foreign investment.(1) A second condition is that the firm can obtain the highest economic rent for its technology assets only by investing abroad in production and marketing under its control. That is to say, the returns from direct investment must be higher than the returns from exporting its products or directly licensing its technology assets.

Why should a firm be able to obtain a higher economic rent from direct foreign investment than from licensing? The first point to be made is that only some of the firm's technology assets can be transferred to an independent foreign company. Patents, trademarks, and product- or process-specific know-how are generally suitable for transfer via licensing, but the firm's skills in management, organization, finance, marketing, and the creation of new technology are so intimately bound to its personnel and ongoing operations that they can seldom be transferred in their entirety to outsiders.(2) And yet the acquisition of the full economic rent in the marketplace may depend on the blending of all technology assets into a single package.

Second, the licensor firm shares economic rent (monopoly power) with the licensee firm. Indeed, the motivation of the licensee is to strengthen its own competitive advantage through access to the licensor's technology. Moreover, the licensee at some future time may directly reduce the licensor's economic rent by entering third markets as a competitor.

Third, licensing a nonaffiliated firm carries a potential opportunity cost, since it may preclude the licensor firm from entering the country market at a later time with exports or direct investment. In some instances, third markets may be closed off to the licensor. For example, one U.S. company granted a British firm a license to manufacture and sell its products in Britian, and, in addition, the exclusive right to sublicense the U.S. company's know-how to the rest of the world. Within a few years opportunities for the U.S. manufacturer's products appeared in several country markets, but the licensing agreement barred any direct participation by the manufacturer in those markets through exports or direct investment.

Fourth, the returns from a licensing agreement may be judged less certain than the returns from direct investment. Apart from market risk (which also confronts the direct investor), there is the risk of poor licensee performance. Most, if not all, licensing income is contingent on the licensee's ability to manufacture and market products utilizing the licensed technology.

In brief, the international firm may gain a higher economic rent for its technology assets through direct investment (or exports) than through licensing because of imperfections in technology transfer. How, in fact, do U.S. firms view the trade-off between direct investment and licensing?

THE FOREIGN MARKET ENTRY STRATEGY OF U.S. FIRMS

Statistical data as well as the behavior of individual firms indicates a strong preference for direct foreign investment over licensing as a strategy to enter foreign markets. In fact, most international licensing involves the transfer of technology from a parent firm to its controlled affiliates abroad. In 1966, for instance, U.S. multinational corporations (3,400 firms with 23,000 affiliates) accounted for 88 percent of total U.S. net receipts of royalties, license fees, nonfilm rentals, and management service fees. (In addition, 43.5 percent of net receipts from nonaffiliated licensees abroad were attributable to these same corporations.)(3) Again, in 1976 U.S. gross receipts of royalties, license fees, and so forth from nonaffiliated foreign companies were only 20.6 percent of total U.S. gross receipts ($844 million out of $4,088 million).(4) Bearing in mind that parent companies may not explicitly charge their affiliates for technology transfers, this percentage probably overestimates the relative importance of nonaffiliate licensing.

These and other data demonstrate that U.S. firms prefer to have an equity interest in the use of their technology assets in foreign countries. With equity goes not only control over the exploitation of their technology, but also (presumably) a greater capture of economic rent than would be possible from licensing.

This preference is hardly surprising. Few industrial firms create technology to sell as technology. Rather technology is viewed as a source of new products that the firm itself will sell in the market. For the most part, therefore, international technology transfer is undertaken to support the production and marketing of the firm's own foreign affiliates.

More specifically, the strong preference to transfer technology via direct foreign investment is based on:

1. the desire to retain control over technology assets that form the basis of the firm's international competitive advantage – control is wanted not only to maximize economic rent, but to prevent outside firms from obtaining the technology;
2. the complexity of much technology transfer, which complexity demands a sustained, intimate relationship between the donor and recipient firms if transfer is to be effective;
3. the importance of protecting the quality image of the manufacturer's product, particularly when trademarks are transferred abroad;
4. the desire to maintain control over the marketing of products, not only to exploit present markets but to build a market position that will facilitate the introduction of new products in the future;
5. the desire to retain control over technology assets to better support an international logistical system that integrates production and marketing over several countries.

Given this strong preference for direct foreign investment, why do U.S. companies enter into licensing arrangements with nonaffiliate foreign firms? The following reasons are offered by international managers:

1. Import/investment restrictions. The representative international firm may employ licensing as a primary mode of foreign-market entry when its exports and investment are kept out of a country by government restrictions. Commonly, licensing is used to retain a market position built up through exports now threatened by restrictions, particularly when the market is too small to justify direct foreign investment. Licensing may also enable the international firm to replace exports of its end products with exports of raw materials, components, and parts to the licensee.

2. Resource constraints. When exports are ruled out by government restrictions, smaller international companies may choose licensing over direct foreign investment (DFI) because they lack, or are unwilling to commit, the necessary capital and management resources.

3. Political risk. The international firm may turn to licensing when it judges unacceptable the political risks associated with DFI. In those instances, it would ordinarily prefer to export, if possible.

4. Reasons not related to market entry. The international firm may use nonaffiliate licensing to (a) discourage possible infringement or loss of patents and trademarks in a foreign country, (b) acquire technology from the licensee (cross-licensing), (c) obtain extra income on its technology assets in markets it does not intend to exploit through exports or DFI, and (d) develop low-cost sources of raw materials, components, or end products for the firm's operations at home or abroad.

What emerges from this review of the foreign-market entry strategies of U.S. firms is the conclusion that U.S. firms consider nonaffiliate licensing as second- or third-best strategy. In a laissez faire world there would apparently be far less use of nonaffiliate licensing as a mode of foreign-market entry.

HOW U.S. FIRMS PRICE TECHNOLOGY TRANSFERS TO NONAFFILIATE FOREIGN FIRMS

International technology transfer usually occurs over a lengthy period of time. Only rarely is technology sold (assigned) to the licensee in a once-and-for-all sale of patents and know-how. Hence the "price" of a technology transfer mainly takes the form of a series of payments by the licensee to the licensor over time. The benefits to the licensee also occur over time, although the payments and the benefits may not have the same duration. Technology transfer is a process over time because its purpose is to create a technological capability in the licensee that matches, in whole or in part, the capability of the licensor.

Pricing technology flows to affiliate licensees. The pricing of technology flows from a parent company to its foreign affiliates is a question of intracorporate transfer pricing, which is under the administrative control of the parent company. This intracorporate transfer price is seldom intended to reflect the price that would result from "arm's length" bargaining between two independent companies.

Intracorporate pricing may undervalue or overvalue technology transfer as compared to the "arm's length" price. Corporate practices range from no explicit charge for technology transferred to foreign affiliates (on the assumption that compensation will occur in the form of higher income earned by the affiliate) to formal licensing agreements with royalty payments or R & D charges based on a fixed percentage of affiliate sales or income.

Most international firms distinguish between R & D work specifically requested by foreign affiliates and R & D work performed for the corporation as a whole. For the former, affiliates pay the full R & D cost (usually based on time and variable costs); for the latter, affiliates pay royalties, technical service fees, or allocated charges.(5)

To conclude, intracorporate pricing of international technology flows is much more a matter of administrative policy and internal accounting procedures than of economics. Also, at times the transfer price may be deliberately manipulated to minimize taxes or evade exchange restrictions. Consequently, the intracorporate transfer price offers little if any guidance to managers in the pricing of technology transfers to nonaffiliated foreign companies.

The rate-of-return approach to pricing technology transfers to nonaffiliate licensees. In a survey of 240 U.S. manufacturing companies covering all major industry groups and size classifications, Lovell found no agreement on which costs are incurred by a licensing program, and little consensus on how licensing costs and income should be treated in internal accounting records.(6)

International firms do not assign values to the proprietary rights they grant to foreign licensees. It is, of course, difficult – if not impossible – to match R & D outlays against specific operations. But this difficulty, as well as the problem of joint costs, is overcome for domestic operations by treating R & D expenses as an overhead cost that is allocated among different units according to sales or another yardstick. And as we have noted, such allocations sometimes include foreign affiliates. Whatever the reason, the failure to impute values to proprietary rights granted to nonaffiliate foreign licensees means that U.S. companies are assuming (implicitly or explicitly) that R & D costs have been fully covered in other operations.(7) In other words, they are regarding R & D costs as a sunk cost.

Instead of a return-on-investment or asset-return approach to pricing technology transfers, U.S. firms apparently seek to "maximize" the contribution margin between transfer costs and licensing income.(8) I propose to discuss this pricing approach from the perspective of a normative model of price determination via licensor/licensee negotiations.

PRICE DETERMINATION IN LICENSOR/LICENSEE NEGOTIATIONS: A NORMATIVE MODEL

The model postulates that the licensor enters negotiations with a range of possible offer prices for a given technology package while the licensee enters negotiations with a range of possible bid prices. The variables that determine these two ranges are depicted in figure 5.1.

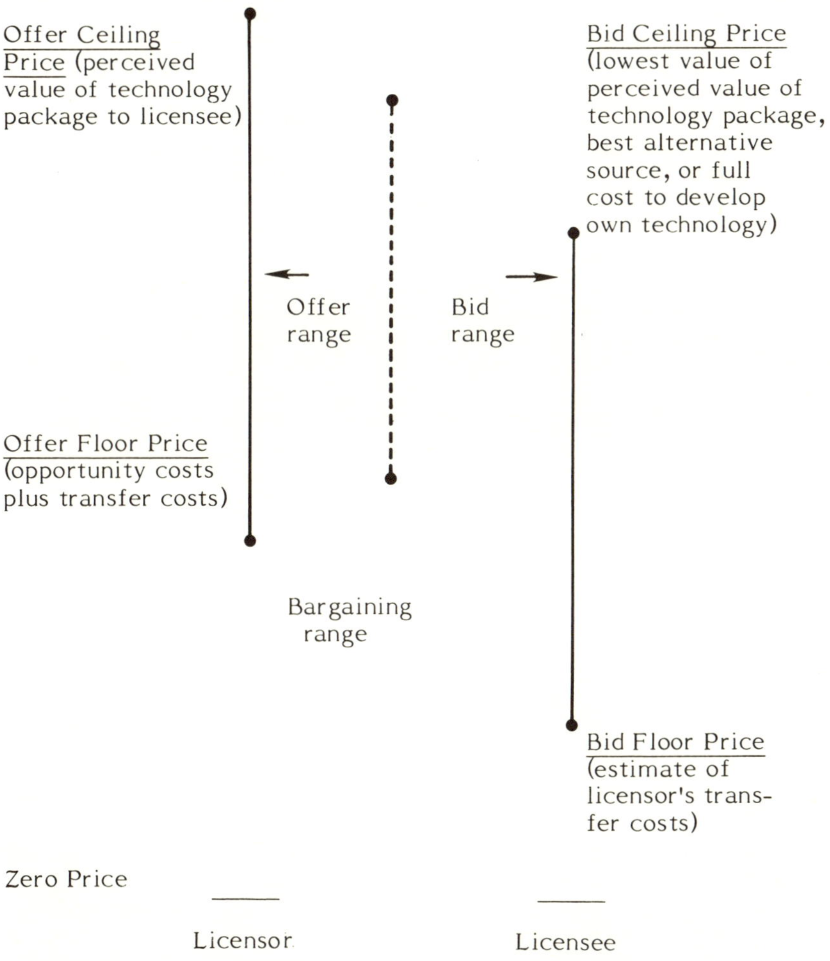

Figure 5.1. Offer and bid price ranges in licensing negotiations.

The floor offer price is the sum of the licensor's opportunity costs and the costs (direct and indirect) that he will incur in transferring the licensing package to the licensee.

The licensor's opportunity costs are the revenues (actual or potential) that are foresaken when he enters a particular licensing agreement. Depending on the circumstances, the licensor may give up actual or potential net revenues from exports, direct investment, or another alternative entry mode. Furthermore, he may lose net revenues in third markets if the licensee enters those markets. A licensor may assess his opportunity costs as inconsequential or so high as to rule out licensing as a foreign market entry strategy. In the latter instance, the licensor's opportunity costs would exceed his offer ceiling – the licensor's perceived value of the technology package to the licensee.

Transfer costs may take several forms. They include all costs incurred by the licensor to establish the licensing agreement and to bring the licensee to the production/marketing stage: market research, negotiators' travel and salary expenses, expenses of inside or outside legal counsel, expenditures on local patent or trademark protection, costs of preparing blueprints and other technical materials, expenses for training the licensee's employees, costs of adapting the technology to the licensee's needs, and the service costs of engineering, construction, and plant installation. In addition, transfer costs include all ongoing costs to service the licensing agreement over its life: travel and salary expenses of supervisory and technical personnel needed to maintain the licensee's operations, allocated salary expenses of administrators and technical people in the home country, policing and litigation costs to protect the licensed rights in the foreign country, auditing and inspection costs, and the costs of procurement or marketing assistance to the licensee.

The licensor's offer ceiling price is his perception of the value of the technology package to the licensee. Although the licensor (like any seller) would like to get as much as he can for his technology, he cannot reasonably expect to get more than its value to the licensee. At the offer ceiling price, the licensor would expect to obtain all the economic rent resulting from the utilization of the technology package.

The licensee's bid ceiling price is determined by the lowest value of three variables. Under no circumstances would the licensee pay more for the technology package than its value to him – the incremental net revenues he would obtain from its possession. But he would be unwilling to pay that much if he could obtain the same technology package from an alternative source or from his own R & D efforts at a lower cost. Depending on the circumstances, these three values may assume different orders. However, it is probable that the full cost of duplicating the licensor's technology package would exceed the other two values.(9) Evidently, if self-development costs were the lowest of the three values (taking into account the time factor), then the licensee would not enter licensing negotiations.

The licensee's bid floor price is his perception of the licensor's transfer costs. (The licensee ignores the licensor's opportunity costs, and in any event does not possess the information to estimate them.) The licensee cannot reasonably expect to get the technology package for less than its transfer costs. At the bid floor price, he would expect to obtain all the economic rent from the utilization of the technology package.

The licensor's offer floor price and the licensee's bid ceiling price together determine the bargaining price range, as is shown in figure 5.1. The final negotiated price will be at a point on the bargaining range.

The bargaining range may change during the course of negotiations for three reasons: (1) a change in the technology package (what is being transferred), (2) new information, and (3) new circumstances.

The bargaining range is most likely to change because of a change in the technology package. That is to say, the package itself will be subject to negotiation. New information generated by the negotiation process is most likely to alter the bid ceiling price, although it can alter the offer floor price as well. Changes in economic and political factors during negotiations can also cause a shift in the bargaining range, particularly in the bid ceiling price.

Licensing negotiations may be viewed as a mix of shifting bargaining ranges and bargaining within each range. The key issues to be resolved in negotiations may be classified into three groups: (1) the content of the technology transfer package (patents, trademarks, know-how, and services to be provided by the licensor to the licensee); (2) the use conditions of the package (rights and obligations of the licensor vis-a-vis the licensee with respect to the maintenance of quality standards, minimum performance, sourcing of inputs from the licensor, sale of licensed products to the licensor, defense against patent or trademark infringement, tax payments, government approval, duration and renewal of the contract, and so on); and (3) the compensation (price) of the technology package (lump-sum payments, royalty rate and base, minimum annual royalty, technical-management service fees, equity, transfer prices of sales to licensee, commissions on sales for licensee, commissions on sales for licensee, grant-back rights on technical improvements, and others).

The many elements that make up these three issue groups may be combined in any number of ways. This flexibility generates many trade-offs that facilitate final agreement. In particular, the technology transfer price may be raised or lowered by changes in the technology package and/or changes in its use conditions. The compensation mix is also variable. A lower percentage royalty rate may be traded off against lump-sum payments, minimum annual royalties, equity, service fees, transfer prices for parts or materials supplied to the licensee by the licensor, grant-back rights, and so on. The royalty base itself may assume many forms.

Flexibility in the compensation mix becomes important to the success of negotiations when royalty rate ceilings are imposed by host governments or when either party insists on a "conventional" rate. It should also be noted that royalty income is the product of the royalty rate and the royalty base (net sales of the licensee). The compensation mix makes the percentage royalty rate (and in some instances royalty payments) an unreliable indicator of the technology transfer price.

Licensing negotiations take place under conditions of partial ignorance and risk for both parties. Terms must be negotiated before the value of the technology package to the recipient can be ascertained in fact. The costs and income of the two parties are contingent on each other's future performance over the agreement's life, as well as the future behavior of economic and political factors. Uncertainty with respect to these contingencies creates enterprise, economic, and political risks. The normative model assumes a risk/return trade-off for both the licensor and the licensee. The higher the risk perceived by (say) the licensor, the more insistent he will be on a higher share of the economic rent. In effect, perceived risk raises the licensor's offer floor price. At the extreme, risk perceptions will eliminate the bargaining range.

A priori there is no reason to assume that the licensor's bargaining power will enable him to obtain the full economic rent on the technology package. Clearly, the stronger the licensor's bargaining leverage, the closer the final negotiated price will be to the bid ceiling price. Conversely, the stronger the licensee's bargaining leverage, the closer the final price will be to the offer floor price. But the actual price is determinable only at the conclusion of the licensing agreement. Hence the negotiated price is a contingent price, dependent on the future manufacture and sale of the licensed product or utilization of the licensed process.

Finally, the successful conclusion of negotiations (marked by the signing of the licensing contract) is only the beginning of an ongoing relationship between the licensor and the licensee. Fundamentally a licensing agreement is a mutual commitment to work together for the benefit of both parties. Thus it is not to the advantage of either side to force a hard bargain that will cause resentment later on. Putting the matter in another way, a licensing agreement will remain viable over the longer run only if both parties share in the economic rent created by the technology transfer.(10)

SOME EMPIRICAL NOTES ON PRICING INTERNATIONAL TECHNOLOGY TRANSFERS BY U.S. MANUFACTURERS

This section reports on some preliminary findings of a current investigation of technology transfer pricing by the Multinational Enterprise Unit of the Wharton School.(11) The mail survey sample comprises thirty-eight U.S. manufacturing firms in several industries and 100 individual licensing agreements of those companies with nonaffiliated foreign companies in forty-one countries. This sample is supplemented by personal interviews with thirteen U.S. manufacturers.

Respondents were asked to rank their reasons for licensing unaffiliated foreign firms. Two reasons were given overwhelming and approximately equal importance: (1) royalty and other income, and (2) entry into foreign markets.(12) These two dominant reasons are complementary rather than alternatives for the majority of firms who view licensing as a profitable way of getting their products into foreign markets. Only a minority of firms regard licensing as simply an opportunity to obtain income without reference to market entry.

Transfer costs. Transfer costs averaged 11.6 percent of total licensing returns in the agreements of the sample firms. These costs included both direct costs (technical services, legal services, marketing assistance, and travel) and indirect costs (legal and general management overheads allocated to the negotiation and servicing of a license). Direct costs were 77 percent of total transfer costs. The average transfer cost was $198,000; in only 9 percent of the sample were transfer costs below $10,000. Opportunity costs were not estimated by the respondents.(13)

These transfer costs should be regarded as estimates rather than cost accounting data. Lovell found that "virtually none" of the sample companies kept segregated records for each type of input and expenses that could be attributed directly or indirectly to their foreign licensing activity.(14) However, our findings do show that firms consider transfer costs when negotiating licensing agreements. Furthermore, transfer costs are frequently significant in absolute dollar amount. Nonetheless, the small percentage that transfer costs bear to total licensing returns demonstrates that licensers do not use a cost-oriented model in pricing technology transfers. And since licensors make no reference to sunk costs in negotiating licensing agreements (although these respondents indicated they would like to), it is also evident (as stated earlier), that they do not use a return-on-asset pricing model.

Pricing technology transfers. How, then, do licensors price their technology transfers? Do they simply price, as is commonly alleged by spokesmen for the developing countries, as discriminating monopolists – charging what the traffic will bear? In an early study Lovell found that the pricing objective of the typical licensor is to get "what the traffic will bear" with "due" consideration for the mutual and long-term interests of the contracting parties. A survey of thirty companies by this writer in 1977 suggested that the term "what the traffic will bear" was synonymous for most respondents with "the value of technology to the licensee."(15) To say the least, the term is ambiguous.

Our current study casts doubt on the monopoly pricing model. Although licensing returns are not based on a markup over cost, we found a positive correlation between total transfer costs and total royalties in the sample licensing agreements.(16) This result is inconsistent with the discriminating monopolist model: monopoly pricing would bear no systematic relationship to transfer costs.(17)

Responses to the question, "How do you set 'price' or 'return' of an agreement?" support the positive correlation between transfer cost and royalty income. By far, the most important criteria cited by respondents were (1) "licensee's market share or sales," (2) "industry norms," and (3) "the amount of technical and other services provided."(18) In the language of the normative model, the first criterion bears directly on the offer ceiling price, and the third criterion, on the offer floor price. The second criterion (industry norms) reflects the presence of competition among licensors. For only about one-quarter (26.8 percent) of the sample licensing agreements did respondents indicate that the technology in question was available solely from their own firms. Since managers tend to think well of their own firms, this self-assessment may well be modest. More to the point, for almost three-quarters of the sample licensing agreements similar technology was available from one or more alternative sources. This finding suggests that licensors are commonly restrained in their pricing by an awareness of what other firms are charging for the same technology. Apart from competition, "industry norms" may also reflect a general awareness of industry practice and the precedents set by a firm's previous licensing agreements that define an "acceptable" level for royalties and other forms of compensation.

Compensation mix. The returns for the majority of the sample licensing agreements take the form of royalties and front-end fees. Technical assistance fees and net income from sales of materials or products to the licensee are fairly common (both occurred in twenty-two of eighty-two licensing agreements), but other forms of compensation are far less common. In only 5 percent of the agreements was there any equity compensation. Thus it would appear that the most frequent compensation trade-off in negotiations is between lump-sum fees and royalties.(19)

Empirical support of the normative model. These research findings support the normative model of technology transfer pricing, at least on the licensor side. Determined in negotiations, a technology transfer price will most probably be a price that substantially exceeds transfer costs but is less than a monopoly price. Monopoly pricing is constrained by the common existence of alternative supply sources, by industry and company practice, and by a recognition that the long-term viability of a licensing arrangement requires that both parties share the economic rent.

Once again, it should be noted that the licensor's transfer cost does not include opportunity costs or R & D costs. To which degree the contribution margin of a particular agreement represents a compensation for these two costs and to which degree it represents pure economic rent can only be ascertained by empirical investigation, if at all. By the same token, it would be erroneous to consider the entire contribution margin as pure economic rent.

SOME IMPLICATIONS FOR PUBLIC POLICY

The most important implication of this examination of international technology transfer pricing is that host governments can do very little directly to lower the price of technology imports. Nor can home governments. And the same can be said for international technology codes.(20)

The effective role of a host government in the negotiating process that determines the pricing of technology transfers is mainly limited to increasing the quantum and improving the quality of information possessed by the licensee. With more and better information, the licensee can better assess his technology requirements, the value to him of individual technology packages, and the alternative sources of comparable technology. In contrast, the imposition by a host government of constraints on the licensor (such as percentage royalty ceilings, duration of the agreement, trademark rights, and so on) can be nullified by adjustments in technology packages (for example, less technical service), adjustments in compensation arrangements (for example, greater lump-sum payments), or the refusal of foreign firms to enter licensing agreements.(21) In sum, the effectiveness of regulatory measures to lower technology transfer prices (apart from their administrative effectiveness) is highly questionable. Even if they were effective, the opportunity cost in lower availability of desired technology can be high.

This analysis indicates that whether the price is "high" or "low" for a given technology package depends on the relative bargaining power of the two parties. Our preliminary evidence suggests that this bargaining power is only marginally influenced – if at all – by the nationality of the licensee. The question of whether the price of technology is too high for the developing countries is, therefore, meaningless in market terms. The only general statement that can be made is that the cost of a given technology transfer is too high if it can be obtained at a lower cost in any other way.

Since a licensing contract is entered voluntarily by both sides, a reasonable conclusion is that both sides gain (or, more accurately, expect to gain) from the arrangement. In retrospect, one or both sides may have made a bad bargain, but such is the risk of decisions made under conditions of partial ignorance and uncertainty.

Circumstantial evidence suggests that licensees in the developing countries enjoy greater bargaining power today than in the past as a consequence of the greater number of alternative sources of technology, better information on those sources, and improvements in negotiation skills that come with experience. Licensees can also take advantage of the fact that licensors seldom consider R & D costs and opportunity costs in determining their floor offer prices. As is well known, this practice has opened U.S. firms to charges from organized labor and others that they transfer technology abroad at too low a price.

NOTES

(1) Horizontal foreign investment occurs when the investing firm enters foreign countries to manufacture the same products as it manufactures at home.

(2) Such skills can be transferred, in part, to outside firms under management and service contracts. The critical transfer resources are most commonly persons with the desired technical and managerial knowledge.

(3) U.S. Senate, Committee on Finance, Implications of Multinational Firms for World Trade and Investment and for U.S. Trade and Labor (Washington: U.S. Government Printing Office, 1973), p. 600.

(4) U.S. Department of Commerce, Bureau of Economic Analysis, June 1977. In 1966 the percentage was 26.6.

(5) Michael G. Duerr, R & D in the Multinational Company (New York: Conference Board, 1970), pp. 65-74.

(6) Enid Baird Lovell, Appraising Foreign Licensing Performance (New York: National Industrial Conference Board, 1969), p. 31.

(7) Ibid., p. 70. Lovell believes this practice reflects the idea that proprietary rights are not "invested assets" in the same sense as money and tangible property. But companies sometimes do capitalize proprietary rights as an equity contribution to a foreign joint venture.

(8) "Maximize" is placed in quotations marks because the licensor may not know enough about his incremental costs and income to truly maximize. Indeed, almost two-thirds of Lovell's sample companies made no attempt to determine net profits or returns from foreign licensing activity. See ibid., p. 55.

(9) It would also take much more time than the acquisition of the technology package from a licensor.

(10) In terms of the normative model, the licensor's long-run profit-maximizing price will be less than his offer ceiling price. Similarly, the licensee's long-run profit-maximizing price (or cost) will be higher than his bid floor price.

(11) The researchers are Farok Contractor and Franklin R. Root.

(12) Other reasons cited were (in rank order) exports of components or materials to licensee, ways to get other business in the licensee country, exports of finished goods to licensee, access to technical improvements made by the licensee, and low-cost source of components or products.

(13) However, respondents estimated lost export sales to the licensee country for nineteen licensing agreements. The mean value of this opportunity cost was $1.8 million.

(14) Lovell, Appraising Foreign Licensing Performance, p. 59.

(15) Enid Baird Lovell, Foreign Licensing Agreements II, Contract Negotiation and Administration (New York: National Industrial Conference Board, 1959), pp. 51-52.

(16) The correlation coefficient was 0.57, significant at the 1 percent level.

(17) We are planning to test for national differences in contribution margins and for relationships between contribution margins and the content of technology packages. The monopoly pricing model implies national differences (because of differences in bargaining power) and no correlation between package content and price.

(18) Each of these three criteria was cited by more than half the respondents. The services criterion was cited by 86%.

(19) Royalty rates in 62 licensing agreements ranged from 1 percent to 10 percent, with a mean of 3.9 percent and a standard deviation of 1.8 percent. Therefore, approximately two-thirds of the rates fell between 2.1 percent and 5.7 percent. The mean life of the agreements was 12.5 years.

(20) These statements should not be interpreted as a general denial of the usefulness of technology control systems. My concern here is only technology pricing. I am not concerned with other control objectives relating to issues such as appropriate technology, unbundling, unfair practices, and the patent system.

(21) In our survey sample, eighteen of thirty-seven companies stated that government-mandated royalty rate ceilings had a "significant impact" on the decision to license in a country. Twelve of the thirty-seven countries said that such constraints made them seek alternate channels of revenue, such as front-end fees, management fees, and other overheads charged to the licensee.

Comments

Donald L. Guertin

I am very pleased to have the opportunity to comment on Professor Root's paper. In my remarks I will first comment on some important points made in Professor Root's paper, including points based on Exxon's experience, and, secondly, comment on the implications for public policy noted by Professor Root, again drawing on some Exxon experience.

COMMENTS ON SOME POINTS IN THE PAPER

MNCs develop technology primarily for their own use. This is sometimes lost sight of in discussions on codes and the transfer of technology. As technology is generally developed by MNCs for proprietary reasons, it must be protected by the developer through the use of patents. Governments which accept the concept of private property and patents generally will not force transfer of technology because of the proprietary nature of the technology and also because they view technology as a national asset. Exxon experience demonstrates this point. In Exxon, most research and development is sponsored by affiliates to meet their needs. The corporation sponsors basic research. Expenditures in 1978 of about $300 million were charged through a Standard Research Agreement at cost. The table below indicates the approximate percentages of R & D in different major areas.

Field	Percent
Oil and Gas	40
Synthetic Fuels	10
Advanced Energy Forms	10
Other	20

Licensing income is used to help reduce the cost of this R & D. In my judgment, Professor Root's study casts doubts on the monopoly pricing model. The criteria used by Professor Root include the "licensee's market share or sales;" "industry norms;" and "the amount of technical and other services provided."

What has been Exxon's experience in the area of refining technology, using criteria similar to those above? Generally, technology is available from many sources, so there is a competitive market.(1) In evaluating market prices for technology it is of course essential to take into account the impacts on licensing income of the extent of use of the technology. Fees per unit volume of product will probably be lower for large scale licensees. Compensation to Exxon for technology may be either a paid-up license or an annual fee based on capacity. Prices exceed transfer price but are less than monopoly prices as indicated in Professor Root's study.

Professor Root states that "Whether the price is 'high' or 'low' – depends on the relative bargaining power of the two parties," and further that "Price is only marginally influenced – if at all – by nationality of license."

Exxon's experience in licensing supports this point. There are a variety of sources for refining technology. Bargaining strengths are of course very important. In the case, for example, of Venezuela, Exxon's strength is based on technology. The agreement with Venezuela after nationalization included an agreement on crude oil prices – Venezuela's strength is, of course, crude; an agreement under which Exxon would supply technology (including personnel); and payment for facilities at book value. Examples of technology where the petroleum industry has a strong bargaining position include offshore exploration and production; synthetics – liquid fuels from coal; and advanced energy sources – solar.

COMMENTS ON IMPLICATIONS FOR PUBLIC POLICY

Professor Root notes that, "Host governments can do very little directly to lower the price of technology imports. Nor can home governments. And the same can be said of international technology codes," and further that "The effective role of a host government – in negotiatory process – is mainly limited to increasing the quantum and improving the quality of information possessed by the licensee."

Exxon's experience in the area of refining technology demonstrates that the market and value of technology determine the price and not specific government actions.

Efforts at educating users are very important and in this regard screening processes are regarded as useful by governments. Encouragement of technical training is of course always beneficial. Training in MNCs like Exxon, incidentally, contributes to development of staff competence, thereby developing the overall skill level in a country. Mikesell's paper on technology transfer in a recent Fund for Multinational Management Education study develops this point in some depth.(2)

Efforts to develop codes, in my judgment, will not contribute to decreased prices but can have some benefits: by helping educate suppliers and recipients; highlighting key interests of both; and providing an overview of what both parties should keep in mind in the technology transfer process. The UNCTAD code exercise has had the above benefits but there are still significant points of disagreement.

Points of disagreement in the negotiation of the UNCTAD code include:

- Whether or not affiliate-parent relationships should be covered. They should not, in the views of most MNCs.
- Responsibilities of suppliers to meet host country objectives. At this time this responsibility is too sweeping and it would not be possible for an MNC to broadly agree to sweeping language.
- Choice of dispute settlement mechanisms (national or international). MNCs believe that the parties to an agreement should be free to designate ways of settling disputes and that these approaches should include third party arbitration.
- Guarantees. While recognizing responsibilities, MNCs must be realistic in the guarantees they can provide and licensees must recognize their own responsibilities, as well as the provisions of the contract which provides certain know-how and guarantees for a particular change.
- National law. MNCs believe that host countries should provide national treatment to foreign investors and act in accord with international law.

NOTES

(1) W.L. Nelson, Guide to Refinery Operating Costs (New York: Petroleum Publishing Company, 1976).

(2) Raymond F. Mikesell, "International Transfers of Technology in the Extractive Industries," in Public Policy and Technology Transfer: Viewpoints of U.S. Business (New York: FMME, 1978), vol. 2, pp. 69-121.

6 Taxation and Technology Transfer

Frances Stewart

At both academic and policy levels, there is growing interest and concern about the nature of technology transfer to the Third World, and with policies which might improve the process from a Third World point of view. Among policy instruments taxation has been, perhaps, relatively neglected.(1) This chapter aims to explore how the tax instrument might be used in parallel with and as a complement to other methods of control.

Potentially, the quantity, terms, and nature of technology transfer may be affected by almost any tax – since any tax that affects the conditions of production or consumption affects the conditions of TT, directly or indirectly. Taxes on consumption affect the pattern of what is produced and consequently the nature of technology transferred, while taxes on production, such as payroll taxes for social security, affect the choice of technique. But this paper concentrates on the taxes that most directly affect the process of TT from overseas, as against taxes that affect the conditions of production and consumption generally and hence the process of TT indirectly. The taxes most obviously involved are direct taxes (corporation, income, and withholding) on income generated by foreign-owned assets; the tax treatment of royalty and other know-how payments remitted overseas; tariffs on imported machinery and other inputs; and the tax treatment of management services and other technical services from overseas. Technology transfer from overseas may occur through direct investment in branches or subsidiaries of multinational corporations, joint ventures, licenses, management and technical services to local companies or machinery (and other input) sales. Payments for TT may take a correspondingly

*In writing this paper I have greatly benefited from discussions with Mr. N.P. Singh of the Department of Science and Technology, Delhi. I am also grateful to C.V. Vaitsos for helpful information.

varied form; payments for one type of service may be disguised as another. Consequently taxes and tariffs related to all forms of TT need to be considered.

Consideration of the tax system as an instrument of control is somewhat complex for a number of reasons. One is the elusiveness and shiftability of the target, which means that the intended control often eludes the policymakers' grasp. There are various types of shiftability, all of which need to be comprehended and preempted. First, there are economists' concepts of tax shiftability, that is, that taxes are shifted forward (or backward) as a result of changes in prices and production so that the consumers of the product, or the suppliers of inputs, bear the tax rather than those for whom it is intended. For example, let us assume that the aim is to tax the process of TT by taxing the profits of a multinational corporation's subsidiary. If the tax is shifted forward, by increasing the price of the product, the consumers (local or foreign, depending on whether the product is consumed locally or exported) will bear some or all of the tax, rather than the MNC. If it is shifted backward, then suppliers of raw materials may bear the tax. In either case the suppliers of technology, for whom the tax is intended, will not bear it. The evidence on shifting is unclear.(2) It has been suggested that it is less in more open and competitive economies (Prest, 1975). In what follows this type of shifting will be ignored. Were it of major significance, it could invalidate some of the argument. A second type of shiftability might be described as the fungibility of payments for TT. Firms have considerable choice in the form (and quantitites of various forms) in which they remit payments for technology. For example, subsidiaries of MNCs may choose how far they remit payments as profits, royalties, management fees, or as inflated prices for inputs. Taxation of one form (for example, profits) may cause a shift to payments in other forms. Third, there is a form of geographic shiftability in payments for international TT. Because of the international nature of the operation, firms are often in a position – particularly if they are free to choose their transfer prices for intrafirm transactions – to choose in which country they wish to declare their profits. This makes it difficult for one tax authority to tax the process of TT. The multinational nature of TT causes further problems. Payments for technology transmitted from one nation to another enter the jurisdiction of at least two tax authorities. Taxes levied by one authority may be nullified in their effects by the tax system of other countries. Hence in formulating policies, the less developed country's tax authority needs to know and allow for the tax system, and particularly the double tax arrangements, of the other countries involved. What makes this particularly complex is that the tax arrangements vary between technology-supplying countries, so that arrangements that are appropriate to the tax arrangements of one supplying country may be inappropriate to those of another. The country that receives technology from both and that, for reasons of simplicity and equity, wishes to adopt a uniform treatment of TT payments to both is then in an impossible quandary. A further complication is that the objectives of technology control vary between countries, and may be in conflict.

OBJECTIVES OF TECHNOLOGY CONTROL

There are four broad areas of technology control, each of which may be affected by the tax system: the quantity of TT, its terms, its form, and the characteristics or quality of the technology transferred. There are potential interactions between the four, and there may be potential contradictions in objectives in relation to them. Taxes designed to secure one set of objectives may then impede others. It is necessary to consider objectives in relation to the four categories to provide a context in which the use of the tax system for TT may be discussed.

Quantity. There is no consensus on objectives in relation to quantity. On the one hand, those who believe that industrialization is necessary to development, and that imported technology is essential for industrialization, would aim to maximize the transfer of technology from abroad, if this does not involve too great a sacrifice of other objectives. This, broadly, is the view of those who emphasize the advantages of being a latecomer, notably Gerschenkron. This view was held by many less developed countries when they acquired political independence: industrialization based on imported technology became a prime aim. But partly as a reaction to the consequences of this strategy, the view has been challenged from two angles. The dependency school believes that technological dependence is of fundamental importance to the general dependent relationship between the center and the periphery; to reduce dependency, delinking between North and South is required and a significant element of this delinking consists of technological delinking. A second challenge comes from the alternative technology school, encompassing Schumacher and Illych, which argues that the process of development based on imported advanced country technology has numerous undesirable consequences, and that an alternative technology is required. While in practice both the dependency and the alternative technology approaches agree on the need to slow down the import of technology from overseas, there are significant theoretical differences between them that are of relevance to technology policy. The dependency school emphasizes above all the need to delink from the center, but does not object to the characteristics of the imported technology, while the alternative technology school emphasizes above all the undesirable characteristics of the imported technology and would not object to continued import of technology with the desired characteristics.

The terms of TT. The terms of TT for most Third World countries consist of the price of the imported technology. (It is worth noting that Third World countries differ in this respect from developed countries that enjoy a two-way trade in technology, for whom the relevant price is the ratio of the price of technology acquired to that sold.) The price is not simply the overt payments for royalties, technical services, licenses, and so on. Intrafirm payments for technology may appear as royalty payments or as profits remitted, or as inflated transfer prices, or as management fees. Moreover, restrictions on activities imposed as part of the process of TT – for example, on exporting to third parties – are also part of the price of technology imported. Similar disguised payments and restrictions may be imposed on joint ventures or on local firms. For both these categories the price charged for the imported capital goods, or for parts or for managements services, may include an element of payment for technology. (Kopits (1976b) has shown that the breakdown of intrafirm payments between royalties and transfer prices can be explained empirically with reference to tax incentives for remittances to appear as different types of payment. Work by Vaitsos, Chudnovsky, and Lall, among others, has indicated the substantial magnitude that such disguised payments may take. This question of disguised payments – which for brevity we may describe as the transfer price problem, but which in fact encompasses much more, since it includes inflated capital goods prices for "arm's length" transactions, and the pricing of management services and so on – raises some difficult and interesting questions. First, what is the "proper" breakdown between the different types of payment? It is arguable that there is no "proper" breakdown because of the problem of the unallocable overheads involved, and because of the large monopolistic elements involved in the market for technology. The knowledge market has to be imperfect if it is to remain a market. But if it is imperfect and if there is no reasonable way of allocating costs of production of the knowledge, then any price is somewhat arbitrary. Moreover, the technology transfer package is often associated with other services that are also monopolistically controlled, such as access to capital, to markets, and so on. Hence it would not be justified to attribute all excessive (above-normal, if these could be identified) profits to technology, because these other elements are also responsible. Yet while these other elements have some independent influence they may also be derived from technological factors. It is thus virtually impossible for LDCs to separate out the technology payments from other payments. This suggests that from a tax point of view they should be treated uniformly.

If, as just argued, there are strong monopolistic elements in the market for technology, then it might be possible to improve the terms, from an LDC point of view, by taking a stronger bargaining stance. The Andean Pact and UNCTAD have concentrated on improving the terms in this way, and some notable improvements have been secured, suggesting that the market is indeed monopolistic in this way. The terms on which an LDC enterprise acquires technology consists of the price paid for it, including all the disguised payments just referred to. But the terms on which the LDC as a country acquires the technology consists of the price paid for the technology less the tax paid to the LDC's treasury. A third price is the price received by the technology suppliers, which consists of the net price (less LDC tax) paid by the LDC, adjusted for taxation in the (developed) country of origin. Thus there are three different, and relevant, prices of technology with tax wedges between them. Tax policy may alter the terms by altering the size and nature of the wedges. It will be helpful to the later argument to set this out algebraically.

Let Pe be the price the LDC enterprise pays for the imported technology.

$$Pe = r + m' + p' + k',$$

where r = royalty payments; m' = payments for management fees over and above the competitive price; p' = payments for technology 'disguised' as profits; k' = payments for technology disguised as inflated input prices. (As noted above, it is difficult both theoretically and practically to know how far some of these payments are due to technology, and how much to other things, but this does not matter very much for this part of the discussion.)

The price of the technology paid by the LDC country, $P1 = Pe - T1$,

where T1 is the taxes paid to the LDC on the payments for technology

$$T1 = t^{r} \cdot r + t^{m} \cdot m' + t^{p} \cdot p' = t^{k} \cdot k'$$

where t^r is the tax paid on royalties (normally a withholding tax); t^m is the tax on management services from abroad (normally a withholding tax)

t^p is the tax on profits made by foreign owned enterprises (normally a corporation tax plus a withholding tax on profits remitted overseas);

t^k is a tax on inputs from abroad (normally a tariff).

The terms on which an LDC country acquires technology may therefore be altered by altering T1 for any given gross price, and this may occur through changes in tax rates, and/or changes in the weight of different types of payment for technology where the different types of payment are subject to different tax rates (that is, changing the ratio of r:m':p':k'), where

$$t^r + t^m + t^p + t^k$$

Let Ps be the price received by the technology supplier:

$Ps = P1 - (T1 + Ts)$, where Ts = tax paid in the source developed country. In many cases Ts is related to T1 through double tax agreements or unilateral relief given by the treasury of the country of origin for taxes paid abroad. The precise relationship is complex, varying from country to country and according to the type of payment. For example, where a developed technology-supplying country gives tax relief on a tax credit basis and where the LDC tax rate is no higher than the developed country tax rate, changes in T1 due to changes in corporation tax will be offset by changes in Ts in the opposite direction. In such cases the price received by the technology supplier will not be affected by changes in the taxes paid in the LDC. But for other payments – tariffs, for example – no relief is given. And in countries where tax relief for taxes paid abroad is provided on an exemption basis, Ts will be zero irrespective of Tl, (for those taxes for which complete exemption is given).

The relationship between Tl and Ts is of significance in two respects. First, the incentive or disincentive effect on technology suppliers of changes in the terms of transfer depends on how Ps changes, not how Pe or Pl changes. Yet many LDC policies are directed at Pe or Pl. Second, tax receipts by the LDC, and consequently the net cost to the country for any gross price of technology, depends on the weight of different types of payment in the total and the tax rates applicable to the different types. Given that foreign technology suppliers have considerable freedom to choose the form in which payment appears, then the incentive to the supplier to show technology payments in a particular form will be of significance to the terms of transfer. These incentives depend not only on the different tax rates on the different payments in the LDC, but also on how the taxes paid in the LDC affect taxes due in other countries, because the net incentive consists in the relative tax due on different types of payment in all countries.

Assume that a technology supplier has complete freedom to choose between receiving payments as royalties, as profits, or as inflated input prices, and assume that net tax due is the only consideration. Then to minimize tax due the firm will choose according to whether

$$\left\{t_1^r + t_s^n\right\} \gtrless \left\{t_1^p + t_s^p\right\} \gtrless \left\{t_1^k + t_s^k\right\}$$

where subscript l indicate payments in the LDC, subscript s tax payments in the technology-supplying country. In practice other considerations obviously also enter, while firms have some but not unlimited freedom to choose – as is shown by the fact that most technology suppliers receive payment in more than one form.

In considering LDCs' objectives in relation to the terms of transfer, it is clear that a major aim is to improve the terms on which the country acquires the technology. However, there might be some conflict between this aim and the quantity objective insofar as (1) the countries concerned wish to increase the quantity transferred; and (2) the terms received by the technology suppliers are relevant to the quantity transferred (to be discussed later). Such conflicts are likely to be greatest if the improvement in terms occurs through changes in gross price or changes in net price with no corresponding changes in foreign taxes due; least (and sometimes nonexistent) if the changes can be secured by increasing local taxes in such a way that foreign taxes fall by a corresponding amount.

The form of TT. Technology may be transferred in various forms, ranging from direct overseas investment by multinational corporations to the sale of machinery to local firms. LDC policy objectives toward the form of transfer also vary. Some countries welcome private overseas investment for bringing management, capital, markets, and marketing skills as well as technology. Others resent the foreign ownership and control involved and insist on some degree of local participation. There are all sorts of direct fiats by which the desired degree of local participation may be acquired, for example, by insisting on at least 51 percent local ownership. Exchange control regulations can also act as a powerful incentive. The tax system may be used to provide an incentive to technology suppliers to provide technology in one way rather than another.(3) For example, licensing becomes more attractive than equity participation if the tax system (taking into account the home country tax as well as that of the host country) is relatively harsh in its treatment of profits and profit remittances as compared to its taxation of license fees. Tax holidays related to profits are likely to encourage transfer of technology via foreign direct investment, as compared with licensing.

Characteristics of TT. It is now generally acknowledged that foreign technology has tended to be over-capital-intensive, and large scale in relation to that appropriate to LDCs; that imported inputs have been used too much and local linkages have been too few; and that the products produced with the imported technology have been over-sophisticated and designed for high-income consumers (see Stewart, 1977). One objective of technology control, then, is to secure a more appropriate technology. This objective involves (1) a more appropriate choice from within the available range; and (2) more R & D and adaption related to the needs and conditions of LDCs. The tax system may affect both by providing an appropriate set of incentives. The characteristics objective may be in conflict with some of the earlier objectives. Increased quantity may not be consistent with improved characteristics, which is why the intermediate technology school opposes increased quantity; much depends on the type of technology involved. A further conflict would arise if improved terms of trade made inappropriate technology cheaper and therefore encouraged its use.

Another (related) objective vis-a-vis technology is that of promoting technological independence for Third World countries. This may be wanted in itself, as a means of acquiring general economic independence, in order to improve the bargaining position of the country and therefore the terms of transfer, or to secure more appropriate technology. Policies toward TT may aid the independence objective by discouraging transfer of technology and protecting local technology, and by being selective about technology transfer, importing only technology that is likely to help and not compete with local technological development. However, the latter is probably better secured by administrative means rather than through the tax system.

Again, there may be some conflicts between objectives. Particular policies may promote some and hinder others. Consideration of tax policy therefore needs to consider the effects of policies on each of the objectives.

THE EXISTING TAX SYSTEM

Countries differ in their definition of different types of income, in tax rates, in treatment of depreciation, in incentives provided, in the way in which they deal with the "transfer price" problem, and in their treatment of income earned overseas. Moreover, there are large numbers of bilateral tax treaties, particularly between developed countries, that affect the way income flows between countries are treated.(4) It would thus be a mammoth task to describe all the relevant parts of all countries' tax systems. This section attempts to provide a very brief overview of the salient features in LDCs and DCs, in order to be able to highlight the main ways in which the systems might be improved so as to assist in technology control.(5)

LDC Tax Arrangements

Most LDCs impose a corporation tax on the profits of subsidiaries of multinational corporations.(6) Rates vary typically between 30 and 50 percent (see table 6.1), although tax haven countries (such as the Bahamas and Bermudas, Panama, and Liberia) levy very low or zero taxes. Generally, LDCs impose a uniform tax rate on all profits, retained or distributed. Profits remitted are subject to a withholding tax; again, rates vary, as table 6.1 shows. In most countries interest and royalty payments are deducted from income before calculating profits liable to corporation tax, and these payments are then subject to withholding tax on remittance. Some countries (this is a provision of the Andean Pact, Decision 24) disallow deductions for intrafirm transactions, regarding these as a means of tax avoidance. Others limit the amount deductible. Table 6.2 illustrates provisions in a few countries. Rates of withholding taxes on royalties vary greatly between countries, some levying very low (or zero) rates, and others rates as great, or greater, than the rate of corporation tax. In 1968 fifteen countries taxed royalties as highly as corporate income, while twenty-eight countries levied withholding taxes below the corporation tax rate. In twenty countries the difference was more than 20 percent. From a regional point of view it appears that withholding tax rates tend to be highest in Latin American countries and lowest in African countries.(7) Table6.3 provides some more up-to-date figures for a few countries. It should be noted that there is a significant difference between allowing deductibility and levying a withholding tax equivalent to the rate of corporation tax, on the one hand, and disallowing deductibility and treating royalties as corporate profits, on the other. In the former case the payments are solely subject to the withholding tax; in the latter they are subject to corporation tax plus the withholding tax – typically 15 to 20 percent but sometimes much higher – on remitted dividends. Depreciation allowances are deductible before calculating taxable profits. Varying conventions determine the depreciation allowances. In some cases generous allowances form part of the tax incentive for technology transfer.(8)

Table 6.1. Rates of Tax on Company Income

	Company Tax Rate	Withholding Tax on		
		Dividends	Interest[a]	Royalties[a]
Industrial Countries (1968)				
Austria	44	5	0	0
Belgium	38.5	15	15	5
Canada	50	15	15	15
Denmark	36	5	0	0
France	50	5	15	0
Germany	52.5	15	0	0
Italy	43	5	38	0
Japan	35	10	10	10
Luxembourg	40	5	0	0
Netherlands	46	5	0	0
Norway	30	10	0	0
Sweden	40	5	5	0
Switzerland	7.2	15	0	0
U.K.	45	15	0	0
U.S.	48	15	0	0
Developing Countries (1970s)				
Argentina	33(45)	29.5	39.7	39.7
Brazil	30(47.5)	25	25	25
Chile	52	40	37.5	37.5
Colombia	40	20	20	20
Costa Rica	20-45	15	15	20
Dom. Rep.	10-38	18	18	18
Ecuador	33.3(11.1)	44.4	40	40
El Salvador	2.5-15(38)	28	38	38
Guatemala	5-48	12	10	10
Honduras	5-40	15	5	10
Jamaica	45-50	37.5[a]	0	0
Mexico	5-42	15-20	35	42
Neth. Ant.	34[a]	0[a]	0	0
Nicaragua	6-40	0	30	30
Panama	20-50	10	45	22.5
Peru	20-55	30-40	37	37
Puerto Rico	24-47	25	29	29
Trin. & Tob.	50	25-30	30	30
Uruguay	25	20	35	40
Venezuela	15-50	15	50	50
Iraq	15-50	0	0	0
Israel	47[a]	30[a]	47	47
Kuwait	0-50	0	0	0
Lebanon	30	10[a]	10	10
Saudi Arabia	1-5(25-45)	0	0	0

Table 6.1. Rates of Tax on Company Income (Cont.)

	Company Tax Rate	Withholding Tax on		
		Dividends	Interest[a]	Royalties[a]
Taiwan	0-35	35	15	15
Hong Kong	17	0	0	0
Indonesia	20-45	20	20	20
S. Korea	20-40	20	5	5
Malaysia	45	40[a]	40	40
Pakistan	60	13.8[a]	60	0
Philippines	25-40	15	35	35
Singapore	40	40[a]	40	40
Sri Lanka	60(66)	33.3	39.3	39.3
Thailand	20-30	15	25	25
Algeria	60	18	18	18
Kenya	45(52.5)	12.5	12.5	20
Liberia	10-45	15	15	15
Morocco	40-48	15[a]	0	0
Nigeria	40-45	15[a]	20	0
Tanzania	50(57.5)	15	12.5	20
Zambia	45	20	0	0

[a]1968

Note: Figures in parenthesis apply to nonresidential Companies.

Source: 1968 figures Kopits (1976a)
1970s figures Lent (1978)

Tax incentives. In order to attract foreign investment most LDCs provide incentives of various kinds. Apart from tax concessions, these include guarantees about convertibility and exchange rates, tariff protection for markets, and tariff remittances on inputs. Tax incentives include tax holidays, investment allowances and investment grants, decelerated depreciation, loss carry-over provisions, and exemption from stamp taxes, sales taxes, and so on.(9) Some tax concessions are given on a discretionary basis to particular companies and are subject to negotiations about the nature of the investment between the company and the host government. Others are given on a nondiscretionary basis to all firms that fall into specified categories. "Pioneer" status, which normally involves tax holidays and often other concessions, is conferred on a discretionary basis in some countries, a nondiscretionary basis in others. The most widely given direct tax concession is a tax holiday for five-plus years. In Reuber's (1973)

Table 6.2. Deductibility of Royalty Payments in Selected Countries When Paid to Nonresidents

Deductibility Denied	Deductibility Limited	Deductibility Limited and Depends on Government Approval of Transfer Agreement
Brazil-if nonresident is controlling corporation or interest	Brazil-limited to 5 percent of sales if not paid to a principal shareholder	Mexico-limited to 1.5 to 3.0 percent of net sales
Argentina-if nonresident is a related party	Argentina-limited to an "arm's length" amount if not paid to a related party	India-limited to 5 percent of sales if not paid for the acquisition of capital rights
	Philippines-deductible if not excessive	Guatemala-limited to 15 percent of gross sales
	Ghana-limited to a fair and reasonable amount	

Source: Carlson and Hufbauer (1976).

Table 6.3. Withholding Tax Treatment of Royalties Arising in Selected Developing Countries and Paid to Nonresidents not Maintaining a Permanent Establishment in the Developing Country

Country	Tax on Gross Royalty (percent)	Tax on Gross Royalty Less Expense Deduction (percent)
Brazil	25.0	
Chile	37.5	
India[1]	40.0	
Israel[2]	25.0	
Mexico[3]	5.0 to 42.0	
Pakistan	60.0	
Peru	37.0	
Philippines	35.0	
Argentina		41.0, with expense deduction of 50 percent of royalty
Ghana		50.0, with deduction for reasonable expenses
Panama		45.0, with expense deduction of 50 percent of royalty
Uruguay		38.0 rate reduced to 19.0 percent if expenses incurred directly associated with royalty
Venezuela		15.0 to 50.0, with expense deduction of 20.0 percent of royalty

[1]India taxes royalties paid to foreign companies under approved agreements made after March 31, 1975.
[2]A recently signed treaty between the U.S. and Israel provides for a rate of 15 percent on industrial royalties. The treaty has not yet been ratified.
[3]Plus a sales tax of 4 percent of gross royalty.

Source: Carlson and Hufbauer (1976).

survey, nearly half the projects undertaken by firms from OECD countries in LDCs were given tax holidays the mean length of which was five years in all areas apart from Latin America, where it was seven. All thirteen countries surveyed by Lent (1967) provided tax holidays ranging from two to twenty-five years. The report of the U.N. tax experts noted a wide variety of tax holidays.(10) They reported, "In general, the trend among developing countries is towards increasingly flexible or 'tailored' tax holiday incentive schemes. Another tendency is towards increasing regional uniformity as illustrated by common incentive schemes in effect or under development in Central America, the Andean countries of South America, the Caribbean Free Trade Area and the East African Community" (p. 100). However, Shoup (1976, annex c) notes a lack of progress in limiting tax-incentive competition within regional groupings.

Two other important types of tax incentives are generous tax treatment of investment, and special tariff arrangements. The former may take the form of accelerated depreciation and/or of the provision of investment allowances or grants, permitting more than 100 percent write-off of the cost of capital equipment. In Reuber's sample more than a quarter of the cases received accelerated depreciation allowances. Lent reports investment allowances in Jamaica, India, Kenya and Uganda, Malawi and Turkey, and grants in Morocco.(11) Tax incentives related to investment expenditure – whether in the form of grants or allowances – are normally given in some proportion to investment expenditure. In a few countries, grants are also related to the employment generated by the investment; for example, in Puerto Rico, Carne (1968) reported, grants of 6 to 15 percent of fixed investment were given depending on the region and number of employees, plus $250 per employee. (In some cases, tax holiday provisions are also related to employment and to the regional contribution of the investment.)

Special tariff arrangements may take the form of protection against imports in the home market and tariff reductions of inputs, such as imported capital equipment and raw materials. Both are widely used. In Reuber's sample (1973, tariffs were imposed or already in existence in about 45 percent of the cases, and in a similar share quotas were imposed or in use. The average level of the tariffs was 68 percent. In over half the cases concessions were provided on imported capital equipment and the concessions averaged 40 percent. In over a third of the cases import concessions were offered on imported parts (from an average of 50 percent to roughly zero), and on raw materials (from 38 to 5 percent). Duty remissions were also offered on exports in a third of the cases for which data was provided. While the precise figures vary, Reuber's evidence appears to be similar to that established by other observers, for example, Lent, who argues that "Remission of import duties is generally thought to be of greater potential value to most manufacturing businesses than income tax concessions".

<u>The transfer price problem</u>. Insofar as LDCs make a serious attempt to deal with this problem, and to establish that MNCs' transactions occur

at a "fair" price for tax purposes, the effort is generally made on a transaction-by-transaction basis. According to replies to a questionnaire to the U.N. tax experts (Sixth Report), special regulations to deal with transfer prices are in being in Ghana, India, Pakistan, the Philippines, and Sri Lanka, with India having the most detailed legislation concerning reapportionment of profits. The transaction-by-transaction assessment is in general conducted on principles similar to those used by the United States, namely the "arm's length" method, based on comparable uncontrolled transactions between third parties, cost plus or resale price. Ghana adopts a general apportionment method in which profits declared for Ghanaian tax are required to be the same ratio to total group profit as the Ghana turnover bears to group turnover.

Developed Country Tax Arrangements

As I have already suggested, developed countries' treatment of income flowing from LDCs for TT is of critical importance in establishing the net effect of any LDC tax provisions. Income that passes through more than one tax authority is potentially subject to more than one lot of taxation. The principle of avoiding double taxation – which might be said to form part of the religion of taxation – has led to unilateral tax relief and to bilateral tax treaties with the main purpose of avoiding double taxation. Bilateral tax treaties between developed and developing countries remain somewhat rare; the main form of relief from double taxation occurs through the unilateral relief system adopted by the developed countries. Broadly, DCs have adopted two different types of unilateral relief: the exemption method and the credit method. (Table 6.4 lists the countries adopting the different methods.) According to the exemption method, income that has already been taxed overseas is not subject to home-country corporation tax. But no exemption is given, unilaterally, in relation to withholding taxes on dividends. According to the credit system, taxes paid abroad on subsidiary profits are credited against liability to home-country corporation tax. Credit is given for withholding taxes, and, in most cases for "underlying tax," that is, for corporation tax.

As far as corporation tax is concerned, the difference between the two systems of relief depends on the rates of tax in the countries concerned. With the credit method the MNC pays the higher of the two tax rates. With the exemption method, the MNC pays the host-country (that is, LDC) rate. Thus if the LDC tax rate is higher than the DC tax rate, the tax paid should be the same in both systems. However, in those circumstances there would be a major incentive for the corporation to redirect its profits to some third low-tax (tax haven) country. With the exemption method of relief there is always an incentive to "earn" profits in tax havens; with the credit method there is an incentive if the tax due is greater in the LDC than the DC.(12) Deferral advantages may also be obtained by the use of tax havens, irrespective of the relative rate of tax in DC and LDC.

The position is complicated, as far as the credit method is concerned, by the form of corporation tax system in use, and how the credit system is operated. For example, the two-tier or imputation corporation tax systems may effectively provide less credit than the classical uniform rate system.(13) Whether or not companies are permitted to offset one type of tax payment against another (for instance, withholding tax on royalties against corporation tax) and whether credits are calculated on a per source, per country or an overall basis(14) also helps determine the net effects of tax changes in any one country on the overall tax position of the MNC, and consequently on its incentives.

Table 6.4. System of Unilateral Relief Applied to Dividend Income Received by Parent Company from Foreign Subsidiary

Country	System
Australia	Credit
Belgium	Credit
Brazil	Exemption
Canada	Credit
France	Exemption
Holland	Exemption
Italy	Credit
Japan	Credit
U.K.	Credit
U.S.A.	Credit
W. Germany	Credit

Source: Adams and Whalley, 1977, p. 59.

A third system of relief has been suggested,(15) the deduction method. This would permit companies to deduct foreign taxes as costs in computing taxable income. Profits after payment of foreign taxes would be subject to source-country taxation.

Royalties. In the absence of double tax treaties, royalties received within DCs are subject to tax, but withholding taxes are normally credited against tax liability.

Transfer pricing. Most DCs have some sort of regulations to prevent tax avoidance through the use of transfer pricing. But they appear to differ in the extent and toughness with which these are enforced. It seems to be agreed that the United States is toughest, applying "arm's length" regulations as is possible and appropriate in the particular case, on a transaction-by-transaction basis. Tough procedures adopted by DCs can have important implications for LDCs, if they avoid the use of tax havens for declaring profits en route from LDCs to DCs. By preventing this, the DC authority may in certain cases increase taxable profits and revenue in the source LDC. It has been suggested that DCs should introduce an accrual basis of taxation of subsidiaries which would (if administered effectively) eliminate the incentive to declare profits in low-tax countries. Such a system, unless accompanied by tax-sparing agreements, would effectively also eliminate much of the effect of LDC tax incentives.

Effects of DC Tax System on Incentives for TT to LDCs

The effects depend on (a) the system of unilateral double tax relief adopted; (b) the burden of taxes in the DC relative to those in LDCs and how the system of credit is calculated; (c) tax allowances in the DC for investment in LDCs; and (d) bilateral tax agreements.

With the exemption system of relief, any tax incentives provided by LDCs increase the after-tax profits of the companies. Where the tax burden in the LDC is notably higher than in the country to which the profits are to be remitted, then tax reliefs down to the point – but not below – at which the taxes due are the same in both countries will increase companies' after-tax profits. However, below that point, and for reliefs in any LDC where the tax burden is less than in the relevant DC, tax holidays/incentives will ultimately increase the tax revenue of the DC, rather than the after-tax profits of the LDC. Most LDCs fall into this category, since typically their tax rates are below those in DCs. However, insofar as DC taxation occurs on a remittance and not an accrual basis, tax holidays and other allowances will benefit the companies to the extent of the deferral in the time at which tax becomes due. But the greater beneficiary will be the governments of the DCs. Branches are subject to tax on an accrual basis, and under various regulations, especially for U.S. companies, so are some subsidiaries.

Some countries, notably Japan, have "tax-sparing" provisions in bilateral tax agreements with LDCs. These agreements work in such a way that the tax spared by various incentive schemes counts as if it had been paid for purposes of calculating tax credit. The tax-sparing provision means that even with the credit system of taxation, tax incentives provided by LDCs benefit the companies and not the governments of DCs. While the U.K. has a few such agreements, the United States has resisted entering into them.

Tariff relief by LDCs (and/or tariff protection) increases the before- and after-tax profits of subsidiaries and branches. Since tariffs are not counted as creditable taxes, these reliefs do increase after-tax profits of the companies, irrespective of the country to which the profits are remitted.

For the most part, countries provide investment allowances exclusively for investment in their own country. But West Germany and Japan also give allowances on various types of investments in LDCs. Whether these allowances are effective in terms of increasing after-tax profits of companies who qualify for them depends on whether any DC taxes are due (against which the allowances may be claimed) after credit is allowed for LDC tax. For high-tax LDC countries no credit may be due;(16) for low-tax countries – or those giving generous investment incentives – the allowances may increase after-tax profits. By reducing the liability to tax in DCs, such allowances are likely to increase the net effect of tax reliefs in LDCs in terms of companies' after-tax profits.

Table 6.5 summarizes the effects of tax incentives given by LDCs on company after-tax profits and on DC revenue under various types of tax arrangements. Viewed from the point of view of the tax reliefs given by LDCs that effectively increase the after-tax profits of companies, the worst situation is where the LDC tax benefit simply and immediately increases the DC tax revenue; this only occurs in cases where taxation is levied by the DC on an accrual basis, and where the LDC tax is below that of the DC. This would be the situation for various types of branch investment. The second worst situation is where the LDC tax is below the DC tax, where DC taxes are levied on remittances and relief is given on a credit basis, and no tax-sparing agreements are in force. This is likely to be the situation for much U.S. and some U.K. overseas investment, which, as table 6.6 shows, together accounted for two-thirds of total DC investment in LDCs in 1971.

Table 6.5. Effect of LDC Tax Incentives in Various Situations

DC tax relief	Exemption		Credit/ Remittance		Credit/ Remittance, Tax Sparing		Credit/ Accrual		Deduction	
LDC	DC-T	P	DC-T	P	DC-T	P	DC-T	P	DC-T	P
1. tax rate < DC tax	0	+	+	i^d	0	+	+	0	+	+
2. tax rate > DC tax	0	+	0	+	0	+	0	+	+	+

Key: DC-T = Developed country tax revenue
P_d = After-tax profits of company
i^d = Interest gain on deferral of tax due

Table 6.6. Foreign Private Investment in LDCs by Country (percent)

	1971	1967
U.S.	52	55
U.K.	15	16
France	5	6
W. Germany	3-4	3
Switzerland	3-4	4
Canada	3-4	3
Japan	3-4	1

Note: Total book value of foreign private investment, as estimated by the U.N. Department of Economic and Social Affairs.

Source: Lall and Streeten, 1977, p. 6.

ASSESSMENT OF EXISTING ARRANGEMENTS IN RELATION TO TT OBJECTIVES

The taxation of income arising from international technology transfer may be assessed from many point of view – from that of maximizing "world" income and output, from that of equity(17) or of tax neutrality and efficiency, from the point of view of maximizing national welfare of the various countries involved, and so on. Moreover, the power to effect changes and therefore the changes appropriate vary according to which governments are involved. To further complicate matters, there is an element of gamesmanship in the question, since tax changes by one authority may induce subsequent and offsetting changes by other authorities.(18)

This chapter is concerned with assessing the system and exploring changes from the point of view of how LDCs can better secure their objectives (as discussed in the first section) in relation to TT. Questions of maximizing world output,(19) national and international equity and so on, are therefore not considered in themselves, but only if instrumental to the objectives of control of TT.

Incentives and the Quantity of TT

A major aspect of the tax system, as described above, is the use of various tax incentives by LDCs aimed at increasing the quantity of technology transferred. Empirical investigations into the effects of these incentives have shown that they have been relatively ineffective, while they have been costly both in terms of revenue foregone and in implications for the characteristics of TT.

Investigations in Nigeria (see May, 1965; Hakam, 1966; and Phillips, 1969) suggested that tax incentives were of significance only in a minority of cases. In Hakam's inquiry, 16 percent of firms mentioned them as a factor in locating in Nigeria. In another it was mentioned by one-third of the companies, but ranked first by only two companies, as table 6.7 shows.

Table 6.7. Nigeria: Rankin in Determining Investment Decisions (number of times mentioned)

	1st	2nd	3rd	4th	5th
Market conditions	27	14	-	-	-
Government attitudes	12	5	12	6	6
Tax incentives	2	17	18	4	-
Raw materials	-	3	7	20	11
Cheap labor	-	2	4	11	24

Source: Phillips, 1969.

Hughes and Seng (1969) come to even stronger conclusions with respect to foreign investment in Singapore: "The most striking conclusion of the survey was that the foreign investors almost without exception, stated that taxation concessions played no role at all in bringing them to Singapore" (my underlining). Hinricks (1974) found little relationship between tax incentives and new investment in Guatemala. King (1970) concludes, "In summary, neither of the tax exemption schemes that Mexico has practised seems to have been economically very significant or a necessary part of her industrial growth." For Jamaica, Chen-Young (1967) concludes, "This direct contribution to the manufacturing sector in terms of investment, salaries and wages, employment and value added have not been particularly significant." In a study of Argentina (Herschel, 1965), tax exemptions were cited as a factor influencing investment decisions in only 6 percent of the cases.

Reuber's survey (1973) suggested that financial incentives "have had some effect on decisions about where to locate projects among the LDCs. The most important of these seems to be tariffs and quotas on competing imports, concessions on imports of inputs, and tax concessions." Financial incentives were of greatest significance for export oriented projects. One survey(20) of twenty LDC governments and a variety of investors showed that tax concessions were viewed as the most significant influence over investment decisions, according to the responses of the host governments, and were not mentioned at all as a significant factor by the investing firms. The costs of the incentives in terms of foregone revenue vary. Lent (1967) found costs ranging from 2 to 13 percent of tax revenue.

For a variety of reasons it is not surprising that the tax incentives in LDCs(21) have been relatively ineffective. First, the DC tax system is such that tax holidays/incentives in LDCs may not reduce the total tax payable by MNCs, but rather may increase the share of tax received by the DC tax authorities, although they may also provide some deferral advantages.(22) Second, the complexity and uncertainty of the incentives reduce their effectiveness as does high risk in the market environment (Mellors, 1973). Third, the motivation and behavior of MNCs may be such as to give minor significance to tax concessions in determination of behavior. A priori, the same incentives can be shown to lead to very different effects according to the model of investment behavior adopted.(23) In the context of LDCs it would appear that the preservation (and extension) of market shares may be a predominant motive, so that after-tax profits become a constraint rather than an objective.(24) Fourth, multinational corporations' ability to manipulate transfer pricing and so on means that tax rates on declared profits are not of great significance since, to an extent, companies can choose the level of their declared profits.(25) Fifth, tax competition between countries has eroded the relative advantage offered by the concessions of any one country. If one country on its own offered substantial concessions it might succeed in attracting investment away from other similar countries, but if similar countries all offer incentives there will be little net effect on the total flow, which will depend on other factors. The competitive offering of tax concessions, although notoriously a feature of LDCs, is not confined to them. DCs also offer increasingly attractive tax allowances (and/or reduced rates of tax) in the hope of stimulating investment. Individually, countries find themselves in a situation of oligopoly with a kinked supply curve. If one country on its own abandons concessions, it risks losing most of its new investment to other countries who maintain their concessions. If it increases concessions, other countries will follow. Individual countries then find it necessary to match the concessions of others, fearing that otherwise they will lose their share of the total flow, but taken together the countries as a group lose revenue with little gain in quantity of TT. The main effect of the concessions is loss of revenue to the LDC governments, the gains from which are shared between the companies and DC governments according to the DC tax system in force.

The surveys just described, indicating the relative ineffectiveness of tax incentives in altering the quantity of TT, have been primarily concerned with the various concessions related to taxation of corporate income. It seems to be agreed that other incentives – in particular, protection of the domestic market, concessions on tariffs of imported inputs, and provision of infrastructure – are more cost-effective (see, for example, Reuber, 1973, and Lent, 1967).

Incentives and the Quality/Characteristics of TT

Tax incentives. The incentives related to the quantity of investment tend to encourage capital intensity of investment and thus be in conflict with desired characteristics. Modifications of these allowances so that they are related to employment might therefore be preferable. Tax holiday incentives have less distorting effects, although insofar as their value is related to the profitability of the project and total profitability is related to capital intensity then, relatively, they are greater for capital-intensive projects. They (along with accelerated depreciation) tend to favor shortlived projects.(26) Reinvestment of earnings is more likely to be encouraged by investment allowances than by tax holidays.

Tariff provisions. Heavy tariff protection of domestic markets tends to encourage an inefficient pattern of production and also capital intensity of product and technique (see, for example, Little, Scitovsky, and Scott, 1970). Tariff concessions on imported inputs, particularly on capital equipment, encourage capital intensity and also minimization of local value added.

The incentives provided for the import of technology, through the tax system and in other ways, have tended to inhibit the development and use of local technology, with consequences for the characteristics of technology available as well as for technological dependence. The characteristics of imported technology tend to be inappropriate to LDCs and, rather, in line with the resource availability and income levels of the advanced countries.

Taxation and the Form of TT

Some incentives are neutral between different forms of technology transfer. However, many of them appear to be designed to encourage an inflow of foreign direct investment, and thus favor that form. For example, some pioneer incentives are only effectively available for foreign enterprises. But where corporate taxation is heavy or where there are restrictions on remittances, TT via licensing to third parties is encouraged. Chown (1974) argues that the tax system may provide an inducement to joint ventures and/or to licensing arrangements in certain circumstances. For example, "an American parent may be better off having a 49.9 percent of a venture with an effective tax shelter than having 70 percent without a tax shelter" (p. 76). (This is to avoid U.S. legislation designed to prevent the use of tax havens as tax avoidance devices.) "Where the ultimate ownership is in, or can be transferred to, a low tax jurisdiction, licensing will usually result in a substantially lower global tax charge than a joint venture" (p. 81). "In a licensing agreement with an arms length licensee, the licensor's share may often be extracted at considerably lower tax charge"(p. 69); this is in circumstances where royalties in intrafirm transactions are nondeductible, but royalties at "arm's length" are subject to low withholding taxes.

Tax Arrangements and the Terms of TT

In the discussion of objectives I showed that the terms on which an LDC acquires technology are the gross price for the various services less the taxes paid in the LDC. The terms may be improved, therefore, by reducing the gross price or by increasing the taxes paid. Much work has concentrated on improved bargaining and hence improving the gross price; this chapter examines the possibility of changing the net price by increasing the tax. Whether or not this will cause a reduction in the total flow depends (1) on whether the market for technology is competitive or not. If the price is oligopolistically determined – the outcome of a process of bargaining – then some reduction in price may be obtained without reducing the quantity. This is the view on which the improved bargaining strategy is based. It also depends (2) on whether a change in the LDC tax results in a corresponding change in DC taxation and no change in after-tax profits, or whether it leads to a corresponding change in after-tax profits, which depends on the DC tax system. As I have already argued, for much U.S. overseas investment, which forms the bulk of the total flow, any change in LDC taxation is likely at least partially to be offset by a change in DC taxation and not to affect company after-tax profits proportionately.

For any gross price of technology – which we will take as given in total, although the type (that is, whether it appears as a royalty payment or as high input prices, or as declared profits) may vary – the total LDC tax revenue is the combined tax on all the types of payment for TT. Simplifying to three types (and ignoring underinvoicing of exports and management fees), from the first section,

$$T1 = t^{r} \cdot r + t^{p} \cdot p' + t^{k} \cdot k'$$

and may be altered by changing effective tax rates (through changes in actual rates, or in allowances) and changing the ratio in which the types of technology payments are made.

Forgetting about foreign taxes, for the moment, the existing tax arrangements affect the terms (1) through tax holidays, and so on, that reduce the effective rate of corporation tax; and (2) through the possibility of manipulating the type of payment – which arises because different types of payment are subject to different effective tax rates, and because the administration is unable to enforce effectively a fair "arm's length" form of payment. Both factors are present in the existing tax arrangements of many countries.(27) It is apparent from empirical investigation that a substantial amount of price manipulation takes place. Part of this manipulation is a response to nontax factors,(28) such as restrictions on profits or remittances, or desire to show "normal" profits for the sake of good relations with the country and with trade unions. However, part is due to the big discrepancy between taxation of different items, which then provides an incentive for making payments appear in one form or another.

Generally the tax rate on declared profits – which includes both corporation tax and withholding tax on dividends – is greater than the withholding tax on royalties. Whether it is greater or less than the tariff on imported inputs depends on their level, but it is likely to be substantially greater where concessions are given on tariffs on these imported inputs. It is likely, then, that there will be an incentive in many cases to disguise profit payments as royalty payments and/or inflated prices for imported inputs. The incentive to do so is likely to be greater where low or zero tariffs are charged on imported inputs. The incentive is likely to be less the lower the effective rate of tax on profits owing to tax holidays, and so on. These effects may be reinforced by nontax considerations.

These arguments apply to the incentives for MNCs to try and minimize total tax payments to the host country. But what matters to the company is overall taxation; the net incentive then depends on how the income is taxed after it leaves the host territory. To the extent that profits may be rerouted(29) to tax havens with near-zero tax rates, there is a strong incentive to do so, so long as the extra taxes incurred by increasing prices of inputs and royalties do not offset the reduction in taxes on declared profits. To the extent that underdeclared profits in the host country simply result in overdeclared profits in the home country, the incentive depends on the relative taxation in the host and home country. Where the DC tax rate is higher than the LDC – as it often is – then the incentive would be to overdeclare profits in the host LDC rather than the home DC. Only where MNCs have unused tax credits in the DC – for example, because of heavy R & D and investment expenditure – would there, in these circumstances, be an incentive to overdeclare in the DC and underdeclare in the LDC.(30)

These conclusions apply where an exemption system of tax credit is used. But where a tax credit system is used, then profits earned in the LDC are ultimately subject to the higher of the two tax rates, whether they are overtly declared in the LDC, or overtly declared in the DC. However, deferral advantages are likely to lead to a preference for declaration in the LDC where the tax rate there is lower. But again, much depends on how other taxes are affected, in particular, withholding taxes on dividends and royalties and tariffs on inputs. Withholding taxes on dividends and royalties are normally credited against tax due on these payments in the home country, but tariffs paid receive no credit.

What general policy conclusions can be drawn from these various factors? The net incentive to manipulate payments is complex, depending on the nature of the tax system in DC and LDC, the balance of unused tax credits in each, the system of double tax relief in use, and the possibilities of making use of tax havens. However, despite this complexity, policies that might be adopted to counteract such manipulation are relatively straightforward in principle. Two approaches are possible: administrative changes, and changes in the tax base. Probably most effective would be simultaneous action on both fronts.

The administrative approach involves toughening up tax administration so as to avoid transfer price and royalty manipulation by enforcing "arm's length" prices. This is the policy adopted by developed countries and a few developing countries. The weaknesses of the approach, taken by itself, are first the requirements made on the administrative system, and second the problems of defining and identifying "arm's length" prices. Methods of identifying "arm's length" prices are prices of comparable transactions between unrelated entities, cost plus, or resale price. All present problems (discussed fully in Musgrave, 1972). In many cases there is no "comparable transaction." Moreover, there may be cost savings in transaction within a firm which make the "comparable uncontrolled price" irrelevant. A study (O'Connor and Russo, 1973) of the U.S. administration of Section 482 prescribing the regulation of "arm's length" prices showed that the cost-plus interpretation of the U.S. authorities varied between cost plus 20 percent to cost plus 150 percent. Thirty-six percent of the 145 allocations surveyed used none of the three prescribed methods, but a fourth method, described as "some appropriate method . . . other than those described," to be applied where none of the previous three can be "reasonably applied." Some of the respondents questioned about the administration of Section 482 replied that its use was "strictly a horse trade" and "arbitrary."

The problems in identifying "arm's length" prices are particularly marked for technology transactions. This is because in one way each transaction is by its nature unique. Moreover, the cost of producing the technology is difficult to determine. It is difficult to allocate R & D costs to a particular development. Even if allocated, the marginal cost is likely to be very low in relation to the average. Since the "cost" has normally been recovered in DCs before the technology is transferred to LDCs, some have argued(31) that royalties from LDCs represent pure profit and should be taxed as such. But others believe that the cost of developing new technologies should be written off against taxation of royalties. The guidelines produced by the U.N. experts on tax policy recognize this principle but accept the difficulties of allocating costs in practice, and so recommend that the writing off of costs be taken into account by a reduced tax rate on the gross royalty.(32)

Another approach is to change the basis of taxation in such a way as to reduce the incentive for price manipulation or, if it does take place, to reduce its effects in terms of lost tax revenue.

Formula apportionment taxation. Taxation based on formula apportionment taxes company profits in any country on the basis of what the profits would be if they bore the same proportion to overall company profits as does the company's employment or sales or turnover or investment in the country, or some weighted average of the possible variables to worldwide company employment or sales or turnover or investment. There is then no incentive to manipulate transfer prices for the sake of avoiding taxation within the country that is using the formula apportionment basis.(33) Problems about this approach are: (1) Which is the "correct" basis to use? Since the ratio of profits to any of the variables is likely to vary, there is no "correct" formula, if the aim is to secure the best proxy for "real" (although undeclared and unknown) profits. According to how the formula is calculated, companies will have an incentive to increase or reduce their contribution in relation to the different variables. In choosing the formula, therefore, attention needs to be paid to which aspects of company activity the country wishes to encourage and which to discourage. (2) There are difficulties in obtaining information about the worldwide profits, turnover, and so on essential for the application of the method.(34) Disclosure requirements vary between countries; some companies are likely to have headquarters in (or to shift them to) countries in which disclosure requirements are inadequate for a proper application of the method.(35) It has been suggested that this type of taxation is not possible without international disclosure requirements and international tax administration.(36)

While this conclusion may be exaggerated, it does seem that current disclosure requirements are likely to be inadequate for satisfactory operation of a formula apportionment approach. Nonetheless, it might be an improvement on the current system, and the adoption of the approach might itself lead companies to improve disclosure to avoid excessive taxation. Formula apportionment deals with the problem of transfer price manipulation in intrafirm transactions, but not with transactions between unrelated enterprises. When technology is transferred to third parties (or joint ventures), taxation may be avoided by charging inflated prices for machinery and inputs, supplied along with the license for know-how on which royalties are paid.

<u>Uniformity of tax treatment</u>. An alternative approach is for tax authorities to treat all relevant items uniformly. This approach avoids any loss of revenue, whatever the manipulation of prices. Thus if royalties are treated as corporate profits for tax purposes, then whether the payments are declared as royalties or as profits will not make a difference to tax revenue. It is for this reason that the Andean Pact adopted this provision. Similarly, tariff rates on imported inputs should be in line with the taxation of corporate profits and royalties. One way to achieve required uniformity is to bring the tax rates in line so that the combined tax rates on each item are broadly similar. This would mean setting tariff rates on imported items at the same rate as the sum of corporation tax plus withholding tax on remitted profits; and royalties would either be amalgamated with profits for tax purposes, or the withholding tax on royalties plus any other taxes (such as turnover taxes, levied in, for example, Algeria), should be set at the same rate. Another way of achieving uniformity would be to redefine the tax base so that all the relevant items were included – the corporation tax would be replaced by a tax on profits plus royalties plus remitted management fees plus imported items less exports.(37) The tax would be similar to a turnover tax that excluded payments for local labor and exports. The

deduction of exports would eliminate the incentive to underinvoice exports for tax reasons.

This uniformity of tax treatment would not mean that manipulation of payments would stop. Incentives in other parts of the world – for example, in relation to DC taxation – for particular types of payment would still exist, and other motives for manipulation would as well. But it would mean that the consequent deterioration in terms of TT would be prevented, which is a major reason for concern about such manipulation. The uniformity approach would reduce countries' freedom to alter the relative taxes on different items. This would probably not matter much, since to a large extent the freedom is illusory, given manipulation possibilities that often make the use of relative tax rates to achieve policy ends ineffective. In any case, countries have tended to use the relative tax rates in the wrong direction, encouraging over-capital-intensive and import-dependent types of project. Moreover, governments would maintain the ability to alter relative costs by use of subsidies, which is normally a much more effective means of achieving the desired consequences.

It would also certainly be objected that taxing royalties at the same rate as corporate profits would involve an unjustified burden of taxation on royalties, which according to some, for example Carlson and Hufbauer (1976), are already taxed very heavily by LDCs in relation to the net of cost return. Their argument assumes a high cost element in technology supplies to LDCs, yet in practice most of the technology supplied to LDCs has already been written off against taxation by royalties paid within DCs. Hence royalty payments from LDCs represent a taxable surplus. Any argument about the need for high surplus on some technology items to finance the failures and provide a general incentive to R & D is inappropriate in the context of LDCs. DC R & D is undertaken for DC markets rather than LDCs, and much of it results in technology that is inappropriate for LDCs, and that in many ways the LDCs would be better off without (see the discussion in Stewart, 1977). In any case, under the present system payments described as royalties and taxed accordingly are not the sole return for technological transfer; the gross return may include some of the profits on overseas investment and some of the price of parts and equipment supplied. Thus to assess whether the technology is being fairly taxed requires an assessment of taxation of all these elements.

The uniformity approach would deal with the question of the terms of TT to unrelated enterprises, since the royalties they pay and the imported items they purchase would be subject to the same tax rate. It is important that any change in tax system not simply result in a change in form of TT as a tax avoidance device – unless this change in form is desired for itself. Hence any tightening up in taxation of MNCs' royalties – for example, by forbidding deduction – needs to be accompanied by raised rates of withholding tax on licenses between third parties.(38)

Rate of tax. Because of the double tax provisions, it is probably pointless for LDCs to charge tax rates less than DC tax rates for investment from countries using the credit method, and which do not have tax-sparing agreements. If countries wish to give additional incentives they should take the form of provision of infrastructure or direct subsidies – for example on employment – which do not simply increase the DC tax take. It is often argued that despite the relative uselessness of tax incentives, it is difficult for any one country to remove them and/or raise their rate of tax because they fear they will lose investment to neighboring countries who do not follow. "Perhaps the most potent argument against dropping income taxation concessions in Singapore is that since all the other Southeast Asian countries have them, and are still in the process of extending them, Singapore cannot afford to be without them because it might then on a superficial view appear not to welcome foreign investment as much as surrounding countries" (Hughes, p. 187, in Hughes and Seng, 1969). This is a powerful argument for joint LDC policies. But even where harmonization of taxation is proceeding, it is normally on the basis of common incentives, rather than common abolition of incentives. Thus while joint action is to be preferred, a second-best individual LDC action may be appropriate. In the light of the overwhelmign evidence about the relative ineffectiveness of tax incentives, particularly in LDCs where risk is high,(39) individual action to raise tax rates to the level of those current in DCs would not seem to be a high-risk strategy, particularly if in part the revenue were used to finance infrastructure and pay direct subsidies to employment and local research and development.

The administrative and taxation approaches to the problem need to be pursued together. It has been argued (Kopits, 1976a) that "with the exception of the United States, most countries have neither the willingness nor the administrative machinery to enforce arms length prices. . . ." As we have seen, there are problems in administration even for the United States. International auditing would be a substantial help, but without that it is utopian to expect most LDCs to succeed by administrative action alone. The taxation approach – whether by formula apportionment or the uniformity approach – would eliminate the tax loss from transfer price manipulation and also reduce the incentive to practice it. If it were introduced together with an increase in tax rates to DC levels in those countries where they are currently below them, and an abolition of most tax incentives, a substantial improvement in the terms of trade of TT could be expected. Moreover, there should also be an improvement in the quality of TT, since the current incentives favor capital-intensive and inefficient projects.

Administrative action would still be required to deal with those cases where transfer price manipulation is practiced for nontax reasons, for example to avoid restrictions on remittances.

The quality of technology could be improved by using the revenue gained to subsidize projects with the desired characteristics, to provide suitable infrastructure for appropriate projects, and to promote local R & D. There are a number of ways the tax system could be used to promote local R & D. One possible policy would be to have a lower rate of tax on royalties paid within the country, but differential tax rates would tend to give rise to manipulation of payments for tax avoidance. Consequently, direct subsidies are to be preferred.

Another possibility is to provide special tax allowances for local R & D. Peru has introduced an R & D tax (2 percent of net income before taxes of industrial enterprises), which is refunded to firms who use it for research and development; the remainder goes to an industrial technology research fund (see Sagasti, 1975). Apparently this has succeeded in raising expenditure on R & D, according to a sample survey showing much higher expenditure in Peru than in Ecuador or Colombia (Mytelka, 1978).

SOME CONCLUSIONS

This discussion has been directed at considering ways in which LDCs might use the tax system to improve their control over TT. It has assumed that the general tax environment remains the same. A quite different – and much more radical – approach would be to consider how the world system of taxation of multinational income flows might be reformed. National tax systems make little sense in the context of the multinational operation of much world economic activity. Quite apart from the possibilities of concealing and disguising information that occur especially in international transactions, the impossibility of truly allocating many overheads makes it impossible to arrive at a "true" national profit for multinational activities. To avoid double taxation, as income is taxed by more than one authority, or zero taxation as income slips between all authorities, a single international system of taxing all worldwide profits of multinational activities is required. The problem of how to divide the revenue between the various countries in which the companies operate would remain, but it would then be clearly seen for what it is – a somewhat arbitrary matter. The distributional consequences – what Musgrave and Musgrave (1972) call "inter-country equity" – would automatically be at the center of the discussion, in contrast to the situation today, in which these consequences are the net and often unintended outcome of a complex situation.

An international system of taxation of multinational activity may be the eventual outcome of a series of less ambitious changes, which would in themselves help LDCs in the control of TT. International agreement on disclosure requirements, on definitions, on exchange of information, and on international auditing of multinational company accounts would provide LDCs with the information they need to enforce a rational system of taxation. To some extent there is an identity of interest between host and source countries in terms of securing information and preventing major abuses of the system, such as the use of tax havens. The identity of interest, if it exists at all, is between the tax authorities in the two sets of countries. But governments represent a wider constituency than that, and consequently it may well be naive to expect any active support by DC governments for improved disclosure intended primarily to help LDCs, which may be against the interests of DC companies. In line with this, arrangements for mutual exchange of information between tax authorities in DCs and LDCs appear to be often weak or nonexistent.

While international reform and/or united LDC changes may be best suited to securing a more rational system of taxation of multinational activities, and of improving TT, this discussion has suggested that there are all sorts of ways in which individual LDCs, acting on their own, might use the tax system to improve the terms, the characteristics, and probably – if they wish – the quantity of TT. Recently, there have been moves to develop a code of conduct for international taxation (e.g., Shoup, 1976). In the field of taxation, it is probably overoptimistic to believe that a voluntary code of conduct can be a substitute for appropriate legislation.

NOTES

(1) UNCTAD's "Major Issues Arising from the Transfer of Technology to Developing Countries" deals only indirectly with the question of taxes. Vaitsos (1974) shows how the tax system may give rise to transfer pricing problems but is less explicit in policy towards taxation. The main policies of the Andean Pact have been related to ownership and control rather than taxation, although Decision 24 of the Andean Pact deals with taxation of royalties. Taxation is dealt with by Shoup (1976), and by the work of the U.N. Experts on Tax Treaties between Developed and Developing Countries, as well, of course, as in the very large literature directly concerned with fiscal matters – but in this literature problems of control of technology transfer to LDCs normally enter only indirectly.

(2) Kryzaniak and Musgrave found over 100 percent shifting in the United States. Others have found much lower shifting, the extent varying with the degree of concentration (Gordon, 1967). "As in industrial countries, the weight of expert opinion on the incidence of the corporation tax in developing countries is inconclusive" (Lent, 1978).

(3) Chown, in a book (1974) "intended for the executives of, and professional advisers to, businesses which have tax problems or opportunities in countries other than their own," argues that "apart from being an alternative to direct operations, licensing is often a powerful tool for international tax planning, often providing a means of transferring profits for tax purposes from one jurisdiction to another" (p. 3).

(4) See for example table 3, p. 60, in Adams and Whalley, 1977.

(5) For detailed descriptions see Chown (1974), the documentation produced by the International Bureau of Fiscal Documentation, and the reports of the U.N. Experts on Tax Treaties between Developed and Developing Countries.

(6) Lent (1978) describes the many forms of corporate taxation in developing countries. As is clear from his description, the discussion here presents an oversimplified view.

(7) See Kopits (1976a), table 2.

(8) Branches of MNCs are normally subject to the same corporation tax as subsidiaries but in most cases their remittances are not subject to withholding taxes. Developed countries' taxation on profits of branches is levied on an accrual basis (i.e., when made) and not a remittance basis, as with subsidiaries. Branches are not generally permitted to make deductions for intrafirm payments, such as royalties, in calculating profits. In general, it is believed that it is advantageous, for tax reasons, to operate overseas as a subsidiary as against a branch, although this view has been challenged (see Adams and Whalley, 1977, n. 5.7, p. 51). The bulk of overseas operations are through subsidiaries – in 1971 the U.K. worldwide net profits of branches were reported as 111 million, as against a reported 546 million in worldwide net profits of U.K. subsidiaries. Some countries (e.g., India) insist on local incorporation.

(9) For a full (though now a bit dated) description see Lent (1967); see also Reuber (1973), ch. 3, Heller and Kauffman (1963), the Third Report of the meeting of U.N. Experts on Tax Treaties Between Developed and Developing Countries, and for particular countries Ross and Christensen (Mexico), Chen-Young (Jamaica), Phillips (Nigeria), Hughes and Seng (Singapore), Taylor (Puerto Rico) and in Kleinsorge (Panama), Bird (Colombia), Tansi (Ecuador), and Herschel (Argentina).

(10) U.N. (1972): "Among the countries that typically offer full tax holidays are the Bahamas, Ethiopia, Guyana, Laos, Malaysia, Nigeria, Pakistan, Philippines, Republic of Korea, Thailand, Trinidad and Tobago, Saudi Arabia and Zambia. Partial tax holidays are offered by Ecuador, India, Israel and Sudan. Another pattern for tax holidays is illustrated by the examples of Argentina, Barbados, Costa Rica, which offer tax holidays in which the benefits decline over time. Still another pattern is the individually negotiated agreement (régime fiscal de longue dureé) for industries of special importance with lesser benefits available under general provisions for other investors. This pattern is illustrated by Burundi, Cameroon, Dahomey, Gabon, Guinea, Niger and Togo . . . The extent of the benefits may be fixed or may vary with the quality and quantity of the economic contribution made by the 'approved' enterprise." (p. 100)

(11) With every budget tax rates and allowances may be changed. The examples cited here all date back some time, and should be regarded now as illustrative rather than descriptive of the current position in particular countries.

(12) In each LDC if credit is calculated on a per country basis; overall, if credited on an overall basis.

(13) This is a consequence of the U.K. system introduced in 1973. But the pre-1965 U.K. system provided more credit than the post-1965 classical system.

(14) In the United States before 1976 companies could select whether to use a per country or overall basis. The per country option was removed in 1976. In the U.K. credit is given on a per source basis.

(15) See Musgrave (1969 and 1978). The deduction method is in use in some countries for taxation of branch profits.

(16) Jenkins and Wright (1975) show that, in the special case of oil, substantial excess tax credits had been earned by U.S. companies in the 1960s. But foreign tax credit provisions for gas and oil were restricted in 1975. For all U.S. corporations, U.S. taxation of foreign income in 1972 was estimated to be 5 percent (Musgrave, 1978).

(17) Musgrave and Musgrave (1972) distinguish three concepts of equity: international equity, which requires that the taxation of any company or individual should be the same irrespective of the source of income; national equity, which requires that equal domestic tax should be levied irrespective of source; internation equity, or the equitable division of tax revenue from a given multinational transaction between nations. The various concepts of equity are related to concepts of tax neutrality, which may also be variously interpreted according to whether it is a matter of neutrality in relation to activities of firms of the capital-exporting nation, or of activities within the capital-importing nation.

(18) Hamada (1966) shows how the question of taxation of multinational firms may be viewed as one of levying the "optimal" tax from a national point of view, and that this may induce reactions and counterreactions from other tax authorities in much the same way as is the case with the "optimal" tariff.

(19) It is my view that maximizing world "output" makes no sense as an objective for any one country or a group of countries so long as there is no international mechanism for securing international redistribution of income.

(20) H.J. Robinson, cited in Reuber, 1973.

(21) Estimates for the impact of taxation on flows of overseas investment to DCs vary, but in general it appears that the impact is greater. See the summary in Kopits, 1976a.

(22) Hufbauer and Nunns (1975) estimate that the cost to the U.S. Treasury of deferral of U.S. tax on undistributed earnings of foreign subsidiaries was $590 million in 1974. The advantage to the investor was substantially greater than this, because it also includes deferred LDC withholding tax on remittances.

(23) This has been illustrated very clearly in attempts to assess the impact of investment incentives in the U.S. and U.K. See the varying estimates in Fromm (1971); also Eisner (1969) and Corner and Williams (1965).

(24) Horst (1974) summarizes alternative theories of the firms as applied to the MNC, but Stevens (1974) argues that "even at the level of pure theory, differences in objectives may not lead to differences in behavior, or if they do the differences in behavior are far too subtle to be detected with existing data."

(25) Thus Horst (1971) explores a model in which the main impact of changes in tax rates is on transfer pricing, and argues that "a country with high tariffs should be quite willing to raise its tax rate above the general norm – tax revenues would be increased with no effect on the behavior of firms selling to the domestic market." The question of transfer pricing is treated in a similar way by Robbins and Stobaugh (1973).

(26) See Levy and Sarnat (1974) for an algebraic demonstration of these conclusions.

(27) See Kopits (1976a), table 4, and UNCTAD (1977) for summaries of the evidence.

(28) Discussed in Vaitsos (1974), Lall (1973), Kopits (1976a), and UNCTAD (1977).

(29) The extent of such rerouting is unknown, but that some takes place is evidenced by the very large number of companies in tax havens. According to Chown (1974), in the Bahamas in 1973 over 12,000 companies were incorporated. (In the U.K. in 1966 there were 11,000 paid-up registered companies and 400,000 private companies: Adams and Whalley, 1977, pp. 131-32).

(30) This is how Vaitsos (1974) explains the overinvoicing he finds.

(31) See the discussion of the U.N. Experts on Tax Treaties Between Developed and Developing Countries.

(32) Carlson and Hufbauer (1976) argue that LDC taxation of royalties is excessive because the withholding tax is based on the gross payment.

(33) A formula apportionment basis is adopted by some states in the United States for company transactions across states. The proposal to introduce it for the taxation of multinationals is discussed by Group of Eminent Persons, The Impact of Multinational Corporations on Development and on International Relations, ST/ESA/6 (New York, 1974).

(34) As Shoup (1974) has suggested, there may be no unique worldwide profit, given fluctuations in exchange rates.

(35) In some countries companies are required to publish their consolidated world accounts, but this is rare outside West Germany, the U.K., and the United States.

(36) One should distinguish between standardized international disclosure requirements and international tax administration. The former may be a realistic aim in the medium term.

(37) Suggested in ILO (1972), Technical Paper 17, which suggests excluding exports from the tax base. There is a similarity between this suggestion and the proposal to replace the U.K. profits tax by a value-added tax that would have included imports and excluded exports. However, the proposed U.K. value-added tax would have included taxation of wages paid in the U.K. See Richardson, 1964.

(38) And by prohibition of capitalization of know-how fees (as in Decision 24 of the Andean Pact), which is a tax avoidance device.

(39) Kopits (1976a) concludes his survey of taxation of MNCs thus: "Where risk is likely to overwhelm the role of tax differentials in the firms' decisions, host countries (particularly less developed ones) would want to refrain from offering generous tax concessions; instead they should attempt to raise badly needed tax revenues from direct investment from abroad . . . and to stabilise the overall investment climate over the long run."

REFERENCES

Adams, J.D.R., and Whalley, J. 1977. The international taxation of multinational enterprises. London: Allen and Unwin.

Bird, R.M. 1970. Taxation and development: Lessons from Colombian experience. Cambridge: Harvard University Press.

Carlson, G.N., and Hufbauer, G.C. 1976. Tax barriers to technology transfers. U.S. Treasury Department, O.T.A. paper 16.

Carne, R. 1968. Investment stimulating methods in Israel and in the world. Economic Quarterly 16.

Chen-Young, P.L. 1967. A study of tax incentives in Jamaica. National Tax Journal 20.

Chown, J.F. 1974. Taxation and multinational enterprise. London: Longman.

Corner, D.C. and Williams, D. 1965. The sensitivity of businesses to initial and investment allowances. Economica 32.

Dunning, J, ed. 1974. Economic analysis and the multinational enterprise. London: Allen and Unwin.

Eisner, R. 1969. Tax Policy and Investment Behavior. American Economic Review 59.

Fromm, G., ed. 1971. Tax incentives and capital spending. Washington: Brookings Institution.

Gordon, R. 1967. The incidence of the corporation income tax. American Economic Review.

Hakam, A.N. 1966. The motivation to invest and the locational pattern of foreign investment in Nigeria. The Nigerian Journal of Economic and Social Studies.

Hamada, K. 1966. Strategic aspects of taxation on foreign investment income. Quarterly Journal of Economics.

Herschel, F.J. 1965. Fiscal policy for economic growth in Latin America. Baltimore.

Heller, J. and Kauffman, K. 1963. Tax incentives for industry in less developed countries. Cambridge: MIT.

Hinricks, H. 1974. Tax reform and fiscal harmonization: Guatemala. In Fiscal policy for industrialization and development in Latin America, ed. D. Geithman, Florida.

Horst, T.O. 1971 The theory of the multinational firm: Optimal behaviour under different tariff and tax rates. Journal of Political Economy 79.

Horst, T.O. 1974. The theory of the firm. In Dunning (1974).

Hufbauer, G.C., and Nunns, J.R. 1975. Tax payments and tax expenditures on international investment and employment. Colombia Journal of World Business.

Hughes, H., and Seng, Y. eds. 1969. Foreign investment and industrialisation in Singapore. Australian National University Press.

International Labour Organisation. 1972. Employment, incomes and equality: A strategy for increasing productive employment in Kenya. Geneva: ILO.

Jenkins, G.P., and Wright, B.D. 1975. Taxation of income of multinational corporations: The case of the U.S. petroleum industry. Review of Economics and Statistics 57.

King, T. 1970. Mexico, industrialisation and trade policies since 1940. Oxford: Oxford University Press.

Kopits, G.F. 1976a. Taxation and multinational firm behavior: A critical survey. IMF Staff Papers 23.

Kopits (1976b). Intrafirm royalties crossing frontiers and transfer pricing behaviour. Economic Journal 86.

Kryzaniak, M. and Musgrave, R.A. 1963. The shifting of the corporation income tax. Baltimore: Johns Hopkins Press.

Lall, S. 1973. Transfer-pricing by multinational manufacturing firms. Oxford Bulletin of Economics and Statistics.

Lall, S., and Streeten, P. 1977. Foreign investment, transnationals and developing countries. London: Macmillan

Lent, G.E. 1967. Tax incentives for developing countries. IMF Staff Papers 14.

Lent, G.E. 1978. Corporation income tax structure in developing countries. IMF Staff Papers 25.

Levy, H. and Sarnat, M. 1974. Investment incentives and the allocation of resources. Economic Development and Cultural Change 23.

Little, I.M.D.; Scitovsky, T.; and Scott, M. 1970. Industry and trade in some developing countries. Oxford: Oxford University Press.

May, R.S. 1965. Direct overseas investment in Nigeria, 1953-1963. Scottish Journal of Political Economy 12.

Mellors, J. 1973. Corporate taxation and the location of overseas direct investment: A pilot study. Discussion Paper in International Investment and Business Studies, No. 5, University of Reading.

Musgrave, P. 1969. United States taxation of foreign investment income: Issues and arguments. Cambridge: Harvard University Press.

Musgrave, P.B. 1972. International tax base division and the multinational corporation. Public Finance 27.

Musgrave, P. 1978. Taxation and American investment abroad: The interests of workers and investors. Forthcoming.

Musgrave, R.A. and P.B. Musgrave 1972, "Inter-nation Equity" in Modern fiscal issues: Essays in honor of C.S. Shoup, ed. R.M. Bird and J.G. Head.

Mytelka, L.K. 1978. Licensing and technology dependence in the Andean Group. World Development, 4.

O'Conner, W.F., Russo, S.M. 1973. A Study of Corporate Experience with Section 482. The European Tax Review.

Phillips, A.O. 1968. Nigeria's experience with income tax: A preliminary result. Nigeria's Companies Income Tax. Nigerian Journal of Economic and Social Studies.

Phillips, A.O. 1969. The significance of Nigeria's income tax relief incentive. Nigerian Journal of Economic and Social Studies 20.

Prest, A.R. 1975. Public finance. 5th ed. London.

Reuber, G.L. et al. 1973. Private foreign investment in development. Oxford: Clarendon Press.

Richardson, G. (Chmn). 1964. Report of the committee on turnover taxation. Cmd 2300.

Robbins, S.M., and Stobaugh, R.M. 1973. Money in the multinational enterprise: A study in financial policy. New York.

Robinson, H.J. 1961. The motivation and flow of private foreign investment. International Development Center, Stanford Research Institute.

Ross, G., and Christensen, J.B. 1959. Tax incentives for industry in Mexico. Cambridge: Harvard University Press.

Sagasti, F.R. 1975. The ITINTEC system for industrial technology policy in Peru. World Development 3.

Shoup, C.S. 1974. Taxation of multinational corporations. In The impact of multinational corporations on development and international relations: Technical papers taxation, ST/ESA/11. New York: United Nations.

Shoup, C.S. 1976. Proposed code of conduct on international income taxation for transnational corporations operating in developing countries with correlative action by home and host governments. Consensus proposed for Aspen Consultation, December 1976, auspices of Carnegie Center for Transnational Studies.

Stevens, G.V.G. 1974. The determinants of investment. In Dunning (1974).

Stewart, F. 1977. Technology and underdevelopment. London: Macmillan.

Tansi 1969. Tax incentives and economic development: The Ecuadorian experience. Finanzarchiv.

Taylor, M.C. 1957. Industrial tax exemption in Puerto Rico. Madison: University of Wisconsin .

Taylor, M.C. 1966. Tax incentives for development: Panama's experience. In Public finance and welfare: Essays in honour of W.C. Macy, ed. P.L. Kleinsorge.

United Nations. 1974. Guidelines for Tax Treaties between Developed and Developing Countries, ST/ES A/14, New York: United Nations, Dept. of Social and Economic Affairs.

United Nations, Tax Treaties between Developed and Developing Countries
1969 1st Report: Sales no. E69 XVI 2
1971 2nd Report: Sales no. E71 XVI 2
1972 3rd Report: Sales no. E72 XVI 4
1973 4th Report: Sales no. E73 XVI 1
1975 5th Report: Sales no. E75 XVI 1
1977 6th Report: Sales no. E77 XVI

United Nations Conference on Trade and Development. 1972. Major issues arising from the transfer of technology to developing countries. TD/B/AC 11/10.

United Nations Conference on Trade and Development. 1977. Dominant positions of market power of transnational corporations: Use of the transfer pricing mechanism. ST/MD/6.

Vaitsos, C.V. 1974. Intercountry income distribution and transnational corporations. Oxford: Oxford University Press.

Comments

Louis T. Wells, Jr.

I had a little difficulty deciding what Frances Stewart's major subject was. There seemed to be two possibilities. Her title indicates that she intended to write about how a country can use the tax system to affect technology transfer. On the other hand, the majority of the recommendations in the paper were addressed, with a much narrower focus, on how a country should tax multinational enterprises. Multinationals are only one source of technology, of course. A focus on multinationals does not disturb me. I think that the reasons why that emphasis might be appropriate will become clearer in my comments.

I also found it difficult, in reading the paper, to determine whether she preferred one set of recommendations over the other. One group of recommendations would make every effort to move to an international system of income allocation for taxing multinationals. Under that proposal, countries would agree on a method of assigning the global income of multinationals to individual tax jurisdictions. On the other hand, she seems to recommend the uniform tax system as the appropriate method to deal with tax problems of developing countries. I really like her first set of recommendations better.

Professor Stewart started the process of outlining the steps necessary to move toward an allocation system, but I suspect that we might have to add more steps. First, countries are unlikely to reach agreement on international allocation until there is a greater feeling of symmetry among the parties to an agreement. Feelings of symmetry arise when each country sees itself as both exporter and importer of foreign investment. Once a number of countries reach that state, then the problems of determining a formula for income allocation ease somewhat. Inside the United States, the states have allocation systems. But in that case a sense of symmetry is present. Once that sense is obtained, the exact formula for allocation is less critical. The world is moving in that direction, as Japanese and European firms invest in the United States. A sense of symmetry among the OECD countries might be not as far away as one would have thought ten years ago.

The second step that would be necessary before we can move very far in that direction is perhaps also on the horizon. Multinational enterprises themselves must feel the need for an agreement. We may well move in that direction fairly rapidly as national authorities become much better at policing the transactions of their multinationals and of others' subsidiaries within their borders. As sophisticatoin increases, multinational enterprises begin to face double taxation, for example, when the United States imposes one transfer price and the host country imposes another transfer price on the same transaction. When the two authorities cannot agree, the enterprise pays taxes on the same profit in both countries. Eventually multinational enterprises themselves are likely to favor an allocation mechanism to avoid the double taxation that results from the conflicting actions of individual jurisdictions.

The alternative proposal, the uniform tax, is unlikely to be adopted in developing countries. Almost all would be unwilling to surrender their ability to use the tariff system as a method of granting protection to local industry. Perhaps they should not be so insistent in the matter, but the chances of their moving in this direction seem remote.

The emphasis in the paper on taxing the multinationals, rather than on all the other tax methods that have been proposed to influence technology, was appropriate because one can really hardly envision the domestic tax system being designed primarily for its effects on technology transfer. In the case of the multinational, this concern might be more important, but there, the likely principal concern will be cost. The tax system, as Professor Stewart has pointed out, has relatively little effect on the quantity of technology transfer, except for export industries. The domestic tax system is usually overloaded with objectives, besides that of affecting technology transfer. In addition to the protection of local industry, it is burdened with income distribution, revenue collection, and locational impacts.

Professor Stewart has suggested that developing countries set their corporate income tax rate close to the corporate income tax rates of the capital-exporting countries that are most important to them. In practice, what often happens is that the country attempts to adjust the domestic corporate income tax by adding a withholding tax on dividends, such that the total effect of the two is close to the home country income tax. This approach enables the country to separate domestic taxation from that imposed on the multinational enterprise. This approach is, I think, appropriate. It is only a minor correction to what is proposed in this paper.

A few countries have tried to differentiate the tax they impose on foreign investors by the source of the foreign investment. Panama and Liberia both have tried to impose what has been called a "soak-up" tax, in an effort to tax the foreign investor up to the point that the tax credits in the home country were used up. That would be the extreme of capturing all of the taxes that one could from the home country. In both of those cases, it has proved administratively unworkable. I believe that both countries have abandoned the effort.

Professor Stewart points out that the tax system has relatively little effect, according to empirical work, on transfer decisions. The studies she points to have looked largely at income tax holidays as a means of affecting the international transfer of technology. There are some exceptions where the tax system has apparently been somewhat effective in influencing technology transfers. First, the tariff system clearly has some effect, even where taxes are unimportant. A number of studies have pointed out that the choice between labor-intensive and capital-intensive technologies is influenced by the degree of competition that a manager faces. A higher tariff in many cases decreases competition. This leads to the choice of a more capital-intensive technology than would have occurred if the protection were absent. Second, there is some indication that the tariff rate has a quite direct effect on the choice, by raising the price of capital equipment. This is the first effect that an entrepreneur faces from the tax system when he is designing his plant. What little evidence is available suggests it enters into his calculations.

There is another means of influencing technology choice that has not been used very widely, but which has been quite effective in one country: the differential excise tax, in which the excise tax on certain products is based on the process used in the manufacture of the products. This system can be applied only in cases where there is a clear break in technology from labor- to capital-intensive. Indonesia has used this effectively in two cases. The excise tax on handrolled cigarettes is much lower than the excise tax on machine-rolled cigarettes. The result is to encourage handrolling of cigarettes, a labor-intensive technology, and to discourage machine-rolling of cigarettes. A similar tax has been applied to soft drink bottling.

In spite of the studies on the effects of tax holidays on investment decisions, there is not much evidence about the effect of tax holidays on the transfer of technology for export-oriented manufacturing investments. The research that has been done has been almost entirely on firms oriented toward production for the domestic market. It may be that tax holidays are more effective in attracting foreign firms to a particular country for export operations than for domestic market operations. One can at least hypothesize that in these cases countries really are competing one against the other for this kind of investment, because investors have some choice of where they will go. In that case it may well be that the tax holiday influences the location of investment, but not the total quantity available to all developing countries. If that is the case, regional agreements limiting tax holidays are likely to be productive.

There were other tax proposals that Professor Stewart quite intentionally did not deal with, although they affect the quality of the technology transferred. There are a number of proposals, for example, that would modify the icome tax base according to the company's wages bill or its depreciation bill. My guess is that such policies would have very little influence on the technology actually chosen by most investors in developing countries. One very important reason for the lack of influence in many developing countries, and in some European countries as well, is that the rules in the income tax law have relatively little to do with what is actually paid as income tax. Entrepreneurs know that the tax settlement at the end of the year is a negotiated tax payment. Minor changes in the tax law have little influence on how much taxes are actually paid. Thus entrepreneurs simply ignore small changes in rules in making their decisions. As a result, I suspect that if one wants to use the tax system to influence technology flows in ways other than having an impact on the cost of technology, one has to make fairly radical changes. There are other reasons. My experience in negotiations is that most foreign investment negotiators do not understand tax issues well. The negotiator sent out to the developing country has a very rudimentary understanding of his own home tax system or of the tax system of the host country. The negotiator tends to behave in many cases more as a "trophy collector" than as someone trying to maximize after-tax profits. He wants to obtain tax holidays, even though they may mean little to the firm. If he gets them, he can go back to his boss and show what a good negotiator he was. It seems that the tax experts tend to be called in later, after a basic agreement has been reached; they modify as necessary to minimize taxes after all the other decisions are made. Given the usual behavior of and lack of information available to the negotiator, only a very dramatic change of taxes will have an impact on the firm.

There are indeed a few benefits that a firm gets from a tax holiday that one ought not to ignore, although I still think they are relatively unimportant in the decision-making process. If an American investor reinvests profits in the country where the project is located he obtains more money to invest under a tax holiday than without one, since the U.S. tax liability is postponed. Second, companies have been able to use the increased cash flow resulting from tax holidays to repay debt obligations without incurring a home country tax.

7 Restrictive Business Practices and International Controls on Transfer of Technology

Michiko Ariga

Commitment of the member countries of the United Nations to the maintenance of world peace through expansion of international trade necessarily draws attention to trade barriers established not only by governments but also by private enterprises. In this connection, attempts have been made to reduce or eliminate such trade barriers under the auspices of the United Nations.

The General Agreement on Tariffs and Trade that took effect on January 1st of 1948 is a landmark successful attempt of member governments to tackle with trade barriers by countries.

Efforts were also made in the United Nations to eliminate the various practices of private enterprises that were believed to affect international trade, leading in 1948 to the proposal to establish the International Trade Organization. Due to the failure to obtain the required number of ratifications, this never took effect. Efforts continued by the Committee of Experts on Restrictive Business Practices of the Economic and Social Council (ECOSOC) of the United Nations, which in March 1953 submitted the Draft Articles of Agreement on Anti-Monopolies, but again these failed to win agreement.

These failures in the early postwar period do not mean that these efforts were fruitless. On the contrary, they gave rise to an important recognition: before a far-reaching international system can be adopted to control restrictive business practices that act as trade barriers, many other things must be done, including the identification of what restrictive business practices in international trade are, and the enactment of some sort of regulatory legislation by each member country on restrictive business practices. For without general agreement on the principle of fair and free competition within a country, promotion of international trade among countries of the world will only be a political ploy, and not based at all on the philosophy of the New Economic Order declared by the General Assembly of the United Nations in 1972.

International transfer of technology is a business practice with substantial implications for international economic growth and the welfare of peoples. The importance of technology in the modern world – in the promotion of employment and export, enhancement of living standards, and development of national economic growth – has become widely recognized, and the need to obtain technologies from outside sources to speed up the pace of national economic development of a country has also been recognized. The proposed International Trade Organization of 1948 and the Draft Articles of Agreement on Anti-Monopolies of 1953, although they failed, both recognized the importance of the transfer of technology and therefore the necessity of controls on restrictive business practices that are likely to occur in the context of a transfer of technology. A rather idealistic or hasty approach to these efforts prevented their realization; there were wide differences among member countries in the political, legal, economic, social, and cultural spheres, and there were laws controlling restrictive business practices in only a few of the signatory countries. With a world population of about 3 billion living in almost 170 large or small independent, sovereign countries that differ greatly in their levels of economic and educational development, a fair international flow of technology – in spite of the vital role it can play in furthering development – cannot be expected to occur as easily as some would desire.

The owner of a technology that is usually the fruit of study and research carried out at great expense generally maintains a policy not only recovering the money defrayed but also of earning additional profits on the technology by using the technology or licensing it to others. Since the owner of a technology is generally a business enterprise that pursues its business interests not only domestically but also internationally, technology tends to be used as leverage, either to maximize profits or to protect or strengthen competitiveness.

The leverage function becomes stronger when the position of the owner of technology is such that he enjoys a legal monopoly by virtue of a national patent system. Although the patent system of one country may differ from that of another, a patented product and a patented process to manufacture a product under one country's patent system enjoys a legal monopoly right of limited scope. The owner of know-how – a technology undisclosed to others – may also enjoy a somewhat monopolistic position, or at least better position than others. Even today, when communications are widely and easily available, patented technologies that enjoy a legal monopoly and technologies that are being kept secret cannot be lawfully available to third parties without the owner's consent. Here is where the licensing contract or agreement comes in. It stipulates terms of licensing whereby the owner of the technology concerned may find maximize profits as a business enterprise. Clauses imposing restrictions on the activities of a licensee that go beyond the limited scope of the patent-defined right of a legal monopoly or the justifiable scope of the owner's rights to know-how are likely to be included as terms of licensing.

Even in a country that has legislation controlling restrictive business practices it is rather difficult to decide whether or not a restriction in a patent licensing contract goes beyond the limited scope of a patent right, because legislation on restrictive business practices is a relatively new phenomenon – mostly postwar – while a patent system is sometimes a centuries'-old phenomenon, during whose existence a patent owner might have enjoyed a monopoly without ever having the reasonableness of strong and broad patent protection challenged. Thus it appears that there is some conflict between a legal monopoly grant and control of restrictive business practices.

This situation becomes more difficult and complicated at the international level. A patent system is primarily a nationally established one, with the exception of the recent phenomenon of the European Patent System. Today legislation on restrictive business practices is mostly found in the developed countries.

Since international trade has come to be undertaken on global basis, providing trading opportunities to enterprises of every country of the world, the transfer of technology has become an important practice, enabling enterprises that lack useful technologies to participate in international business. This is all the more true of countries that are endeavoring to develop their national economies by promoting national production and employment and increasing living standards. The World Intellectual Property Organization (WIPO) is working to provide a model law of industrial property, and in addition to the increasing number of national regulatory legislative successes in many developing countries, UNCTAD and the Transnational Corporations Commission of the United Nations are formulating international controls on restrictive business practices in relation to technology transfer.(1)

Since the endeavor by UNCTAD is on a world scale, I will focus here on the items of restrictive business practices UNCTAD proposes to bring under control, and discuss possible difficulties that have to be understood and overcome before controls on restrictive business practices vis-a-vis exploitation of patents and know-how in the international arena can be successfully implemented.

Any effective international control of unreasonable restrictive business practices linked with the transfer of technology must be based on the fullest understanding of the different realities that exist in the countries of the world.

WHAT IS MEANT BY "RESTRICTIVE BUSINESS PRACTICES"?

The term "restrictive business practices" is relatively new. However, forestalling, engrossing, or hoarding were penalized in Roman days because an enterprise conducting such practices made money at the expense of the consuming public, through the private control of a market. Now various other types of restrictive business practices, including monopolization, price fixing, boycotting, refusals to sell, and tie-in sales have come under attack by those seeking to maintain a free and fair market order or free and fair competition in market.

How a restrictive business practice is treated differs from one country to another. In a common-law country such as England, a restrictive business practice occurring in a transaction that was based on a contract freely entered into was sometimes invalidated by the courts because the restriction was considered "unreasonable" from the standpoint of equity. Another way a restrictive business practice in a contract can be considered is from the standpoint of public policy or public interest. The term "public policy" means the policy of a country that is identified by its political and economic objectives. Once national objectives are defined, then a restrictive business practice in the private sectors that contravenes any of the national objectives will be unenforceable. This is the extended concept of equity, with the growth of integration of separate small communities into a solid state.

In a country that has a controlled type of economic system, a practice forcing outsiders to abide by an established cartel will not be considered restrictive, and outsiders who do not comply with certain cartel agreements will be regarded as doing business against the public policy of the country.

In a country whose policy is to expand exports by domestic enterprises, an export cartel whose quota system restricts the export volume of member enterprises will be considered contrary to the national interest.

On the other hand, the market-economy countries believe that free and fair competition within their markets is the backbone of their political democracy. A restrictive business practice that deprives one party to a transaction of freedom to do business will be considered illegal if it affects competition in a market. A business practice that distorts the smooth and free functioning of a market mechanism is condemned as restrictive and in contravention of the public policy.

The term "restrictive business practice" covers a wide range of business practices, including self-restriction of business activities, restrictions on the business activities of the other party to a transaction, and practices restricting or adversely affecting the business activities of a third party. All three types, generally speaking, have negative effects on the economic development of a country that takes freedom of doing business as its basic political as well as economic policy. However, what types of business practices falling under each of the above categories are condemned as restrictive will depend on each country's stage of economic development and national objectives.

Types of Restrictive Business Practices

Monopolization. The restrictive business practices most generally condemned are monopolization and cartelization. Monopolization is a business practice whereby an enterprise, by itself or in combination with others, controls a market in terms of the price of commodity concerned or its volume of market supply, which indirectly affects the market price or the quality of the commodity. A cartel agreement is an agreement entered into by enterprises with the purpose of manipulating the market mechanism so that an artificial market price can be produced. Monopolization and cartelization are condemned because these practices are conducted at the expense of consumers. If the price or quality of a commodity is manipulated by enterprises, technology development cannot be expected to occur, because the enterprises are safeguarded against the pressure of competition and thereby may be allowed to idle along with their less efficient and old-fashioned production facilities. Economic development through expansion of production, inducement of new technologies, increase of employment, and increase of general income will be hampered by monopolization and cartelization.

In Japan, a monopolistic enterprise is subject to divestiture of some part of its business. In the United Kingdom a monopoly will also be made subject to divestiture. In the United States, although attempts to monopolize are covered by controls, the status of monopoly as such is not yet under specific statutory control; but there are cases in which an enterprise maintaining its monopolistic position was judged in violation of Section 1 of the Sherman Act.(2)

Similarly, the merger of companies or an acquisition of the assets of a company in the same line of business is made subject to control in the United States, the United Kingdom, West Germany, and Japan. Although mergers as such are generally permissible under national law, it may, in certain cases, lead to the attainment of a monopolistic position, giving the merged company a market-controlling power similar to that of a monopoly. Stockholding and interlocking directorates may help achieve the same position, and therefore such practices are subject to control in some developed countries.(3)

A monopoly that may be the end result of fair competition is nevertheless susceptible to public control because it has the power to maintain control over a market, thereby suppressing competitive challenges.

In almost all countries that have national legislation on restrictive business practices, controls on monopolization and cartelization are based on varying legal concepts and procedures. West Germany and the European Economic Community both control practices that are tantamount to abuse of the dominant market position by an enterprise.

Cartelization. A cartel agreement is, it is generally considered, an agreement in restraint of trade, an agreement among enterprises on price, volume of production, volume of supply, or market division, as a result of which the market concerned is controlled. A horizontal agreement among enterprises – in other words, an agreement among enterprises in competition with each other – usually constitutes a cartel agreement in so far as it distorts a market mechanism and thereby produces an artificial market condition.

Under some countries' statutes, not only horizontal agreements, in which the restrictions on each of the participants are similar in nature, but also vertical agreements among enterprises who are not in competition with each other (for instance, an agreement of manufacturers and wholesalers of some commodity) are treated as cartel agreements. In such cases the restrictions on the participants differ in nature, but the combination of such restrictions may result in an overall restraint of trade in a market.(4)

Unfair business practices. Business practices such as refusal to sell or boycotting, dumping, discrimination, resale price maintenance, tying, exclusive dealing, deceptive representation or misrepresentation, and industry spying, all of which adversely affect a free and fair competitive market order and the interests of consumers, are generally called restrictive business practices even though some of them do not involve the imposition of restrictions upon a particular enterprise. Some countries statutorily define such business practices as illegal,(5) and some countries condemn them generally as unfair business or trade practices, without defining them specifically.(6)

In the course of the economic development of one country, business practices that are criticized as inimical to wholesome economic development are identified, and the types of practices that must be made subject to control increases. When the laissez faire doctrine prevailed in France and England, a freely-entered-into contract containing restrictions on one of the parties to a transaction was upheld, but later (as was mentioned earlier) equity considerations made such restrictive provisions invalid and unenforceable, on the grounds that they are unreasonable from the standpoint of the right to pursue a profession. In other words, unreasonable restrictions on one other party to a transaction result in a distortion of fair competition. Business practices that distort fair competition are found not only in a contract involving several enterprises whose interests are primarily opposed, but also in the unilateral practices of enterprises. Refusal to sell and misrepresentation are two examples of the latter. Restrictions provided for in a contract or unilaterally self-imposed are considered restrictive not only because they are likely to adversely affect the fair competitive order but also because the repetition and accumulation of such effects lead to a monopoly.

PATENT LEGAL MONOPOLY AND CONTROL OF RESTRICTIVE BUSINESS PRACTICES

The number of countries that have specific legislation on restrictive business practices is reportedly thirty-six.(7) Each of the OECD member countries except Italy – namely the Netherlands, Belgium, Luxembourg, West Germany, France, Denmark, the United Kingdom, Ireland, Finland, Norway, Sweden, Switzerland, Spain, Portugal, Austria, Canada, the United States, Japan, and Australia – has specific legislation on restrictive business practices. In non-OECD member countries such as Israel, Korea, India, Mexico, Brazil, and Sri Lanka legislation on restrictive business practices has been enacted. In the Rome Treaty establishing the European Community there are specific provisions controlling restrictive business practices, and these provisions apply to all the EC member countries, including Italy.

Although the national legislation of each country differs from that of others in various respects, due to differences in legal, political, economic, and social structure, it is generally true that these countries prevent in some way or other a substantial restraint of trade, either by monopolization or cartelization, and some business practices considered unfair, such as resale price maintenance and boycotting.

As was mentioned earlier, a law under which a patent is issued generally gives to the grantee of a patent a monopoly on the working of the patented invention. It has been asked whether there is a conflict between controls on the restraint of trade and authorization of a legal monopoly through the issuing of a patent. Some countries, such as West Germany, the United Kingdom, and Japan, solved this question by inserting specific provisions into their legislation on restrictive business practices.(8) In the case of the United States, where antitrust legislation (the Sherman Act) was enacted as early as 1890 and has no particularly relevant provision with respect to the scope of the legitimate exploitation of a legal patent monopoly, there is an accumulated case law.

Although the groundrules on the scope of legal patent monopoly are generally laid down by the patent law itself, and it is recognized as legitimate by some legislation regarding restrictive business practices, the increase in sophistication in exploiting a patent monopoly has come into conflict with the enforcement of controls on restrictive business practices, which may also be strengthened as the economy of a country develops. Therefore this area of conflict will always be a national legal problem.

A country whose patent law grants a legal monopoly in working the patented invention, but is not clear about the legitimate scope of the patent right, is encountered with a question, must a licensee accede to and comply with an obligation imposed upon him by the licensor which is in the interest of the licensor but not in the interest of the licensee? The answer should be ascertained from the fundamental purpose of the law. Generally speaking, the legal monopoly right accruing from a patented invention encompasses the exclusive right to manufacture and sell a patented product or the exclusive right to use the patented process to manufacture a product for a certain limited period of time, as provided for by the patent law, in the entire territory of the country. (The patent laws of some countries recognize patentability for composite matter, plants, and designs. To simplify the discussion here, only patented products and patented process will be dealt with.) The importation of a product similar to the product on which a patent has been obtained in the importing country by the importer himself is recognized in most countries as a lawful exercise of the importer's patent right, but some countries do not regard it as part of the patent right unless the importer-patentee himself is actually working the patent, in other words, manufacturing the patented product in the importing country.(9)

As far as a patentee or patent owner is concerned, he is in the legal position to exercise his exclusive right against any infringer of his patent. Only a licensed party can be safe from the patentee or patent owner's attack of infringement.

In granting a license, a patentee does so on certain conditions. The terms of licensing will necessarily include provisions relating to the period, the territory, and the type of working of the patented invention, and also to whether the license is exclusive or nonexclusive. Since a patent is private property, the patentee is entitled to demand royalty from his licensee, thus enabling the patentee to obtain a pecuniary reward for his patented invention. Beside the terms or conditions flowing from patent law, other terms or conditions may be included in a license contract for the sake of the patentee's interest, reflecting his competitive position as an enterprise in a market. If a patent owner himself works the patented invention himself, then he has an interest in restricting the licensee as to the amount or volume of production or the number of uses of the patented process to manufacture the product. The Act against Restraint of Competition of West Germany and the Japanese FTC's Antimonopoly Act Guidelines for International Licensing Agreements provide that these restrictions are within the exclusive right of a patent right, although they are not explicitly provided for in patent law.(10) As to whether a patent owner may lawfully restrict the price of a product manufactured by licensed technology, an affirmative answer would be dangerous, since although the "GE Doctrine" of 1926 of the U.S. Supreme Court has not yet been overruled, the field in which the doctrine may be applied has been very much narrowed by the same Supreme Court.(11)

Noncontesting of the licensed patent is a frequent condition in license contracts. This practice has been challenged rather recently in the United States.(12) The state's promise to its people to grant a patent to a truly patentable invention and give the patentee the power to exploit his monopoly legitimately is broken by a mispatented invention, and since such mispatentee exercises a patent monopoly power not originally entitled to him, the licensee, after terminating the license, should be able to contest the patentability of the invention concerned.

Other restrictions of various types imposed upon a licensee, which do not flow, theoretically speaking, from the exclusive right of the patentee but only help restrict the freedom of the business activities of the licensee, are matters to be considered in the light of each country's controls on restrictive business practices, because such restrictions may affect the smooth functioning of the market mechanism of the country.

A license contract sometimes includes restrictions applicable to the licensor himself by virtue of an agreement with the licensee. If such a restriction affects the fair competitive order of the country, it would also be subject to controls on restrictive business practices.

Since both the licensor and the licensee in a license contract are usually business enterprises engaging in a similar industry, and therefore are competitors, they are prone to make an anticompetitive arrangement, often in the form of a cross-licensing scheme, whereby a substantial restraint of trade is achieved in violation of prohibitions against cartels.

So far, this discussion has dealt with a license solely within a country, in connection with the possible conflict between a patent law and the control on restrictive business practices. When the setting is international – that is, an agreement between a domestic licensee and a foreign licensor – the problems that arise are much more complicated and difficult, due to the differences in the patent law and in the legislation or regulations on restrictive business practices of the countries concerned. It is most important to note that international legal conflicts can arise only between countries that have differing specific legislation on patents and on controls on restrictive business practices. Without harmonization or cooperation based on mutual understanding of the countries' legal systems, this conflict will never be resolved enough to ensure a smooth international transfer of technology.

RESTRICTIVE BUSINESS PRACTICES IN THE INTERNATIONAL TRANSFER OF TECHNOLOGY

Parties to an international transfer of technology are usually private enterprises, one having a technology and the other desirous of obtaining it. The discussion in this section will mainly focus on this situation. There are, however, cases where one party to the transaction is a state industry. Since a state industry in such a case is engaging in an entrepreneurial business, there is no difference in its business status from that of a private enterprise. Therefore, a state industry engaging in a transfer of technology should also be subject to controls on the restrictive business practices it employs.

A transfer of technology between private enterprises is usually made by a written contract, wherein various factors culminating in the actual transfer of the technology are set forth in provisions that are to be complied with either by the recipient enterprise alone or by both recipient and supplying parties. In order to reach an agreement between parties whose interests in the transfer of technology are different or even opposing, coordination or compromise of such interests must to be achieved. From the standpoint of the theory of freedom of contract or laissez faire, the compromise or coordination made between the parties should be honored, allowing no one to interfere. In the domestic framework, a contract containing provisions pertaining to restrictions involving a transfer of technology may be subject to public interference from the standpoint of equity, public interest, or public policy, through administrative or judicial procedure, in accordance with the controls on restrictive business practices, making any restrictive provisions unenforceable if they are found to be unjustified or unreasonable. In an international sphere, where no supranational controls on restrictive business practices have yet been established, this sort of legal solution cannot be obtained. Furthermore, even if each of the countries of the parties to a technology transfer has legislation on patents and on restrictive business practices, application of such controls on restrictive business practices to an international transfer of technology by one country may affect the economic interests of the other (foreign) party, thus giving a rise to a problem of extraterritorial application. If both countries apply their controls independently, matters will be even worse. An international agreement of some sort between the countries concerned will become necessary.

It is the reality of today's world, however, that many developing countries are most interested in receiving technologies from developed countries, but do not have national controls on restrictive business practices or even patents laws. Those countries may not need national controls on restrictive business practices because of their economic underdevelopment, but they are very much concerned with restrictive business practices in an international contract, particularly pertaining to the transfer of technologies. How should this situation be comprehended? Complaints about the restrictive business practices of foreign enterprises may be in line with the assertion of the economic sovereignty of such developing countries. Then to solve this problem, an international agreement of some sort by both developed and developing countries would be required to promote the economic development of developing countries and worldwide trade.

In the days of colonization, the so-called international transfer of technologies was seen in various industries, but it was of an entirely different nature from that which is sought today. Then there was no recognition of the independent interests of the peoples who have since gained sovereignty by becoming independent countries. The gap in technological development between the countries who transferred their technologies in their own interest and the countries to which technologies were brought has been surprisingly wide, because thus far the transfer of technologies has not contributed to the real and general enhancement of the technological capabilities of the developing countries. Natural resources and manpower, abundant in the developing countries, were utilized for the benefit of the developed countries, even though the natural resources were the property of the developing countries and their people, and ignoring the fact that these people were entitled to sufficient education to enable them to develop the technological capabilities that would surely serve to spur the economic development of their country.

The wish to narrow this gap as quickly as possible has been strong on the side of developing countries, and the concern of the technology-receiving countries has been keenly felt with respect to the restrictive provisions included in international transfer-of-technology contracts made between private enterprises – (developing-country) domestic enterprises and (developed-country) foreign enterprise. Developed countries with advanced technologies have come to realize their responsibilities in regard to this situation. Both factors have contributed to the opening of international fora of various sorts to discussions of, in particular, restrictive business practices involved in international transfers of technology from the standpoint of the New International Economic Order, which was declared by the United Nations General Assembly Resolution of 3201 (S-VI), May 1, 1974.

The representative forum where this discussion is occurring with the aim of solving these problems is UNCTAD.

It was in 1972 that the OECD issued a report on the restrictive business practices involved in the transfer of technology among member countries and in international trade. With respect to the impact of restrictive business practices employed by enterprises of one member country on other member countries, and with respect to so-called extraterritorial application of the legislation on restrictive business practices by one member country, which may affect certain economic policies of other member countries, the OECD adopted a resolution as early as 1967 (on the basis of its grand objective to promote a fair and free international trade) intended to promote cooperation on this matter.(13) When the activities of multinational enterprises in world trade became of ever greater concern to member countries, the OECD, after considerable study, issued its 1976 guidelines addressed to multinational enterprises.

Although the actual and real developments of international controls on restrictive business practices can be seen, at least among OECD member countries, this is not enough to solve the problems of the same nature arising in relation to trade between enterprises of developed and developing countries. In this connection, the UNCTAD effort can be said to be a true and comprehensive one. An appropriate approach to the problems of international control of restrictive business practices in the transfer of technologies would be to consider the practices taken up in the UNCTAD forum as the activities to be brought under an international control. In parallel with the UNCTAD effort, the Commission on Transnational Corporations of the ECOSOC is working on a code of conduct for transnational corporations, which naturally involves restrictive business practices in general, including those particularly seen in the transfer of technology, but I think that it is more suitable to discuss the restrictive business practices enumerated in the UNCTAD forum for possible international control than those pointed out by the Transnational Corporation Commission, because the UNCTAD enumeration is more comprehensive in its applicability to enterprises.

UNCTAD'S INTERNATIONAL CODE OF CONDUCT FOR THE TRANSFER OF TECHNOLOGY

It is interesting to note that in the UNCTAD forum where efforts are being made to formulate a possible Code of Conduct on the Transfer of Technology, Group B (consisting of the developed countries other than the socialist countries) presented fourteen items of restrictive business practices, while the Group of 77 (the developing countries) enumerated twenty restrictive business practices.(14)

The number of items presented by each of the two groups is the result of the reconsideration of items they previously presented. At the meeting of the fourth session of the intergovernmental working group of experts, Group B added one item to its originally submitted items, making the total number fourteen, and the Group of 77 reduced its originally submitted forty items to twenty. In the case of Group B, the addition of one item to the list is a real addition, while in the case of the Group of 77 the large number of reductions is not considered to mean that the items withdrawn from the list were viewed as unnecessary; on the contrary, the items withdrawn from the list of restrictive business practices were either incorporated into other items by modifying the text or transferred to the chapter dealing with matters of guarantee, which is still under discussion.

According to the Tenatative Composite Draft Text of the Proposed International Code of Conduct on Transfer of Technology, annexed to the report (TD/AC. 1/11) mentioned above, (15) it appears that the restrictive business practices listed are items agreed to be taken up for control by both groups, but it also seems that further effort of clarification and discussion is needed before the final text will be agreed on.

Items agreed on by Group B and the Group of 77 are as follows:

1. Restriction after expiration of arrangement
2. Payments after expiration of industrial property right
3. Exclusive grant-back provisions
4. Challenge to validity
5. Sales or representation arrangements related to competing technologies or products
6. Restrictions on research
7. Restrictions on use of personnel
8. Price fixing
9. Restrictions on adaptations
10. Exclusive sales or representation agreements
11. Tying arrangements
12. Export restrictions
13. Cartels or patent-pool or cross-licensing agreements
14. Restrictions on publicity
15. Limitations on volume, scope, etc.
16. Use of quality controls
17. Obligation to use trademarks
18. Requirement to provide equity or participate in management
19. Unlimited or unduly long duration of arrangement
20. Limitations on use of technology already imported.

It is not my intention to criticize the itemization of restrictive business practices by both groups. When groups of countries differing in economic, technological, and legislative development sit around the table in an attempt to tackle restrictive business practices, some degree of illogicality cannot help but creep in. This illogicality (if any), however, must be straightened out in order to reach a sincere and effective agreement for possible international controls on restrictive business practices in transfer of technology.

Technologies to be transferred include patented inventions (including utility models), industrial designs, unpatented know-hows, and even management technology. A trademark is a mark affixed to a product to certify its quality by the manufacturer of the product and distinguish one product from another. Since a trademark on a product certifies the quality of the product, patented or not, it naturally relates to how the product bearing the trademark is manufactured, in order to maintain the established quality with which the trademark is associated. Therefore, the licensing of a registered trademark involves the transfer of pertinent technology to manufacture the similar product on which the licensed trademark is to be affixed. So when the transfer of technology is discussed, trademark licensing is also a subject of negotiation. Under the trademark law of any country, a registered trademark enjoys a legal monopoly to use of that trademark for a limited period of time, although it is renewable.

When the legal monopolies under patent law and trademark law are combined, it is easy to see that the economic power of the licensor of both rights is stronger than that of the licensee. We must bear in mind, however, that these two legal monopolies are inherently different from each other in character and scope. As long as a product is the original trademark owner-manufacturer's, importation of that product in another country cannot be prohibited in the case when the licensee (or assignee) markets a similar product with the same trademark. The reason for this is that the objective of trademark law is to guarantee the quality of the product to which the trademark is affixed.(16) Misappropriation of a registered trademark in the licensee's country by another manufacturer of a like product is of course an infringement of the trademark law of the country of the licensee.

Now let us turn back to the list of restrictive business practices.

Items 2 and 12 are restrictions that obviously do not flow from the legal monopoly recognized in most countries having a patent law.

Item 1 seems to refer to an arrangement for the transfer of technology, which may be a patent and/or know-how. A patent license cannot survive the patent's limited period of life. Know-how also loses its value as such when it becomes knowledge in the public domain. Although the nature of the restrictions are not specified, it is clear at least that Group B and the Group of 77 have agreed that any restriction extending beyond the life of a technology transfer agreement and effective even after the expiration of an arrangement is a restrictive business practice. There are circumstances, however, in which a clear-cut decision may not apply, particularly in the case of a know-how license involving a restriction that calls for keeping the know-how secret several years after the license and royalty payments have expired. Rule-of-reason applicability will have to be considered here.

The items aside from 8, 15, 19, and 20 restricting activities of a transferee are clearly not restrictions within a legal patent monopoly. By such restrictions a transferror uses his advantageous position vis-a-vis a transferee in his own business interest. Even if such restrictions are accepted, wholeheartedly or grudgingly, by a transferee, they may affect the national interest of the technology-receiving country by thwarting its competitive order. For this reason it is imperative for a technology-receiving country to have national regulatory provisions able to say that such restrictive business practices are unjustifiable and condemnable.

Items 8 and 15 need some discussion. The power to fix the price of the product of a licensee-manufacturer is not within the patent right. However, there are circumstances in which the licensor's power of price-fixing might be considered reasonable, and these need to be distinguished. The wide dissemination of a patented invention by licensing it would surely aid the speedy technological development of the country, the purpose of the patent system of any country. But since licensing is the legal method of creating competitors of the licensor, it would be understandable if the patent owner did not welcome the licensing of his patent. Although a licensor usually charges royalty from his licensees, the cost of the licensees' patented product, in which royalty payment is included, may not necessarily be higher than the cost of the licensor's product, because the licensee might absorb the royalty cost through rationalization. Therefore, price-fixing restrictions by a manufacturer-licensor on the licensed product would be considered reasonable, as striking a balance between a wider dissemination of technology and a restriction on the inherent free pricing activity of a licensee.(17) If a patent owner is not importing the patented product or manufacturing it for sale in the technology-receiving country, the fact that he is licensing the patent does not create his competitor, and therefore there is no reasonableness to allowing him to fix the price of his licensee's product. Actual situations are not as simple as this, however. To produce a sophisticated product, for example, a television receiver set, several patented inventions are very often used, with some of them belonging to one patent owner and others to other patent owners. If a licensor allows a patent to be used in manufacturing a sophisticated product in a process also using other patents belonging to other patent owners, neither he nor the other patent licensors are in a reasonable position to fix the price of the product. By the same thinking, where a patent is used in only a small portion of a product, the licensor may not fix the price of the licensee's product.

All this relates to price fixing by a patent licensor with regard to the price to be charged by the licensee, in other words the first sale by the licensee. If the restriction relates to fixing the price of a patented product on the distribution level, that is, wholesaling and retailing, such a restriction cannot be reasonable, since the licensee is not party to transactions at the distribution level. Thus although the reasonableness of price-fixing restrictions by a licensor upon a licensee is theoretically recognized, actual circumstances to which such reasonableness apply will be very narrow.

Item 15 is "limitations on volume, scope, etc." A patent license is primarily intended to permit the licensee to use the licensed patent to the extent authorized by the license under the licensor's legal patent monopoly. Therefore, theoretically, at least, limitations on volume, scope, or field of use are generally considered to be within patent right.(18) Therefore, for this sort of limitation to be treated as a restrictive business practice, the technology-receiving country must first have appropriate regulatory provisions in force.

Items 14 and 16 again concern trademark licensing. A trademark established under a national law, although it enjoys a legal monopoly, differs from a patented technology in the characteristics of the protection given it. It is a quality certificate, a guarantee to the general public by the producer of the product to which the trademark is affixed. Since a trademark usually ensures (by renewal of registration) an unlimited period of protection, an established trademark can strongly enhance the marketability of a product. Obtaining a license on a patented invention for use in producing a product plus a trademark to affix to that product gives the licensee a better marketing position. However, it may eventually create an undesirable situation for the licensee, in that he will find himself the licensor's subcontractor, and discover that he is only publicizing the name of the licensor without making the business his own. Whether to get a trademark license from the technology licensor or not is an important choice for a technology licensee. Once a trademark is licensed, the licensee cannot help but accept controls on product quality and restrictions on publicity imposed by the licensor. It is generally the licensor who insists on the use of the trademark, since this is in the interest of the licensor and his world business strategy. Rejection of these obligations means no technology licensing at all. If the technology is a patented one, a patent license only might be obtained under compulsory licensing regulations of the technology-receiving country, but no such measure would be available for an unpatented technology.

Item 19 may relate to a situation where an unpatented technology, that is know-how, is involved. In most technology transfers involving licenses for patented technology, know-how – certain secret technologies necessary in the realization of the patented technology in practice – is included. Regardless of whether licensing of know-how is arranged independently or together with a patent, the advantage over competitors is prolonged as long as the know-how is kept secret by the parties.

The problem with know-how is not only that its precise delineation is not clear, due to the fact that it is not patented, but also that its real technological value is not known to the licensee until it is completely disclosed to him – after the license contract is signed. Furthermore, there exists the danger of disclosure of the know-how by a third party. A would-be licensee must therefore be careful in negotiating a license, particularly if it involves payment of royalty. Apart from the method of payment for a know-how license, the character of payment – as royalty or whatever – must be considered, so that unexpected difficulties that arise after the license contract is signed can be properly resolved. In sum, in obtaining a know-how license it would be better to consider it as the buying of desired know-how; then the negotiated price can be paid either in a lump or in installments, in accordance with whatever sort of agreement is reached.

When integrated technologies are transferred in the form of a full turnkey or product-in-hand system, these sorts of know-how licensing problems may not arise. It is very doubtful, however, that turnkey or product-in-hand systems can by themselves serve to teach the people of developing countries the vast amount of technology (including know-how) incorporated and integrated in the system. I must add here that although I admire the efforts of countries that are utilizing turnkey or product-in-hand systems to narrow the time lag in their economic development, I place much more importance on countries that are at the same time deploying a vast amount of their expenditures for general education, primary as well as higher, to upgrade their peoples' technological capabilities so that they can develop indigenous technologies as well as digest high-level technologies from developed countries, to make the economic growth of their countries sure and steady. Item 19 reflects the self-help intentions of the developing countries.

Item 13 is different from the other items. Of concern here is a division of the world market that affects the exports of individual countries and international trade.

A cross-licensing arrangement by which a division of the market is achieved is made between enterprises in different countries with different technologies that are equally useful in manufacturing their competitive product. Establishing an import barrier against the inflow of the competitive product by receiving an exclusive license from the competitor will be a sure method of protecting both parties' respective markets. However, this practice, whereby a national patent system is flagrantly abused, is contrary to the objective of free international trade.

Technology-pooling practices can be found in both the domestic and international arena. Depending on the purpose and practice of such a pool, it is welcomed or condemned. If it serves to gather technologies pertinent to an industry, with licensing of these technologies open to any applicant (just like a public library), such a pool is welcomed. But if its licensing is done selectively according to an agreement among the pool members, then the goal of elimination of competition is apparent, and the pool will therefore, be dealt with as a restrictive business practice. A division of a market can be achieved by cross-licensing and/or a patent pool.

The legal nature of the proposed code of conduct on transfer technology is still undecided. It is interesting to note that Group B proposed that there be a separate chapter on the responsibilities of technology-receiving and technology-supplying enterprises, while the Group of 77 proposed that there be a chapter on guarantees. When a transfer of technology is desired from one enterprise to another, it is normal practice to formulate a contract stipulating, in addition to the primary positive matters relating to the transfer of the technology, other matters relating to the responsibilities of the technology-supplying enterprise and the guarantees offered by the recipient enterprise. Whatever wording is used, the point is not only to avoid disputes between the parties that might arise later over unsatisfactory or noncompliance in regard to what either party is required to perform in the contract, but also to clarify the extent of damages or liabilities for both parties. Therefore, it is wise to have such provisions.

However, the meaning of the words "responsibilities" and "guarantees" in an international transfer-of-technology contract differs, in that the former apparently refers to the responsibilities of the parties themselves only, while the latter implies something more, which goes beyond the competence of any particular party. The role of governments in an international transfer-of-technology contract seems to be what is in the minds of the drafters of UNCTAD's proposed code of conduct.

Payments of royalties abroad have direct bearing on the balance of payments of a country. Therefore, some countries have an approval system on international transfers of technology, and thereby undertake a selective introduction of technology. In some technology-receiving countries there is a procedure for scrutinizing contract provisions before they are put into effect, so that any restrictive business practice prohibited in that country will not be agreed to and practiced. After a green light is given to an international technology transfer, the parties can go ahead with the contract.

Nonperformance by either of the parties of any part of a contract thus approved may finally be brought to litigation by one or the other party under the civil law of the supplying or the receiving country. Take payment of royalties, for example. The approval of a contract, implying a guarantee of such payment abroad, by the government of a recipient country should, of course, not be taken as a guarantee given to the supplying enterprise – it is given to the recipient enterprise. If the recipient enterprise becomes unable to buy the necessary foreign currency for payment of royalties abroad due to its own bad business, the possible recourse of the supplying enterprise might be to a damage suit, usually following termination of the contract. This is a kind of business risk that any enterprise might encounter. In connection with restrictive business practices, the business risk that parties to an international transfer-of-technology contract may encounter will be serious, particularly from the legal aspect. This problem will be discussed by describing the development of enforcement of the Japanese Antimonopoly Act in the following section.

THE JAPANESE EXAMPLE

I would like to call the reader's attention to Section 6 of Japan's Antimonopoly Act of 1947 as amended in 1958 and 1977. As is well known, the Antimonopoly Act was enacted in compliance with a directive of the General Headquarters of the Occupation Forces, and was a transplanting of the United States' antitrust acts in an ideal and more stringent form. The act was significantly amended in 1953, however, to ease its provisions in line with the real economic characteristics of Japan, and was again amended in 1977 to strengthen it in order to deal with changes in industrial structure so as to allow more effective enforcement of the act. Section 6 deals with international contracts or agreements. It runs as follows:

> Sec. 6 (Prohibited international agreements)
>
> (1) No entrepreneur shall enter into an international agreement or an international contract which contains such matters as constitute unreasonable restraint of trade or unfair business practices.
>
> (2) An entrepreneur who has entered into an international agreement or an international contract shall, in accordance with the Rules of the Fair Trade Commission, file a report thereof with the Commission, accompanied by a copy of the said agreement or contract (in the case of an oral agreement or contract, a document describing the contents thereof), within thirty days as from the execution of such act.
>
> (3) The provisions of the preceding subsection shall not apply to an agreement or contract regarding a single transaction (excluding such transactions of which the delivery of the goods extends over a period of one year), or to an agreement or contract merely creating an agency in business matters (excluding an agreement or contract containing conditions that restrict the business activities of the other party).

Section 6 covers all types of private international agreements and contracts, including cartelization, technology licensing, money borrowing, agency licensing, trademark licensing, and joint ventures involving Japanese and foreign enterprises. It also covers both incoming and outgoing transactions.(19)

Up to fifteen years ago Japan lived on exports of products made by using raw materials imported from abroad. In order to produce exportable products Japan needed foreign technologies, because it was far behind the technological standard of the United States and the European developed countries. Its financial position was as miserable as could be, due to the devastation of war. To export was imperative for Japan, not only for reconstruction but simply to survive as an independent country. But it faced two problems: (1) how to get technology from abroad, and (2) how to earn the money it needed to buy necessary raw materials. To cope with these problems Japan had to select the most effective technology on one hand, and earmark limited moneys for the importation of materials for important industries on the other. The government took a major responsibility in this regard and exercised strong control over industry. The Foreign Exchange and Foreign Trade Control Act (1949) and the Foreign Investment Act (1950) were the statutory basis of this control.

Section 7 of the Foreign Investment Act provides that in the case of technical assistance supplied by a foreign company, the Minister of Finance and the Minister of International Trade and Industry shall make public the kinds of technology the country desires to obtain.

The attitude toward proposed technical assistance contracts gradually changed in the direction of liberalization. In its first ten years, the Foreign Investment Council had the authority to grant approvals, selected from among many proposed contracts only those technologies that were regarded as contributing positively to the improvement of Japan's balance of payments and the development of important industries. From 1959 on the scope of technology to be introduced was widened, and only technologies that, it was feared, would adversely affect the Japanese economy were not allowed. At present the approval system itself, although it still remains on the book, is no longer used.(20)

Although the procedures for finalizing technical assistance contracts have been almost entirely liberalized, the two laws provide that approval of a proposed technical assistance contract by the Foreign Investment Council does not necessarily mean a clean bill of health as far as the Antimonopoly Act is concerned. In other words, although today no approval is needed for a technical assistance contract, it is still subject to scrutiny under the Antimonopoly Act.

The purpose of the issuance of the FTC's Guidelines mentioned earlier, was to refresh the memories of Japanese enterprises regarding the antimonopoly regulations concerning international technical assistance contracts and, furthermore, to help them avoid stipulating restrictions that fall under the heading of unfair business practices. The Guidelines deal only with a single licensing contract involving a licensor and a licensee. In the case of a complex international agreement constituting a cartel by way of international cross-licensing, or a joint venture scheme in which transfer of technology is involved, the several interrelated agreements will be considered as an integrated whole under the substantive provisions of the Antimonopoly Act.

Let us now turn back again to Section 6 (1) of the Antimonopoly Act. The proper and authentic interpretation of this provision is that it prohibits not only a contract that at the time of its conclusion is found to contain provisions that constitute unreasonable restraint of trade or unfair business practices, but also that it prohibits a contract found years after to contain such provisions.

All international agreements or international contracts that are entered into must be filed with the Fair Trade Commission, the enforcement agency of the Antimonopoly Act. The commission scrutinizes international agreements or contracts filed with it on the basis of the then-current market conditions of competition. For example, if the commission finds that a technical assistance contract contains clauses that restrict the activities of the Japanese licensee and that are likely to impede fair competition in the market concerned, it will advise the Japanese party to modify or delete such provisions so that this unfair business practice will not be included. Because of jurisdictional difficulties, the addressee of such advice is the Japanese firm, the licensee; it is liable to prosecution for violation of Section 6 (1), by being a party to the contract in question. If the addressee does not follow the advice of the commission, for example, because of inability to renegotiate with the foreign licensor, the commission will take formal action against the alleged violation of the act in accord with the procedures provided for in the Act, and will issue an order for the elimination of the restrictive practice. It is rather rare for the commission to take formal action on an international contract containing a restrictive clause. Advice from the commission – in other words, administrative guidance – is usually complied with by the parties involved. Of about 6,000 international contracts filed in 1977, about 1,400 related to technical assistance, and the commission gave administrative guidance on nearly 30 percent of these contracts.

Some readers will no doubt wonder what happens when the Japanese firm accepts the commission's advice despite the opposition of the foreign licensor. An administrative formal action can be taken only against a Japanese firm; therefore an order not to maintain the disputed clause is issued to him only and not to the foreign party.(21) This can mean a difficult legal situation for both parties, particularly at the civil law level.

Now, what about an international contract in which the commission did not find any restrictive provisions – in other words, where the commission, after scrutinizing a contract filed with it, did not say anything? This nonaction of the commission does not mean that it has given a binding approval, or that it is unable to take later action if circumstances require it. This is a very important point of Section 6. Since a contract will be in existence for a certain period of time – five, ten, or more years – during this time the market conditions of competition will change, and certain provisions of the contract that were originally considered not in restraint of competition may become restrictive on the licensee and may affect fair competition in the market. When such a situation occurs, it is the duty of the commission to interfere with the contract to eliminate the restrictive provision. Thus parties to an international contract cannot be completely safe from challenge under the Antimonopoly Act as long as the contract is in existence. Uncertainty about a challenge from the Antimonopoly Act may be very troublesome to the parties, particularly to the licensor. This is, however, unavoidable, since the market factors that constitute the premise on which a particular technical assistance contract was entered into may change in the course of time, and may make certain provisions in the contract in violation of Section 6 (1).

An illustration of such troublesome situations is the fact that technical assistance contracts that the parties intended to renew are currently subject to administrative guidance of the commission, which wants the modificaton or deletion of certain provisions in the original contract, although at the time of the initial filing nothing was said by the commission. More sophisticated cases, such as joint venture agreements setting up a jointly owned company in Japan, to manufacture and sell products using the technology of the foreign party and the manufacturing, marketing, and management ability of the Japanese party, have now given rise to an antimonopoly problem. Fiduciary relationships between parties in an international private contract could not be countenanced by the Antimonopoly Act, which is a public law prohibiting a private monopolization, unreasonable restraint of trade and employment, or unfair business practices – expressing the public policy of maintaining a fair and free competitive order in the domestic market. Here again, the difference among legislations on restrictive business practices in different countries with respect to international private transfer-of-technology contracts calls our attention to the OECD's resolution of 1973 on cooperation in the field of control of restrictive business practices. In order to have cooperation among countries in this field, each government must first represent a sovereign state willing to take full responsibility for domestic as well as international economic order.

CONCLUSION

I may have spent too many pages in describing the Japanese experience with respect to the control of restrictive business practices in the transfer of technology. I have merely tried to indicate that any international control of restrictive business practices in the transfer of technology will be unsuccessful unless each country (1) establishes its own economic development objectives, which may vary from one country to another; (2) perceives some distance between its present economic condition to the expected goal; (3) decides ways and means to attain the goal, such as the enhancement of prerequisite general educational standards in order to enable its people to absorb selected necessary technologies from abroad; and (4) prepares legislative measures for safeguarding its business from those restrictive business practices that are likely to affect the development of its economy.

The itemization of restrictive business practices by the Group of 77 in UNCTAD is considered exhaustive, but all of these items may not be vital to every developing country. However they serve as a good guide for a developing country on how to control restrictive business practices related to the transfer of technology. The most important thing of all, however, is for a country to have a definite idea why, from the standpoint of its own social and economic conditions, certain types of restrictions should be treated as restrictive business practice.

The New International Economic Order declared in the United Nations will surely call on developed countries to contribute to the development of developing countries in various ways. I am convinced, however, that the mere submission of complaints to the United Nations by developing countries will not solve the social and economic problems those countries face, because restrictive business practices employed in the international transfer of technology cannot be legally dealt with unless the country making the complaint has national legislation controlling such practices.

NOTES

(1) The relation of industrial property, particularly as regards patents, to restrictive business practices has been the theme of many studies. The report of the Committee of Experts on Restrictive Business Practices, issued September 27, 1972, on "Restrictive Business Practices Relating to Patents and Licenses," focused on these practices' international implications.

(2) Out of many cases the following are cited: Standard Oil Co. of New Jersey v. United States, 211 U.S. 1, 1911; United States v. Aluminum Company of America, 148F(2d)416, 1945; United States v. Griffith, 334 U.S. 100, 1948; American Tobacco Co. v. United States, 328 U.S. 781, 1946.

(3) The United States, West Germany, Canada, the United Kingdom, Japan, the EEC.

(4) For example, U.S. Sherman Act, Section 1; EEC, Article 85. In Japan only a horizontal cartel agreement is prohibited by the second part of Section 3 of the Antimonopoly Act, as an unreasonable restraint of trade. A vertical agreement in substantial restraint of trade constitutes a monopolization violating the first part of Section 3 of the act.

(5) For example, Section 19 of Japan's Antimonopoly Act, FTC regulation no. 11 of 1953, designating generally applicable unfair business practices, plus several regulations designating unfair business practices in specific industries.

(6) Section 5 of the Federal Trade Commission Act of the United States.

(7) According to the report of the Japan FTC of April 1977, twenty countries out of the twenty-four member countries of the OECD, plus sixteen nonmember countries in Asia and Latin America, have legislation on restrictive business practices; the EEC, EFTA, and Caribbean communities also have regulatory provisions in the treaties establishing each community.

(8) Japan: Antimonopoly Act, Sections 23 and 100, FTC's Antimonopoly Act Guidelines for International Licensing Agreements May 24, 1967 (hereinafter Guidelines). West Germany: Act against Restraint of Competition, Sections 20 and 21. United Kingdom: Restrictive Trade

Practices Act of 1956, as amended in 1968, Sections 6, 8(4); Patent Act of 1949, Section 57; Monopolies Act of 1948, Section 20(1).

(9) While almost all developed countries' patent laws provide that the importation of the patented product is within the patent right, some developing countries do not.

(10) Act against Restraint of Competition of West Germany, Section 20(1), proviso; the Guidelines of FTC, Japan, Section 3. Although the Guidelines deal with restrictons in international licensing, the Japanese FTC's Antimonopoly Act enforcement against restrictions in purely domestic licensing is similar in thinking.

(11) Japan FTC's Guidelines do not say that a price-fixing restriction upon a licensee of a patented product is safe, while the Act against Restraint of Competition of West Germany exempts such restrictions, in Section 20(2) of the act.

(12) Lear, Inc. v. Adkins, U.S. Sup. Ct., 395 U.S. 653 19.

(13) The so-called notification system is adopted under which a member country notifies of its action taken or to be taken on a case wherein a foreign enterprise is involved to the member country to which such foreign enterprise belongs. In 1973 the system was further extended to include complaint procedure.

(14) See the report of the Intergovernmental Group of Experts on an International Code of Conduct on Transfer of Technology on its fourth session, held at Palais des Nations, Geneva, from October 31 to November 11, 1977, TD/AC. 1/11.

(15) Ibid.

(16) Centrafarm BV and Adriaan de Peijper v. Sterling Drug Inc., Court of Justice EC, Case No. 15/74, October 31, 1974. Parallel import, of a genuine product bearing the same trademark does not constitute an infringement of a registered trademark in Japan and the United States.

(17) See Section 20(2) of the West German Act against Restraint of Competition.

(18) Section 20(1) of the West German Act against Restraint of Competition; Section 3 of the Guidelines of Japan. In the United States a restriction on field of use has come under challenge based on the Sherman Act; see Committee of Experts, "Restrictive Business Practices," p. 47.

(19) For a historical view of the development of Antimonopoly Act enforcement of international licensing, see Michiko Ariga, "Regulation

of International Licensing Agreements under the Japanese Anti-monopoly Law," in Patent and Know-how Licensing in Japan and the U.S., ed. T. Doi and W.L. Shattuck (1977), pp. 278-319.

(20) Japan today still depends on foreign technology, however. Royalties paid abroad still exceed the amount of royalties received from foreign licensees.

(21) Novo Industri A/S v. Fair Trade Commission, Supreme Court decision of November 28, 1975.

Comments

Miguel S. Wionczek

A report published in 1953 by the Ad Hoc Committee on Restrictive Business Practices of the U.N. Economic and Social Council complained that

> governments that subject restrictive business practices in domestic trade to some measure of governmental action or scrutiny have generally abstained from inquiry into the existence or nature of practices involving international trade; the few cases where such inquiry was made covered only limited parts of the foreign trade of individual countries.(1)

Noting the similar lack of interest in that subject in the interwar period among international organizations, the report added that

> no private body and institution appears either to be continuously engaged in the systematic collection of data concerning the extent of restrictive business practices.

As the paper submitted to this conference by Dr. Ariga demonstrates, the situation is very much different today. Moreover, it is quite probable that in the near future, before the end of 1979, a worldwide agreement will be reached on policing – whether mandatorily or voluntarily – a rather impressive list of restrictive business practices considered undesirable in the international technological transactions that are part and parcel of international trade.(2)

The fact that restrictive business practices have increasingly become an important policy issue for governments and international agencies is the result of two clearly contradictory trends.

First, government policies, especially in major industrial countries, by encouraging and permitting the concentration of economic power at the national and international levels, have inevitably facilitated the growth of market power and have provided ample opportunities for increased and effective use of restrictive business practices by private enterprises.

Second, concurrently with the expanded use of such practices, particularly by transnational enterprises, it has become widely recognized that restrictive practices affect the interests not only of their home countries of large enterprises but the interests of other countries as well.

While currently the subject of restrictive business practices is being treated in many places, both in and out of the United Nations, it is quite probable that the treatment it has received within the UNCTAD negotiations on an international code of conduct on technology transfer offers better operative possibilities than other attempts, at such a code, because nowhere outside the UNCTAD draft code has an effort been made to specify and exemplify all major restrictive practices currently in use. The progress achieved in that respect by the UNCTAD code negotiators should not obscure, however, the scale of the disagreement that still exists between the developing world and the Western industrial countries.

At two sessions of the U.N. conference on a code, agreement has been reached (as of February 1979) on the provisions relating to exclusive dealing, exclusive sales or representation agreements, and payments and other obligations after expirations of industrial property rights. Moreover, with respect to the criteria to be applied, agreement has been reached on including the notion of restraint of trade and of adverse effects on the international flow of technology, particularly when restrictive practices hinder the economic and technological development of acquiring countries. But behind the unwillingness of the industrial countries to include in the list of restrictive practices six items that appear in the draft code, elaborated by the Group of 77 (limitations on volume and scope of production, use of quality controls, obligation to use trademarks, requirement to provide equity or participate in management, unlimited or unduly long duration of arrangement, and limitations upon use of technology already imported), lies a series of major disagreements between the two major negotiating groups.

According to an UNCTAD Secretariat document that presents to the UNCTAD V (1979) a review of major issues under negotiations in the field of the transfer of technology and the industrial property system,(3) the following difficulties with respect to the chapter on restrictive practices in the draft code still persist.

1. While fourteen restrictive business practices, agreed by all parties, are considered by the industrial countries all-inclusive by virtue of being anticompetitive in nature, the Group of 77 insists on the inclusion of the six additional practices mentioned above, independent of whether they are anticompetitive or not, because they are regarded as being, for one reason or another, unfair to the acquiring party or adverse to economic and technological development in the wider sense. Such is particularly the case of the obligation to use trademarks, which not only strenghtens the monopoly patent but is directly responsible for creating outward-directed consumer preferences.

2. The question of the coverage and scope of the practices is far from settled. The industrial countries would like to refer in the title of the chapter dealing with them simply to "restrictive business practices," while the Group of 77 explicitly suggests their regulation by proposing a general title, "The regulation of practices and arrangements involving the transfer of technology."

3. The industrial countries continue to insist that it is necessary not only to enumerate restrictive practices but to determine in real life whether their effect is actually harmful. Consequently, they want to introduce in this, as in other fields covered by the code, the notion of the "rule of reason" or "public interest." The Group of 77, however, objects to the inclusion of this concept in the definition of restrictive practices, on the grounds that application of the "rule of reason" gives rise to a real danger of arbitrary interpretation of restrictive practices by the same parties that originate such practices, be they private parties or their governments.

4. Finally, on the grounds of "unequal partnership," the Group of 77 defends the right of competent authorities of the technology-acquiring country to disregard, in exceptional circumstances, the restrictive practices proscribed by the code, provided that on balance and on development grounds there will be no adverse effect on its national economy. Industrial countries do not accept such an exception clause in the code.

In brief, while the shape of the treatment that the issue of restrictive business practices will receive in the UNCTAD code is slowly emerging, the process of international negotiation has been much less smooth than is suggested by Dr. Ariga's paper. Moreover, it would perhaps never have reached such a detailed level were it not for the expertise available within the Group of 77 as the consequence of the recent introduction in many less developed countries of restrictive business practices legislation. Such legislation arose mostly from the double concern (1) about the structure of economic power in itself, resulting from historical developments and economic dependence on the advanced industrial economies; and (2) about the ability of mostly foreign enterprises with dominant power to use it to the detriment of the broadly defined national interest, by overpricing goods and technology and controlling the production and distribution of products. It is no accident that these concerns found their reflection in the most detailed ever presentation of forty restrictive practices in the early draft code elaborated by the Group of 77. As Dr. Ariga correctly observes, the reduction of this original list to twenty practices between 1975 and 1978 should not be understood as the admission by the Group of 77 of the irrelevance of half of the practices originally detected by its experts. On the contrary, for the sake of arriving at a mutually agreeable consensus the "missing" twenty were incorporated in the shorter basic list or transferred in new legal language to other parts of the draft code.

It would be unrealistic and naive to assume that any sort of international regulation of such practices, including one that may emerge from the UNCTAD code on the transfer of technology once it is adopted, will by itself take care of the concentration of economic power and its encroachment on the development process in less developed countries. Following Dr. Ariga, one has to stress that international regulation of restrictive business practices will fail unless the individual countries (1) set up their own development strategies; (2) design and implement policies aimed at increasing their domestic economic and technological capability; and (3) introduce national legislation against restrictive business practices, defined more broadly than is the case in the similar legislation in force in the industrial countries or their groupings, such as the European Economic Community.(4)

NOTES

(1) U.N. Economic and Social Council, Restrictive Business Practices, Analysis of Governmental Measures Relating to Restrictive Business Practices, Official Records: 16th sess., Suppl. 11A (New York: United Nations, 1953).

(2) In fact a set of worldwide voluntary guidelines on restrictive business practices was adopted in April of 1980 under the auspices of UNCTAD. Eds.

(3) UNCTAD, Restructuring the Legal and Juridical Environment: Issues under Negotiations - Technology, 5th sess., Manila, Philippines, TD/237, January 29, 1979, Geneva.

(4) UNCTAD, Control of Restrictive Business Practices in the European Economic Community, Report by the UNCTAD Secretariat, TD/B/608, Geneva, 1977.

Comments

Enrique Aguilar

The following comments will focus attention on the legal and economic implications that derive from international technology transfer agreements, with special emphasis on restrictive business practices in licensing transactions in developing countries.

On this basis it will provide a different perspective than does the excellent paper prepared by Michiko Ariga, yet it will draw it on the global situation and on the existing international system that aims at controlling restrictive practices.

BACKGROUND

As world economic and development structures have changed, technical knowledge has become an increasingly significant factor in the development process of nations. This becomes particularly important for the developing countries that most heavily rely on the acquisition of proprietary know-how from foreign sources.

In recent years the developing countries have become better prepared to deal with the complexities of international licensing and have developed, with the assistance of international organizations (UNIDO in particular), a system and a coordinated approach to orient and regulate the acquisition of foreign know-how.

For this purpose national agencies have been created with the necessary authority and governmental support to rationalize licensing procedures, and to acquire technology in terms and conditions most convenient to the recipient country's economy.

From the outset it should be stated that developing countries will in the future need to increasingly rely on and spend more for the purchase of foreign technology, mainly because technological requirements will increase (both in cost and sophistication) in close co-relation with the industrial development of the country. As a result, the transfer of technology from developed to developing nations is going to play an even greater role in international economic relations.

Technology, however, very often involves a high price not only in terms of direct payments and royalties, but also as a result of contractual provisions that recipient companies in the developing countries are forced to accept in order to attain specific objectives.

Many licensors of "technological knowledge" – including patents, trademarks, process organization, trade secrets, and the like – have managed to unduly use that knowledge as a lever to impose obligations, restrictions, and controls over production on the recipient country that otherwise would not be possible.

Many firms from industrialized countries that enjoyed technical advancement and expertise have in the past been able to impose contractual provisions that not only perpetuate noncompetitive market structures but also – more important – tend to harm the internal economy of the recipient country.

The conditions under which the commercialization of technology operates on a world scale have posed multiple legal, economic, and political questions; the awareness of its implications among the recipient developing countries explains the active role the United Nations and the academic community as a whole have taken in regard to these questions.

Another area of concern about this situation, which can be seen as a new variant of potential market domination, is growing on the international level. Both developed and developing countries are seeking ways to limit the influence and control of the business interests of other nations.

In order to deal with this problem, several governments of developing countries have enacted specific legislation beginning in early 1970.

It should be mentioned that more than 90 percent of the technological inflow (and corresponding payments) and direct foreign investment into Third World nations originates from the advanced countries. The United States by and large has been one of the world leaders in technological development and, likewise, U.S. enterprises have been prominent in licensing and controlling technology.

In this context the U.S. antitrust laws provide one source of control over restrictive trade practices contained in contracts with foreign companies. Antitrust principles have been applied to such agreements in the past, but from the point of view of developing countries the exact scope and limitations of those principles are not clear.

Under U.S. law, technological knowledge is considered intellectual property and has special status and protection. Certain forms, such as patents or trademarks, do provide "legal monopoly" to their holder. However, even this grant protection is strictly limited in terms of scope, field of use, duration, and so on.

It emerges from this that a legal monopoly cannot be used to obtain greater control or to work against antitrust principles. From the point of view of government authorities it becomes essential to properly define the scope of legally protected monopolies with as much accuracy as possible, since restrictive terms outside the "protected area" may be construed as violations of antitrust laws.

The Mexican experience has shown that more than 75 percent of technological licensing agreements do not involve patents, and fall within the category of know-how licensing contracts. Extensive research conducted at the international level demonstrates that in the future the trend will be to rely more on know-how licensing and to minimize the use of patents as a main object (with exceptions in certain fields, such as pharmaceuticals).

In this connection it is pertinent to point out that secrets and unpatented technological know-how cannot obtain legal monopoly protection, and in practice this becomes a fundamental issue that calls for careful examination.

A basic area of concern is that the "nonprotected category of technology" is only protectable by contractual provisions of nondisclosure and/or nonuse. Under a similar analysis, restrictions in such contracts cannot be used to produce controls that go beyond the necessary scope of protection.

The basic relationship between technology licensing and antitrust regulations is that control of technology on the one hand and legal monopolies on the other cannot be used to give the holder more rights than are specifically defined in legal statutes. Likewise, neither contracts for technology nor specific business arrangements, such as joint ventures or wholly owned subsidiary arrangements, can be used to overcome these limitations and to create anticompetitive market structures that are otherwise illegal.

Although this remains a highly technical and specialized field of activity, governments of developing countries must carefully study the implications that derive from contractual arrangements in the legal and economic spheres.

To illustrate this point, I will refer to the Mexican experience in dealing with licensing of nonpatented technology that quite frequently called for an excessive nondisclosure (secrecy) period on the part of the recipient – either an indefinite period or an obligation that extended beyond twenty years.

Mexican patent law grants a maximum of only fifteen years of protection. It is clear that the licensor is seeking to acquire substantially more as a result of contractual provision that what he would acquire by obtaining a patent. In other words, monopoly rights are sought through contractual arrangements, even though nonpatented know-how does not acquire legal monopoly protection.

Dr. Ariga's paper goes into considerable detail in enumerating the types of restrictions that contravene national objectives and that could be made unenforceable. From the point of view of a developing country these restrictive practices have a different connotation.

First of all, it has to be recognized that the scope and nature of technological requirements extend beyond the specific transfer of patented or nonpatented technologies, which, in turn, implies a greater involvement of the supplier of technology in the planning and execution of projects.

RESTRICTIVE PROVISIONS IN DEVELOPING COUNTRIES

As was indicated previously, considerable progress has been achieved by government agencies in developing countries in the application of guidelines and criteria for the examination and approval of contractual arrangements.

International organizations, including UNIDO and WIPO, have also contributed in the preparation of guidelines for developing countries in this area. On a more general level, the debate under the auspices of UNCTAD on the adoption of an international code of conduct for the transfer of technology contributed to what may be callled the general educational level. Nevertheless, it should be stressed that the most important knowledge is emerging from the practical work and empirical data accumulated by the developing countries themselves.

To illustrate this I will deal primarily with three issues among those that concern government agencies in developing countries: confidentiality, access to improvements (grant-back), and tie-in provisions.

Confidentiality

Issues pertaining to confidentiality in contractual agreements deserve to be considered at two different levels: that of the recipient enterprise in the developing countries, and that of the overall interest of the developing country.

With respect to the recipient enterprise, the issue of confidentiality largely relates to specific provisions in the technology contract; whereas with respect to the recipient developing country the issues relate primarily to the question of sharing and disseminating information and experience with the objective of broadening the area of technological choice, know-how, and expertise about the acquisition of foreign technologies in various production sectors.

A confidentiality provision often implies for the recipient enterprise that the technological know-how and information disclosed shall continue to be the property of the foreign supplier, even to the extent that the know-how cannot be exploited after the expiration of the agreement or outside the field of use so defined, or even of the territory agreed upon between the parties.

The issue of confidentiality could also imply that the know-how cannot be freely communicated to third parties and, furthermore, could extend to contractual obligations for the recipient enterprise and its employees to prevent and safeguard the use of this information.

All or some of the above possibilities could be in direct conflict with the developing country's national interest, for obvious reasons.

Access to Improvements and Grant-Back Provisions

Through the examination of a large number of contracts entered into by developing countries, it has been found that grant-back provisions have been introduced requiring the recipient company to assign back to the licensor any new patents, improvements, or technical developments achieved by the licensee during the life of the agreement.

In this connection, the analysis of grant-back provisions can be divided into three parts:

1. Provisions according to which the licensee is obliged to inform the licensor of all the knowledge and experience the licensee has acquired in connection with the goods and services covered by the contract
2. Provisions that oblige the licensee to assign the rights (patent rights or rights arising from application thereof) related to any improvement, invention, or application of inventions that the licensee has made
3. Provisions that oblige the licensee to grant to the licensor a license on any improvement, invention, or application of invention that the licensee has made.

According to the basic criteria followed by many developing countries, these provisions may not work against the interest of the enterprise or the developing country, if the supplier of technology is placed under the same kind of obligations and if the obligations of the supplier and the recipient company are properly balanced. Furthermore, these criteria also hold that the obligations of the parties are "properly balanced" if the obligations to both parties are similar in kind and established under reciprocal basis in relation to:

- royalty payments
- duration of obligation
- the territory in which improvements will be exploited by either party
- the degree of exclusivity applied to either party.

Grant-back provisions are closely related and can have a determining effect on the access to technological improvements by the recipient developing country. This is a most important issue and needs to be specifically provided for in any technology contract.

Tie-in Provisions

Tie-in provisions imply that the recipient enterprise that wishes to acquire certain technology or goods from the supplier (licensor) is also required to purchase other products from the same licensor.

Such clauses are often considered violations of various antitrust legislation; for example, the Clayton Act, Section 3, focuses attention on the sale of commodities "patented or unpatented for use, consumption, or sale within the United States." On the other hand the Sherman Act can be applied to sales outside the United States and tie-ins may be construed to be violations where the United States' commerce might be affected.

The general criterion is that tie-in clauses are unreasonable whenever a party has sufficient economic power with respect to the tying product to appreciably restrain free competition in the market for the tied product. The legality of this is usually established when the tying product is patented; however, when no patent is involved but only a trade secret or process know-how, this requirement may be harder to prove. Through national legislation in many developing countries, tying clauses are per se not acceptable. The experience of some countries shows that in the majority of cases it is possible to eliminate tie-in provisions from the contract. However, the alimination of a contractual provision may in itself not be sufficient to prevent a wholly owned subsidiary or a licensee with a majority foreign participation to de facto exercise this requirement.

The question of tie-in clauses relating to the exclusive supply of intermediate products or components from the licensor has been the focus of considerable attention, and the question of "transfer pricing" has figured prominently in recent literature. It is obvious that tie-in clauses may not be desirable and could constitute a serious disadvantage to the licensee in a developing country.

From the point of view of a developing country, it is necessary to ensure that domestic manufacture is utilized to the maximum, on an economical and commercial basis. This would reduce the magnitude of the problem and would avoid a common tendency on the part of licensors, especially in the field of engineering products, to limit domestic manufacture in developing countries to assembly or semi-assembly operations for excessively long periods. However, even where an acceptable program of domestic integration is defined, the problem of pricing imported components and materials remains, and negotiations in this regard could center around certain specific areas, such as intermediate products and components bought out by licensors; and, where components are manufactured by the licensor, the cost of such components should be the cost at which the components are priced in the next stage of production in the licensor's plant plus any handling and other costs that may be involved. A solution in terms of contract provisions may perhaps be that (1) the licensee shall be free to obtain such items from any source, and that if the licensor supplies such products and components, he shall (2) supply such items at internationally competitive prices; (3) deduct the price of such imported components from the royalty base; and (4) shall to the extent possible accord a most-favored-licensee clause in relation to the cost of such products.

CONCLUDING REMARKS

It is quite obvious that the implications of restrictive business practices in licensing agreements are quite complex and that in order to deal with these problems the efforts of the developing countries should not be limited to legal scrutiny but should also incorporate economic and technical considerations, to adequately assess the effect of these restrictions on the recipient enterprise and on the economic and industrial development of the country.

From the point of view of the developing countries, the policy rationale in providing intellectual property protection and for regulation in the field of technology transfer should not be viewed in isolation from the economic and technological policy of the country.

The presence of restrictive clauses in international agreements could hamper the economic and technological development of the recipient nations, and the relations between North and South. On the other hand, one can appreciate the need to understand that technological contracts are excellent vehicles for strengthening the technological capabilities of the recipient country. Finally, we believe that these agreements are to be evaluated on the basis of economic and commercial criteria to properly ascertain the magnitude of problems and to create and maintain a receptive climate for technology and foreign investment in the developing countries' economies.

8 Conflicts in the International Transfer of Technology: A U.S. Home Country View

Robert G. Hawkins
Thomas N. Gladwin

Controversies over technology transfer have many manifestations in technology-exporting countries. Because of its multifaceted nature, policies toward and control systems over technology transfer in the major industrial countries are difficult even to identify and describe, much less analyze.

This chapter adopts a narrower focus. It is concerned only with controversies and control systems surrounding the outward transfer of technology from the United States. While this has the practical advantage of limiting the task of the paper, this delimitation does not greatly harm its generality and relevance. For one thing, the United States remains the world's single largest national source of technology for international transfers. Despite the fact that foreign firm activity in the United States has grown dramatically in the 1970s, the flow of outward direct investment by U.S. firms remains about three times that of inward foreign investment into the United States, and U.S. receipts from royalties and license fees (associated with the transfer of U.S. technology abroad) are over ten times U.S. payments to foreigners for the same purposes.(1)

In addition, among the predominantly technology-exporting countries the controversies and policy control issues regarding technology transfer concern almost exclusively outward technology transfer, with relatively few conflicts on the technology import side. For example, the United States permits free entry of foreign firms except in a few strategic industries (communications, defense, and nuclear technology) and has limiting restrictions in some financial service industries such as banking and insurance. Most other conflicts associated with inward

*The research underlying this paper was partially supported by the National Science Foundation (Grant No. RDA 76-20257), and the NYU Project on "The Multinational Firm in the U.S. and World Economy."

technology transfer fall under the issues of U.S. safety, consumer protection, environmental, and product quality standards, which may in fact be viewed as technology transfer issues. But heretofore these conflicts have been relatively few and minor.

Finally, the other major advanced countries that have substantial transfers of technology abroad have experienced many of the same types of conflicts as has the United States, and in many cases have considered or adopted similar policies or control systems.(2) Although the United States leads most of the technology-exporting countries in the breadth of issues over which conflict has developed, there are similar examples in at least some other countries. The United States is thus a useful case study.

This chapter adopts a broad definition of technology and technology transfer. Here technology is considered to be specialized knowledge pertaining to the production of goods and services in organized economic activity, including the knowledge (skills) to manage a set of interrelated, technical processes.(3) The technology may be embodied in products, or it may be disembodied in technique, organization, or process. It may be patented, but it need not be. Technology transfer occurs when technology previously applied or developed in one country is applied in a second country. Technology transfer may involve new (latest) or old (mature) technology. Technology transfer may occur through the international transfer of documents (blueprints, formulas, plans, and other communications on technique or organization), people, or products embodying the technology.(4)

This broad view of technology transfer introduces two important requirements for this chapter. First, the conflicts and policies covered must include those dealing with the export of technology-intensive products as well as those concerned with the export of processes, formulas, techniques, and other vessels for transferring disembodied technology. Second, since a substantial majority of commercial technology (and approximately 85 percent of the R & D) is produced by private firms in the United States, the activities of U.S. companies in producing or licensing production abroad fall under the purview of the present study, as well as the policies of the U.S. government relating to its own technology holdings. Thus our concern includes conflicts, government policies, and potential controls affecting the outward transfer of technology via exports, through the operations of U.S. firms abroad (including licensing, management contracts, turnkey plants, and controlled subsidiaries), and the transfer of government-owned technology.

The chapter is organized as follows. The next section develops a simple conceptual framework to outline the linkages between technology transfer and the achievement of perceived national objectives of the home country. The nature of the objectives are examined, including potential conflicts and trade-offs among the objectives themselves. But the major intent is to identify how unrestricted outward technology transfer may compromise one or more national objectives, or the objectives of certain important national interest groups, and thereby lead to conflicts and agitation for control systems.

The following five sections examine in some detail various broad types of conflict in the United States over outward technology transfer. The coverage does not pretend to be exhaustive, but hopefully all those areas are included where the greatest governmental control over technology transfer activities has been exercised or is currently being sought. In many areas, the interference with or concern over technology transfer is an incidental result of policies introduced for other purposes. These are mentioned as well. The types of conflict dealt with concern technology transfer to communist countries, transfer of nuclear technology, technology transfer affecting domestic jobs and production, environmental preservation, and human rights. In each case the sources of conflict are traced, the U.S. policy actions or restrictions that have been implemented or proposed are presented, examples of the conflicts are identified, and government policy or interest group positions are assessed as to their effectiveness in accomplishing the stated objectives and their costs in terms of other objectives foregone. Since there are several other areas of conflict and policy which bear upon, but are incidental to, the technology transfer debate, these are briefly identified in a further section. The final summary section offers some suggestions on resolving home country conflicts.

NATIONAL OBJECTIVES, TECHNOLOGY TRANSFER, AND SOURCES OF CONFLICT

Conflicts among objectives are endemic in a pluralistic society. It is thus not surprising that conflicts arise over the desirability of international transfer of a nation's technology. This section attempts to develop a simple conceptual framework both to explain the source of conflicts over policy and to assess the effectiveness of home country control systems.

National Objectives

Any nation's objective function is a multifaceted one, with variables that are interdependent, contributing to or detracting from overall national welfare. A further complication is that the well-being (and objectives) of any particular group (such as organized labor, stockholders, or Hispanics) may conflict with national (aggregate) well-being or objectives. National objectives are a mixture of economic, strategic and security, and social-political goals, some of which are abstract and some concrete, some quantifiable and others not practically measurable, some relatively permanent and others transitory and shifting as the visibility and urgency of interest groups waxes and wanes.

At the risk of oversimplifying, table 8.1 identifies commonly recognized U.S. national objectives, most of which are revealed through the articulation and implementation of national policies – that is, they are "revealed" objectives. They are grouped to facilitate discussion in ensuing sections.

Table 8.1. A Catalogue of National Objectives Related to International Transfer of Technology

1. Primarily Economic Objectives

   a. High and sustained economic growth in real income per capita
   b. High domestic employment levels
   c. Relative stability of average prices
   d. Equitable income distribution
   e. Maintenance of effective competition
   f. Limited and smooth structural adjustment in the economy

2. Related Economic and Social Objectives

   a. Preservation and improvement of environmental quality
   b. Consumer protection

3. Strategic and National Security Objectives

   a. Minimize the offensive military capabilities of actual or potential enemies of the nation
   b. Increase the defensive and offensive military capabilities of actual or potential allies

4. Objectives Relating to Humanitarian Sentiment and a Peaceful World Community

   a. Enhance the economic development of poor countries
   b. Enhance humanitarian and democratic behavior in other countries

In line with the nation's basic political and ideological commitment to private property, individual rights, and a democratic political process, the economic objectives of the United States are, as is obvious, heavily laced with the private enterprise ethic and with skepticism about government regulation of private economic decisions. But the growing involvement of government in economic decisions over the past five decades is reflected in the increased emphasis that has been placed on constraining the free discretion of private firms in decisions involving technology transfer and foreign investment. Both of these tendencies are reflected in the objectives in table 8.1.

Among the predominantly economic objectives, the goals of economic growth and full employment are firmly based on legislation and federal policy.(5) The same is true of price stability. Likewise, the commitment to maintain effective competition and affect the distribution of income has a multiplicity of legislative and judicial roots.(6) As for the objective of limiting and smoothing structural adjustment in the economy, there are several clear indications of its existence and importance. Among them are the adjustment assistance provisions of the Trade Reform Act of 1974; the escape and countervailing duty clauses in earlier trade legislation; the federal commitment to taking over bankrupt private railroads in the Conrail system; and a multitude of credit guarantee, loan, and other programs designed to shield individual companies (for example, Lockheed and Chrysler) or industries (such as textiles and steel) from the costs of adjusting to domestic or international competition.

In the last two decades, additional national objectives have come to the forefront under the banner of "social responsibility." These are combinations of social and economic objectives that recognize that national output valued at private market prices is not a uniformly valid measure of national welfare, given the important external costs (and benefits) involved in the production and consumption process. These diverse objectives are identified under the labels of environmental preservation and consumer protection.(7)

A third broad group of objectives has to do with the national security and defense of the nation-state. Those related to technology transfer policy are of two types. The first is the objective of minimizing the military – and possibly the economic – prowess of actual or perceived enemies or likely enemies of the country. The other is the objective of maximizing the resistance (defensive or offensive) capabilities of third nation (allies) which share the concern over the military position of potential enemies.(8)

The final group of objectives is a potpourri of additional social or humanitarian concerns about the human condition in the rest of the world, which may potentially be reflected in national policies on international commerce and technology transfer. The improvement of the economic condition (nutrition, housing, education) of people living abroad, especially in poor countries, is a national objective of industrial countries, as revealed in economic aid, trade, international relief, and other programs and policies. Recently concern has been extended to individual rights and democratic principles abroad, as evidenced most notably in President Carter's "human rights" campaign. Such objectives, toward foreign governments as well as domestically, are often ambiguous and difficult to define operationally. In addition, their coverage may be ill-defined, perhaps extending only to direct U.S. government actions affecting human rights conditions in other nations, or extending further to include such things as minimum occupational safety requirements or even minimum wages, as has been proposed by some groups. Despite the ambiguities, there can be little doubt that this is an increasingly important component within the broader objective function of several home countries, including and perhaps especially the United States.

There may be conflicts or inconsistencies among the various groups of objectives, and indeed even among specific objectives within one group. For example, domestic economic objectives may be harmed by policies to restrict international commerce with perceived enemies – an action deemed to contribute to national security objectives. Similarly, the denial of technology transfers to exercise leverage on a country to improve its human rights performance may reduce employment at home. But conflicts may also occur within a group of objectives: more competition may reduce employment and accelerate structural change; tighter environmental regulation may mean higher inflation rates. In short, some of the diverse national objectives may have complementary relationships with others, while some have competitive relationships. A problem facing policymakers is to understand the trade-offs and complementarities, in an uncertain policy environment with imperfect information.

Linkages Between Technology Transfer and National Objectives

Whether unrestricted or controlled technology transfer contributes most to national economic objectives depends on three important factors (among others). One is obviously the perception of what the national utility function is, that is, the weights given the various components discussed above. The second is the nature of the linkages between technology transfer – or its denial – and one or more of the national objectives: more simply, which objectives are affected by technology transfer, and by how much? A third factor, which conditions the strength of the linkages and thus governs the effectiveness of any control system, is the nature of the international environment in which the linkages exist and the policy is carried out.

Establishing restrictions or control systems for technology transfer by a home country must be expected to make a net contribution to national welfare through one of the following results, as table 8.1 suggests. First, it may have a direct positive economic impact on the home country; second, it may limit (relatively) the military or strategic capabilities of countries perceived to be hostile; and third, it may induce potential recipient countries of the technology transfer to adopt economic, social, political, or military positions more in keeping with the humanitarian or other interests of the home nation. The econmic linkages will be examined first.

Economic linkages

It is only recently that economic theoreticians have begun to examine the effects of international technology transfer (or its restriction and taxation), and the conclusions from most such models are ambiguous, depending upon the relative magnitudes of various parameters.(9) In some instances taxes on outward technology transfer will increase the technology's price, thus making the home country better off than with free transfer. And whether labor or capital or both are benefited by unrestricted transfer also depends on the specific situation.

In addition to the exclusive focus of such trade theory models on one or at most two economic objectives – real national income and its distribution between factors of production – they have at least two other shortcomings as a guide to technology transfer policy. One is that they assume that structural change or adjustment in the economy is costless, or can be ignored. Second, they ignore any interaction between the future stock of technology, which is influenced by current levels of R & D, and the international transfer of technology. The latter may affect the expected return on R & D, and thereby affect its volume.

A more complex set of linkages is thus needed to understand even the economic results of technology transfer. A simplified illustration of those linkages is shown in figure 8.1. This framework ignores, temporarily, the influence of technology transfer on the humanitarian or security objectives of the nation. Nor does the framework deal with the intertemporal structure of the relationships among the variables.

The major direct economic effects are denoted by the triple lines running through the center of the chart. Starting with outward technology transfer at the far left, if it is effectively utilized abroad, foreign production will be initiated utilizing the technology and a stream of financial returns will be generated for the home country. These financial returns may take several forms, such as license fees or royalties. In the case of a multinational firm, the return may be realized as dividends remitted, income transferred through rigged transfer prices among affiliated subsidiaries, or a buildup of assets abroad.

Some of the foreign production may displace production that was previously supplied from the home country. Thus home country exports may be displaced, or domestic production for the local market may be replaced by imports produced with the exported technology. This, in turn, throws an adjustment burden on the domestic economy. Resources in the affected industries must become unemployed, transferred into other more competitive industries, or receive lower factor returns. Meanwhile, a different group in the economy (stockholders and executives of multinationals, licensors, or the government) receives a flow of real income from abroad from fees, royalties, or dividends. Whether the loss to the group of displaced factors due to lost domestic production in the affected industry is greater or less than the gain to the recipients of the returns on the technology transfer is impossible to tell a priori. But the fact that the gainers and the losers are almost always different groups presents a definite conflict over technology transfer policy, even on economic grounds.

It should be noted that the gross returns on the technology transferred, represented by the upper triple lines, are not of the same magnitude (or time pattern) as the gross losses in domestic production represented by the lower triple line. Any difference suggests a net (positive or negative) balance of payments effect, requiring adjustment in the terms of trade (relative prices or the exchange rate) in the long run, which itself will have real income and adjustment cost implications.

The size of the returns and the domestic production displaced depend on two other factors illustrated in the figure. The first is the

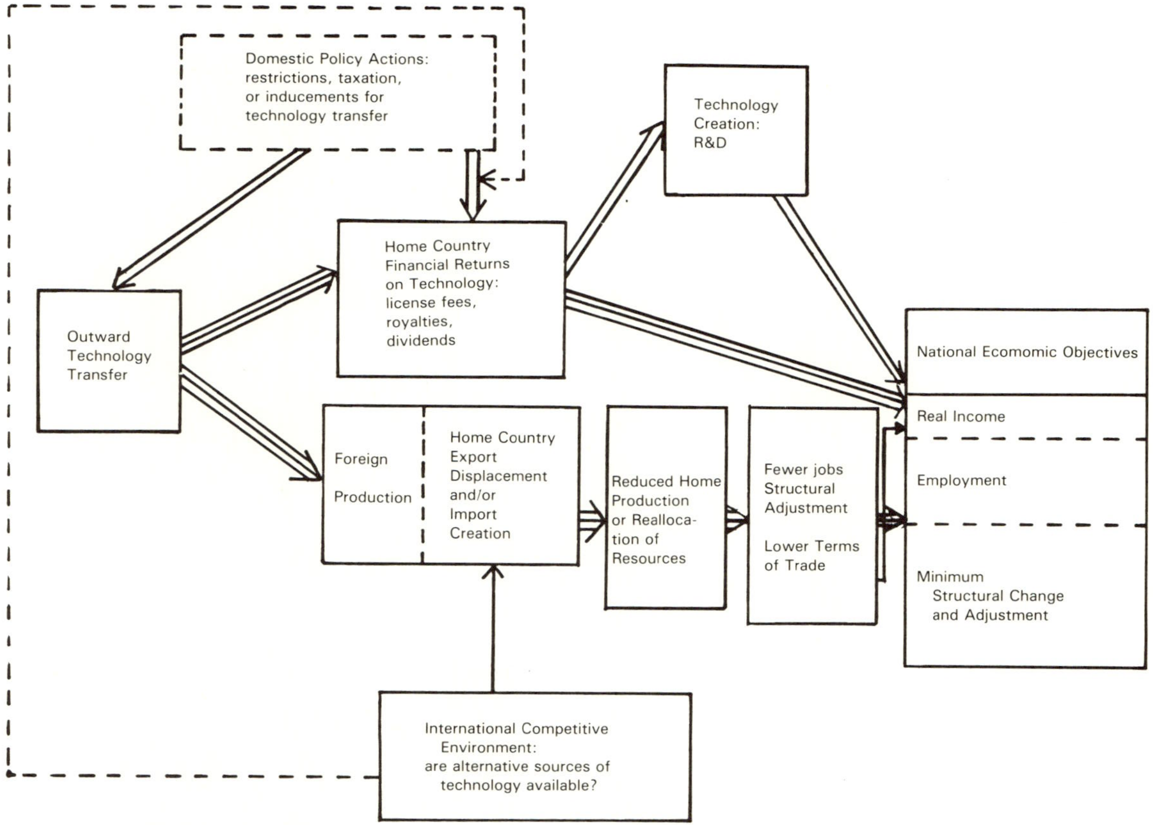

Fig. 8.1. Linkages between technology transfer and home country economic objectives.

home country policy posture toward outward technology transfer. If the home country has a monopoly position in a technology, and the demand for the technology is inelastic in the recipient country, limitations (short of prohibition) on the volume transferred or taxation of the transfer may raise the total returns to the technology. This prediction is parallel to the optimal tariff/tax arguments on imports and exports, and would suggest the need for varying levels of tax or restriction, depending on the elasticity of demand for the technology in potential recipients and the degree of monopoly held in the technology. In short, for a large country that can influence the price of technology in the rest of the world, optimal policy to maximize its economic welfare will involve graduated optimal taxes or limits on technology outflows.(10)

The second factor is the international competitive situation in the technology. This external situation influences the level of optimal tax or restriction to maximize the gross return to the nation. A closely substitutable technology available from another country will make the optimal restriction on its export zero; exclusive monopoly position in the technology will make the optimal tax dependent only upon the foreign elasticity of demand. This influence is illustrated in figure 8.1 by the dotted line from the "competitive environment" at the bottom to "policy actions" at the top.

More important, the availability of closely competitive technology from other countries will determine whether an actual loss in domestic production in the home country can be attributed to the transfer of the technology to a foreign nation. If similar technology can be obtained from a third country (or from sources within the recipient country), some or all of the domestic production would be lost even if the technology transfer from the home country did not take place. In such a case, the linkage denoted by the triple line between the transfer of technology and foreign production is broken. The importance of alternative sources is, again, an empirical question and cannot be determined a priori. The existence of significant foreign alternatives to home country technology, of course, makes restrictions on technology transfer ineffective or counterproductive on economic grounds alone. In the worst case, the gross return on the technology is not received, while the domestic economic dislocation and adjustment is incurred anyway.

One linkage is denoted in figure 8.1 between the return from technology transfer and home country technology creation. If the return on technology transfer raises the perceived expected total return to R & D activities in the home country, the future "stock" of technology may be increased, resulting in more (or faster) innovation, leading to higher national income. Little is known about the magnitude of these linkages, however.(11) Indeed, a counterargument might be that the displacement of domestic production, with the attendent adjustment costs, might stifle R & D activity. This linkage is not shown on the chart and is one that we do not find very compelling evidence for.

The economic effects of technology transfer outlined here would normally involve both positive and negative aspects for one or more national economic objectives. The net balance of economic benefits and costs cannot be predicted without reference to the specific situation. Perhaps the most important determinant of the net balance is the status of alternative sources of the technology abroad. This could depend on previous government policy toward outward technology transfer, and perhaps becomes an economic argument for a control system. In any event, the transfer of technology is likely to affect domestic income distribution; that is, the relative economic status of different groups in the economy, in the short run and the long run.

Other Economic Goals

Aside from the objectives of real income growth, limited and smooth structural adjustment, and maintenance of domestic employment levels, table 8.1 identified several other national economic objectives (price stability and maintenance of effective competition), and two related social and economic objectives (environmental preservation and consumer protection). Technology transfer would likely have selective and unpredictable effects on these variables. One can imagine a situation in which outward technology transfer permits a firm to recover R & D expenditures over a larger volume of output, thus lowering overhead charges on domestic sales and lowering inflationary pressures. Or a company may, by transferring technology abroad, solidify its monopolistic position in worldwide markets, leading to upward movements in prices and reduced competition in the home market in the long run, due to a broader worldwide base for its monopoly. Each case is situation-specific (or built by alternative assumptions) and generalizations are not possible.(12)

Likewise, technology transfer abroad would normally have no impact on environmental preservation and consumer protection at home. Thus in most cases the control of technology transfer for such motives would be for the sake of residents in the recipient country, and would fall under the "humanitarian sentiment" objective in the national welfare function, to be discussed below.

There is, however, an important set of exceptions. Technology transfer does affect the national objectives of environmental preservation and consumer protection when the technology transferred results in transborder pollution or has other harmful transboundary effects. This is obviously important in ocean pollution and air pollution in specific instances, and could be used as a national interest argument in imposing home country monitoring over or limitations on the type of technology transferred.

It should be noted, however, that the immediate motive for transferring certain types of "dirty" technology from home countries may be to escape tight environmental or consumer protection regulations.(13) These situations are particularly difficult to analyze, since several domestic objectives may be harmed. Control of technology transfer in such situations becomes uncertain and subjective.

National Security Objectives

The limiting of technology tranfer to a potentially unfriendly country raises two essential questions. Is the control of the transfer effective in denying the technology to the unfriendly country? And does control of the transfer harm (or contribute to) other goals of the home country? The answer to both depends largely on the international competitive environment for the technology. Are alternative foreign sources available?

This can be illustrated with the aid of figure 8.2. If there are no alternative sources of supply of the technology (if the lower box is empty), denying technology to the unfriendly country would contribute to the national security objectives of the home country, but would also harm the economic growth objective, and perhaps other economic objectives such as price stability. Thus even if the technology transfer control is effective, there is normally still an economic cost to be traded off.

If the technology has already been transferred to a third country (upper box), either by a multinational based in the home country or by license from the government or a private entity of the home country, the home country restriction must then also extend to the licensee or producer in the third country if it is to be effective. This, of course, represents the "extraterritorial" application of home country restrictions, a controversial aspect of the U.S. Export Control Act and of many licensing agreements for arms and nuclear technology. If this transfer from third countries is not controlled, the home country objective of increasiong national security is not achieved, while the harm to the national economic objectives occurs anyway.

The same result occurs when there are willing alternative suppliers of similar but indigenous technology in third countries, as represented in the lower box. In such cases the home country control is not effective, which means that the cost in economic objectives is incurred while the benefit of increased security is not. In such cases, the home country's only recourse is to use moral suasion or economic threat to induce independent third countries to withhold the technology as well – illustrated by the broken line on the lower-left in the chart. To oversimplify: technology transfer controls over technologies shared by several countries will only be effective when there is effective international cooperation.

Humanitarian Motives

A final motive for controlling technology transfer may be to induce potential recipient countries to take actions – economic, social, political – that the home nation finds desirable. In large measure this involves the imposition of some home country values on other nations. Its success depends on the leverage that the home country exercises over the potential recipient.

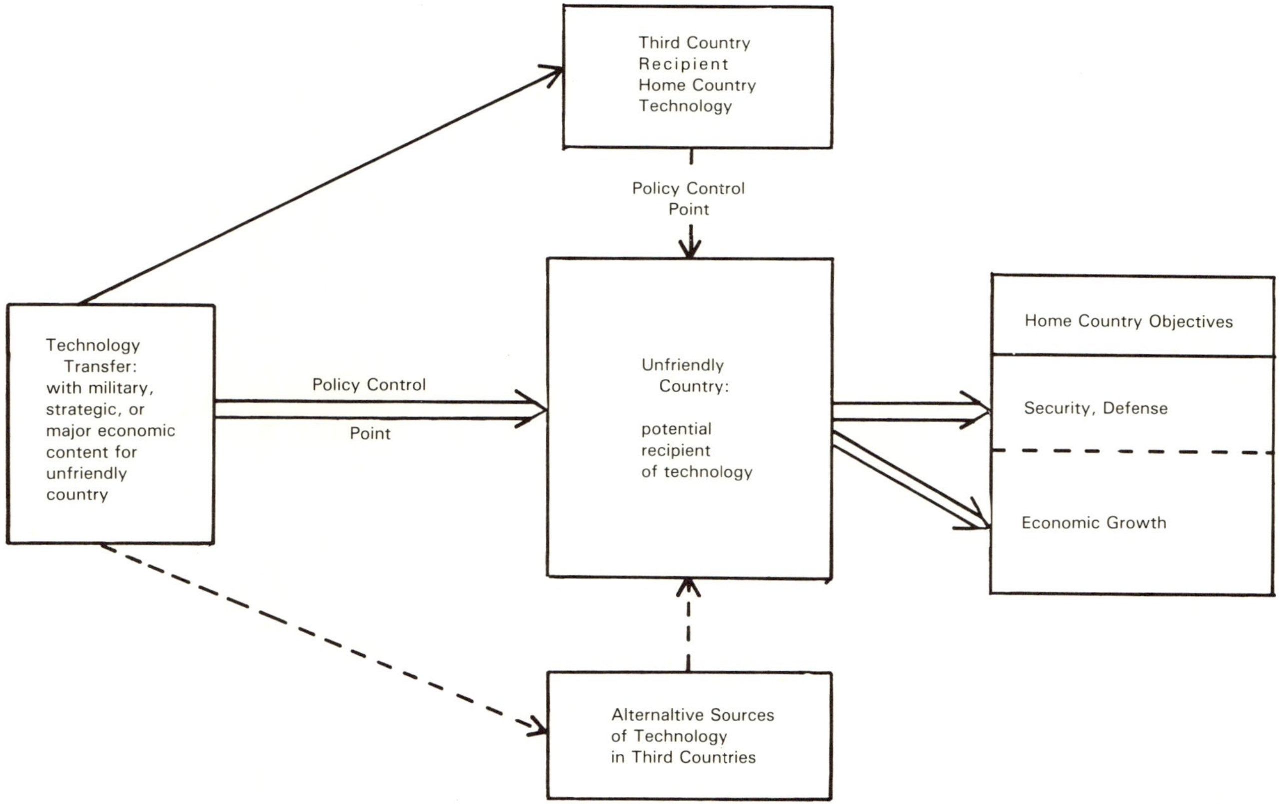

Fig. 8.2. Linkages in technology transfer control for national security purposes.

The linkages are illustrated in figure 8.3., which is similar to figure 8.2 in that the availability of alternative sources (lower box) and the capacity to control the technology already transferred to third countries (upper box) determine whether the home country policy will be effective in influencing the potential recipient to change its policy.

If it does change its policy, thereby contributing to the gratification of home country humanitarian objectives, and the technology is subsequently transferred, then the home country incurs no economic cost in terms of foregone real income. In such a case, the upper double line on the far right would remain, and the lower double line would disappear.

But if the potential recipient is willing to forego the technology and retain the policies repugnant to the home country; if alternative sources of foreign technology are available to the recipient; or if the home country's control over the retransfer of its technology from third country recipients is weak, the situation is reversed. The home country incurs the economic cost (the lower double line on the far right remains), but receives no contribution to its worldwide humanitarian motives (the upper double line disappears). Thus the status of alternative sources of technology to which control systems are applied is again a primary determinant of the effectiveness of the control system and the benefit-cost relationship for the national objectives of the home country.

Multiple Conflicts

The foregoing indicates that, even at a simplified abstract level, policies toward international technology transfer are inherently subject to conflict. This is true because any nation has multiple national objectives, some of which are competitive and not complementary. And groups within the society have interests identified more closely with some objectives than others. Thus, no policy posture can satisfy all groups in the economy.

Among the more obvious and serious conflicts identified are the following:

1. Conflicts within a class of objectives – especially among economic goals. Technology transfer may adversely affect some workers in an industry while benefiting stockholders and managers, and even the economy as a whole.
2. Conflicts between classes of objectives. Technology transfer may detract from economic goals of real income and employment but benefit the objectives of national security or humanitarian behavior abroad.
3. Conflicts among groups in the economy. Workers and stockholders may view technology transfer in one way; the military establishment, environmentalists, or consumer advocates in an opposite way.

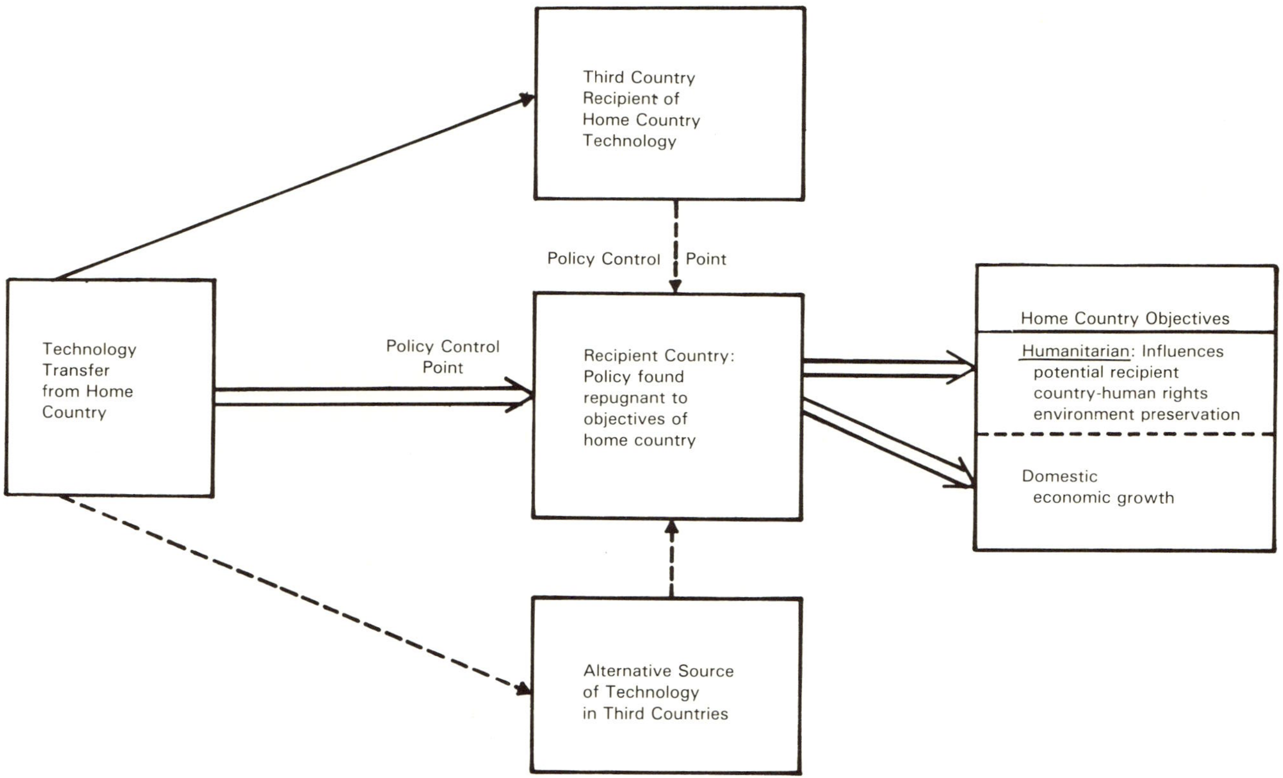

Fig. 8.3. Linkages in technology transfer control for humanitarian purposes.

For all of these types of conflicts, and for any assessment of the effectiveness of technology transfer controls, the availability of competing technologies from other nations is a critical determinant. Also, to be effective, a home nation must control not only its technology transfer, but also the technology it has already transferred to third countries.

NATIONAL SECURITY AND TECHNOLOGY TRANSFER

The area of perhaps the most direct conflict among national objectives is the export of technology to potentially hostile nations.(14) The conceptual framework just discussed predicts that restrictions on technology transfer will be effective for the stated objective of harming the military potential of the target countries only if two conditions are met: (1) the target countries are not able to develop their own indigenous technologies, and (2) the technologies are not available from alternative nation suppliers. The latter can be assumed only if the controlling country has a world monopoly in the technology, or if there is international action by all potential suppliers to deny the technology transfers.

The potential costs of the control include a loss of real national income, and perhaps less desirable income distribution in and adjustment burdens on the domestic economy. If alternative suppliers are available and provide the technologies, the economic costs are absorbed without the objective of the policies, and thus any positive contribution to national welfare, being realized.

U.S. Policy Measures

The first U.S. policy effort to restrict exports and technology transfers to perceived enemies was the Trading with the Enemy Act of 1917, passed after the United States had entered World War I. Section 5(b) of this act gave the president broad and flexible powers to control international transactions during that "war emergency," and after the amendments to the act in 1933, during "national emergencies." Pursuant to the act, the Treasury Department issued Transaction Control Regulations that authorized the scrutiny and licensing of all exports to designated countries, according to both the class of good and the country of delivery. Foreign Assets Control Regulations were also promulgated to prohibit almost all contact between the United States and some foreign countries and to block the assets of those countries in the United States. In recent years, the United States has prohibited, under the act, imports from and exports to North Korea, Vietnam (North and South), Cambodia, Cuba, and (until 1972) China. Controversies arose between the United States and some of its allies due to the extraterritorial application of the act to subsidiaries of American companies operating in third countries. These created several causes celebres among host countries for U.S. multinational firms, such as the Fruehauf and GE-Bull cases in France and the Ford Motor, B.F. Goodrich and Alcan cases in Canada.(15)

The era of maximum or near-maximum U.S. technology controls directed toward the communist countries was ushered in by cold war legislation – the Export Control Act of 1949. This declared that U.S. policy was to utilize export controls (over products and technical data) to the extent necessary to exercise the necessary vigilance over exports from the standpoint of their significance to the national security.(16) The act was administered by the Office of Export Control of the Department of Commerce, which was authorized to establish a licensing system designed to prevent the shipment of any goods or technical data that would make a significant contribution to a communist country's military potential. Two kinds of export licenses were made available: (1) a "general license" covering a specific list of "nonstrategic" commodities that could be exported to specified destinations under specified conditions without the need of special permission in each instance; and (2) a "validated license" required for the sale of all goods (that is, sensitive items) not covered by the "general license." At the peak of the cold war, the Department of Commerce had a "positive" commodity control list of about 1,000 items that were considered strategic and normally not exportable to communist countries. About the only product considered nonstrategic at the time, according to one observer, was bubblegum.(17) After a period of general relaxation of validated license coverage in the late 1970s, restrictions were reestablished by presidential mandate over a broad range of technology-intensive products (and grains). These new restrictions were directed only toward exports to the Soviet Union in response to its invasion of Afghanistan.

U.S. policy also attempted to remove alternative sources of technology potentially available to communist countries. The United States led the effort to impose a collective embargo of strategic exports by all Western countries. The Consulative Group-Coordinating Committee (COCOM) was established in 1949, its members fourteen of the fifteen NATO countries (Iceland excluded), plus Japan. Control lists of embargoed goods were established by consensus, designed to maximize the harm to the offensive capabilities of the Eastern bloc. But from the beginning, the requirement of consensus for inclusion on the COCOM list conflicted with the U.S. wish for more extensive controls, and the COCOM list was less extensive than the unilateral U.S. control list.(18) Any member of COCOM, however, could veto a sale by another member if it could show that the sale might compromise a Western defense advantage. The cooperative COCOM embargo deteriorated in the 1950s and 1960s, as Western European nations pressed for, and won, exemptions from the COCOM list. By the 1970s COCOM had become entangled in competing national export interests, and today it is generally limited to controlling exports of strategic goods, that is, the inner core of technologies with obvious potential military significance.

The legislative underpinning for U.S. participation in COCOM was the Mutual Defense Assistance Control Act (Battle Act) of 1951. This law specifically put pressure on other nations to observe the COCOM embargo on the sale of strategic goods to the Soviet bloc. Any nation violating the embargo would be denied U.S. military, economic, and financial assistance. The Battle Act explicitly embargoed any "shipment of arms, ammunition, and implements of war, atomic energy materials, petroleum, transportation materials of strategic value, and items of primary strategic significance used in the production of arms, ammunition, and implements of war to any nation or combination of nations threatening of the United States, including the Union of Soviet Socialist Republics and all countries under its domination . . ."(19)

Because of a growing concern about the economic (as well as military) strength of the communist bloc, the Export Control Act of 1949 was extended and amended by Congress in 1962 to include goods primarily of economic significance. But the cold war policy of containment began to break down in the 1960s. The deterioration of COCOM, the steadily worsening balance-of-payments position of the United States, and the growing concern of U.S. exporters about the burdens, uncertainties, and lost market opportunities entailed in strategic export controls led to political pressure for liberalization. The costs to U.S. welfare from economic losses were perceived to be rising at the same time that alternative sources of technological exports to the communist bloc made the attainment of the objective of the containment policies highly uncertain.

The result was the Export Administration Act of 1969, which replaced and modified the 1949 act. The new act declared, "It is the policy of the United States both (A) to encourage trade with all countries with which we have diplomatic or trading relations, except those countries with which such trade has been determined by the President to be against the national interest, and (B) to restrict the export of goods and technology which would prove detrimental to the national security of the United States."(20) The act retained the entire mechanism of controls and regulations governing the licensing of exports of goods and technical data, but restricted "validated license" only to those items of "military potential."

U.S. export controls were gradually relaxed under the new law in 1970-71, but corporate and congressional dissatisfaction with the pace of liberalizatino led to passage of the Equal Export Opportunity Act of 1972. This amendment to the 1969 act further recognized the major constraint on the effectiveness of restrictions on technology transfer when alternative sources of technology are available. It abolished export controls on all goods that could be purchased without restrictions from suppliers outside the United States, except in those instances where the absence of control would be detrimental to U.S. national security. At the time the amendment was passed, 550 items were subject to unilateral control by the United States (that is, were not on the COCOM list); by the end of 1972 there were only 73.

The Export Administration Act was further amended in 1974 to formalize and strengthen Pentagon influence over the license-granting procedure. The amendments of 1974 specifically authorized the secretary of defense to "review any proposed export of goods or technology to a country for which a validated license is required by law and, whenever he determines that the export of such goods or technology will significantly increase the military capability of such country, to recommend to the President that such export be disapproved."(21)

In addition to direct controls over the export of technology and products, restrictions on payments or extensions of credit may indirectly impede outward technology transfer to the Eastern bloc.(22) A Depression-era statute, the Johnson Debt Default Act of 1934, placed restrictions on the export of private U.S. capital to foreign governments that are in default on obligations to U.S. citizens or to the U.S. government. Originally designed to apply to Russia because of its nationalization of U.S. companies in the 1920s, the act was extended in the late 1940s to cover the Soviet satellite nations that nationalized U.S. property. Amendments, exemptions, and loopholes discovered over the years have made the act relatively obsolete. Of greater importance today are the constraints placed on Eximbank financing of U.S. exports. The Foreign Assistance and Related Agencies Appropriation Act of 1965 made it illegal to lend Export-Import Bank (Eximbank) funds to any communist country unless a specific determination had been made that such a loan would be in the national interest. A 1968 amendment to the Export-Import Bank Act (of 1945) prohibited the Eximbank from financing exports to any country that aided another country engaged in armed conflict with the United States. This policy was subsequently modified by the Export Expansion Finance Act of 1971, which limited the ban on Eximbank credit support for exports only to those nations directly engaged in armed conflict with the United States (that is, North Vietnam). The 1974 amendments to the Export-Import Bank Act established a new set of specific restrictions. This Stevenson Amendment established a ceiling of $300 million on extension of credit to the Soviet Union. Credits, in addition, could not be extended for production of fossil-fuel energy resources, and funding related to research and exploration of fossil fuel resources could not exceed $40 million.(23) The $300 million ceiling could be exceeded only if the president determined that a higher limit was in the national interest and if Congress concurred. In sum, the Congress, through its limitations on Eximbank lending, has been a restraining factor in the expansion of U.S. East-West trade. U.S. companies, in many cases, have been forced to initiate deals through their subsidiaries in Western Europe, where official credit support programs have been more liberal.

Assessment of the Controls

An evaluation of the U.S. and COCOM system of export controls over technology transfers to the communist countries is difficult. First, were the adversary nations actually deprived of strategically (and economically) important technologies? Or did they, for example, develop them on their own? And were alternative suppliers sufficiently ready to supply technologies that the United States unilaterally withheld, such that the U.S. policies were not effective in denying the technology? And even if the technologies were effectively denied and the military (and economic) capabilities of the adversary countries harmed, did the welfare benefits of that policy – to be optimal – at least offset the net losses in economic or other welfare components to the United States?

Systematic evidence on each of these issues is sparse, while an overall evaluation of the degree to which the trade-offs chosen were optimal is, no doubt, impossible. Much of the evidence is anecdotal. There is, however, a widely held view that U.S. restrictions were not very effective. Yet the trade embargoes that have received full COCOM adherence technologies such as high-speed computers or sophisticated microcircuitry with obvious implications for advancing offensive weapons systems) have surely hurt the Soviet Union in limited areas. Least-cost solutions have been denied the Soviets, and the acquisition of such technologies has been delayed, limiting strategic capabilities at least to some extent.

Aside from the effective denial or delay of these inner-core strategic technologies (upon which there has been COCOM consensus), U.S. controls may have had little effect on the transfer and advancement of technology, particularly with respect to the Soviet Union. Some technologies have been readily and easily obtained from non-U.S. sources. In addition, Soviet efforts at self-sufficiency have been stimulated, and satisfactory solutions have been receiving high priority in most areas bearing upon military strength. The director of Strategic Technology and Munitions Control in the U.S. Defense Department notes,

> It is unrealistic to expect that a system of export restrictions can prevent the USSR from eventually acquiring any level of technology that the West has developed. Inadvertent leakage, clandestine acquisition, and indigenous development will combine to assure that this eventually takes place. The process cannot be halted; it can only be retarded. Thus the true measure of the effectiveness of controls over technology is how long the catch-up process takes. On that basis, the present system scores well, for in a number of critical technologies, the United States has consistently maintained a lead of two to five years over the USSR, and in some cases the margin is even wider.(24)

Another possible benefit to the United States is that export controls have succeeded in inducing communist countries to adopt behavior more acceptable to the United States. Have the delays and higher costs caused by export controls caused embargoed countries to change their ways? Real concessions on noneconomic issues from the East Bloc as a result of U.S. use of controls for leverage seem to have been modest at best. Closing or opening the U.S. economic door has not induced internal reforms or polycentrism within the socialist world, and has not let to the overthrow of any communist regimes. Yet changed behavior might be attributed to Yugoslavia, the People's Republic of China, and perhaps others. To what extent these changes or divergences in behavior result from U.S. policy is impossible to determine. But some are skeptical. As one observer expressed it, "Obviously, the Soviets will not trade their political soul for a mess of capitalist porridge."(25)

While the results of technology transfer controls on the strategic capabilities and degree of belligerence toward the United States of communist countries have been mixed, the costs in U.S. economic welfare have evidently not been trivial. Yet the full cost can hardly be estimated. One result has been that potential markets for U.S. companies have been lost, with the attendent domestic economic losses in real income and employment. It is clear that the United States has unilaterally deprived itself of trading opportunities with the East Bloc. It has also diverted socialist trade from American producers to those from other countries following less restrictive control policies. But there is no very reliable estimate of the economic costs of the U.S. controls.(26) Even in the absence of export controls, however, it is certain that a variety of structural, ideological, and policy obstacles would have remained to impede a rapid expansion of U.S. East-West trade during the past decade. It is possible, therefore, that even though the success of the policy has been mixed, the economic losses for the United States in the past have also not been great.

The Future

The need for overhaul of U.S. export and technology transfer policies has been stressed. The search in Washington for a coherent policy in recent years, however, has generally been hindered by disagreements and conflicts over the sale of technology. Those involved range from the International Longshoreman's Union to the John Birch Society. But the debate between two groups with direct interests in the matter – the military on the one hand and private companies seeking markets on the other – dramatized the trade-offs between the welfare loss due to the fact that free technology transfer helps make more powerful enemies, on the one hand, and the economic benefits that would result from free transfer on the other. The nature of this trade-off was dramatically revealed in President Carter's cancellation of Sperry Rand's license to export a Univac computer to the Soviet Union in 1978, a decision that was eventually reversed in 1979.

The Defense Department asked its Defense Science Board in 1974 for a report on U.S. technology exports. The report issued in 1976 by the Board's task force, headed by J. Fred Bucy, Jr., president of Texas Instruments, Inc., called for a significant change in the criteria utilized in making export control decisions. The report noted, "On the one hand the list of controlled products was too long and its administration too cumbersome to business. On the other, not enough attention went to maintaining our lead time in critical areas of design and production knowhow."(27) The report recommended the easing of licensing procedures, paperwork, and controls on the sale of commercial endproducts, but also the drastic curbing of U.S. exports of sensitive production skills to the Soviet Union.

Presidential Directive No. 18, in 1977, outlined a national strategy for the use of economic power toward the Soviet Union:

> American policy in the economic sphere should reflect the mixed cooperative-and-competitive nature of the U.S.-Soviet relationship and the distribution of economic and technological resources between the two countries. Also, the United States must take advantage of its economic strengths and technological superiority to encourage Soviet cooperation in resolving regional conflicts, reducing tensions, and adequately achieving a viable arms control agreements . . . In short, economic capabilities and economic relations must serve the basic U.S. foreign policy objectives of encouraging East-West cooperation, containing Soviet expansion, and promoting American values.(28)

Presidential Review Memorandum 31 of 1978 confirmed the change in approach. Efforts in the government were made to streamline decision making on technological exports so as to focus on and make the export of critical technological processes conditional on altered Soviet behavior. The National Security Council was given a much larger role in export control committees. A new Ad Hoc Technology Transfer Group was also created within the Security Council structure to deal with the broad policy issues involved in the implementation of the new plan, "a sign that 'economic diplomacy' may become a bigger part of America's policy towards the Communist bloc."(29)

It appears, then, that the use of export controls as a political lever gained support within the U.S. government in the late 1970s, if not to harm the adversary's military capabilities, then at least to induce less belligerent behavior. This movement was dramatically intensified by the U.S. denial of most exports to the Soviet Union in 1980 in retaliation for the Soviet intervention in Afghanistan. Yet the lowering of export barriers toward mainland China also reflects a selective policy based on the realities of technological gaps.

Despite low odds of success, more strategic use of controls "to signal, to bargain, or to do some domestic politicking" can be expected in the years ahead, especially toward the Soviet Union.(30) The United States is unlikely to return to the general "denial" approach of the 1960s, yet is also unlikely to yield to one of laissez faire. Given the new desire of the government to capitalize on U.S. economic and technological advantages in its relations with adversaries, the outlook is rather for a more flexible and perhaps erratic approach that will open and close the door to high-technology exports in line with day-to-day changes in world politics based on transaction-specific assessments of national-security-versus economic-welfare trade-offs.

CONTROLS OVER NUCLEAR TECHNOLOGY

An area closely related to security that has witnessed controls over technology transfer and continued policy conflict is nuclear technology. According to Senator Abraham Ribicoff, "The world may be headed toward new Dark Age in which plutonium replaces gunpowder as the explosive of choice, and warfare and terrorism take on a nuclear dimension."(31) The U.S. response to such dire warnings has been to place more stringent transfer conditions in recent years on the outward flow of "sensitive" nuclear materials, equipment, and services. But the issues and conflicts in the nonproliferation imbroglio have been extremely complex and have attracted much attention.(32)

The transfer of nuclear technology, as is shown by the conceptual framework presented in figures 8.2 and 8.3, involves at least three conflicts. First, the economic growth or national income objective involved in the sale or licensing of nuclear technology to foreign countries may conflict with national security objectives, should the technology be transformed into military potential and, in turn, directly or indirectly aid the strategic capabilities of belligerent countries; but restricting the transfer may conflict with the humanitarian motives that seek to speed the economic development of poor countries, to which nuclear power obviously contributes. The national objectives of injecting appropriate safeguards against radiation escape, both for home country safety, in case of cross-border radiation drift, and for humanitarian reasons, to protect the nationals of the recipient country, may conflict with the humanitarian motive of development enhancement. This complex series of trade-offs in objectives is further complicated by the rapid growth of alternative national sources of nuclear technology, which makes the unilateral achievement of some objectives highly problematical. The search for an acceptable national trade-off between these conflicting objectives continues.

In the earliest days of civilian atomic energy (the late 1940s and early 1950s), U.S. policy severely restricted exports of nuclear materials and technical know-how, reflecting in large measure U.S. self-sufficiency and a desire to protect its monopoly position. In December 1953, however, President Eisenhower launched his "Atoms for Peace" program, which marked the beginning of two decades of aggressive U.S. assistance to other nations in the development of civilian nuclear power in return for guarantees that nuclear capabilities would be used only for peaceful purposes. Bilateral Agreements for Cooperation were typically negotiated between the United States and the purchasing nations in this regard. In 1957, the United States helped to create the International Atomic Energy Agency (IAEA), which instituted a system of international safeguards involving regular reports and inspections to forestall diversions of nuclear materials from civilian to military purposes. In 1968 the fear of proliferation of nuclear weapons led the United States to join 102 other nations in the Nuclear Weapons Non-Proliferation Treaty (NPT), which took force in 1970. The signatory states without nuclear weapons agreed not to develop or aid in the development of nuclear weapons or explosives and to put all their nuclear facilities under IAEA international safeguards. The treaty also committed the signatories to "the fullest possible exchange of equipment, materials, and scientific and technological information for the peaceful uses of nuclear energy," and guaranteed all treaty signers equal access to all forms of peaceful nuclear technology.(33)

The IAEA and NPT were apparent successes during the early 1970s, but that appearance was shattered in 1974 when India, which had not signed the NPT, exploded a "peaceful" nuclear device using plutonium derived from a research reactor supplied by Canada.(34) That explosion indicated that nuclear weapon capabilities were within the grasp of many Third World nations. And 1974 also witnessed the Arab oil embargo and the quadrupling of oil prices, which precipitated a mad rush for alternative sources of nuclear power. West Germany contracted with Brazil to provide it with two 1,325-megawatt reactors, the technology for enriching uranium to provide the fuel, and the reprocessing technology for extracting plutonium from spent fuel.(35) France agreed to sell Pakistan a reprocessing plant for separation of fissionable materials, an agreement that was later canceled. Neither purchasing country was an NPT member, and each agreement would have provided the technology to create weapons-grade nuclear material. These events dramatized the problem of nuclear proliferation and illustrated that formal international mechanisms for technology transfer control had failed to curb proliferation of the two most sensitive technologies inthe nuclear fuel cycle, enrichment and reprocessing plants. Assumptions about nonproliferation by international agreement, which had guided U.S. nuclear export policy for two decades, began to be reassessed. Alternative national sources of the technology had become a reality, which placed U.S. military and humanitarian objectives in doubt. As the conceptual framework predicts, once effective alternative suppliers become available, technology transfer control systems leave the nation with the costs of the controls (to national income) without the benefits of the objectives of the policy.

But convinced that the commercial objectives of the United States had dominated and undermined security concerns by permitting the transfer of nuclear technology in the first place, the U.S. Congress began working on a variety of bills designed to impose more stringent controls on additional commercial sales of U.S. reactors and to restrict government sales of enriched uranium to other nations. The bills proposed in the second session of the Ninety-fourth Congress, according to one observer, were

> based upon the mistaken assumption that the United States was in a position to determine unilaterally how nuclear energy would be developed around the world. In reality, enactment of any of these measures, thereby reducing the role of United States as a nuclear supplier, would likely have decreased U.S. influence on the development of an international nonproliferation policy.(36)

While most of these bills failed to pass, the International Security Assistance and Arms Export Control Act of 1976 was enacted. It added a new provision to the Foreign Assistance Act of 1961 barring foreign aid to any nation that delivers or receives "nuclear reprocessing or enrichment equipment, materials, or technology" unless it agrees to place all its "nuclear fuel and facilities" under the IAEA safeguards system.(37)

Pressures from Congress and then presidential candidate Jimmy Carter prompted the Ford administration to take a leadership role with the so-called London Suppliers Conference, begun in 1975. Nuclear supplier governments agreed to search for a consensus on a code of conduct for nuclear exports. Meeting annually under strict conditions of secrecy, these governments did in fact eventually hammer out a common export policy. In January 1978 the participating governments published guidelines for safeguard conditions, urging restraint on militarily sensitive violations. President Fored announced in October 1976 (a week before a presidential election) that "avoidance of proliferation must take precedence over economic interests" and that reprocessing should be deferred until "there is sound reason to conclude that the world community can effectively overcome the associated risks of proliferation."(38) The United States would thus defer the commercial reprocessing and recycling of plutonium, and all nations were asked to join a three-year moratorium on the transfer of reprocessing and enrichment technologies.

As a presidential candidate, Jimmy Carter indicated that curbing the spread of nuclear weapons would be one of his highest foreign policy priorities, but rejected the idea of a total export moratorium on nuclear technology as ineffective and unworkable due to the availability of alternative sources abroad.(39)

The Carter administration, once in office, sought a complex mix of controls and incentives for nuclear nonproliferation, which were embodied in an administration bill and eventually passed both houses of Congress in 1978 as the Nuclear Nonproliferation Act of 1978.

The overall thrust of the policy was to reduce the likelihood, at home and abroad, of widespread use of plutonium. Underlying the policy was the belief that the premature diffusion of plutonium technology would reduce barriers to the misuse of commercial nuclear energy, erode the international system of safeguards, and thus threaten the "delicate political balance of the internatinal nuclear regime."(40) The administration also apparently believed that there was still time available to forge an international antiproliferation consensus, and hoped that alternative proliferation-resistant technologies were available. The preeminent policy objective was to slow the spread of nuclear weapons capabilities – preferably to zero – and to minimize, and keep under control, any destabilizing impact of the diffusion of nuclear technology. The strategic objective was thus emphasized, while the domestic economic objectives of national income and jobs and the humanitarian objective of faster economic development for developing countries were deemphasized.

The key provisions of the 1978 act were: (1) to set strict criteria to be followed by the Nuclear Regulatory Commission (NRC) in the granting of export licenses for nuclear materials; (2) to provide for a cutoff of nuclear exports to any nations that newly develop or test atomic weapons or violate international safeguards; (3) to direct the president to seek negotiations with other nations to establish an international nuclear fuel stockpile to which a country following nonproliferation rules could turn if it were cut off from its regular supplier; (4) to grant the president power to override the NRC on an export-license denial if he deems it in the national interest (but Congress, in turn, could override the president and block the export); and (5) to require, eighteen months after the bill became law, that foreign countries receiving nuclear materials from the United States agree to "full-scope safeguards," including inspection of all their nuclear facilities, or risk a ban on any nuclear trade with the United States.(41)

What, then, have been the trade-offs and effects of this American policy? First, it is clear that significant alternative sources of nuclear technology exist. Not only has the United States lost its monopoly in most nuclear technologies, it is apparently even losing its comparative advantage in several specific technologies.

A major result has been the diversion of nuclear export business to countries with more reliable and less restrictive policies, thereby imposing the domestic economic cost of the controls on the U.S. economy and contributing to the loss of U.S. technological leadership. U.S. firms have gradually lost their overseas market share. Ten overseas orders for reactors were received by U.S. firms in 1974, five in 1975, one in 1976, none in 1977. Fifty-six percent of all foreign orders went to U.S. firms in 1974, 50 percent in 1975, 17 percent in 1976, and 0 percent in 1977 (in 1977 U.S. firms lost twelve potential orders).(42) Although U.S. restrictions are not the only reason for the decline in U.S. sales, Westinghouse and General Electric have lost out while European firms such as Framatone, Kraftwek Union, ASEA-Atom, and Brown Boveri have gained.

There are other indirect effects of the U.S. policy that no doubt compromise its long-term effectiveness. U.S. policy has damaged allied harmony and stimulated efforts abroad to reduce dependence on the United States. U.S. pressure on France, West Germany, and Japan to neutralize alternative sources of technology has been openly protested and resisted, and Europe and Japan have departed from the restrictive U.S. posture on the commercial reprocessing of spent fuel, uranium enrichment technology, and plutonium-fed fast breeder nuclear reactors.

Another cost of U.S. policy has been a considerable loss of goodwill from developing countries, but this loss probably also provided some developing nations with a strong incentive to set up their own enrichment programs, with or without U.S. assistance or technology. Nations such as Brazil, South Korea, Argentina, Pakistan, Egypt, Israel, and Taiwan have come to perceive the U.S. measures as an attempt to perpetuate their dependence on foreign technology.(43) Thus rather than achieving humanitarian objectives for development and a peaceful world through the restriction of nuclear technology exports, the policy of restrictions may in fact be counterproductive, contributing to rivalry and the deflection of scarce resources needed for development of independent, indigenous nuclear establishments.

The policies and tactics adopted by the United States may also have weakened the existing fabric of global nonproliferation institutions. U.S. restraints on the transfer of sensitive technologies have violated the spirit of the NPT, which guarantees signatory countries nondiscriminatory access to nuclear technology for peaceful purposes. And the United States unilaterally refused to certify IAEA inspections as satisfactory safeguards for U.S. nuclear materials exports. International confidence in the IAEA has dwindled and the good faith of the major supplier nations such as the United States is questioned by importing members.(44)

In summary, the nuclear or plutonium horse is obviously already out of the stable. As with almost any technology, the transfer to or replication of nuclear technology by other countries could not be permanently controlled or forestalled. Once effective alternative sources exist, the costs of restrictive controls by any one country escalate rapidly. But the capacities of international or multilateral institutions will always be limited. Much is heard about the need to develop an international consensus on the further spread of nuclear weapons and how to manage the nuclear fuel cycle – but this will take considerable time, and time may be running short in relation to the problem. Many dream of technological fixes, such as foolproof, proliferation-resistant nuclear technologies – but will they appear any time soon? In the absence of effective institutions for international safeguards, international consensus on the undesirability of nuclear dangers, and proliferation-resistant technologies, the United States had no choice but to attempt leadership.

U.S. policy, in an international environment short on consensus and long on alternative foreign suppliers, is at that stage, and carries significant economic costs in lost sales and domestic jobs. And, as the arguments presented indicate, U.S. policy has limited neither nuclear technology transfer nor its foreign development, because a strict, comprehensive international regime of nuclear and plutonium controls is apparently not feasible. At this stage, the economic costs of the U.S. control system are being borne by the United States without obvious significant contribution to its noneconomic goals.

DOMESTIC JOBS AND TECHNOLOGY TRANSFER

The issue of the displacement of home country jobs by the foreign activities of a nation's firms has become a public one in the 1970s, not only in the United States but in other home countries as well.(45) While it is difficult to disentangle the effects of technology transfer per se from the consequences of other aspects of multinational corporate activities abroad, the protagonists calling for control systems lay a good part of the blame for the perceived adverse employment effect on the outward transfer of technology developed in the United States. As one eminent labor spokesman put it:

> But the major factors [causing a deterioration in the U.S. competitive position and loss of American jobs] have involved the direct transfer of U.S. technology and knowhow – through foreign subsidiaries of U.S. firms, patent agreements, licensing arrangements, and a variety of joint venture affiliations of U.S. firms with foreign companies.(46)

Technology transfer, as suggested in the framework developed earlier, is indeed an essential ingredient in evaluating the impact on domestic jobs of foreign operations of U.S. firms. But if alternative technologies existed abroad, and were equal to or superior to U.S. technologies, then U.S. production and jobs would be displaced in any event – with or without the foreign activities of U.S. firms. The labor market is affected, but the cause is not the transfer of U.S. technology, but the development of alternative technologies abroad. However, in instances where superior technology and know-how remain with U.S. firms, that technology is exploited abroad via direct investment or licensing to secure markets that otherwise could have been served by production in the United States, and thus the effect on the domestic is <u>caused</u> by the technology transfer.(47)

The conflicts concerning jobs and technology transfer obviously involve the distribution of the national income. Some groups lose and some groups gain within the domestic economy. The losing groups, mainly production workers and their representatives in organized labor, have attempted to transfer the conflict from the marketplace (the labor market and product market) to the political arena with a call for different governmental policies toward technology transfer (and imports). Governmental control systems for the technology transfers themselves are sought.(48)

The several economic analyses carried out to estimate the impact of foreign investment and technology transfer by U.S. companies have yielded several widely accepted conclusions. The first is that there are both jobs created and jobs destroyed.(49) Whether the total of jobs created exceeds the number of jobs destroyed is still a matter of debate, but the typical estimate is that the net balance – whether positive or negative – is relatively small compared to total unemployment or other structural changes in the U.S. labor markets.(50) But the skill mixes of the jobs created and jobs destroyed are significantly different. Jobs created involve a higher share of professional, managerial, and clerical jobs than exists for the economy at large, while the jobs displaced have a higher share of semiskilled and unskilled production and operative jobs than for the economy at large.(51) The jobs created are thus relatively more skilled, with higher average pay and less unionization, than the jobs eliminated. The problem is adjustment in the domestic labor market. Finally, the adjustment, when completed, leads to a lower share of the national income going to production labor than would have gone without the technology transfer and investment abroad.(52)

U.S. Policy

To date, the issue of the effect of technology transfer on domestic jobs has been largely a matter of conflict between labor unions and multinational corporations, but has not involved the application of government control systems. This government policy of relative laissez faire stems, it could be argued, from the belief that enhanced aggregate national income and positive contributions to U.S. humanitarian objectives to further economic development abroad outweigh the economic costs of adjustment, including jobs lost at home.

But the issue has been pushed with intensity by organized labor, which estimates that 500,000 U.S. jobs were lost in 1966-69 due to U.S. international trade, and that a significant share of the loss could be attributed to the operations – technology transfer – of U.S. multinationals.(53)

Labor spokesmen also allege that permissive U.S. government policies have encouraged the transfer of production facilities and technology to foreign locations. Among these are inappropriate (liberal) U.S. taxation of foreign source income.(54)

A second issue is that some of the political and commercial risks of foreign investment can be insured at indirectly subsidized rates through the Overseas Private Investment Corporation. A third involves "unfair" U.S. international trade policies, including the Generalized System of Tariff Preferences, enacted in the Trade Reform Act of 1975, which permits the import of a substantial list of products from developing countries at preferential (or zero) tariff rates, and Sections 806.30 and 807.00 of the U.S. Tariff Schedule, which permit American companies producing or assembling products in developing countries for export to the United States to pay U.S. import duties only on the value added abroad to U.S. raw materials or components. Both of these are presumed to encourage technology outflows by U.S. companies.

The combination of these policies, says organized labor, has stimulated U.S. companies to transfer to low-wage countries operations with technologies which have high labor-capital ratios, and thus to reduce the demand for American production workers. The results are the closing of plants in the United States, layoffs, and swelling unemployment.(55) Furthermore, the adjustment assistance programs of the federal government are viewed as wastefully inadequate. The adjustment assistance provisions of the Trade Expansion Act of 1962 and their liberalization in the Trade Act of 1974 do not apply to the foreign operations of U.S. companies unless domestic jobs are lost due to imports from those operations. Second, the success in actually accomplishing adjustment when disruption due to imports occurs – that is, reabsorption of displaced workers into different and more productive jobs – has not been good. Labor's answer to these inadequacies is to block or control the technology transfer causing the loss of jobs.

The Conflicts

The conflicts of interest outlined above, while they have not yet resulted in explicit government control systems, have had two manifestations. One is a concerted attempt by organized labor, spurred on by the unions directly affected, to bring about a major change in U.S. policy toward the outward technology transfer and investment by U.S. companies. The second is the direct use of bargaining power by some unions with companies to influence their foreign investment decisions. The latter remains relatively rare.

The most systematic legislative proposal of the AFL-CIO was the Hartke-Burke Bill introduced in both houses of Congress in 1972.(56) This bill, which has been reintroduced in revised form in most sessions of Congress since 1972, addresses most of the difficulties in U.S. policy as perceived by U.S. trade unions. It would have:

1. eliminated the preferential tax treatment for foreign earnings by removing the deferral provision and tax credit provision in the U.S. revenue code and allowing only a deduction against earnings for foreign taxes paid;
2. eliminated the preferential tariff treatment under sections 806.30 and 807.00 of the Tariff Schedule (both of which would reduce the general incentive to invest in foreign facilities);
3. monitored and restricted foreign investment by U.S. companies whenever U.S. employment would be decreased by this investment;
4. limited foreign licensing of U.S. technology when it adversely affects domestic production by rendering unenforceable the U.S. patent on such technology; and
5. imposed quotas on the import of any product for which the ratio of imports to domestic production rose above a base-period level.

Despite the fact that the Hartke-Burke Bill has never been reported out of committee in either house of Congress, it was still high on the list of priorities of the AFL-CIO in 1979. In addition, pressure through the executive branch led to the inclusion of the elimination of the tax-deferral provision on foreign earnings in the administration's tax reform proposals of 1978.

Another tactic used by labor in a few specific instances is to attempt to protect U.S. jobs through contractual arrangements won in the collective bargaining process or through threat against U.S. operations of a company investing abroad or transferring technology.(57) These direct actions, for which a legal basis has been at least partially established,(58) are of course designed to use the union's bargaining power with companies as a private control mechanism over technology transfer, but the practice is evidently not widespread.

Protecting domestic jobs and production against the transfer of technology is thus an issue, and one that is escalating. Thus far it has not yielded direct governmental control mechanisms in the United States. While these are still being debated, the national policy continues to reflect the view that the economic benefits outweigh the costs of adjustment and that other domestic redistribution policies – such as unemployment compensation – are adequate for protecting workers displaced. Whether this policy will be altered through the political process is difficult to forecast.

CONFLICTS OVER ENVIRONMENT, HEALTH, AND SAFETY

Another debate regarding outward transfer of technology and its control has centered on protecting the environment and/or human health and safety in the recipient country (figure 8.3). Should the United States, in pursuit of its humanitarian objectives, attempt to impose its health and safety standards on the rest of the world? Is it fair to send, often to Third World "pollution havens," production technologies or products deemed hazardous or unfit for domestic use? Some consumer and environmental groups, along with a few federal agencies, have come to believe that the United States should play the role of "environmental policeman of the world." Several controls on technology transfer have thus been proposed, and a few imposed, in the name of bringing about an "environmentally oriented" U.S. international economic policy.(59)

Legislation to limit the freedom of U.S. companies to market abroad products labeled hazardous and restricted in the United States has been proposed throughout this decade. Affected industries have typically argued that their export marketing and foreign production activities are in conformity with the laws of the foreign importing nations, that many U.S. health and safety regulations are unwarranted, and that some products, such as pesticides, are so needed abroad that the benefits of use far outweigh any harm that might result.(60) These views have usually prevailed, with the result that the general rule applicable to exports of products or processes subject to federal product, safety, and environmental legislation is that they need not conform to the requirements of the U.S. statutes, but that they must be labeled to show that they are intended solely for export and conform with local regulations. In terms of the potential trade-offs set out in earlier in this chapter, U.S. policy faces two conflicts. In restricting the transfer of product or process technology on environmental/humanitarian grounds, economic costs in national income and employment may arise. If alternative sources of supply are available, the positive contribution to U.S. humanitarian objectives is not realized either. And banning such transfers may detract from the objective of increasing economic development in poor countries.

U.S. Policy

While U.S. policy has traditionally imposed relatively few controls, the picture has recently begun to change. The U.S. Toxic Substances Control Act (TSCA) of 1976, for example, requires exporters to notify the Environmental Protection Agency (EPA) administrator of any exportation of a toxic chemical. The administrator, in turn, must notify the importing country that data on the chemical is available, or, as the case may be, that regulatory action has been taken in the case of substances found to present unreasonable risks. TSCA exempts chemicals for export from premarketing testing requirements, however, unless they "will present an unreasonable risk of injury to health within the United States or to the environment of the United States." Such a finding was made in the case of chlorofluorocarbons, which, via overseas emissions, could deplete the ozone layer, thus damaging the United States; on those grounds export and foreign production of the substance were banned.(61)

A second example involves interpretation by the Consumer Product Safety Commission (CPSC) of the Federal Hazardous Substances Act to authorize the commission to prohibit the export of banned products that were previously allowed in domestic commerce. It thus banned exports of childrens' sleepwear treated with the flame retardant Tris, a substance found to cause cancer in laboratory animals.(62) Export controls for pesticides were also strengthened in 1978, when Congress passed amendments to the Federal Insecticide, Fungicide, and Rodenticide Act of 1972. The amendments require closer monitoring, informative labeling, and mandatory warnings to foreign recipients of pesticides considered dangerous in the U.S.(63)

Not only is U.S. policy in a state of flux, it contains several inconsistencies and contradictions for which resolution is sought. More than 500 consumer, drug, pesticide, and chemical products have been banned from the U.S. market by the EPA, CPSC, and Food and Drug Administration (FDA). However, many substances (for example, DDT, cyclamate food sweeteners, and food dye Red No. 2) are still exported.(64) A Carter administration interagency working group, in attempting to develop policy on this issue, identified the following conflicting statutes: "eight laws permit export of banned products, and conflicts arise out of three laws that require approval by a foreign government; three that require notificatoin of the foreign power; and, at the other extreme, two laws that give the regulatory agency authority to ban such exports; and two laws that ban such exports."(65) Committees in both houses of Congress have held hearings with a view to harmonizing the multiple and inconsistent regulations and legislation. While the various proposed changes in U.S. export policy have not yet been made, the trend in legislative concern is clearly toward imposing stricter U.S. controls over exports of hazardous product and technologies. This will make it more difficult for U.S. companies to compete with foreign companies not encumbered by such restraints, which suggests that the United States is willing to take the economic cost involved. Whether other supplying nations will follow a similar policy, thus at least realizing the direct objectives of the policy, seems doubtful, at least in the near future.

Environmental groups are concerned about technology transfers taking the form of facilities and processes which may be damaging to the environment. Several groups have attempted to establish a "global reach" for the National Environmental Policy Act (NEPA) of 1969, which requires "all agencies of the Federal Government" to write detailed environmental impact statements (EIS) in advance of "major Federal actions significantly affecting the quality of the human environment." Controversy has raged over whether the EIS requirement should apply to federal export licenses, permits, approvals and other export-related actions on the part of agencies such as the Nuclear Regulatory Commission, Export-Import Bank, Agency for International Development, and Overseas Private Investment Corporation.(66)

A few court cases, mainly involving nuclear and pesticide technology exports, have succeeded in forcing companies to prepare EISes. In 1977, the White House Council on Environmental Quality (CEQ) concluded that the EIS requirement of NEPA applied to all significant effects of proposed federal actions on the quality of the human environment, including those within the jurisdiction of other nations and those outside the jurisdiction of any nation.(67) Under such a policy, environmental impact statements might be required for most exports and foreign investments by U.S. companies. But opposition from several U.S. agencies and private industry led the Carter administration in 1978 to drop the idea that an EIS be filed with <u>all</u> applications for export licenses, but that "abbreviated environmental reviews" be required only for exports of nuclear reactors, potentially toxic products or facilities that could pose serious health risks, and for "certain Federal actions" that could damage the environment of Third World countries or endanger crucial global resources.(68) Still pending, however, is a Natural Resources Defense Council suit against Eximbank to require an EIS prior to its financing of exports.(69)

A final issue concerns restraints on technology transfer involving the location of hazardous industries and technologies abroad to circumvent restrictive environmental or occupational health laws at home.(70) Charges that U.S. companies selectively export hazardous industries to developing nations (Maryland Public Interest Research Group) or locate labor interim plants in countries with lax worker-protection regulations (National Institute of Occupational Safety and Health) have led to call for controls over foreign investment and imports from such countries.(71) And legislative attempts to impose new restrictions have not as yet been successful. If even such locational shifts do accelerate, more stringent regulation would protect relatively few U.S. jobs and little national income. Other nations do not so regulate these external activities of their companies, and unilateral U.S. action would only be marginally effective. And to the extent it is effective, it would involve some cost to the humanitarian economic development objectives in developing countries.

The logic behind such restrictions is much clearer if health or environmental effects spill over across national boundaries and compromise U.S. objectives of national environmental preservation and safety. Controls are obviously justified, for example, where the effects of a transfer would clearly and significantly damage U.S public health or environmental quality. Here the so-called effects doctrine of jurisdiction applies, as pronounced by Judge Learned Hand in the landmark Alcoa case of 1945: "it is settled law that any state may impose obligations, even upon persons not within its allegiance, for conduct outside its borders that has consequences within its borders which the state reprehends."(72) The problem, however, is that it is often very difficult to demonstrate direct effects that reach back to the United States, though many expect "more and more examples in an increasingly impacted biosphere."(73) Even in the absence of international spillovers, restraints such as environmental impact assessment requirements may be warranted in instances where the risks of disaster are very high or where the recipients do not possess the capacity to fully understand the nature of the hazards involved. In January 1976, for example, the Eximbank authorized a loan and guarantees to enable the Philippines to buy a nuclear reactor. No environmental assessment was performed. If it had been made, its disclosures surely would have included the fact that the reactor was to be sited in an earthquake zone near five volcanos.(74)

Beyond such extreme situations, there appears little or no justification for controls on outward technology transfer. Their effectiveness would be limited, since other nations with similar technologies do not usually restrict exports, even when they are restricted at home. Second, the economic loss in the U.S., as a result of the previous point, will occur without the offsetting benefits of the objectives of the policy. When the environmental or health impacts are purely domestic and the importing country is informed of the risks, then extension of American standards seems unwarranted.

HUMAN RIGHTS CONFLICTS AND TECHNOLOGY TRANSFER

Like environmental protection, the issue of human rights is tangential to control systems over technology transfer. But in this case, the sole justification for control of outward technology transfer is to induce a change in or make it costly to carry out the domestic policies of the potential recipient country, policies that have no direct and immediate effect on the economic well-being of the home country. There are thus no direct economic benefits for the imposing country – only economic costs. Whatever contribution they make to national welfare comes through the humanitarian objective, as illustrated in figure 8.3, and this contribution arises only if the potential recipient country changes its behavior in the desired direction. As with the previous discussion, the leverage that one nation has over the behavior of the object government depends on the uniqueness of the technology (that is, the availability of other national alternatives) and the strength of the demand for the technology by the country.

No justification for restraints on outward technology transfer from the United States has been the object of more passionate advocacy or scornful criticism than that of promoting human rights.(75) The issue is laden with ideology and propaganda, and often shrouded by a lack of real information. What restrictions have been imposed by the human rights movement on technology transfer? What results have they achieved? Have the benefits in U.S. social objectives of humanitarian behavior abroad outweighed the economic costs of income and jobs lost to alternative sources of technology?

Jimmy Carter did not discover human rights. A moral tone has dominated American diplomacy, if not policy, since the Declaration of Independence presented an appeal for universal inalienable rights. President Carter, however, was responsible for making human rights a central theme ("the soul") of American foreign policy. He pledged to "restore the moral authority of this country in its conduct of foreign policy" and pledged an "undeviating commitment" to the advancement of human rights abroad.(76)

The popular movement concerned with cruelty, injustice, and oppression is pervasive in scope, involving many national and international public-sector and private-sector organizations – including the White House, U.S. Congress, United Nations, Organization of American States, Interfaith Center on Corporate Responsibility, Committee to Oppose Bank Loans to South Africa, Amnesty International, and several foundations, student and civil rights groups. The movement is motivated by a mix of religious, civil, ethnic, racial, economic, and humanitarian concerns.

Actions by Private Groups and the U.S. Government

The tactics impinging on technology transfers by multinationals are as varied as the groups and motives involved. The common object of nongovernmental action has been to influence management's thinking and decision making at the corporate headquarters level by raising the perceived costs and reducing the perceived rewards associated with transfers to "repressive nations," in other words, to accomplish a self-imposed control of technology outflows. Consumers in the United States have been urged to boycott a wide range of products and services in the name of human rights (such as ITT and Gulf Oil products because of the firms' involvement in South Africa). There have been campaigns to force universities and churches to divest themselves of stock in companies operating in nations violating human rights. And there were more than 150 human rights resolutions voted on at annual shareholders meetings from 1970 to 1978 calling for companies to stop doing business (making loans or transferring technology) with repressive governments. However, very few of these resolutions passed.(77)

U.S. government actions have not directly prohibited or controlled technology and transfers in the name of human rights except in cases of trade embargoes (such as Rhodesia and Cuba) or trade with the Soviet Union. Several governmental actions or proposals, however, indirectly affect technology transfer by affecting the financing or risk of U.S. trade or foreign investment. For example, beginning in 1973, ever more forceful provisions in the foreign assistance laws have made first economic, then military assistance conditional on acceptable human rights practices.(78) Aid cuts under these provisions have been made for several countries, including Argentina, Brazil, and Chile.

Likewise, efforts to restrict Eximbank lending or guarantees for U.S. exports were partially successful in 1977 in requiring adherence to the "Sullivan Code" concerning the promotion of racial equality in South Africa for any recipient.(79) And since 1975, issuance of U.S. company investment abroad against political risk by the Overseas Private Investment Corporation was to be denied, under the Harkin Amendment, for countries that were "gross and consistent" violators of human rights. There have been complaints, however, that OPIC has not adhered to this restriction.(80) Furthermore, the U.S. Congress, in 1976, wrote instructions into the law authorizing the U.S. contributions to the World Bank and other multilateral aid agencies that instructed the U.S. directors to vote against aid to countries grossly violating human rights.(81) Other attempts to "politicize" lending by multilateral agencies on human rights grounds have been largely unsuccessful.

The most direct U.S. governmental involvement in technology transfer controls arises due to restrictions or embargoes on trade, including high-technology goods, with specific countries charged with violating human rights. While the most effective embargoes are multilateral, the United States has at times taken unilateral actions. A voluntary embargo against the sale of military weapons to South Africa was adopted by the U.N. in 1963, and at the same time the United States imposed its own mandatory embargo on exports of arms and ammunition, and all goods and technical data to the police and military of South Africa and Namibia.(82) After the Unilateral Declaration of Independence by Rhodesia in 1965, the United Kingdom persuaded the U.N. to impose selective mandatory economic sanctions in 1966 and subsequently more comprehensive mandatory sanctions in 1968, banning virtually all commodities originating from or destined for its former colony. In unilateral actions, the United States in 1978 slapped a trade and military aid ban on Uganda, tightened its controls on exports of crime-control and detention equipment to all countries, and, in the wake of Soviet trials of dissidents and U.S. newsmen, cancelled (later to be reversed) the sale of a Univac computer to TASS (the Soviet news agency), and placed controls on oil equipment exports to Russia – all in the name of human rights. Finally, one other trade-related restriction that should be noted is the Jackson-Vanik Amendment to the U.S. Trade Act of 1974. Directed primarily at the Soviet Union, the amendment prohibits most-favored-nation status for any "nonmarket economy country" which denies or unduly restricts the rights of its citizens.

The Impact of Human Rights Restrictions

To be effective, technology transfer controls for human rights purposes must induce the other governments to change their policies. It is most likely to do so if the technology is essential, and unavailable from other sources. And even if it is successful, controlling nations must still assess if the humanitarian gain exceeds the economic costs. Overall, the evidence so far is not very promising for the human-rightists. On the other hand, thousands of political prisoners have been released since early 1977, thanks in part to U.S. efforts, in such nations as Indonesia, South Korea, the Philippines, Poland, Morocco, Bangladesh, Pakistan, Tanzania, and the Dominican Republic; emigration has increased from the USSR; Bolivia, Peru, and Ecuador have moved toward elections; and El Salvador and Haiti have agreed to visits by the Inter-American Commission on Human Rights.(83) But evidence that restraints on technology transfer have had major effects in pushing governments along the path to broader observance of human rights is difficult to find. The social and economic costs of repression are apparently not yet greater than the perceived costs of relieving repression to such governments. And the availability of alternative technology from countries that do not share the U.S. fervor is quite extensive, making unilateral actions of doubtful effect. Thus the economic cost to the United States of its human rights sanctions have not been trivial in the form of lost incomes, profits, taxes, and jobs. In addition, the United States has lost substantial good will among many developing nations. And the question ultimately remains whether blocking technology transfer on human rights grounds actually improves the lot of the repressed, or, through harming economic growth of the recipient nation, increases the resolve of the repressive government to be "independent."(84)

SUMMARY AND IMPLICATIONS

This chapter has identified an abundance of governmental and private control systems, laws, and practices that limit or reduce the incentives for technology transfer from the United States. This is ironic, because the United States remains the world's major capital and technology exporter, and is viewed by many as a nation whose economic and international trade policy is built on the liberal tradition of laissez faire. The "controls" extend to technology embodied in products as well as disembodied technology. It applies to technology and know-how owned largely by private firms in the United States, but also extends to some technologies over which the government itself has proprietary control.

The coverage is not complete, since peripheral areas that may affect technology transfer – such as the application of antitrust laws, limitations on arms sales, the Foreign Corrupt Practices Act's criminalization of foreign payoffs, and the anti-Arab-boycott legislation, have not been dealt with specifically. Yet the five policy areas surveyed concern much potential technology transfer from the United States.

The structure of significant controls over U.S. technology transfer, in place or seriously sought, is summarized in table 8.2. Included are the nature of the control measures, the objectives of the measures, the groups in the country seeking such measures, and the U.S. national objectives that are likely to be harmed by the controls. Finally, some indications of the nature of the effectiveness of the controls are given, based upon the foregoing discussion.

Table 8.2. Characteristics of Actual or Proposed U.S. Controls Over Technology Transfer

	1. NATIONAL SECURITY
Objectives Pursued:	Minimization of enemy military capabilities Security for U.S. allies
Groups Seeking Controls:	Defense Department U.S. Congress Conservative political groups White House
Control Measures (in place or sought):	Prohibitions on trade with certain belligerent nations Scrutiny and licensing of strategic exports Collective embargo (COCOM) on certain strategic exports Restrictions on payments or extensions of credit Constraints on Eximbank financing
National Objectives Compromised:	Domestic employment Income per capita Balance of payments East-West cooperation
Effectiveness (actual or potential):	Moderate with respect to the inner core of strategic technologies, upon which there has been COCOM consensus; otherwise limited due to alternative sources, inadevertent leakage, clandestine acquisition, and induced indigenous development

Table 8.2. (Cont.)

	2. NUCLEAR TECHNOLOGY
Objectives Pursued:	Nonproliferation of nuclear weapons Security for U.S. and allies
Groups Seeking Controls:	White House U.S. Congress Peace and arms control groups
Control Measures (in place or sought):	Bilateral agreements for cooperation IAEA international safeguards Denial of aid to nations not complying with IAEA safeguards system Supplier's code of conduct for nuclear exports Nuclear Nonproliferation Act of 1978 (stringent export licensing for nuclear materials, prohibitions on nuclear exports to violators of international safeguards, full scope safeguards requirements for recipient nations, etc.)
National Objectives Compromised:	Income per capita Domestic employment Energy supply expansion abroad Allied harmony and Third World goodwill
Effectiveness (actual or potential):	Moderate with respect to code adopted by major supplying nations; otherwise limited due to alternative sources, reduced U.S. influence, and reduced confidence in existing IAEA safeguards

Table 8.2. (Cont.)

	3. DOMESTIC JOBS
Objectives Pursued:	Domestic employment Income distribution Minimum or slow structured adjustment
Groups Seeking Controls:	AFL-CIO Some national unions A few industry leaders A few congressional leaders
Control Measures (in place or sought):	Removal of Sections 806.30 and 807.00 of Tariff Schedule Removal of deferral and credit for taxes on foreign source income Screening and control of new foreign investment or technology transfer, including licensing Removal of OPIC and Eximbank support for technology transfer activities
National Objectives Compromised:	Income per capita Potential impact on R&D and growth Economic development abroad
Effectiveness (actual or potential:	Limited, alternative sources of technologies available abroad in most instances

Table 8.2. (Cont.)

	4. ENVIRONMENT AND HEALTH
Objectives Pursued:	Environmental, consumer, and worker protection abroad and at home Protection of "global commons"
Groups Seeking Controls:	White House Council on Environmental Quality U.S. federal agencies: EPA, FDA, CPSC Many national environmental groups Some congressional leaders
Control Measures (in place or sought):	Mandatory warnings/notification to foreign governments Closer monitoring and informative labeling Outright export bans on hazardous products Environmental impact statement requirements on U.S.-government-assisted projects abroad Tariffs or bans on imports of hazardous products from nations with lax laws Denial of governmental assistance to exports of workplace or environmental hazards
National Objectives Compromised:	Income per capita Domestic employment Economic development abroad
Effectiveness (actual or potential:	Limited due to alternative sources and counterproductive effects of exporting "pettifoggery"

Table 8.2. (Cont.)

	5. HUMAN RIGHTS
Objectives Pursued:	Improvement of human rights conditions abroad Promotion of democratic values
Groups Seeking Controls:	White House/State Department U.S. Congress Civil rights groups Religious and academic groups
Control Measures (in place of sought):	Nongovernmental actions: boycotts, stock divestitures, proxy resolutions, etc. Denial of bilateral economic and military aid to nations violating rights Denial of Eximbank and OPIC assistance to nations violating rights Denial of multinational foreign aid and credits via U.S. role in international financial institutions Trade embargoes (Rhodesia, South Africa) Denial of most-favored-nation status
National Objectives Compromised:	Income per capita Domestic employment Favorable foreign relations
Effectiveness (actual or potential):	Limited due to alternative sources and counterproductive effects

The controls stem from a diversity of motives, many of which have found expression in the last fifteen years. Their diversity, inconsistency, and lack of coherence is due to the multiplicity of national goals, the fact that technology transfer often involves a redistribution of income or welfare in the transferring country, and the legislative and regulatory process in the United States, which responds to the shifting interests and influences of groups affected by technology transfer.

The simple conceptual model outlined earlier provides the basis for defining the motives and assessing the effects and effectiveness of controls over outward technology transfer. Critical to understanding U.S. controls is the fact that the nation has multiple and sometimes inconsistent objectives. Some are economic – real income, employment, distribution of income, avoidance of structural adjustment. Some are strategic – national security, limiting the military potential of antagonistic nations. And some are humanitarian – economic growth for poor countries, health and safety abroad, environmental preservation, and the like. The effectiveness of any controls over technology transfer depends on the degree of monopoly the home nation has in that technology. Readily available substitutes from other national suppliers limit the leverage that a nation has through the exercise of controls, and raise the likelihood that economic or other costs will be borne without reaping the benefit in terms of other objectives. To be effective when there are alternative sources, international agreements are required to impose similar control systems, an area in which the United States has been a leader with respect to technology transfers of a strategic nature to communist countries.

The United States has attempted to control technology transfer for three purposes. The first was for defense and strategic reasons – to limit the military capabilities of the Soviet bloc, the People's Republic of China, Vietnam, North Korea, and other communist and hostile regimes. It is this area also in which the United States has been most active in attempting to achieve commonality in policy with other industrial countries. National security has also been the major motivation for U.S. policy toward nuclear technology transfers, although other national objectives enter that area as well, such as concerns over human safety and environmental pollution.

A second purpose of control systems over technology transfer has been to induce other nations to take certain actions that the United States, for humanitarian or domestic economic motives, perceives as beneficial. This explains a broad range of controls or limitations over technology transfers and other activities, especially by U.S. companies, designed to protect foreign workers or consumers (whether the nation sees fit to protect them or not), to protect the environment abroad or the "global commons," to induce a foreign government to take other action sought by the United States (for example, to convince Arab nations to end the boycott of Israel and companies doing business with Israel).

A third purpose for which controls over technology have been proposed, but not adopted, is the strictly domestic economic one of affecting income redistribution and jobs, and avoiding economic adjustment in the home economy to changing international competitive conditions. There remains a serious attempt, by organized labor, to install a technology transfer surveillance and control system of a type already existing in some other technology-exporting countries.

Have interferences with technology transfers been effective in accomplishing the objectives they seek? While it is impossible to estimate with assurance what would have happened if policy had been different, much of the evidence presented above suggest that for most types of control systems the effectiveness has not been great and in some cases even counterproductive for the objective sought. The ineffectiveness is most obvious for cases in which alternative national suppliers of that technology do not share the U.S. national objectives and are thereby unwilling to adopt similar control systems. Clear examples are U.S. measures that limit technology transfer on the basis of environmental, worker, or consumer protection; human rights violations; armaments trade; and, recent, nonproliferation of nuclear technology. The technologies in these cases are not monopolized by the United States, and the willingness of alternative suppliers to share U.S. objectives is quite limited. A result is that the United States has often accepted the domestic economic cost of the controls while not realizing much contribution to the national objectives that produced the controls. The direct cost are thus borne by workers, managers, and stockholders whose incomes are suppressed by the controls, while the advocates of the controls succeed in winning the controls but often not the objectives the controls seek.

Perhaps one area where the objective of controls has been realized has been in limiting technology transfer with military capabilities to communist bloc countries. But even here the assessment must be equivocal, since restrictions on trade and technology transfers have, no doubt, spurred the Soviet Union and other potential recipients to greater self-sufficiency and independence, and even competitiveness with U.S. technology. In any event, the shrinking range of technologies in which the United States has a clearcut superiority has increasingly limited the effectiveness of U.S. restrictions even in this area.

Despite the lack of clear contribution to net national benefit from controls over technology transfer, it is not likely that governmental and pressure group interferences with the outward technology transfer process will disappear. The conflicts among priorities are not going to vanish, and may well intensify. America's penchant for misguided idealism and counterproductive controls on technology transfer appears to be here to stay.

NOTES

(1) U.S. Department of Commerce, Survey of Current Business (various issues).

(2) See Robert G. Hawkins and Bertram Finn, "Regulation of Multinational Firms' Foreign Activities: Home Country Policies and Concerns," Journal of Contemporary Business, Autumn 1977, pp. 7-30.

(3) This definition of technology follows closely that in Richard R. Nelson, M. Peck, and F. Kalachek, Technology, Economic Growth, and Public Policy (Washington: Brookings Institution, 1967), pp. 7-8.

(4) See G.R. Hall and R.S. Johnson, "Transfers of U.S. Aerospace Technology to Japan," in The Technology Factor in International Trade, ed. R. Vernon (New York: National Bureau of Economic Research, 1970), pp. 305-58.

(5) The most obvious codification of these objectives is the Employment Act of 1946 and its amendments. Annual reaffirmations of these objectives can be found in the Economic Report of the President to the Congress and the Report of the Council of Economic Advisors.

(6) The concern about maintaining competition received legislative sanction in the Sherman Antitrust Act (1890) and the Clayton Act (1914). Policy objectives concerning income distribution must be gleaned individually from the Internal Revenue Code and various pieces of social legislation, ranging from the Social Security Act (as amended) to legislation covering minimum wages, unemployment compensation, grants-in-aid, and the like.

(7) Some of these measures and objectives are much older, such as the Pure Food and Drug Act and several measures governing occupational safety and labeling of products. But the scope of such regulation has been greatly expanded in the recent laws and regulations of the Environmental Protection Agency, the Federal Trade Commission, and the Consumer Protection Agency.

(8) These derivative objectives obviously do not include perhaps the most important aspect of national security: the military establishment maintained by the nation itself.

(9) The recent theoretical models are critiqued in Thomas Pugel, "International Technology Transfer and Neoclassical Theory: A Survey," unpublished, N.Y.U. working paper no. 78-96 (1978). See also Rachel McCulloch and Janet L. Yellen, "Technology Transfer and the National Interest," unpublished working paper, Harvard Econmics Department, 1978.

(10) This argument has been developed in McCulloch and Yellen, "Technology Transfer and the National Interest"; Ronald Jones, "The Role of Technology in the Theory of International Trade," in Vernon Technology Factor, pp. 73-92; and Carlos Rodriguez, "Trade in Technological Knowledge and the National Advantage," Journal of Political Economy, February 1975, pp. 121-37. For a comparison, see Pugel, "International Technology Transfer."

(11) The empirical evidence on these linkages is not very compelling. Little research has yet been produced on whether technology transfer (in any or all forms) raises the perceived return on total R & D activities. There is some evidence that higher levels of R & D do lead to more innovation and fast productivity growth, and hence higher national income. See Richard B. Freeman, "Investment in Human Capital and Knowledge," in Capital for Productivity and Jobs ed. E. Shapiro and W. White (Englewood Cliffs, N.J.: Prentice-Hall, 1977), pp. 96-123. Additional evidence is reviewed in Rachel McCulloch, Research and Development as a Determinant of U.S. International Competitiveness (Washington: National Planning Association, 1978).

(12) For an analysis of these effects of multinational firms' operations, see Thomas Parry, "Competition and Monopoly in MNC Operations," in Economic Issues of Multinational Firms, ed. R. Hawkins (Greenwich, Conn.: JAI Press, 1978).

(13) Such cases are analyzed in Ingo Walter, International Economics of Pollution (New York: John Wiley, 1975), Thomas N. Gladwin and Ingo Walter, Multinationals Under Fire: Lessons in Conflict Management (New York: John Wiley, 1980).

(14) For general reviews of policy in this area, see Samuel Pisar, Coexistence and Commerce: Guidelines for Transactions between East and West (New York: McGraw-Hill, 1970), and Marshall I. Goodman, Detente and Dollars: Doing Business with the Soviets (New York: Basic Books, 1975).

(15) See Gladwin and Walter, Multinationals Under Fire.

(16) Export Control Act of 1949, Section 2 (63 Stat 7).

(17) Goldman, Detente and Dollars, p. 49.

(18) See Thomas A. Wolf, U.S. East-West Trade Policy (Lexington, Mass.: Lexington Books, 1973), chap. 6.

(19) Mutual Defense Assistance Control Act of 1951, title 1, Section 101 (22 U.S.C. 1611).

(20) Export Administration Act of 1969 (PL 91-184), Section 3.

(21) Export Administration Amendments of 1974 (PL 93-500), (9, 88 Stat. 1552), adding 50 U.S.C. App. 2403 (h).

(22) For a review see Lawrence T. Brainard, "Public and Private Credit Policy in East-West Trade," Law and Policy in International Business 7 (Fall 1975): 1169-86.

(23) See William F. Jones, Jr., "Export-Import Bank Act Amendments of 1974," Law and Policy in International Business, 7 (Summer 1975), pp. 947-58.

(24) Maurice J. Mountain, "Technology Exports and National Security," Foreign Policy, no. 32 (Fall 1978), pp. 101-2.

(25) Samuel P. Huntington, "Trade, Technology and Leverage: Economic Diplomacy," Foreign Policy, no. 32 (Fall 1978), p. 73.

(26) For an assessment of the impact of formal U.S. controls see Thomas Wolf, "The Impact of Formal Western Restraints on East-West Trade: An Assessment of Existing Quantitative Research," in Tariff, Legal and Credit Constraints on East-West Commercial Relations, ed. John Hardt (Institute of Soviet and East European Studies, Carleton University, Canada, 1975).

(27) "Selling Technology," Wall Street Journal, August 8, 1978), p. 20.

(28) Huntington, "Trade, Technology, and Leverage," pp. 64-5.

(29) "Economic Warfare?" Economist, August 26, 1978, p. 32.

(30) Klitgaard, "Sending Signals," p. 103.

(31) "Why the Nuclear Power Race Worries the U.S.," Business Week, August 23, 1976, p. 68.

(32) For a small sample, see Richard K. Betts, "Paranoids, Pygmies, Pariahs and Nonproliferation," Foreign Policy, Spring 1977, pp. 157-183; Adlai E. Stevenson III, "Nuclear Reactors: America Must Act," Foreign Affairs 53, (October 1974): 64-76; George H. Quester, "Can Proliferation Now Be Stopped?" Foreign Affairs 53 (October 1974): 77-97; Abraham A. Ribicoff, "A Market Sharing Approach to the Nuclear Sales Problem," Foreign Affairs 54 (July 1976): 763-87; Paul L. Joskow, "The International Nuclear Industry Today," Foreign Affairs 54 (July 1976): 788-803; Lincoln P. Bloomfield, "Nuclear Spread and World Order," Foreign Affairs 53 (July 1975): 743-55; William O. Doub and Joseph M. Dokert, "Making Nuclear Energy Safe and Secure," Foreign Affairs 53 (July 1975): 756-72; Joseph S. Nye, "Nonproliferation: A Long-Term Strategy," Foreign Affairs 56 (April 1978): 601-23; William O. Doub and Eugene R. Fidell, "International Relations and Nuclear Commerce:

Developments in United States Policy," Law and Policy in International Business 8, no. 4 (1976): 913-62; and Ted Greenwood, Harold A. Feiveson, and Theodore B. Taylor, Nuclear Proliferation: Motivations, Capabilities, and Strategies for Control (New York: McGraw-Hill, 1977).

(33) See Thomas G. Schelling, "The Promise and the Curse," Saturday Review, January 22, 1977, p. 30.

(34) Tad Szulc, "How India Got the Bomb," New Republic, July 22, 1978, pp. 19-21.

(35) See Norman Gall, "Atoms for Brazil, Dangers for All," Foreign Policy, Summer 1976, pp. 155-201.

(36) William O. Doub and Eugene R. Fidell, "International Relations and Nuclear Commerce: Developments in United States Policy," Law and Policy in International Business 8, no. 4 (1976): 927.

(37) Ibid., p. 923.

(38) See Victor Gilinsky, "Plutonium, Proliferation and the Price of Reprocessing," Foreign Affairs 57 (Winter 1978/79): 377-78.

(39) Joseph S. Nye, "Nonproliferation: A Long-Term Strategy," Foreign Affairs 56 (April 1978): 607.

(40) Ibid., p. 609.

(41) "Bill for Strict Control on Nuclear Exports, Sought by Carter, Is Approved by Senate," Wall Street Journal, February 8, 1978, p. 7.

(42) Don Cook, "How Carter's Nuclear Policy Backfired Abroad," Fortune, October 23, 1978, p. 126.

(43) John R. Redick, "Regional Restraint: U.S. Nuclear Policy and Latin America," Orbis 22 (Spring 1978): 161.

(44) David J. Rose and Richard K. Lester, "Nuclear Power, Nuclear Weapons and International Stability," Scientific American 238 (April 1978): 45.

(45) Two examples of the issue in Europe were the preinvestment assurances by Volkswagen and Volvo that home country employment would not be reduced as a result of their U.S. assembly operations.

(46) Nathaniel Goldfinger, "An American Trade Union View of International Trade and Investment," in American Labor and the Multinational Corporation ed. D. Kujawa (New York: Praeger, 1973), p. 41.

(47) The "jobs impact" of multinational firms activities has been thoroughly analyzed. See, for example, Thomas Horst, "American Exports and Foreign Direct Investment," Discussion Paper No. 362, Harvard Institute of Economic Research (May 1974); Robert Stobaugh, "U.S. Multinational Enterprises and the U.S. Economy," in Kujawa, American Labor, chap. 6; and R. Hawkins and M.J. Jadel, "U.S. Jobs and Foreign Investment," in D. Kujawa, International Labor and the Multinational Enterprise (New York: Praeger, 1975), chap. 4.

(48) This point is elaborated in Stephen Magee, "Jobs and the Multinational Corporation: The Home Country Perspective," in The Economic Effects of Multinational Corporations, ed. R. Hawkins (Greenwich, Conn.: JAI Press, 1979).

(49) Findings of a positive linkage between foreign production by U.S. affiliates and U.S. exports to that country (and thus U.S. jobs) can be found in Horst, "American Exports," Hawkins and Jadel, "U.S. Jobs"; Stobaugh, "U.S. Multinational Enterprises"; and R.E. Lipsey and B.Y. Weiss, "The Relation of U.S. Manufacturing Abroad to U.S. Exports," Proceedings of the American Statistical Association, Business and Economics section, 1969.

(50) The estimates have ranged from a net creation of 600,000 U.S. jobs by foreign firm activities, made in Stobaugh, "U.S. Multinational Enterprises," to a loss of 500,000 U.S. jobs made in Stanley Ruttenberg and Associates, Needed: A Constructive Foreign Trade Policy (Washington: AFL-CIO, 1971). These estimates were for the late 1960s. The true net effect is obviously somewhere between these extremes, and gets close to zero if one assumes that only 15 to 25 percent of U.S. firms' production abroad involves a true transfer of technology that is not readily available from foreign sources – i.e., that 15 to 25 percent is not "defensive investment." See R. Hawkins, "The Multinational Corporation: A New Trade Policy Issue in the United States," in The United States and International Markets, ed. R. Hawkins and I. Walter (Lexington, Mass.: Heath, 1972), chap. 7.

(51) For estimates along these lines, see Jose de la Torre, R.B. Stobaugh, and P. Talesio, "U.S. Multinational Enterprises and Changes in the Skill Composition of U.S. Employment," in Kujawa, American Labor, Chapter 7; and R. Hawkins, "Jobs, Skills, and U.S. Multinationals," Testimony before the Subcommittee on International Economic Policy of the House Committee on International Relations (Washington: U.S. Government Printing Office, 1976).

(52) See R.H. Frank and R.T. Freeman, Distributional Consequences of Direct Foreign Investment (New York: Academic Press, 1978); Peggy Musgrave, Direct Investment and the Multinationals: Effects on the U.S. Economy (Washington: U.S. Government Printing Office, August 1975); and Magee, "Jobs and the Multinational Corporation."

(53) See, for example, Elizabeth Jager, "U.S. Labor and the Multinationals," in Kujawa, International Labor, chap. 3. The same argument was made in R. Barnet and R. Muller, Global Reach (New York: Simon and Schuster, 1974), chap. 9.

(54) For a more detailed review of U.S. policy toward foreign operations of American companies, see R. Hawkins and B. Finn, "Regulation of Multinational Firms' Foreign Activities: Home Country Policies and Concerns," Journal of Contemporary Business, Autumn 1977, pp. 7-30; and C. Fred Bergsten, T. Horst, and T. Moran, American Multinationals and American Interests (Washington: Brookings Institution, 1978).

(55) See Gus Tyler, "Testimony," in American Jobs and Foreign Investment, Hearings before the Subcommittee on International Economic Policy of the House Committee on International Relation (Washington: U.S. Government Printing Office, 1976).

(56) For an analysis of the provisions of the bill, see "The Foreign Trade and Investment Act of 1972," Kujawa, American Labor, chap. 5.

(57) For example, see Kujawa, American Labor, chap. 3, and Kujawa, International Labor, chaps. 5 and 10.

(58) See Duane Kujawa, "Collective Bargaining and Labor Relations in Multinational Enterprise: A U.S. Public Policy Perspective," in Hawkins, Economic Effects, chap. 2.

(59) For an early survey see Eugene V. Coan, Julia N. Hills, and Michael McCloskey, "Strategies for International Environmental Action: The Case for an Environmentally Oriented Foreign Policy," Natural Resource Journal 14 (January 1974): 87-102.

(60) See "Export Ban on Banned Products Proposed," Chemical Week, March 31, 1971, p. 15.

(61) "Administrative Survey: October 1976 to September 1977," Law and Policy in International Business 10, no. 1 (1978): 401.

(62) "Agency Reverses Stance on Tris-Treated Exports," Wall Street Journal, May 8, 1978, p. 40.

(63) "New Tighter Controls on U.S. Pesticide Exports," NRDC World Environment Alert 2 (October 1978): 2.

(64) "Products Banned in U.S. Being Exported," Chemical & Engineering News, November 6, 1978, p. 19.

(65) "Double Standard on Exports," Chemical Week, July 19, 1978, p. 19.

(66) See "Renewed Controversy over the International Reach of N.E.P.A.," Environmental Law Reporter 7: 10205.

(67) "U.S. Agencies Resist Environmental Limits Overseas," New York Times, January 19, 1978, p. A7; and "International Application of NEPA Explained by CEQ Spokesman," Pollution Control Guide, April 17, 1978, pp. 198-99.

(68) "U.S. Proposes Extension of Environmental Rules to International Arena," Business International, March 10, 1978, pp. 75-6. See Alan Tonelson, "Exports: Stumbling over Past Priorities," The Inter Dependent, November 1978, p. 6.

(69) Tom Bethel, "Exporting Pettifoggery," Harper's, October 1978, pp. 34-5.

(70) The "pollution haven" issue has been examined conceptually in Thomas N. Gladwin, Environment, Planning and the Multinational Corporation (Greenwich, Conn.: JAI Press, 1977); Thomas N. Gladwin and Ingo Walter, "Multinational Enterprise, Social responsiveness, and Pollution Control," Journal of International Business Studies 7 (Fall/Winter 1976): 57-74; Charles Pearson and Anthony Pryor, Environment: North and South - An Economic Interpretation (New York: Wiley-Interscience, 1977); Ingo Walter, "Environmental Control and Patterns of International Trade and Investment: An Emerging Policy Issue," Banca Nazionale del Lavoro Review, March 1972, pp. 3-27; and Ingo Walter, International Economics of Pollution (London: Macmillan, 1975).

(71) See Sara Hansard, "Concerns Said to Evade U.S. Health Laws," International Herald Tribune, January 14, 1975; "Hazardous Industries Flee to Developing Countries," Not Man Apart, February 1975, p. 4; and "Exporting Pollution: What Doest It Cost U.S.?" Not Man Apart, mid-February 1976, p. 3.

(72) United States v. Aluminum Company of America, 148 F. 2d 416, 443 (2d Cir. 1945).

(73) "A Growing Worry: The Consequences of Development," Conservation Foundation Letter, January 1978, p. 6; and George D. Appelbaum, "Controlling the Environmental Hazards of International Development," Ecology Law Quarterly 5, no. 2 (1976).

(74) Tom Wicker, "Looking Before Leaping," New York Times, June 20, 1978, p. A17.

(75) For a broad survey of human rights conflicts encountered by multinationals see Thomas N. Gladwin and Ingo Walter, Multinationals Under Fire: Lessons in the Management of Conflict (New York: Wiley, forthcoming).

(76) See Robert Keatley, "Human Rights and Diplomatic Pitfalls," Wall Street Journal, March 22, 1977.

(77) See Eleanor Craig, A Shareholders' Manual: For Church Committees on Social Responsibility in Investments (New York: Interfaith Center on Corporate Responsibility, 1977); "Corporate Social Responsibility Challenges – Spring 1977," Corporate Examiner, March 1977; "Corporate Social Responsibility Challenges – Spring 1978," Corporate Examiner, March 1978; and Valerie Heinenen, Church Proxy Resolutions: January 1977 (New York: Interfaith Center on Corporate Responsibility, 1977).

(78) For legislative histories, see Foreign Affairs and National Defense Devision, Congressional Research Service, Library of Congress, Human Rights in the International Community and in U.S. Foreign Policy, 1945-76, Prepared for the Subcommittee on International Organizations of the Committee on International Relations, House of Representatives, July 24, 1977 (Washington: U.S. Government Printing Office, 1977); and Center for International Policy, Human Rights and the U.S. Foreign Assistance Program (Washington: Center for International Policy, 1977).

(79) "House Approves Curbs on Credit Guarantees Involving South Africa," Wall Street Journal, June 5, 1978, p. 6.

(80) See Center for International Policy, "OPIC: Insuring the Status Quo," International Policy Report (Washington: Center for International Policy, September 1977).

(81) Tom Wicker, "The High Cost of Repression," New York Times, December 2, 1977, p. 275; William Goodfellow and James Morrell, "Why Chile Doesn't Lose When It Loses Aid," Philadelphia Inquirer, July 28, 1977, p. 11A; and Center for International Policy, "World Bank Sets $2.9 Billion in Loans to Human Rights Violators for Fiscal Year 1979," Research Study (Washington: Center International Policy, 1977).

(82) Kathleen Teltsch, "U.N. Council Orders Arms-Sale Sanctions Against South Africa," New York Times, November 5, 1977, pp. 1 and 7.

(83) Sandra Vogelgesang, "What Price Principle? U.S. Policy on Human Rights," Foreign Affairs 56 (July 1978): 822-23.

(84) Karen Elliot House, "Uneven Justice? U.S. Officials Worry About Inconsistencies in Human Rights Plan," Wall Street Journal, May 11, 1978, p. 1.

Comments

Bernard Mennis

The paper by Robert Hawkins and Thomas Gladwin is ambitious and impressive not only in terms of its length, but also with respect to the obvious determination of the authors to analyze and evaluate governmental controls over the export of technology explicitly within the context of national objectives. Too many efforts too often are content merely to describe the nature of the control systems that are in place, from a legal and/or historical orientaiton. Hawkins and Gladwin aim to go beyond this minimum by asking not only what controls have been introduced, but in addition why they have been, what purposes they are intended to serve, and whether they actually achieve these purposes. Given this approach, the great length of the piece is virtually guaranteed by the need (1) to present a framework of "national objectives" against which the analysis and evaluation of technology transfer control systems can be played; and (2) to provide as case study data a record of recent policies and practices that follow from these national objectives. The paper in fact is structured along these lines, with the first part devoted to "developing a simple conceptual framework to outline the linkages between technology transfer and the achievement of perceived national objectives of the home country," whereas the larger second part "examines in some detail various broad types of conflict over technology transfer in the United States." It should be noted that Hawkins and Gladwin limit their investigation to only one home country, the United States, and then, with respect to direction, only to the outward transfer of technology, that is, from the United States to other countries.

As is already clear, I strongly endorse the approach they have taken, which attempts to link in an overt way control policies and national objectives, although some reservations do arise in my mind about its implementation in the paper. Table 8.1 "catalogues" what the authors take to be the key national objectives (for the United States as home country) that condition technology transfer policy. Four categories of objectives are identified. My problem with table 8.1 is not that the categories and specific objectives included are not reasonable, but rather that their presentation is not informed by any theoretical conceptualization, nor is the reader told whether this is an exhaustive list or an illustrative one, nor whether the specific objectives (and categories) are of the same "scale" or "kind" (for example, which constitute end goals and which intermediate ones?), nor are any "priorities" or "weights" attached to them (for example, which are core-objectives and which peripheral ones?). Ambiguity is a serious problem here, for the approach adopted by Hawkins and Gladwin requires that great care be used in deriving and articulating national objectives that will be employed as standards against which to measure prospective and proposed restrictions on the export of technology. A simple "catalogue of national objectives" that appears to enjoy "face validity" is insufficient if the authors really mean "to explain the source of conflicts over policy and to assess home country control systems."

The presentation of the "simple conceptual framework" is followed by the attempt to draw certain "linkages between technology transfer and national objectives." Three sets of linkages are examined – economic, national security, and humanitarian. Not too surprisingly, the first of the three is by far the best developed. In a really interesting effort to trace what the consequences regarding national economic objectives would be in the event of restrictions placed on outward technology transfer, Hawkins and Gladwin identify several major variables that would need to be taken into consideration in such a calculation, reveal the really multifaceted nature of the effects of public policies and practices, and indicate just how difficult it would be to trace empirically the cost-effectiveness of taxation or other forms of intervention. This section is both persuasive and very suggestive. I was especially drawn to two implications of the argument. One concerns the need to explicate and empirically ascertain situation-specific parameters that structure the costs and benefits of alternative technology transfer control policies; for example, the international competitive position of the home country in regard to the technology in question (the availability of "willing and able" nonnational suppliers) and the character (elasticity) of demand structures in prospective recipient countries. The second implication concerns the need to think seriously about the future consequences of policies and practices for technology creation, in addition to the more obvious, salient, and immediate consequences for employment, income, balance of payments, and so forth. The discussion contained in this very provocative section encourages modesty as it progressively reveals just how little the current state of the art can tell us about the potentialities of government intervention in regard to technology transfer.

The sections on national security and humanitarian objectives are not as interesting, mainly because they are not very much elaborated. The thrust of the presentation is the not unobvious point that there are definite economic costs to be absorbed when the United States (or any home country for that matter) intervenes in the export of marketable technology for national security or humanitarian reasons and, moreover, the intervention is likely to be unsuccessful if alternative suppliers of the technology are available.

This conceptual section ends with Hawkins and Gladwin concluding that "even at a simplified abstract level, policies toward international technology transfer are inherently subject to conflict . . . because any nation has multiple national objectives, some of which are competitive and not complementary. And groups within the society have interests indentified more closely with some objectives than others." True enough – but that is precisely the problem. The statement is so bland because, not having presented a well-developed conceptualization of national objectives, the authors are unable to state much more than the above, or readily identify where the crucial divisions are likely to be, or why, regarding technology transfer issues. Thus we are left with the considerably less enlightening thought that there will be conflicts of goals within any class of national objectives and also conflicts between classes of objectives – but not which specific goals or classes of objectives in fact have, are, and will be impacting on technology transfer, why, and to what extent.(1)

Following this, Hawkins and Gladwin devote the remainder of the paper to a rather extensive historical investigation of U.S. technology transfer control systems in a variety of areas; particular attention is given there to national security, nuclear technology, domestic jobs, the environment, and human rights. The analysis is detailed and is consistently persuasive. Any attempt at a summary of their findings would be, I believe, rather presumptuous, and therefore none is attempted here. But the conclusions that are drawn at the end are worth emphasizing. Hawkins and Gladwin generally argue that recent U.S. policy on the outward transfer of technology has been both costly in the economic sense and, additionally, ineffective with respect to the attainment of the objectives that motivated the restrictive policies and/or practices. There may be many reasons for this state, including lack of competent implementation. Nevertheless, Hawkins and Gladwin continually stress one factor: that the effectiveness of U.S. policy in restricting or taxing the export of technology "depends upon the degree of monopoly which the home nation has in that technology. Readily available substitutes from other national suppliers limit the leverage a nation has through the exercise of controls, and raise the likelihood that economic or other costs will be borne without reaping the benefit in terms of other objectives." In other words, the availability of alternative national suppliers has undermined the efficacy of U.S. policy, since potential recipients have ultimately been able to attain the technology from competitors not similarly restricted, competitors who thereby gain markets at the expense of U.S.-based firms. The result is that the United States is frustrated in its pursuit of those national objectives for which these restrictions have been imposed, while at the same time it experiences negative feedback in terms of employment, income, balance of payments, and general international competitiveness.

Needless to say, Hawkins and Gladwin are not sympathetic to the use of restrictions on the transfer of technology as a vehicle for achieving typically noneconomic national objectives. But given the legitimacy of these objectives, one is hard put to suggest what other methods would be successful. At any rate, Hawkins and Gladwin do not seem to anticipate a change in what they see as a counterproductive and costly exercise. They conclude the piece on a pessimistic note, saying that "the conflicts among priorities are not going to vanish, and may well intensify. America's penchant for misguided idealism and counterproductive controls on technology transfer appears to be here to stay."

I think it is a mistake here to mix directly in the conclusion "misguided idealism" and "counterproductive controls." Hawkins and Gladwin show rather clearly that the effects of control policies are often counterproductive in regard to both economic costs and goal achievement, but this is not to say that there exists "misguided idealism" in regard to the objectives or motivations upon which the policies are based. I do not take it that the authors mean that the objectives or motivations are frivolous or utopian or violate national interests and values. On the contrary, what seems to be the problem is that the means chosen to implement these objectives are not appropriate because, given certain parameters that obtain, these means will not accomplish the desired outcomes. But this is a serious intellectual as well as political problem, in the sense that at present the state of the art is such that we do not have empirically derived models to guide the translation of objectives into means so as to achieve favorable outcomes, nor do we seem to be able to attain the political commitment necessary for the task, perhaps because of the conflicts among groups that the authors speak about often in the earlier part of their paper. In either case, the problem does not appear to be "misguided idealism."

I believe it also would be worthwhile to step back for a moment from the great detail of the case studies that are presented by Hawkins and Gladwin and cast a somewhat broader net. Such a perspective would underline the fact that, except in a few selective areas, significant technologies are no longer monopolized by U.S. firms. There has been and continues to be a relentless diffusion of technology and the initial source of much of the new innovative technology no longer is (or necessarily is) the U.S. public or private sectors. But beyond the fact that non-U.S. competitors are more frequently innovating significant marketable technology and are willing and able to replicate increasingly standardized technology initially sourced in the United States, in addition the home country governments of these alternative suppliers do not share U.S. national objectives nor are they willing to harmonize their public policy with that of the U.S. regarding technology transfer in particular or international trade and investment in general. Quite obviously, the private gains to be extracted for each foreign country's "national champion" through the exporting of technology are too great, or are perceived to be too great, for harmonization or cartelization on the supply side to be established and endure. The temptation clearly is for each national supplier to resist the extraterritorial embrace of U.S. public policy and take advantage of markets left uncontested as the result of U.S. restrictions on outward technology transfer.

But these endemic tendencies could conceivably be constrained, as often they were in the past, through intergovernmental collaboration in which other governments, for a variety of reasons, were responsive to U.S. government objectives. Put somewhat directly, it is not only the existence of alternative national suppliers that is the source of the American predicament. It is also the unwillingness of the home governments of these suppliers to employ their good offices to persuade their national champions to forego or at least compromise potential gains as a result of U.S. firms being handicapped in order that these governments should maintain (and not risk undermining) good relations with the United States. This is another way of saying that U.S. political capacity (or bargaining power) with respect to Western Europe and Japan has deteriorated to the point that the United States cannot expect "allied governments" to subordinate the economic gains of their firms to international security "high politics." The problem of U.S. policy efficacy did not become acute when there appeared alternative national suppliers of technology. It did become acute when the political capacity of the United States declined, as the political/military/security benefits that it could provide to other governments no longer seemed as significant, and not nearly impressive enough to compensate for the economic and other gains to be realized through developing policies more independent of the United States.

Yet it should be clear that the above is not a late-1970s nor an early-1980s problem for the United States, but rather a late-1960s one, with the 1970s providing clues to the U.S. reaction to the situaiton. First, in a world in which detente with the Soviet Union, China, and other adversaries continues to advance, although of course not exactly in a strictly linear fashion, and the use and utility of military force appears increasingly problematic except in peripheral areas and among secondary countries, there is little incentive for the Western European countries and Japan to subordinate their national economic policy to U.S. desires, however sincere and well-intentioned the latter may be. Short of the emergence of a new international political/security crisis along the lines of the cold war, a crisis perceived by all home countries as seriously threatening their survival and well-being, the situation that existed in the last decade will continue.

Hawkins and Gladwin argue that the lack of a degree of monopoly over marketable technologies is the key to explaining the lack of efficacy of United States government control regarding outward technology transfer. This must be the case once it is admitted that the United States no longer has the capacity to "transfer" its public policy to other host countries and their national firms through intergovernmental collaboration and persuasion. But what is the likelihood of such a degree of monopoly being established in those areas surveyed by Hawkins and Gladwin in which agitation for controls has been evident and strong? The record of the 1970s has not been promising in this respect. To reiterate, the diffusion of technological innovation has been the general rule, whether one turns, following the authors, to weapons and electronics associated with national security pursuits, closely related nuclear technology, industries characterized by standardized technology where the loss of domestic jobs is the main issue, drugs, chemical products including pesticides, and so forth. Even though the United States still maintains a commanding lead, research and development capability is increasingly nationally decentralized among a number of developed market economies. Moreover, it is improbable that a reversal of this general diffusion can be accomplished. And even if by certain unusual circumstances it were, it is likely to be a temporary, short-term phenomenon with little durability. But even if it could be sustained for some period of time, and restrictions over the outward transfer of technology thus could be enforced, is it likely that the U.S. government in those circumstances would and could withstand the internal political pressure of national firms who would be in the intolerable position of not gaining realizable revenue, and/or the external political pressure and resentment of foreign governments who would have carrots dangling before them that they could consume if only they modified their behavior? The prospects are not promising, even in those cases where it is clear that unrestricted transfers could encourage terrible political destabilization or counterproductive consequences for recipients of the technology and their environment.

Given all of this, the United States has been urged to adopt what I take to be the recommendation of economic orthodoxy – exploit the few areas where the U.S. still monopolizes technology or at least commands a relatively leading position. This is the counsel of Hawkins and Gladwin, who argue that control of outward technology transfer, if there is to be any at all, should be in areas "in which the United States has an effective monopoly and for which the foreign demand is price inelastic." My impression is that they imply that this has not been tried. To the contrary, I would argue that, as they themselves have shown, a good deal, if not even a majority, of the discussion regarding controls has been directed at restricting exports of national-security-oriented weapons, nuclear knowledge, and electronics/computers, areas in which if the U.S. does not enjoy monopoly, it commands at a minimum a dominating technological position. In spite of this structural advantage and the massive agitation in these areas, the 1970s has seen a tremendous rise in arms sales deriving precisely from the huge demand abroad for such "products" and the escalating international trade and balance-of-payments problems encountered by the United States, as well as from the persuasion of domestic manufacturers. These three forces have acted to override the attempt to administer export controls in areas where the United States does enjoy competitive advantages, export controls that can be justified on a variety of bases: humanitarian and moral grounds concerning the export of deadly wares, more fequently than not to most unattractive regimes; economic grounds concerning the distortions and dislocations deriving from diverting great amounts of investment capital, production facilities, jobs, and so forth, into nonproductive military items; and political grounds concerning the self-defeating consequences of supplying advanced hardware to repressive, essentially unstable leaders and thereby encouraging general destabilization as the perverse consequences of such exports become manifest (for example, as in the case of the Shah's Iran).

And, of course, if alternative foreign suppliers begin to appear and the degree of monopoly is thereby eroded, it becomes even easier to ignore these pleas for restrictive controls. The dilemma is a terrible one, because very profitable markets exist for, and alternative national suppliers are ready and willing to service, the transfer of powerful destructive weapons, potentially dangerous nuclear material and knowledge, insufficiently tested drugs, and ecologically detrimental agricultural chemicals, in spite of the longer-term negative costs that may exist. Yet is not this situation increasingly endemic to advanced market economies caught up in the grip of short-term economic imperatives?

NOTE

(1) For a recent but only moderately successful example of what could be attempted along these lines (regarding however, U.S. national objectives and policies in the raw materials sector), see Stephen D. Krasner, Defending the National Interest: Raw Materials Investment and United States Foreign Policy (Princeton: Princeton University Press, 1978).

II
Policy Perspectives and Implications

9 Control and Incentives for Technology Transfer: A Multinational Perspective

Robert E. Driscoll
Harvey W. Wallender III

Much of the current debate over the transfer of technology, centering as it does on the control of the transfer process, ignores major factors that shape the role of technology in development and provides little assistance in formulating national policies designed to improve the acquisition and utilization of technology by developing-country firms. The multinational firm, as a principal supplier of technology, contends that control of the transfer process is secondary to the more central concerns of utilizing technology – including local capabilities to receive and adapt; the local legal, political, and economic environment; incentives for transfer; and the local technological infrastructure. Although for many years the developing countries have rejected this point of view and focused on the control of technology transfer through regulating technology supplies or seeking greater access to technology. Many in the Third World now recognize that a key to development is the ability to use technologies already available. Shifts in the national policies of several countries and a new emphasis on building local capabilities to absorb and manage technology are now evident.

As a primary source of technologies, multinational firms have played a central role in the international transfer of technology. They are one of the most important elements of the equation of international transfer currently under analysis and of the debate in their role of relating suppliers and users. However, this debate generally overlooks the needs and abilities of the receiver firms, and the role of the national environment in encouraging technology search, acquisition, and local adaptation and development. By focusing on this neglected area of the debate, much can be accomplished in encouraging improved technology transfers by multinational corporations while at the same time making the application of those technologies more supportive of national development goals.

FACTORS AFFECTING TECHNOLOGY TRANSFER

The Fund for Multinational Management Education (FMME), in conjunction with the Council of the Americas, the U.S. Council of the International Chamber of Commerce, and George Washington University, recently conducted a project in which over 120 U.S. multinational corporations participated in an in-depth review of technology transfer. Three principal conclusions emerged from this four-volume study:(1)

- Private transfer has been responsive to incentives
- Users must have capabilities to absorb technology
- Technology transfer is a function of the environment.

A related project,(2) analyzed user firms from the private and public sectors in Kenya, Tanzania, South Korea, Brazil, and Peru. Case studies based on interviews with sixty-seven firms were prepared, with the objective of examining the main factors affecting the successful absorption of technology. This study identified two major types of external environmental factors that condition a firm's behavior, which are

- the general economic and political environment
- the technological infrastructure and the availability of resources to enable a firm to exploit new technologies.

From these two studies we can see that the same type of factors relating to general political and technological environments affect the decision-making process of foreign and domestic firms in developing countries. Table 9.1 indicates some of the key variables within the environment affecting technology transfers.

Table 9.1. Key Variables Affecting Technology Transfers

General Economic and Political Factors

Incentives and control systems
- government regulations
- market characteristics and competition
- inflation patterns
- balance of payments

Political stability
- turbulence
- predictability of change in environment

Technological Infrastructure Factors

Direct suppliers of knowledge and manpower
- universities

Table 9.1. (Cont.)

Direct suppliers of knowledge and manpower (cont.)
- other industries
- government programs
- foreign firms
- immigration
- specialized training institutions
- equipment suppliers
- licensors
- private volunteer organizations

Local support systems
- financial institutions
- local consultants
- special government assistance agencies
- service companies such as labs, tool and dye shops
- physical infrastructure including roads, water power, communications systems

Diffusion systems
- standards offices
- patent offices
- business and professional associations
- universities and training organizations
- government information offices
- general publications
- free press

General Political and Economic Environment

The legal, political, and economic environments of both supplier and receiver countries are critical variables that inhibit or promote technology transfer and development. Government regulatory policies can either encourage or discourage technology importation; such regulations may make it difficult or prohibitively expensive for firms to seek new technologies. Regulations controlling technology transfer, if properly applied, can strengthen the position of the developing country firm. They can, at the same time, discourage the foreign supplier of technology by making entry costs to the firm too high.

There are various types of incentives that affect decisions of firms regarding technology transfer. The ability of the technology developer to recoup his investment is a key factor in encouraging a firm to undertake further R & D, and to transfer its technology. Government incentives are also central to the decision by a firm to seek new technologies from foreign or domestic sources.

A third type of incentive is the market, which is key to inducing or inhibiting the firm in taking risks necessary for technological improvement. The results of Technology Transfer and Management in the Developing Countries indicate that many developing-country firms see little reason to alter existing management systems or change technology use behavior unless those changes will improve their market position. Local competition may also affect technology transfer. In certain industries competitors are exposed to technology innovations and development through each other and are forced to take risks in order to keep up with the changing market patterns. Firms enjoying a secure market position may have no real incentives for investing in technological change and improvement.

From the perspective of multinational firms, the existing system of commerce in technology is working and has resulted in cost-effective technology transfer to developing countries.(3) Because they reflect better the diversity of users and suppliers of technology, case-by-case adjustments are deemed preferable by the companies and more cost-effective than a centrally controlled system.

Political stability has a great impact on planning and risk evaluation of the individual firms. Political stability seems to relate less to the ideological position of a country's government than to the predictability of change in the environment. In a country such as Tanzania, we find that although technology is not readily available locally, firms are willing to take risks and invest in foreign technologies because they are confident that the government's policies will not change radically or rapidly and that they will be able to continue using foreign technologies. In countries such as Peru that evidence great political instability and turbulence, firms lack confidence in their ability to predict how the environment will change and are unable to evaluate options for improved performance and growth. Political stability and predictability of change in the environment are factors often as pervasive as those of competitive or market forces and are exceptionally important in conditioning the firm and its ability to adapt and plan future growth.

The general economic environment also may encourage or discourage risking limited resources in new technology. For example, the rapid deterioration in Peru's balance of payments position during 1976-77 discouraged many firms from committing themselves to new programs involving long-term technological change. As Peru's foreign reserves continued to dwindle, the government stepped in to control all international travel, purchase of foreign equipment, and other business activities that might negatively affect the balance of payments. As a result, many of the firms that had begun technological programs tied to foreign assistance were negatively affected, and long-term development of the programs was thwarted.

Technological Infrastructure Factors

The effectiveness of technology transfer is greatly dependent on the capabilities of the user to absorb it. These capabilities in part reflect the technological infrastructure and the support in the environment that assists firms in receiving technology. The study Technology Transfer and Management in the Developing Countries suggests that the availability of skilled personnel and physical facilities is second only to the market in attracting foreign technology and encouraging local research and development activities. Indeed, this is perhaps the critical aspect of the transfer question. If developing-country goals are to expand local opportunities and upgrade national capabilities, the availability of trained manpower and managerial skills will be key determinants of the success of such policies.

The local technostructure is an important element in the success of technology transfers. It affects both the ability of the supplier to transmit the technology and that of the user to receive, adapt, and utilize it. Three elements of the internal technical capabilities of developing countries can be isolated: direct suppliers, support systems, and diffusion systems.

Access to direct suppliers of manpower and know-how is a principal concern of any company. As a rule, firms look to local universities and to training offered by public or private institutions as a prime source of their personnel and management resources. Another source of manpower is the general business community. Many firms hire critical personnel away from other firms when they are planning to improve technological capabilities; as a result, other industries and competitors become an important direct supplier of manpower and specific technological systems. In some cases certain government programs provide the type of experience that will be useful to independent firms and they may also be a source of manpower. Firms often have access to sources of technology, both hardward and software, through the purchase of equipment and supplies, through licensing, or through technical assistance agreements.

Support systems are often essential to the firm in acquiring and using technology. Banks, government offices, or other general business institutions are in many cases important sources for financing, counsel, and general information aiding the firm in its technological development. Local consultants provide specialized manpower and system-specific know-how only periodically, primarily because of their limited expertise. Managers in all five countries examined(4) indicated that stronger local consulting systems would contribute significantly to the efficiency of their operations. Many of the managers interviewed draw on whatever local consultants there are, but also rely heavily on foreign consultants. Such consultants are prohibitively expensive in some cases, however.

A national government's commitment to developing specialized assistance organizations may be a significant aid in improving technology transfer. These organizations provide management consultants and specialized technicians for local firms, publish information on technology sources, and offer special seminars and training programs in general and specific management techniques. A notable example of this type of organization is CEBRAE in Brazil. Specialized service facilities such as laboratories and tool and die shops are also important for developing or reinforcing a firm's ability to adapt to changing patterns in the environment.

The physical infrastructure is an element in the local support system that may have a great impact on a firm's ability to grow. Without sufficient power or water supplies, communication systems, or transportation systems, a firm may not be able to implement plans for developing and adapting new technologies.

Diffusion systems are necessary for linking different parts of the technological infrastructure. An active patent office serves indirectly to make more information available to inventors, businessmen, and government technicians. Standards bureaus and standards organizations are vital as guides to firms in developing new products or improving existing products and procedures.

Business and professional associations encourage the exchange of ideas and discussion of new developments, and in some instances they have pooled resources for the financing of specialized studies or information-gathering systems. The effectiveness of these associations as information sources varies greatly among developing countries and industries, however. Findings from our research indicate that in some instances, particularly in countries such as Kenya and Peru, these associations are used more as lobbying groups for dealing with the government. In business associations in Korea and Brazil, there seems to be greater exchange of industry- and system-specific information as well as discussion of political and general environmental trends.

Universities and training centers can stimulate diffusion of information and exchange of ideas. For example, the Korean Institute of Science and Technology (KIST) encourages companies to develop new products and research programs. KIST works closely with universities and advanced schools of engineering and scientific training in Korea, such as the Korea Advanced Institute of Science, and therefore is an important link between the public and private sectors.

Local and foreign journals and general publications were cited repeatedly by local managers as sources of ideas and information on recent industry developments. A free press also seems to be an important aspect of the technological infrastructure. A number of managers indicated they rely heavily on the local press for information on political, economic, and social developments. Confidence in the local press's accuracy in reporting news will encourage firms to plan growth based on readily available information presented in newspapers and journals and through radio and television.

In summary, there are a number of factors relating to the general economic and political environment, and to the technological infrastructure, that affect where multinational corporations seek to invest and transfer technology, how contracts are negotiated, and how technology is received and applied by user firms. The importance of these factors in conditioning technology transfer argues for national policies that reflect the characteristics of individual countries. This type of approach contrasts with many exercises in the international arena, such as UNCTAD's work on a code of conduct for the transfer of technology. An analysis of current policies of several leading Third World nations reveals a greater recognition of the importance of their national environments in conditioning technology transfer, and reflects an increasing concern with programs that will make their political, economic, and legal environment and the technological infrastructure more receptive to technology receipt and application.

A SHIFT IN NATIONAL POLICIES

During the late 1960s and early 1970s developing-country governments felt constrained by the perceived oligopolistic nature of the transfer process. Dominated by a few large firms, the market was seen as skewed in favor of these firms to the detriment of the developing countries. Competition for developing-country markets has grown, however, with European and Japanese firms taking a commanding position against the U.S. multinationals. No longer is any country dependent on a few sources of technology. The multiplicity of sources of technology has encouraged developing countries to shop carefully and compare various technologies with regard to national development goals. The availability of technology from many suppliers has induced developing countries gradually to rely less on legislated or decreed controls on technology transfer, and more on bargaining to acquire technology; stated differently, negotiation and bargaining are used increasingly as mechanisms for obtaining technology on terms acceptable to both the local private enterprise and the national government. Developing countries are acquiring technology through a case-by-case evaluation of its applicability to their own environmental characteristics.

At the same time, however, governments maintain rights of approval of contracts. Nonetheless, the new emphasis is not on approving contracts already negotiated, but rather on participating in the negotiation process, seeking the best possible terms through bargaining with the supplier rather than through fiat. This approach is more in line with the needs of the private sectors of both the supplier and user countries, and can in the long run result both in better agreements and a long-term relationship between the foreign company and the developing country – a relationship desired by the majority of the multinational corporations.

The mere screening of technology agreements, or efforts to control the entry of foreign technologies through international codes of conduct, cannot deal adequately with the development issues of the Third World. These issues affect the motivation of domestic and international firms to undertake the risks involved in technology transfers – both through direct investment and through licensing. While on the surface the problems in international technology transfers to developing countries have been the price, conditions of transfer, and cultural and social impacts, these are by far secondary to the pressing need for increasing employment, for gaining access to foreign markets for exports, and for strengthening the local technical infrastructure.

Current government policies more and more reflect the concerns of foreign and domestic enterprises regarding the general and technological environments, and thereby evidence a more pragmatic approach to technology transfer questions. Governmental initiatives of the early 1970s that aimed at screening technology agreements are shifting in focus. In a seminar held in January 1979 in Mexico City, executives from the Mexican private sector emphasized that the registry system of that country was working successfully to strengthen the local private sector's search for foreign technologies under favorable terms. One of the main reasons cited for this success was the flexibility of the registry in reviewing contracts and in applying Mexico's technology legislation on a case-by-case basis. At that same seminar, a representative of the private sector commented:

> The government . . . has begun to play an important role in helping the user community improve its capability to plan and bargain for technology. It was felt that the Mexican government is working to facilitate interaction with business and is beginning to focus more on improving the process of technology selection than on simply reducing the cost of technology.(5)

Some countries are formulating national development plans that stress strategies for foreign technology acquisition other than simply imposing controls on its importation. National goals are shifting, with increasing emphasis on strategies that directly enhance the abilities of developing countries to develop and utilize technology. As these plans are implemented, their ultimate goals will assist both national and foreign firms in better applying technologies to support national development goals.

Figure 9.1 displays Brazilian technology strategies and development targets. It should be noted that the majority of the strategies are aimed at the underlying constraints within Brazil that affect the application of science and technology to development. The emphasis of these strategies is on applying and using technology efficiently while at the same time furthering national goals.

At the seminar in Mexico City, discussions centered around initiatives that might be taken by both governments and private companies to reduce constraints on the technology transfer process. These initiatives are also displayed in table 9.2. As can be noted, these programs do not relate to the control issue, but rather to the recognized problems of applying technology effectively for maximum support of national development goals.

These new emphases in policies of several developing country governments foreshadow what may be a shift in technology policy focus in much of the Third World. Although there is still much energy spent in trying to control perceived abuses by multinational corporations, there is also an increasing number of countries with policies directed to the underlying environmental factors affecting the application of science and technology to development. In the absence of thorough analysis of the costs and benefits of policies that favor controlling the entry of technology, it is impossible to evaluate them, beyond tentative assessments of their relative value. Nonetheless, the trend that seems to be emerging, at least as evidenced in the new policy emphases of several key Third World countries, is toward policies that reflect the fundamental environmental factors conditioning the application of technology to national development.

Strategies \ Targets	Build Manpower	Improve BOP	Control entry and improve search	Encourage local control	Decentralize development benefits	User firm growth	R&D
Tax incentives for locals	●				●	●	●
Stimulate market demand						●	
Create trading companies		●					
Information services			●	●			
Regional technology programs					●		●
Rural exension centers					●		
National Council for S&T Development							
Singling out appropriate technologies			●			●	
Standards institutions			●			●	
Small/medium industrial transfer channels	●					●	
Private technology centers	●					●	
Local consulting/ engineering firms	●					●	●
Postgraduate training	●						
International cooperation		●					
Stimulate mergers				●		●	
Minority ownership incentives for foreign capital				●			
Limit foreign royalties and fees		●					
Relocation tax incentives for foreign capital					●		
Training tax incentives	●				●		
Limit on payment period for foreign technology		●					

Fig. 9.1. Brazilian technology strategies and their development targets, 1975-77.

Table 9.2. Summary of Mexican Programs, Constraints and Complementary Programs in the Application of Science and Technology for Development

Mexican Programs	Constraints	Complementary Programs
Decentralization of industry	Limited local market demand to pull through science and technology	Foreign and national public/private sector analysis of science/technology issues/programs
Diversifiaction of markets, products	International competitiveness	
Agricultural development	Inadequate conditions for local small and medium-sized firm development	Private voluntary organization/business cooperation
Mexicanization		
Control/limit foreign technology (screening)	Limited foreign small/medium-sized firm participation	Government facilitation programs
Employment generation	Scarcity of information and ability to manage information	Policy centers
Tax incentives/restrictions	Lack of managers, technicians, technically skilled workers, entrepreneurs	Training assistance
R & D centers		Brokering organizations/extension services
Information services (for technology selection)	Conflicting local legislation	
Government purchasing	Poor linkage/diffusion networks	Information management systems
Economically/socially appropriate technology	Limited flexibility of foreign technical supplier	Contracting with foreign firms and institutions for R/D/E and product design
Special studies fund		
Technology centers	Inactive technology user community	Public/private product research
Income distribution		
Integration of science and technology	Limited user diagnostic/planning/negotiation capability	Use of retired corporate executives in training programs
Local small/medium-sized firm development	Limited capital formation	Technology corps

Table 9.2. (Cont.)

Mexican Programs	Constraints	Complementary Programs
	Complexity of transfer and technology process	Multinational corporation training programs, as separate business activity
	Limited technology infrastructure	Improved foreign educational curriculum for Third World needs
	Weak capacity to assimilate technology	
	Poorly integrated technology system	
	Inappropriate product mix for majority of population	

Note: Items are not listed in order or importance. "Complementary programs" are those that could be conducted in Mexico by local and foreign governments and businesses to complement national development objectives.

For most developing countries, multinational corporations remain a key source of needed technologies and know-how, even though many conflicts and policy questions remain unresolved. From the perspective of the multinational corporation, this new emphasis reflects a more pragmatic review of national needs, and is one that will lead to better applications of technology in developing countries. In its December 1978 report(6), the Advisory Committee on the Application of Science and Technology to Development of the United Nations (ACAST) called for "bilateral negotiations with transnationals not only on the transfer of specific technologies, but also on increasing the direct contribution of these transfers to the domestic scientific and technological capability"(7). This report goes on to state:

> Transfer, adaptation, and diffusion policies will have to be designed in such a way that optimal advantage is gained for strengthening the domestic science and technology capability; 1) through affiliates of transnational corporations; 2) through contracting of proprietary technology from independent foreign producers; and 3) through using foreign-originated but freely-available technology.(8)

In recognizing the importance of national economic and technological environments, the report further emphasizes the opportunities for adaptation of transnational technologies to the size of the national market and to the factor endowment of the country. How these adaptations are made will depend on what types of policies are formulated. Multinational corporations can and will adapt technologies to fit varying environmental factors and national development needs if the appropriate incentives are available in the form of adequate return on investment in the technology and its adaptation.

NEW POLICY MECHANISMS FOR TECHNOLOGY TRANSFER

If, as the ACAST report indicates, developing countries will be dependent on foreign sources of technology for some time, policies must be found which seek to maximize the contributions of MNCs to domestic industrial development and science and technology goals. The following paragraphs briefly describe types of programs that can be adapted and expanded to accommodate varying national environments and to improve the application of science and technology to development.

Improved bargaining between developing countries and MNCs will play a key role in the search for ways to maximize these contributions. Developed countries and the United Nations system have a special role to play in raising the skills levels of developing-country negotiators. The U.N. Centre on Transnational Corporations has conducted a series of seminars on negotiations in various regional areas aimed at training government technicians in negotiating skills. These programs must be augmented and extended, perhaps to include the participation of individuals from multinational corporations who have extensive experience in negotiating technology contracts, and can provide practical knowledge of the decision-making process of MNCs.

Private groups as well are involved in negotiation training. FMME has undertaken such seminars in a number of developing countries. The Georgetown University Center conducts a series of negotiations seminars in the United States and abroad each year. Mechanisms must be found to extend these programs.

Training is one of the strengths of the MNC. Companies train numerous individuals at all levels – skilled, technical, and managerial – in their overseas operations. The ACAST report notes that, while such programs in themselves will have "limited value for augmenting the domestic technological capability . . . their indirect effects upon the diffusion of a technological culture should not be underestimated, particularly in those situations where new job opportunities increase the mobility of high-level managerial and technical manpower within the country."(9) Government negotiators should therefore negotiate with MNCs for increased amounts of training to be provided, in order to strengthen the national technological base.

Government negotiators should also bargain for increased domestic inputs into the products of multinational corporations. The creation of supplier industries within the developing country and the provision of technical and managerial know-how to these supplier firms by the multinational corporation can broaden the industrial base of the country, as well as increase the participation of small and medium-sized firms in economic growth and development.

Complementary programs should be designed that will involve domestic and foreign corporations in devising solutions to blockages affecting the implementation of national industrial and technology policies. A key factor in their success is the close coordination between these initiatives and national policies and programs.

Other programs might be developed that would increase the participation of multinational managers in removing some of the bottlenecks in the application of science and technology to development. One of the conclusions from the study Technology Transfer and Management in the Developing Countries was that developing-country firms do not initiate their searches for technology effectively. Through close collaboration between managers of these firms and those of MNCs, new planning, management, and search techniques could be imparted, leading to increased demand on the part of the private sector for new technologies. Such demand is essential if programs aimed at developing indigenous technologies are to be successful.

CONCLUSIONS

Technology transfer only occurs when know-how is transmitted and applied in the productive process. How technology is transmitted and how it is received, absorbed, and used is conditioned largely by the environmental factors that shape the decisions of MNCs on where to transfer and that condition how technology is applied by the user. A review of the factors relating to the general political and economic environments and to the technological infrastructure indicates that the interests of the MNC and the local productive sectors, public and private, overlap in many areas. Their interests appear far more harmonious than is evidenced in the historical emphasis in international discussions and in some national policies on controlling the entry of foreign technology.

National development goals in many Third World countries stress social considerations. However, they also stress economic and industrial growth, which will enable them to better meet the social needs of their countries. This economic and industrial growth is conditioned in part by the characteristics of the economic, legal, and political environment, as well as by the technological infrastructure of the country, that is, by the same factors that determine multinational corporations' decisions to invest and transfer technology.

An increasing recognition of the importance of local environmental factors in conditioning technology transfer is evident in a gradual shift in emphasis in national development policies, which are concentrating less on controlling the entry of foreign technology and more on developing local capabilities to receive and use technology. A review of current policies in Brazil and Mexico, both leaders in the Third World, are evidence of rethinking on the part of policymakers and of the increasing emphasis on shaping local environments that encourage and facilitate economic and industrial growth.

The multinational corporation has much to contribute to that growth, and can provide both the services and the technology needed for achieving many of the development goals of Third World countries. How they contribute to economic growth depends in part on how government negotiators deal with MNCs and how proposals for foreign investment and technology transfer are reviewed. If, on the one hand, policies that emphasize the narrow application of regulations or legislation on technology agreements continue to govern foreign investment or technology transfer, agreements will be more difficult to reach, and the contribution of MNCs to economic growth will be reduced. If, on the other hand, policies are formulated that stress expanding the cooperation between MNCs and developing-country productive sectors, economic growth – through manpower training, local supplier development, and access to developed-country markets – will be spurred and national development goals will be supported. Through a careful evaluation on a case-by-case basis of the potential contribution of multinational investment and technology, government negotiators will be able to maximize the benefits of multinational corporations' presence while simultaneously moving to reach national development goals.

NOTES

(1) See Fund for Multinational Education, Public Policy and Technology Transfer: Viewpoints of U.S. Business, four volumes (New York: FMME, 1978).

(2) See Harvey W. Wallender III et al., Technology Transfer and Management in The Developing Countries (Cambridge: Ballinger, 1978).

(3) This conclusion is supported by over seventy cases prepared for the project on Public Policy and Technology Transfer, which illustrate the flexibility of the market-coordinated system and the many adaptations and adjustments already made in response to various national needs and requirements. See Fund for Multinational Management Education, Public Policy and Technology Transfer, 1978.

(4) See Wallender, 1978.

(5) FMME, p. 22.

(6) ACAST, Report of the Advisory Committee on the Application of Science and Technology to Development (New York: United Nations Economic and Social Council, 1978).

(7) Ibid., p. 38.

(8) Ibid., p. 42.

(9) Ibid., p. 41.

Comments

Robert G. Hawkins

One dimension of technology transfer control systems is who or what applies the controls. Technology transfer may be controlled by the multinational firm which obviously has some discretion over the type, timing, charges (or conditions) and target country of technology transfer. The second is the government (or governmental bodies, which may be more or less coordinated) in the home or source country. Home country control may involve actions to induce or constrain behavior by its private firms, but also involves the behavior of governmental units themselves – both of which may stimulate or retard technology transfer.

A third source of control is, of course, the host government (or governmental bodies). Again, these may be designed to limit, enhance, or redirect technology transfer. A fourth level of control system may be an international body – such as codes or rules developed by the OECD, UNCTAD, or U.N. Transnational Center, which are agreed to by participating members and may involve obligations on companies, host governments, and/or home governments, as the case may be. Alternatively, it may be a set of common rules, agreed upon by a group of host (or conceivably home) countries, that may have the affect of eliminating certain aspects of competition among the members for technology from outside. The Andean Pact in perhaps the prime example.

And finally, a technology transfer control system exists in the form of the market. The existence of alternative suppliers, if they compete and do not cartelize, constrain at least to some degree the discretionary company power over the timing, volume, and price of technology transfer. Likewise, alternative host nation locations constrain the conditions and regulations that any one competing host country can apply to technology transfer.

Several questions present themselves. Which types of control systems are becoming more important? Which are adequate and which inadequate – and for what purposes? And what should be done to strengthen or change the mix of technology control systems, if anything?

The Driscoll-Wallender paper takes a particular tack on these issues, which I generally agree with. They argue that much (or even most) of the world's commercial technology is the property of private multinationals, and that this area of technology transfer is "where the action is." This focus of their paper, which in a broad sense is correct, needs an important qualification. Much technology is not in the hands of completely private enterprise, but is owned by public or quasi-public concerns. The latter is a growing share in total technology transfer internationally. And in the case of some technologies (military or nuclear), the proprietary interests are shared by private and public bodies. This complicated mixture of public and private enterprise in technology ownership is likely to be an escalating source of conflict over the next decade, and is worthy of more attention than this paper has given it.

Granting the paper's contention that technology transfer by transnational firms is "where the action is" currently, we can see that the paper has two basic thrusts. First, technology transfer is a highly variegated phenomenon, with unique company- (or technology-) specific characteristics, and unique host country environments. And because there is so little commonality, the paper comes close to arguing that no general control system can be very successful, perhaps for any objective that it might seek. Ad hoc, case-by-case, flexible, and negotiated terms are the characteristics of the system of controls that Driscoll and Wallender would prefer. Besides, a growing number of the successful developing countries are tending to follow more closely this flexible approach. Indeed, this is perhaps even more generally true than the limited number of cases (albeit important ones) examined in the paper. (One might also include South Korea, the Philippines, and Taiwan, among others, among the group employing ad hoc, flexible negotiation control systems.)

The second contention is that the operation of the market has moved in a way to favor the host countries. Individual boundaries in which negotiated terms of technology transfer are set are determined either by market forces (the existence of alternatives), by law, or perhaps by international codes. Driscoll and Wallender seem to prefer the market solution to the constraints issue, as a matter of principle. They correctly argue that the degree of competition among transnational firms, in almost any well-defined industry, has increased dramatically in the course of the postwar period. This has made the market work more effectively as a control system, and in a way that is more favorable for the interests of technology-importing less developed countries. Indeed, more emphasis might have been given to this point.

On the other hand, there is a contrary force that the paper does not acknowledge, but that is becoming more evident. This is a tendency for transnational firms (some with government equity positions) from different nations to negotiate joint ventures in a third country, in technology-intensive activities in which the participating firms have typically heretofore completed with each other. An issue worthy of study and antitrust concern is whether these joint ventures reduce, or limit the growth of, competition in the technology market, or whether they are a form of back-door cartelization. If they are predominantly the latter, the negotiating leverage of technology-importing developing countries may well be adversely affected.

The policy recommendations of Driscoll and Wallender are basically actions that would improve the workings of the market. They emphasize the diversity of technology transfer modes and conditions, giving prime attention to the importance of the environment in the host country as a determinant of the successful transfer of technology, and the terms at which it is transferred. The specific recommendations are to improve host country bargaining capacity – including the ability to use and choose appropriate technologies and to integrate them into the production structure of the host economy. The observation is made that host countries are becoming smarter in technology transfer negotiations, as well as more flexible in that is acceptable. But further improvements are deemed necessary, and a broader country coverage is sought. Thus the paper recommends:

1. Further improvement in negotiation skills, through training and seminars in those skills by international and governmental organizations, and from transnational corporations themselves (an activity that many companies would gladly forego).
2. That emphasis be given by LDC negotiators in striking deals with transnational firms to obtaining obligations from the firms for more training of local employees, and more local sourcing of inputs. The objective is, of course, to facilitate the transfer of technology to local entities.
3. That transnational firms cooperate in removing bottlenecks in industrial or technological structure, and in the labor force.

There can be little disagreement with such recommendations, all of which are designed to improve the capacity of host LDCs to negotiate and accomplish a successful and complete technology transfer to the local economy. But such recommendations, relying as they do on improving the workings of the market as the principal constraint on the volume and terms of technology transfer, are essentially "status quo" in nature, and unlikely to be satisfying to LDC spokesmen who seek codes of conduct.

While the paper presents a rather orthodox view from the perspective of transnational firms, it is somewhat frustrating at several points. It is inadequate and unspecific about the relative importance of the components of the incentives and the host country environment and technological infrastructure which received much emphasis. The linkages between these infrastructing components and the terms of technology transfer or the need for control systems – are obscure and largely asserted.

I conclude from this failure to specify these linkages that the diversity in host environments and technological characteristics is too great to allow general governmental controls over technology transfers either by host government or international bodies. No two countries can have the same optimal control systems. Further, international controls, if they are to be adhered to or effective, must probably be confined to a few, narrow, and specific items where there is substantial commonality of interest among the host countries, and perhaps between host and home countries. Two areas that might fruitfully be pursued are taxation and transfer pricing practices. If these are correct interpretations of the implications of the paper, then I am in basic agreement with it, with the omissions noted above.

Comments

Endel-Jakob Kolde

In this critique of the study by Robert E. Driscoll and Harvey W. Wallender III, the stress will be on the positive: to build upon their paper, to add to it, with the objective of achieving a more complete elaboration and understanding of the control of technology transfer rather than of finding fault in their presentation. In the context of this bigger effort, the weaknesses of the paper will of course be revealed, but only as inevitable by-products. Within the conceptual framework that the authors adopt the paper is a commendable piece of work. However, I find the authors' framework, restricted essentially to direct-investment-embodied technology, to be too restrictive to allow us to address other types of technology transfer. I shall start by attempting to explain why a broader and more flexible conceptual framework is a necessity, and connect up with the main themes of the Driscoll and Wallender paper at appropriate points as I proceed.

Technology transfer is a very elusive concept. While definable in theory, it evades concrete delineation, which would allow us to examine it as a discrete variable in practice. As a result there is a general tendency to treat the technology not as a distinctly defined factor, but as an ingredient in a definitional compound in which other constituents, such as capital investment, labor, entrepreneurship, and some socio-cultural elements, may occur in varying proportions. Such compound variables provide at best questionable proxies for technology as a distinct phenomenon and at the very least expose the analysis to far greater risks than those that we are accustomed to tolerate in the more mature branches of knowledge.

To say that the paper under review utilizes such compound data can be taken, therefore, both as a commendation and a criticism. It is a commendation if viewed from the perspective of the present state of the art: most studies of technology transfer are based on such compound data. Thus the Driscoll-Wallender paper is quite consistent with the current state of the art. But to let the matter rest here is to ignore the substantive issue. When composite data are used to represent a distinct factor, their limitations and possible inconsistencies deserve special attention if false imageries of scholarly accuracy are to be avoided. Silent assumptions that underlie the composite proxies for the technology are potentially the gravest sources of misinformation about the proxy variables. The same is true about the interdependencies, such as the supply-demand-price equations borrowed from conventional economic analysis. Explicit articulation of the silent assumptions would therefore have added both clarity and convincingness to the paper under review.

NEGOTIATIONS

According to the paper, the main control mechanism for international technology transfer is the process of case-by-case negotiation between MNCs and recipient nations and enterprises. By implication, the negotiation process is to be preferred to the exclusion of alternative transfer systems.

Negotiations are indeed useful in many transfer transactions. My experience suggests that many a developing country might have now possessed certain technologies if they had relied more on negotiations and less on proclamations. But this is not the whole truth. Are negotiations needed in all transfer transactions? Indeed, are negotiations with MNCs capable of meeting all technology needs? An answer to this question requires first an answer to a more fundamental question: Is all technology negotiable?(1) Obviously not. For example, IBM has consistently refused to place its technology on the negotiating table in any country; Coca-Cola packed its bags and left a market of some 600 million people in India rather than share its technology with New Delhi authorities. These are just the most recent headline-making incidents. But they are not exceptional. Many earlier cases extend as far back as Nobel's invention of dynamite, Westinghouse's air brakes, the RCA's color TV tubes.

The hard fact of the business world is that the most desired technology is rarely negotiable. Such technology is transferable only as an inseparable ingredient of foreign investments – it is embodied in the strictest sense of the term. Transfer of such technology is primarily an international direct investment issue, rather than one of technology transfer per se. The paper under discussion implicitly but erroneously assumes that this type of capital-embodied technology is representative of all technology transfers worth discussing.

Public or Nonproprietary Technology

On the other end of the spectrum we have nonproprietary technology, technology in the public realm; it is nonnegotiable for the simple reason that nobody needs to buy it and only the unscrupulous are prepared to sell it. The only cost of acquisition of such technology is that of what is needed to gather it in – perhaps train some people, hire technicians, or spend some time in a public library or on a university campus.

By far the largest share of the technology needed by the developing nations belongs to the nonproprietary category. Negotiations here could take place only if some middleman, such as a consulting firm, slips in and performs the gathering and screening functions. But what such a middleman can offer to negotiate is not the technology as such, rather the assembling and possibly packaging of it for export. The technology itself is there for the taking.

Shelved Technology

The most negotiable technology is so-called shelved technology. Shelved technologies are mainly patents for which the owners have no known application. Such patents come about as unexpected by-products or spinoffs of targeted R & D activities of industrial firms. All firms engaged in R & D accumulate some shelved technology; the big companies have heaps of it.(2)

How shelved technology differs from all other notions of technology is succinctly illustrated by the Boeing Company. Over the years this firm has acquired scores of patents that have no apparent use in the aerospace industry. To dispose of these patents, a special department was created and staffed with market researchers, engineers, and lawyers. The main mission of this department is to scan other industries in search for buyers or licensees for the shelved patents. The success rate of the department has been rather dismal. The same seems to be true in other corporations. When Boeing suffered its serious setbacks in early 1970s, the whole business community in the greater Seattle area joined the company's efforts to market its shelved technology. Parts of this program sponsored visits by regional delegations, consisting of leading businessmen and government officials, to other industrial nations, especially those with a balance-of-payments surplus; thus West Germany and Japan were the principal targets. But despite all the planning and prestige, and most sophisticated salesmanship, the delegations found no takers.

The reason for the lack of success had nothing to do with the price. Even when the delegations tried to give the patents away, with only the condition that the West German or Japanese recipient reciprocate by making an industrial investment in the Puget Sound area (to alleviate the severe business slump there), the response remained negative. No one was anxious to acquire this shelved technology for the simple reason that it lacked economic application. The West German and Japanese industries' response was, in effect, "If you don't know how to employ your patents profitably, how do you expect us to do it?"

These events underscore a very crucial aspect of technology that is too often omitted from the transfer equation – namely, that any given technology, to be economically useful, must meet the test of being an engineering invention and also the value-added test of being capable of converting inputs of a given cost value, say 10, into outputs of a value greater than 10. A technical invention, regardless how ingenious or how unanticipated in engineering terms, is of no economic value unless it can also meet the inputs-less-than-outputs test. In addition, in our economic system the new technology must pass still a third test before qualifying as an economic phenomenon. This third test is that of the competition, of being capable of adding value to inputs more efficiently than other available technologies.

The shelved technology, as a rule, fails either the value-added test or the comparative efficiency test. It remains on the shelf not because of any technical demerits, but because it lacks any known potential for generating positive cash flows if employed.

It is a sad fact that the developing nations are more likely candidates to become buyers of shelved technology than are the industrial countries. When this happens the charge of inappropriate technology derives added credibility, fueling fires that have burned pretty briskly for a number of analogous reasons.

Mature Technologies

The technologies that are truly negotiable in a meaningful sense belong in the more or less mature and diversified (many alternative techniques to make the same product) category, such as textiles, machine tools, plastics, and pharmaceuticals. The category encompasses a very wide range of products and services. However, its great scope is dwarfed by comparison with that of nonnegotiable technology. Thus ignoring the difference between the negotiable and the nonnegotiable spheres in the technology universe puts the subject of controlling technology transfer into a distorted frame of reference.

One cannot, therefore, escape the conclusions (1) that negotiations between MNCs and potential recipients is not a broad enough concept to offer us a control system which would be applicable to the full range of international transfer processes; and (2) that to devise any operationally realistic control systems, it is necessary to differentiate among the relevant types or classifications of technology. The practice of treating technology as an undifferentiated factor violates both scientific and administrative realities.

Systems Framework

At its irreducible minimum, a control system for transboundary technology transfers must be able to cope with the following classificational dualities:

1. public (nonproprietary) versus private (proprietary)
2. embodied (dependent) versus unembodied (independent)
3. hard (engineering) versus soft (managerial).

The economic tests of added value and comparative efficiency, which were discussed above, are presumed to have been satisfied as a precondition.

Since the three classification dualities are not mutually exclusive but overlapping, a systems approach to transfer control demands further differentiation by a three-dimensional matrix, which the lack of space does not permit me to explore in any detail here.

MISTAKEN IDENTITIES

To control technology (to give it desired direction, scope, and composition) we must have a clear concept of the institutions and processes that form the channels for the export-import flows of technology. This is more easily said than done. Unlike goods and capital, technology has no clearly defined, not to mention well-established, market institutions. There are no counterparts to banks, securities brokers, stock exchanges, or export-import merchants in the technology "trade."(3) Perhaps that is why the term "transfer" rather than "trade" has become the terminology in the technology business. Yet despite the lack of any technology trading, despite the absence of distinct international business channels designed for technology flows, both government authorities and economists have attempted to control technology transfer with methods analogous to those used in foreign exchange and foreign trade policies.

Partial success is the most optimistic projection possible for such efforts; and the probabilities for the future are distressing indeed.

Inexhaustible Supply

What sets technology sharply apart from goods and capital is the intrinsic fact that any particular technology, once perfected, has an inexhaustible supply; that is, its "inventory" is not depleted or even diminished when it is transferred (exported) from one party to another. The original owner can still continue to use it at any scale after he has made it available to any number of other users. Therefore, the export (out-transfer) of technology need not cause the source country or company to reduce its own use of the technology while augmenting the technology base of the recipient's firm and/or economy. Indirectly, the original owner's ability to employ the technology may be reduced if other countries use technology acquisition to enter the end product markets and if that market happens to approach its saturation point. In aging fields of technology, such as textiles, steel, shipbuilding, and a few others, this has in fact occurred. However, the main threat to most technologies has not come from international dissemination, but from obsolescence – the emergence of new, economically superior technologies that have rendered the old ones obsolete.

Fictitious Equilibrium

These intrinsic supply characteristics of technology do not permit accumulation of supply, such as we have for goods in the form of inventories and for capital in the form of deposits and claims. As a consequence, attempts to arrive at objective prices for technology on the basis of the familiar supply and demand equilibria are, fundamentally, exercises in futility. For technology such equilibria do not exist. While a demand schedule can be constructed at least in terms of derived data (derived from the input requirements of the end product), the supply schedule defies definition.

The Role of Multinational Corporations

Returning to the transfer channels, I have no quarrel with Driscoll and Wallender's contention that as matters presently stand MNCs serve as the principal channels for technology trade. However, some additional specifics will help to bring out the different ways and combinations in which technology transfer figures in the international activities of MNCs.

The main instruments of technology transfer are direct investment, licensing agreements, patents, management contracts, training and service contracts, and feasibility studies. Except for the last two, none of these instruments is designed primarily for international distributions. Their primary purpose is an economic advantage that a direct investment or licensing agreement makes possible. Technology transfer is inherent in these arrangements, but only to the degree that it is necessary to achieve the objectives of the investing or licensing firm. Hence these channels are not open for technology transfer for other purposes. A wide variety of diverse technologies will be transferred by direct investment only when many multinationals from different industries invest in a host country.

Management contracts are similarly constrained and in reality offer even fewer alternatives than does direct investment, due to the fact that they normally presume the existence of the establishment to be managed (such as a hotel).

Patents may accompany direct investment as integral components, or they may be licensed for use separately. Most developing countries have relied primarily on direct-investment-embodied patented technology.

Trademarks have tended to have a more even split: most of the multinationals typically exploit their trademarks in association with direct investment, while other firms act as licensors for local companies in host countries.

For neither patents nor trademarks is there an institutional channel solely for the purpose of distributing the technology; the distribution and transfer processes are coincidental to the overall objectives of the source companies. Thus despite the official registration and public record of patents and trademarks, their international transfer channels are in substance quite similar to those of unrecorded proprietary technology, such as unpublished production methods and managerial know-how. A somewhat greater degree of flexibility is found in the technological services sold by engineering and consulting firms. However, their contribution is mainly of a one-shot nature – to get a one-time problem solved – and the technology they supply is, so to speak, second-hand, often from the public sphere. Also consulting firms are seldom prepared to sustain an operating entity for an indefinite life, as the multinationals can do.

New Distribution Models Needed

Organizationally neither the multinationals nor the developing countries possess the necessary structures to facilitate an efficient international allocation of technology. This lack has not gone unnoticed. Several proposals to remedy the situation have been put forward. All of the proposals known to me focus rather narrowly on the technology transfer process and do not come to grips with the supply and consumption aspects of the technology market.(4)

Two main themes dominate in the proposals. One advocates the creation of networks of governmental agencies for the monitoring, evaluation, and allocation of technology, both intranationally and internationally. The emphasis here seems to be more on the regulatory than on the marketing and developmental aspects. As such, this approach harbors the dangers of suffocating the links that now play the most vital role in transboundary technology flows.

The other innovative thrust aims more directly at economic objectives. It argues for the creation of developmental institutions whose main concern would be to procure new technology, facilitate technology imports, encourage indigenous technology production, increase the absorptive capacity of the host country's economy for technology, and promote training programs to utilize technology.

One plan of the latter type visualizes a set of interconnected institutions made up of governmental, academic, applied research, social, and business organizations. The coordinated activities of this set would include (1) information monitoring, assimilation, and dissemination; and (2) impact evaluation and facilitative intervention at one or more resolution levels and sectors of the economy and society.

National technology institutions are proposed to be linked internationally to guarantee effective transfer to technology from one economy to the next. Whatever the domestic merits of such national networks of technology institutions, they fail to offer satisfactory solutions to the international difficulties of technology transfer. Their main weaknesses lie (1) in the lack of organizational systems designed for transboundary technology trade, and (2) in the vagueness of how the technology content of investments and trade could be subjected to separate measurement, bargaining, and pricing.

The Capacity to Absorb Technology

In the Driscoll-Wallender paper, the lack of absorptive capacity in technology-importing countries is stressed as a main barrier to greater technology dissemination as well as to international economic progress. The factual record lends much support for this general proposition. However, like most other generalizations, it needs a number of qualifications to reflect its limitations, and there is ample room for reservations on the grounds of the conflicting self-interests of the parties to transfer transactions, leading to conflicting interpretations of the same set of facts. For example, what is interpreted as the inability to absorb a given technology by the potential exporter of the technology may be regarded by the importing nation or firm as the outcome of an infusion of the wrong or inappropriate technology.

The selection and adaptation (qualitative aspects) of technology are, therefore, just as pivotal to the control of the transfer process as are the access to and cost of the technology.

Right or Wrong Technology

The issue of maladaptation or inappropriateness arises mainly from the fact that the objectives and priorities of MNCs and host country governments are not synonymous. Three kinds of conflicts are common:

1. The multinational favoring automation for increased efficiency versus the developing country favoring labor-intensive technology to increase employment (especially true in the agricultural sector).
2. The multinational favoring production of expensive manufactured goods with high profit margins versus the developing country tending to favor inexpensive, low-priced goods.
3. The question of control and ownership of the technological resources involved.

These conflicts are particularly apt to occur in those developing countries that have decided not to imitate American or West European patterns of economic development but to forge ahead on a path built by themselves. Such independent development strategies demand that the developing country define its national priorities and actively review development activities constantly to make sure they are in compliance with the national policy. Often this may involve sacrifices or trade-offs, such as using methods that are less efficient but maximize employment. Invariably such development strategies are cast in the form of a national economic plan. From the multinational company's standpoint, rigid governmental planning and monitoring of business activity represents not only an added risk but may also undermine the financial feasibility of its investment. As a result, new investments may be withheld and the existing ones liquidated. Cooperation rather than confrontation is normally the rational prescription for a mutually satisfactory solution in such situations.

LIMITATIONS OF MULTINATIONALS AS TECHNOLOGY ASSIMINATORS

The fact that MNCs have been a principal source of technology for the developing nations has not been accepted with universal acclaim. On the contrary, the assessment of the multinationals' net effect has become an explosively controversial subject, as is so vividly demonstrated by the ongoing Iranian revolution. To be sure, there are many other factors at play in Iran at this time, but there is no room for doubt that the Khomeini regime regards both MNCs and imported technology as major sources of evil. Khomeini's denouncements have been echoed throughout the world, at various amplitudes. Clearly, MNCs are objects of the grave suspicions of significant factions in many countries.

Leaving to one side the purely ideological and ultranationalistic arguments, the criticisms of the MNCs as technology transfer agents relate to the following:

1. The tendency of MNCs to concentrate their R & D activities in the educationally and technologically most advanced countries, particularly in the headquarters country. MNCs rarely conduct any research at all in the developing countries. The indigenous scientific and technological community of developing countries, small as it is, is thus excluded from participation in the MNCs R & D activities.

2. The multinationals' practice of expanding by acquisition – buying up successful native enterprises – rather than starting new and different entities. Along with other assets, the acquisition also includes the perfected technologies as well as ongoing R & D activities of the absorbed native company. Thus the MNC gains control over what typically has been the most productive indigenous technologies. What happens to the R & D of the absorbed firm afterwards is the most sensitive issue; is it transferred from the developing to the developed country where the corporation's own R & D activities are located?

3. The transplantation of American and other Western products and production processes into non-Western societies without proper regard to the society's economic standards and developmental priorities. The feared results are wasteful consumption and corrosion of indigenous sociocultural structure.

4. The patent-hoarding tendencies of MNCs. In the developing countries' view, the primary objectives of the patent system are to offer incentives for domestic inventors and adapters and to attract inbound direct investment in technologically backward sectors. The multinationals' practice of using the patent system to restrict the use and imitation of patented technologies by local businesses and to strive for a monopolistically segmented market structure is seen as an abuse of the system. To the extent that multinational corporations' patent holdings fall into the shelved technology category, this argument would seem to be fallacious. But shelved technology is not likely to be patented in developing countries. Though no data are available to show the breakdown between economically usable and nonusable patents, an a priori assumption is irresistible: that most MNC's patents registered in

the developing countries possess such economic merit that a real competitive risk exists when the company decides to apply for a patent. What other purpose would justify the costs connected with the processing of the patent? Viewed from the MNC's perspective, patenting is a protective, not productive, process. A patent is taken to protect the owner against patent piracy; it is the owner's cost of inventing the patentable invention and the future income stream that derive from it that are the objective of the legal device we call a patent. When and where a patent will be used are matters that can be decided only after studies of production costs, markets, and competition have been completed.

The criticisms of the multinationals by the developing countries on this score are often quite misplaced. No antisocial and antidevelopmental implication can be imputed to the simple fact that a patent is not exploited. For such criticism to be justifiable there must be proof that the patent is in fact economically exploitable and that the best place to do so is the particular host country concerned.

Thus no definitive statement is possible on the patent-hoarding practice. It can be harmful or it may be necessary. Both the developing nations and the MNCs can be right, depending on the specifics.

SUMMARY

Traditional economic analysis can be dangerously misleading in the study of technology transfer. If the silent assumptions on which such studies are based are explicitly articulated, the lack of any fundamental analogy between technology transfers and goods/payments transfers becomes evident. From false analogies flow false conclusions. Incessant treatment of technology as identical to goods and capital, as a proxy, particularly synonymous with direct investment, has introduced false proxies for technology transfer and, what is worse, is rapidly creating a dangerous fiction of objective truth where none exists.

Technology is not a homogeneous phenomenon. There are different types of technology, each posing fundamentally different problems and demanding different solutions in the international transfer process. Ad hoc negotiations on a case-by-case basis are often useful in the proprietary technology sphere. But even here, case-by-case negotiations cannot be regarded as an adequate control system capable of the screening and direction that is necessary to meet the needs of national development as perceived by recipient nations.

In the nonproprietary sphere, case-by-case negotiation can have but marginal effect. A workable system of control must combine satisfactory mechanisms for all types of technology flows.

NOTES

(1) Negotiability is not to be confused with availability of alternative technology.

(2) Shelved technology is not synonymous with front-end technology, that is, the most advanced state of a given technology.

(3) To regard consulting firms as technology traders is to stretch the meaning of "institution" to an extreme. Rarely is the consulting firm's primary objective technology transfer; rather, it transfers technology as something coincidental to its main purposes.

(4) See E.J. Kolde, "Technology Market and Developing Countries." mimeo, 1979.

Comments

Gabriel Wilner

The contribution of multinational corporations to the development and diffusion of technology throughout the world is undoubted. The priority for the multinational corporation(1) is markets that will utilize the technology produced. Ideal markets are those that can afford to pay for the technology that industrial countries produce, that can adapt technology already produced, that give incentives to sellers of technology, that have sophisticated technological infrastructures, and perhaps, above all, that have legal and economic regimes that will permit the seller to set the terms of trade without any interference.

The point is that such ideal conditions for multinationals have existed primarily in the developing countries. The economic power of multinationals and the eagerness of the developing world to spur on its economic development – particularly in industrial sectors – have made it easy for the seller of technology to set conditions and terms to a buyer whose only wish is to reap the benefits of "progress."

Since the market for technology transfer to developing countries is a seller's market, as is generally admitted, a central question is the best manner for the buyer and the buyer's country to cope with this imbalance in economic power and technological sophistication.

Professors Driscoll and Wallender suggest that the parties concerned with transfer of technology in developing countries should become better negotiators. They point to assistance for this purpose that is presently being given to buyers of technology by various international organizations. In fact, increasing the negotiating skills of buyers of technology seems most useful; they can thus begin to work with existing legal and economic rules to make the best of the existing situation. The current situation is one in which the national and international rules(2) favor the owner of technology, and are, further, premised on the notion that sellers and buyers of technology and their respective countries have roughly similar economic power and technological sophistication.

There is an alternative to attempting to resolve the current economic and social imbalance between the position of the multinational seller and the developing-country buyer of technology by making believe that all that is needed is more sophisticated negotiation techniques by developing-country buyers of technology. This alternative is a modification of a legal regime that at present does not take into account the particular situation of economic disadvantage in which developing-country buyers of technology and their national societies find themselves today. Legal rules and regimes, both national or international, are not immutable. They should respond to economic and social needs as they arise.

The argument made here for changes in legal structure should not be read to mean that such changes need pose any threat to the economic and political underpinnings of the multinational corporation. When economic power is perceived to favor one side of an economic relationship or when enterprises are considered to have too dominant an economic position, the law can be used to redress such an imbalance. This, it would appear, is at the root of economic regulation of business within any national society. The United States antitrust law is a vivid example of such regulation.

There is nothing extraordinary or unusual about national regulations by certain developing countries controlling the entry of foreign-owned technology through commercial transfer-of-technology transactions. Countries such as Japan and Spain have stringent regulations that have a direct effect on the transfer of technology, while other industrialized countries use more direct regulation, such as exchange controls. In the field of competition law, the European Economic Community uses screening in the form of "negative clearance." In the United States, control over certain types of exports of goods and of technology is maintained through a system of special licenses, which is essentially a screening process.

Certain persons, including Driscoll and Wallender, maintain that developing countries cannot afford to regulate technology flows, because to do so will discourage the multinationals from selling technology and, more generally, from investing in the developing countries. These activities of multinationals are perceived as being essential to economic development. Thus the price of technology purchased, the conditions of its purchase, and the cultural and social impact of such purchases are seen as secondary in importance to increasing employment in the developing country, gaining it access to foreign markets, and strengthening its technological infrastructures. In short, the argument seems to be that the transfer of technology at any price and under any conditions is the panacea ensuring economic development.

In fact, paying close attention to all the factors listed above is essential to a developing country if it is to achieve a decent standard of living and social and cultural well-being for its people. There is value in a national legal and economic structure that permits a national society, through its government, to make some determination about the types of technology that should be brought into the national economic and social system and about how the transfer of technology will affect the nation's financial capacities, including its balance-of-payments situation. The decision about whether a particular technology is useful for a particular society ought not to be made by private international economic entities, managed in accordance with a particular global point of view that is principally oriented to economic profit. The multinational corporation may be admirably equipped to determine economic efficiency, and it must be permitted to do so, but within the framework that the political entities – the states – have determined is necessary to foster their broader goals.

National action by developing-country governments to set standards for the inward flow of technology has, in recent times, been accompanied by calls for universal standards. An element of the reason for the interest in universal solution is the perception by developing countries that they must present a unified front against the dominant sellers of technology – the multinational corporations. Decision 24 of the Andean Pact is a regional example of this notion of a uniform position.

In the proposed Code of Conduct on the Transfer of Technology, currently being negotiated in UNCTAD,(3) a reformist tendency can be perceived in the attitude of the developing countries. No great structural changes in the terms of trade are proposed. The code of conduct seeks to eliminate abuses by the economically dominant sellers of technology in their transactions that have an adverse effect on the economic development and national well-being of the technology-acquiring country. In return for some reform of the current totally unregulated international scene, the code of conduct sets forth some limitations on national action in this field. The idea here is that there should also be some international standards to which governments must be subject when they enact and apply laws relating to the transfer of technology. Thus the code seeks to bring about reform of the present international legal structure on technology in ways that would affect all the actors on the technological scene.

On both the national and international levels, the changes in the present legal structure sought by the developing countries, and implemented by some of them, are meant to end a system that essentially favors the seller of technology. These changes will not eliminate the favorable climate that multinationals seek; they will only temper that favorable climate with a similarly favorable climate for ultimately achieving the economic and social well-being of the country of the technology acquirer. National and international regulations need not eliminate incentives for the sale of technology by multinationals, but they may bring down levels of profits and economic domination. Regulation may set priorities that will in fact attract the technology that is necessary for the development of national technological infrastructures in developing countries. This capacity will in turn result in greater ability not only to adapt technologies that are imported, but also to develop new technologies that respond to national needs. None of this can occur if the decisions on technology transfer are made solely in a marketplace where the seller dominates.

The law can serve the evolution of economic and political relationships by providing the means for controlled change without destroying those elements of the relationships that work well. Reform of the current legal rules and institutions, through some regulation of the activities of sellers and buyers of technology, responds to the constructive role that law can play.

NOTES

(1) It should be noted that the term "transnational corporation" is now in more current use, although spokesmen for these enterprises appear to prefer the older term. The term "multinational corporation" is used here, since the paper under discussion uses the term.

(2) The traditional rules relating to technology are basically national rules on the protection of industrial property rights. The international regime, namely the International Convention on the Protection of Industrial Property (Paris, 1884) reinforces these national rules. Further, national rules on contract and unfair competition are meant to give unconditional protection to owners of technological knowledge.

(3) See chapter 14 by A.A. Fatouros, "International Controls of Technology Transfer," in this volume.

10 Technology-Importing National Perspectives

Hubert A. Janiszewski

With the growing trading of technology – including technology protected by various industrial rights such as patents, trademarks, and utility models – governments are recognizing the potential impact and effect that such a "commodity" as technology may have on trade, industrial development, and employment.

In the spirit of protection of free competition and unrestrained trade, governments of such industrialized countries as the United States, Japan, and the EEC member countries have introduced various antitrust measures – for example the Sherman and Clayton Acts in the United States and Articles 85 and 86 of the EEC's Treaty of Rome – that have also found application in the past twenty years or more with regard to exchange of industrial property rights and know-how, and to technology trading generally.

Since the Second World War, the trading of technology has become a more and more important economic activity. Its fast-growing share in world trade turnover reached in 1965 the level of about $2.7 billion, and in 1975 a level of approximately $11 billion.(1)

Out of many reasons and conditions that have created a favorable environment for the drastic growth of technology transfer, at least a few should be mentioned:

1. Tremendous technological progress, particularly in the United States, later also in Western Europe and Japan;
2. Need for the rapid rebuilding of economies and industries of Europe and Japan after the Second World War;
3. Modernization of the industries of the socialist countries and the Soviet Union, particularly strong in the last fifteen years or so;
4. Emergence of a great number of new, independent countries in Africa and Asia, which in certain instances introduced limitations on trading goods and temporarily affected direct foreign investment because of nationalization policies, high import tariffs, and the like.

The emergence of new, independent countries in all parts of the world, which because of their social and economic underdevelopment are called developing nations, also created rather favorable conditions for various forms of technology transfer. Technology transfer was and still is one of the means of industrialization of those countries, which because of the lack or underdevelopment of an indigenous R & D base are dependent on foreign technology either in the form of foreign investments or direct technology flows.

This technological dependence has also been one of the reasons for the increase of technology flow to developing countries, which amounted to perhaps $300 million U.S. in 1965 and reached a figure of over $1 billion U.S. in 1975.(2)

According to UNIDO estimates,(3) direct expenditure in technology by developing countries may reach a figure of about $4 to 6 billion in 1985, and their share in the world flow of technology may increase from the current approximate 10 percent to some 13 to 15 percent in six to seven years, due to increased industrialization efforts and needs, among other things.

It will also be extremely useful to have a closer look at the pattern of technology importation by developing countries. A few of the more advanced developing countries are purchasing the predominant part of technology. In 1972 Argentina imported technology worth $120 million; Brazil $154 million; Mexico, between $150 and 200 million; Spain, over $200 million; and India, around $50 million, together comprising approximately 75 percent of the total technology imports by developing countries.(4)

In view of this pattern of technology flow to developing countries – particularly in Latin America – the governments of these countries were compelled to take a closer look at such substantial payments and outflows of large amounts of foreign currencies, and at the contribution of those flows toward economic and social development of their nations.

The earliest regulation of transactions involving the transfer of technologies resulted, therefore, from the need for effective government regulation of foreign currency payments through, for example, the registration of agreements with central banks in order to effect or authorize transfer of payments to foreign licensors and suppliers.

Only subsequently have some governments decided to look into the substance of technology agreements, particularly from the point of view of the agreements' impact on the national economy on one hand and of their terms and conditions on the other hand.

As is well known, the first developing country to evaluate and scrutinize technology agreements was India, which introduced rather strict administrative guidelines concerning technology agreements as far back as the early 1950s, and also established a special interministerial body to deal with these issues.

India's Industries Development and Regulation Act of 1951 specified the conditions under which so-called technical collaboration may be approved by the government. Later on a specialized body was created, the Foreign Investment Board, to deal with all issues related to technology transfer agreements as well as with foreign investments; as a matter of fact, this still operates today in an expanded and rather complex manner.

As early as 1967, Colombia, because of the necessity of tight control on its foreign currency situation, introduced special legislation to deal with control of technology agreements (Decree 688 and Decree 444) and established a special body called the Royalties Commission to deal with those issues.

The example of Colombia, and studies and investigations carried out by the Andean Pact Secretariat in this field under the leadership of C. Vaitsos, led later to the introduction of the now famous Decision 24 of the Andean Pact, affecting other countries of the group – Ecuador, Chile, Peru, Bolivia, and later Venezuela.

In addition to the rather high cost of technology imports by developing countries – amounting to about 3.8 percent of the total exports to Brazil, 1.8 percent to Colombia, 7 percent to Mexico, over 5 percent to Argentina, and 6 percent to Spain – there were also other conditions that led to direct government intervention and regulation of technology imports:

1. the negative impact of restrictive business practices, particularly export, volume, and field-of-use restrictions;
2. the flow of technology into the most profitable industrial branches in a way not necessarily compatible with the development goals of individual governments;
3. the purchase of technologies inappropriate in terms both of their technical level and their impact on employment; and
4. the negative impact of the uncontrolled flow of technology on indigenous research and development activities.

An example of the amount of restrictive provisions in agreements for technology transfer to developing countries is the following summary of data for Mexico.(5)

Nature of Restrictive Provisions	% of 2,200 agreements
Grant-back provisions	5.8
Tie-ins of various types	3.1
Export restrictions	6.2
Prohibition against use of complementary know-how	0.9
Tie-outs	0.3
Volume restrictions	9.6

The data collected by the Andean Pact, however, suggest that, for example, territorial restrictions on the average were to be found in about 70 percent of the examined contracts of Chile, Colombia, and Peru, and were therefore much more frequent than in Mexico.

Another factor should also be stressed as leading toward introduction of regulatory measures in technology transfer, measures that as a matter of fact were only a logical follow-up to growing government regulations over direct foreign investments and to increased overall economic intervention of the state. This factor is the frequently occurring incompatibility between the goals of national governments and national companies (mostly private) on one hand and the governments and suppliers of technology (mostly transnational corporations) on the other.

The evolution of state or government intervention, particularly strong in developing countries without large industrial sectors, has predominantly led to control or regulation of technology transfer in order to achieve long-term policy goals.

As can be observed, both internal and external reasons and conditions have led to the introduction of government regulatory and controlling measures for technology flow in developing countries.

APPLIED SYSTEMS OF REGULATION OF TECHNOLOGY TRANSFER AND THEIR ASSESSMENT

From more than 100 developing countries at present the following have introduced various systems of control over technology transfer:

Latin and Central America (9). Argentina, Brazil, Peru, Colombia, Ecuador, Bolivia, Venezuela, Mexico, and Cuba.

Africa (4). Algeria, Egypt, Libya, and Tanzania.

Europe (5). Spain, Portugal, Turkey, Yugoslavia, Romania.

Asia (10). Iraq, Pakistan, India, Malaysia, Philippines, Sri Lanka, People's Republic of China, Israel, Republic of South Korea, Republic of Vietnam.

According to new developments and information available at UNIDO, the following countries are also considering the introduction of various regulatory measures in technology transfer: Guatemala, Panama, Sierra Leone, Nigeria, Ethiopia, and Somalia, as well as some others. This trend toward government intervention in this specific field has been very clearly strengthened by the negotiations on the UNCTAD code of conduct for technology transfer, in the course of which the experiences of certain countries, such as Mexico, Brazil, the Andean Pact countries, and India, have been very widely discussed among the governments of the developing countries.

From among all the modes or systems applied to the control of technology transfers, two basic ones emerge most prominently: the legislative and administrative approaches, as well as a mixture of the two in certain countries.

Legislative Systems

The following developing countries have adopted a distinct legislative approach: Argentina, Brazil, the Andean Pact member countries, Mexico, Spain, Portugal, the Philippines, and the Republic of Korea. A legislative approach can usually be characterized by the introduction of specific legislation or a legislative framework that stipulates in greater or lesser detail the conditions that technology transfer agreements must comply with, and that creates high-level government body to implement such legislation.

The systems existing in the above-mentioned countries can even be called the direct legislative approach, since these governments take a direct position on the process of regulating technology transfer, first by introducing special legislation and second by creating a special government body entrusted with its implementation. The legislation establishes basic criteria and conditions under which technology may be allowed into the country, while the administrative body called upon or established by the legislation serves as a watchdog, supervising whether the imposed conditions are fully met.

Usually such a governmental body or agency – often called the Registry of Technology or Department of Technology – in order to fulfill its tasks elaborates and applies specific evaluation criteria for economic, legal, and technical analysis of technology transfer agreements.

It should be emphasized that legislation in the area of technology transfer does not occur in isolation and usually has been introduced as part of a package of legislation regulating foreign investments as well as in some instances industrial property rights (for example, in Mexico, the Andean Pact, and Brazil).

Administrative Approach

Among countries that opt for the administrative system of controlling technology transfer, the following may be included: Cuba, Egypt, Libya, Turkey, Yugoslavia, Romania, Iraq, Pakistan, India, the People's Republic of China, Malaysia, and the Republic of Vietnam.

Unlike the legislative approach, this system does not use legislation specifically dealing with technology transfer and specifying conditions under which technology agreements may be concluded, but rather uses administrative measures and internal criteria, which in many instances are not too well known either to recipients or suppliers of technology.

The administrative approach may establish a government body controlling all technology agreements, as is the case with India, Libya, Cuba, or Malaysia, or those responsibilities may be scattered among different governmental and paragovernmental institutions, as in the People's Republic of China, Romania, Yugoslavia, Egypt, and Iraq. As has been often observed, in countries that opt for the administrative system criteria and conditions for technology inflow may vary not only from year to year, but also depending on which government institution deals with the issue. (This is particularly true in regard to the criteria for approval of contractual conditions and the degree of strictness used in evaluating agreements.)

It should also be mentioned that a mixed approach is used in such countries as Turkey and Malaysia, where existing legislation only implicitly covers technology transfer, while some specialized government institutions deal with the regulation of technology inflows.

It should further be noted that the administrative approach in such countries as Cuba, Romania, the Republic of Vietnam, and the Peoples' Republic of China has been very much oriented toward similar systems introduced earlier in other European socialist countries and in the Soviet Union, where usually one or more specialized foreign trade units implements the planned acquisition of technology needed for the national econmies of those countries on behalf of their governments.

In principle, in all three categories – legislative, administrative, and mixed – central or similarly positioned regulatory agencies have developed over the years a set of internal rules, proceedings, and evaluation systems that either reflect the requirements of existing legislative frameworks or the overall economic and social needs of the country, as expressed by the government's long-term plans and policies.

It is not easy to assess the effectiveness and impact of the three basic categories of technology transfer control systems, and such an assessment should be viewed from at least the following standpoints:

1. attainment of national objectives and goals;
2. strengthening of the negotiating position of recipient companies;
3. increase or decrease of technology supply by major sources;
4. elimination of discrepancies and limitations in the trading of technology (elimination of restrictive business provisions);
5. effectiveness of introduction of regulatory or controlling measures in technology transfer.

The assessment of all three categories will therefore be carried out – in as much detail as possible – according to these criteria.

The attainment of national objectives is a rather broad criterion, and therefore the following may be specifically taken into consideration:

- long-term decrease of dependence on foreign technology and simultaneous building up of indigenous technological capabilities;
- decreasing and more efficient use of foreign exchange reserves used for the purchase of foreign technology;
- opening of markets for indigenous products based on foreign technology;
- increase of employment;
- acceleration of industrial development;
- increase in standards of living.

Assessment and clear evaluation of which of the three types of controlling systems is more efficient with regard to decreasing dependence on foreign technology is at present very difficult or even impossible, particularly since collected and available experience is too recent.

Certain indications in this direction, however, may be gathered from India, which through very direct, strict administrative measures has achieved a considerable level of technical and technological independence.

The experience of countries that adopted the legislative system (particularly in Latin America) is too brief for full analysis, and in any case the average technological level of those countries at the moment of introduction of the control system was considerably higher than that of India at the time it introduced its controls.

A unique situation, on the other hand, can be found in the Republic of South Korea, which has adopted the legislative model, very much oriented toward the Japanese system of close interaction between industry and government, and where technological independence has reached a considerable level.

Some of the effects achieved in the Republic of Korea (up to 1975) are shown here(6):

Contents of Agreements Approved by Government	Frequency(%)
Shortening of contract period in an agreement	24.8
Reduction of technical fees	22.1
Quality guarantee of the products	7.0
Utilization of local raw materials	6.1
Amendment of royalty payment methods	2.9
Shortening of foreign technical experts stay and reduction of their fees	11.9
Ban on using foreign trademark on the local market	10.2
Full export rights for all products	7.9

In terms of a decrease of foreign exchange costs of technology transfer or its more efficient use, both systems seem to be equally useful; however, perhaps administrative measures can be introduced within a shorter period of time and with more immediate effect.

In terms of absolute figures, only in a few cases can a drop in expenditure on technology in the first year or two after the introduction of regulatory mechanisms be observed (Colombia and Peru);(7) the trend of growth is in principle maintained in all the countries surveyed.

According to the former director general of the Mexican National Registry for Technology Transfer, Lic. Jaime Alvarez Soberanis, foreign exchange savings that occurred because of the Mexican law on technology and operation of the registry reached the figure of one year's imports of technology, estimated in 1976 at the level of about $250 million. It should be mentioned, however, that there was no drop in the absolute figures for Mexico's imports of technology, and the accumulated savings have been calculated on the basis of reduced royalty and other payments, elimination of volume and territorial restriction, and other improvements in contractual conditions that can be quantified.

No doubt the administrative system can drastically affect payments for technology in a much more effective way, particularly over a short period of time. However, this may lead to big increases during more lax policy periods, something not necessarily compatible with the long-term policy goals of a given country.

The opening of export markets for products manufactured with imported technology has been achieved in principle in both legislative and administrative systems.

The statistics on India and an analysis made in Argentina by Chudnovsky in 1974(8) prove that both systems of control of technology have been rather effective in this respect, but are not without problems and difficulties in terms of real exports. (Potential export possibilities arising out of technology agreements do not automatically mean real exports of goods.)

Other criteria, such as increased employment, accelerated industrial growth, and increased standards of living, are only partially affected by controls on technology transfer, and depend also on the overall economic policy that usually determines the policy in the area of technology transfer.

In terms of strengthening the negotiating position of recipient companies, it seems that both systems are almost as effective; however, the legislative system offers recipients of technologies, that is, domestic companies, a broader base for negotiations both with suppliers and their own government agencies.

Most of the legislation and adopted legislative systems (Mexico, Brazil, Spain, Portugal) provide a possibility for continuous dialogue between recipient companies and the government, as well as informal expressions of opinion by the government vis-a-vis foreign suppliers of technology.(9)

Available information from the suppliers of technology also indicates their preference for dealing with legislative systems for the following reasons:

- broad legislative base open for negotiation;
- clearly spelled-out conditions;
- clearly indicated responsibilities for all parties.

Naturally, suppliers of technology were not too happy with the introduction of national controlling systems, as this has created new market situations, increased competition, and perhaps lowered profits. On the other hand, most of the suppliers of technology accept new situations and the new rules of the game. It should be mentioned however, that some of the suppliers are rethinking their policies on technology transfer, but the effects of that process will only become visible in years to come.

The overall trend of technology flow to developing countries in the 1970-76 period, both to those who have introduced regulatory systems and those who have not, is increasing at a similar pace, and therefore control over technology transfer has not as yet negatively affected trading in technology.

At this point it is worth mentioning the generally tighter application of antitrust provisions in the United States as well as in Europe on technology transfer agreements, which confirms the prevailing thesis that worldwide control over technology transfer is increasing.(10)

One unquestionable achievement of national technology transfer control policies was the drastic elimination or reduction of various restrictive business provisions in technology contracts.

Statistics available from such countries as India, Argentina, Brazil, Spain, Portugal, and Mexico show very clearly that in this specific field most restrictive provisions have been eliminated, particularly territorial and volume restrictions, tie-ins and tie-outs, and field-of-use restrictions, irrespective of whether legislative or administrative systems have been applied.

It was found, for example, that once the Mexican law on technology transfer and the approval criteria applied by the National Registry became well-known among suppliers of technology, the amount of restrictive provisions in the agreements submitted for approval has diminished very considerably.(11)

Naturally, the total elimination of all restrictive provisions is not possible, due to earlier contractual obligations of technology suppliers, the monopolistic position of some individual suppliers of technology, the patent protection in their possession, and so on, as well as the importance of certain technologies for the economies of individual countries.

In many instances, the legislative system, offering as it does a rather broad base for negotiations, has been found more acceptable by suppliers of technology, particularly when they insisted on the introduction of certain restrictions that, if found to be justified, were accepted by certain government regulatory agencies. Examples of this sort may be found in the cases of Argentina, Brazil, Mexico, Spain, and Portugal.

There is another aspect of the attitude of suppliers of technology toward the introduction of control systems, and that is whether such systems do or do not hinder the supply of the most modern, sophisticated, "front-end" technologies.

This issue is connected with another aspect of any controlling system, namely that of substantial loosening of control over technology by the suppliers.(12)

Judging from available information, the introduction of control systems does not lead to the supply of inferior technologies, but on the other hand definitely contributes to increased competition among technology suppliers. This phenomenon could be treated as a positive result of the introduction of technology control systems by developing nations.

Finally, we should look into the effectiveness of the introduction of control systems at the national level.

Here, the following factors will play an important role: the nature of already-existing strong industry in the country, the political will of the government, and, finally, the extent of the government's understanding and knowledge of the intricacies of technology agreements.

EXPECTED DEVELOPMENTS IN THE REGULATION OF THE TECHNOLOGY TRANSFER PROCESS

As was mentioned earlier, it is not only developing countries that have introduced various regulatory measures in technology transfer, but also highly industrialized nations – the United States, Japan, and the EEC countries.

The example of Japan has had a particularly strong influence on developing countries that desire to effectively regulate the flow of technology.

Observing the developments within developing countries in the last five to eight years, the following three basic conclusions may be reached:

1. There is a definite trend in developing countries to curb the uncontrolled flow of technology.

2. Control policies and measures gradually evolve into complex technological policies, and the focus of government attention moves from, initially, control over technology imports to, ultimately, development of indigenous technological capabilities.

3. It is possible that, gradually, direct control measures for technology transfer by governments will be eliminated or diluted in favor of explicit and complex technological policy, where "regulatory" or "controlling" measures will play a minor or hidden role.

This section of the present chapter will attempt to substantiate these conclusions and to provide certain perspectives on how the goals of developing countries to achieve just and equal terms of technology transfer compatible with their development objectives can be achieved.

In the early 1970s only a few of the developing countries exercised direct governmental control over technology transfer; those were the Andean Group member countries in Latin America, India in Asia, and Romania and, to a degree, Turkey in Europe. In some other countries a certain indirect control has been introduced, usually through central banks, in order to keep track of payments for technology (Philippines, Malaysia, Pakistan, and Egypt).

Innovative approach by the Andean Group and the successful technological policy of Japan have led to a broadening of the concept that technology transfer should be looked at, not only as a commodity transfer, but also as a very important element of control by the supplier companies of recipient licensees, with impacts on balance of trade, employment, R & D activities, and so forth.

Also, at that time various U.N. organizations such as the U.N. Secretariat in New York, UNCTAD, and UNIDO initiated worldwide discussions on the nature of technology transfer, its implications for and effect on the economies of developing countries, the impact of patent laws, and so on, within the context of the New International Economic Order.

These discussions and this political climate have led other countries to follow suit, and regulatory measures have been introduced in the Republic of South Korea (1973), the Philippines (1970 and 1971), Egypt, Algeria, Mexico (all 1973), Argentina (1971), and Brazil.

It should also be mentioned that all these countries have reached a level of industrial development at which the flow of technology not embodied in complete plants or equipment plays a very considerable role in their further technical progress.

Discussions in UNCTAD regarding the Code of Conduct for Technology Transfer as well as at the Second General Conference of UNIDO, which produced the Lima Declaration and Plan of Action with its goal that by the year 2000 the developing countries as a whole should reach 25 percent of world industrial output, intensified the motion toward introduction of government measures controlling technology transfer. These measures were to secure just and equal conditions for technology flow, ensure that imported technology will be compatible with a developing country's overall economic strategy, and contribute to the rapid development of local research and development capabilities.

This trend continues at present and promises to do so in the future, provided appropriate conditions develop in certain countries (see the recent developments in Portugal, Venezuela, or Ecuador).

Another interesting development can be observed – namely, that strict controlling policies introduced by governments gradually evolve into complex technological policies that include not only the regulation of technology transfer but also the regulation of the flow of foreign capital, development and coordination of local R & D, and technological services and information.

Such developments may for example be observed in Brazil, Mexico (the merger of the National Registry for Technology and the Registry for Foreign Investments), Portugal, the Republic of South Korea, the Philippines, and Malaysia, as well as in Venezuela and Peru.

It is perhaps premature to say that all countries will follow this evolution from control systems per se to complex technological policies. But so far no contrary examples can be pointed to, and countries that at present are considering the introduction of control systems are doing so on a much more complex basis (Guatemala, Ghana, Ethiopia, and Iraq).

It seems, therefore, that in the future suppliers of technology will deal with countries with clearcut technological policies, where technological inputs will have to fit into long-term policies and where excessive, one-sided conditions will not be accepted.

Technological policies will also provide for a sufficient information base, which in many instances will provide current information on alternative sources of technology supply that will broaden the options of technology recipients and enable them to make better and more appropriate choices.

It can further be expected that the successful introduction and implementation of such policies will help make industry sufficiently strong in developing countries to operate within the framework set up by such policies, so that control systems in the strict sense will no longer be required.

Of course, this is a rather long process, and may take a decade or two in certain instances. But the example of Japan, which abandoned most of its regulatory and control systems in the past few years, gives rise to hope that a similar trend will follow elsewhere.

At this time we should visualize the following picture of the world trade in technology in the decade to come. One one hand we will have broad antitrust legislation applied to technology transfer agreements in countries that are major supplies of technology, and on the other, individual national regulatory systems (very frequently reflecting the antitrust legislation of supplies of technology) in developing countries.

It is my conviction, after a careful analysis of the national legislation and administrative systems of major importers of technology (see table 10.1) and of the antitrust legislation of major suppliers of technology, that the aims and rules of both sides of the "technology game" are by and large similar.

On top of the very obvious similarities, one might add the conditions currently on the agenda at the UNCTAD discussions on a code of conduct for technology transfer, which, if agreed upon, will provide basic international guidelines for technology transfer to be observed by both suppliers and recipients of technology.

Therefore, taking into consideration the obvious similarities and parallelism of interpretation of restrictive contractual provisions – and those, apart from price, are the most criticized and discussed issues between recipients and suppliers of technology – we may conclude that as the present pattern of technology transfer is developing, suppliers of technology as a group are not meeting with new market situations in developing countries, but rather with an enlargement of markets where certain basic rules of "behavior" have been applied since a certain period of time.

It seems, therefore, that over the long term the technology market will become more "orderly," with clearly defined "basic" conditions under which both suppliers and recipients will operate.

One more issue that often becomes a central issue of a technology control system is the cost of technology.

I have attempted to present some basic figures concerning the totals of technology transfer, as well as concerning the order of magnitude of imports by some developing countries. Those expenditures, when compared with investment needs, seem to be low, but it should not be forgotten that on average \$1 U.S. paid for technology should bring about \$10 U.S. worth of production. In other words, the over \$1 billion U.S. spent on technology by developing countries in 1975 should have

Table 10.1. Restrictive Provisions in Licensing Agreements as Viewed by Legislation in Industrialized and Developing Countries

Type of Restriction	EEC	UK	FRG	Andean Group	Spain	Portugal	Mexico
Territorial restrictions	-	+	+	-	?	$\underline{+}$	-
Exclusivity	-	+	+	$\underline{+}$	+	+	+
Export restrictions	-	$\underline{+}$	+	-	-	-	-
Field-of-use restrictions	?	$\underline{+}$	+	-	-	-	-
Royalties on products not covered by license	-	-	-	-	-	-	-
Royalties from know-how in public demands	-	?	-	-	?	-	-
Grant-backs	-	$\underline{+}$	?	$\underline{+}$	-	$\underline{+}$	-
Volume restrictions	+	$\underline{+}$	+	-	-	-	-
Obligation by license not to use	-	?	-	-	-	-	-
technology after contract expiration	-						
Price fixing	-	-	+	-	-	-	-
Tie-ins	$\underline{+}$	$\underline{+}$	$\underline{+}$	$\underline{+}$	-	$\underline{+}$	$\underline{+}$
Tie-outs	?	$\underline{+}$	$\underline{+}$	$\underline{+}$	-	$\underline{+}$	$\underline{+}$
Restriction on license to acquire		+	-	-	-	$\underline{+}$	-
competitive technology	-	-					

Key:
+ allowed by law
- prohibited
? unclear
$\underline{+}$ allowed in principle – special clarification required

Source: Business International, December 4, 1974.

resulted in over $10 billion in production in 1977 or 1978 at the latest, and that is a very considerable amount.

Most government technology-controlling agencies often scrutinize the price of technology, which in many instances leads to protracted negotiations, the need for additional clarification at the supplier's end, and reevaluation of the input of technology, its market potential, and so on. As a result of government scrutiny, payments are usually reduced (be it royalties or lump-sum payments); however, supplier companies have often felt deprived of their just profits.

In this regard we should take a closer look at the scientific methods that, for example, the Registry of Technology uses in its evaluation of payments, using the share of the licensor in the licensee's profit as a principal measure of payment.(13)

For example, in Mexico (but also in other countries) the licensor's "take" in licensee profits very often was equal to 100 percent or even more! It is clear that in such circumstances the government pressed for a drastic reduction of fees and was very reluctant to approve such agreements.

It is perhaps because of unjust terms and the high cost of technology that new arrangements should be sought to secure even greater flows of technology from industrialized to developing nations.

One such arrangement might be, for example, a special multi-million-dollar fund that will provide financing for technology flows to the developing countries. The fund would on one hand collect information on available technologies and store it in some way, and on the other hand, would collect inquiries about technologies needed.

Once the matching operation is finished, the fund will undertake financing of the transaction, imposing contractual conditions on both licensor/supplier and licensee/recipient.

Provided large financial means are available and sound technologies are offered and sought, such a fund, by the mere scope of its operations, might very positively influence the technology market, particularly in relation to developing countries.

Control systems imposed either by industrialized or developing countries will not by themselves rapidly change the pattern of technology flow in favor of underdeveloped and poor nations. That goal should also be supported by more concrete actions.

NOTES

(1) UNIDO, ID/WG.275/7, May 17, 1978.

(2) All figures include only so-called direct payments for technology and do not cover payments for accompanying equipment, raw materials, components, etc.

(3) UNIDO, ID/WG/275/7, May 17, 1978.

(4) Based on UNCTAD, TD/B/AC.10/Rev. 10; TD/B/AC.II/20; Business Latin America, April 24, 1974.

(5) UNIDO, National Approaches to the Acquisition of Technology, Development and Transfer of Technology, Series #1 (Vienna and New York: U.N., 1977), ID/B/187.

(6) UNIDO, "Recent developments in the regulation of foreign technology in selected developing countries," ID/WG/275/8, May 29, 1978.

(7) See UNIDO, ID/WG/275/8, May 24, 1978.

(8) UNIDO, ID/WG/278/8; and D. Chudnovsky, "Aspectos Economicos de la importacion de Tecnologia en la Argentina en 1972," INTI, 1974, mimeo.

(9) See, for example, UNIDO, ID/187.

(10) See, for example, the third "Draft of EEC Regulations," Les Nouvelles, December 1978.

(11) See Jaime A. Soberanis, "Towards prospects of a future policy for transfer of technology in Mexico", 1977, mimeo.

(12) This issue is discussed at length in Jack Baranson, Technology and the Multi-nationals, where the author expresses his view that licensing to independent licensees – affected by government controlling measures – loosens the licensors' grip over technology.

(13) An explanation of this method can be found in an article by Hubert A. Janiszewski in the December 1978 issue of Les Nouvelles.

Comments

Denis Goulet

Janiszewski's paper is mainly descriptive and historical, and ends with a few classifications and predictions about probable future trends. Several comments cognate to his argument occurred to this reader.

TERMINOLOGY

Because the paper purports to outline technology-importing national perspectives – mainly those of Third World nations – it should have referred, I think, to the main criticisms leveled at the term "transfer." Much of the technology that flows across national boundaries is not effectively transferred, on two counts:

- Control over it is not passed from one set of institutional actors to another, and
- The capacity to take further steps in elaborating technological modifications or improvements is likewise not vested in new hands, in particular indigenous Third World ones.

The term "transfer" also fails to suggest the varied modes and channels by which technology "moves" from one site to another. Although most transactions are commercial, numerous others are made possible by subsidies, direct and indirect. More important, as the World Bank's Charles Weiss has recently observed, "Ways must therefore be found to finance the 'front-end' costs and risks of innovations intended to benefit the poor – costs and risks that in most technical fields are borne by wealthy civilian or military customers whose expenditures are governed by noneconomic motivations."(1)

To offset these omissions and ambiguities, I propose the generic term "circulation of technology," which embraces all types of flows across geographical frontiers.

IMPORT SUBSTITUTION

Janiszewski lists several factors that have led to increases in the quantities of technology purchased by the Third World. I would add to his list the import-substituting strategies made popular fifteen years ago by the Prebisch/ECLA writings and adopted later in many development arenas outside Latin America. The noteworthy point, however, is that dependency was not diminished by import substitution; on the contrary, it increased. And why? Because something akin to a "sequence of dependency" is at work (obviously with much overlap) here. At first Third World nations depend on outside suppliers for capital. Once they can provide their own financing they still rely on them for technology (mainly of the product- and process-embodied kinds). Should they, however, gain some mastery over these, they need still others to bring them managerial or coordinative abilities. Finally, they continue to require competitive access to markets, for raw materials and intermediate goods as well as for outlets for final goods.

Students of the product-cycle theory of business competition will immediately perceive points of similarity. Technologies, like finished products, also have competitive life spans of varying durations. Moreover, technological advantage operates within larger frameworks, each of whose elements is also subject to shifts in comparative advantage. Consequently, technology circulation across national borders needs to be described not in static but in dynamic terms. This Janiszewski's paper fails to do.

THE NEED FOR POSITIVE INCENTIVES

The Janiszewski paper emphasizes the need for regulation of imported technology, yet it makes no mention of the other element of policy: the provision of positive incentives to induce national users of technology to employ what is, or can be, supplied by alternative sources. The experience of many Latin American countries suggests, however, that negative incentives, in the form of restrictions, rarely alter the patterns of technology supply and demand in the absence of parallel positive inducements to make such alterations. The reason is simply that firm managers – even in state-owned enterprises – pursue other goals in their acquisition of technology than those prized by governments. Managers want reliability, punctuality, and marketability in their technologies, as well as good credit terms from sellers. Accordingly, even if national policymakers attach a premium to saving foreign exchange, the same priority is not compelling to firm managers. As a result, they need to be induced, by positive incentives (tax breaks, financing, free information, and so on), to look for alternative suppliers. In the absence of such positive stimuli, the imposition of restrictions by governments will frequently lead to circumvention of controls on the part of regulated firms, by means of a black market in technology, for instance, or duplicate bookkeeping to disguise technological payments.

SELECTIVE LINKAGE AND DELINKAGE

In his discussion of technology transfer agreements Janiszewski warns of the danger of falling into excessive dependency. It would have been helpful, I think, to invoke the strategy concept in which the pursuit of reduced dependency as an objective is now frequently couched. This is selective linkage and delinkage, with the positive and negative impulses emanating from "metropolitan" (center) institutions as they impinge on the "periphery."

One illustration that never did get implemented is a case involving the Mexican government and Nabisco several years ago. The company sought permission to build an additional plant on Mexican soil. As a counter-proposal the government urged Nabisco to utilize its R & D abilities to develop a protein-enriched cracker that it would mass-produce and sell cheaply to poor people. In exchange, the firm would be allowed to raise the price of its luxury items manufactured in other plants – Oreo cookies, Fig Newtons, Shredded Wheat, Ritz crackers, and so on. The offer was refused by the company, not because it would not be lucrative, but because it would shift decisional control regarding important elements of production to actors outside the company. This aborted case is cited solely to illustrate, in concrete terms, what is meant by selective linkage and delinkage; this is precisely what the Mexican government sought to do in this instance.

ACTORS IN THE TECHNOLOGY ARENA

Following convention, Janiszewski has restricted his discussion of actors in technology trade to selling and buying firms and host and home country governments. Two additional actors are vital: consultants and financing institutions (for those transactions where third party subsidies play a role). Were one to consider other "affected" parties as well, one would also have to consider the interests of labor unions, local associations, and the like.

CONTROL AS PART OF A PROCESS

Janiszewski does well to describe controls on the flow of technology as a first step usually leading, by a process of evolution, to what he calls "complex technological policies." His assertion that control and regulation may eventually be relegated to a minor role is congenial to the point I have made above about the need for positive and negative incentives.

UNANSWERED QUESTIONS

The paper adumbrates, but never explicitly raises, three important questions needing clarification. These are:

- What is a fair price for imported technology? How, precisely, is R & D amortized and how can the costs of research failures be equitably distributed over the entire gamut of technology users?
- Are monopolistic financial rewards, in the form of exclusive rights to royalty payments on patents, truly necessary to stimulate invention? What merits are there to the generalized Third World opposition to current international legislation on industrial property? Which important productive activities could be duly satisfied by new modes of circulation of nonproprietary technology?
- What is the comparative utility of the different models of a code of conduct advocated by many both in regulating and in fostering a more developmental utilization of technology?

A GENERAL REMARK

Janiszewski introduces a helpful classification of control systems as either legislative or administrative (or mixed). Yet a certain antiseptic quality marks his treatment of these systems. In my view, the most effective control system for a country is to have a clearly articulated vital nexus, that is, an explicit normative linkage among its preferred values, its development strategies as supportive of those values, and its criteria for problem solving in any given arena (in this case in technology policy), as derived both from development objectives and overarching values.(2) During the Chinese Cultural Revolution, eloquent expression was given to the principle of the "vital nexus" in the following slogan:

> Values command politics,
> Politics command economics,
> Economics commands technique.

It is my contention that to the degree that a nation is clear about the values it wishes to preserve and promote in its development efforts, and to the extent that it is able to translate the commitment to these values into development strategies congenial to them, to that same degree will it have the instruments it needs to derive criteria of technology choices that will enable it to link and delink selectively with those technologies, and technological institutions, that favor its purposes.

Obviously, not all national objectives are equally worthy – whether from the point of view of sound development, human welfare, autonomous cultural maturation, or ecological health. The point, of course, is that technological arenas are not value-free; on the contrary, therein are joined basic value questions lying at the heart of competing visions of development and social organization. For this reason it behooves both analysts of technology circulation and prescribers of solutions to display the kind of "normative diffidence" for which Ronald Dore so eloquently pleads in his recent essay on another value-laden question, basic human needs.(3)

NOTES

(1) Charles Weiss, Jr., "Mobilizing Technology for Developing Countries," Science 203 (March 16, 1979): 1084. Emphasis mine.

(2) A detailed explanation of the "vital nexus" is found in Denis Goulet, The Uncertain Promise: Value Conflicts in Technology Transfer, New York: IDOC/North America, 1977. The "vital nexus" construct allows on to tackle the value questions head on without, however, prescribing normative outcomes for specific societies.

(3) Ronald Dore, "Scholars and Preachers," IDS Bulletin 9, no. 4 (June 1978): 12-15.

Comments

Bernard Mennis

This informative paper by Janiszewski makes several contributions toward increasing our limited understanding of international technology transfer and its regulation by recipient countries, in particular by developing countries.

First, it provides by way of introduction some very interesting descriptive data on the international trading of technology as it has developed over the last several decades. Janiszewski wants to show that there has been and continues to be, especially in the last five years or so, a tremendous growth in technology transfer. For me, this ongoing growth implies increasing international technological interdependence, in spite of the great amount of discussion devoted to the concept and development strategy of national self-reliance. And as the number of suppliers of technology has come to increase, as firms based in Europe and Japan join those based in the United States, transfer patterns linking suppliers and recipients (both developed and developing countries) have emerged that are quite differentiated and complex. This of course is not to say that all the links are of equal importance, but simply to note that the present variety of national suppliers of technology has complicated the picture on the exporting side of the equation and has thereby also influenced the flexibility and bargaining situation (and leverage) of the importing side.

Particular attention is paid by Janiszewski to "North-South" technology transfer. He tells, for example, which "Southern" countries are large purchasers of technology. Unfortunately, his analysis does not move beyond the data to react to some of the implications of his findings. It should have been emphasized that even though North-South trading in technology is rapidly expanding, its shape remains very concentrated, with only a few of the developing countries accounting for a disproportionate share of the purchases. Furthermore, these countries tend to be among the most advanced of the developing countries, the so-called upper tier countries. This suggests, as some have argued, that development encourages demand for technology and technological interdependence. It also suggests the need to consider whether and how this relationship, if generally empirically valid, conditions the regulation of the international trading of technology by recipient governments regarding both its quantitative and qualitative aspects. But these possibilities are not taken up by Janiszewski.

Janiszewski does go on, however, to briefly summarize the history of national regulation, stressing, quite rightly, the leading roles played by India and the Andean countries. Yet again, the follow-up discussion is less than satisfying because it goes no further than presenting essentially a chronology of what has been done. It seems at least as important to ask the why of what happened in regard to regulation strategies. Janiszewski opts for a safer course and sticks close to the facts, but thereby deprives the reader of his interpretation of the record. And this is disappointing because, on the basis of his obvious command of the issues, it is clear that the interpretation, if stated, would be very well informed. What we have, then, is a summary presentation of the history of national regulation of technology, but one that is less provocative than it could or should be.

Following this, Janiszewski distinguishes and assesses two general "applied systems of regulation of technology transfer" – legislative systems and administrative ones. He describes each, indicates through illustration which governments have adopted each approach, and evaluates the appropriateness of each for the national objectives of recipient countries and the corporate objectives of supplier firms (usually transnational business enterrises). There is no need here to summarize the argument, except to note that Janiszewski appears to find the legislative approach to be the more successful one for both recipient governments and supplier firms, in the main because of the greater clarity and consistency of administration achieved regarding regulations covering technology transfer.

Once again I found the discussion useful, but not ambitious enough in pursuing questions that arise directly from the analysis. For one thing, Janiszewski does not explain why certain countries have favored one approach while others favored the alternative. Closely related to this is the need to spell out the parameters that condition the success (or failure, for that matter) of each approach. In other words, under what conditions are these alternative systems of regulation effective, and do they have any connection to the propensities by which countries make a choice between alternatives? Among the conditions that could impact on both the choice between and effectiveness of the legislative and administrative approaches might be the level of development of the recipient country, the growth rate and structural characteristics of its economy, the nature of its polity, the product or product line in which the technology is embodied (if it in fact is), the structure of the industry of the supplier firms, precedents in related industries where similar technologies are present, precedents in the same technology deriving from the experience of other comparable importing countries, and so forth. What I am suggesting is that, given the limited empirical extent and mediocre quality of the state of the art, it is necessary and desirable to encourage speculation that attempts to interpret the seemingly discrete national regulative responses that have been described.

I believe it is especially vital to push ahead because strategies of technology regulation and control often are "transferred" internationally among importing countries as a result of the phenomenon known as the "demonstration effect." But obviously just because a particular approach may appear to work for one country does not mean it will work equally well elsewhere, if it will work at all. To understand whether and how a regulation strategy can be transferred successfully, one has to understand many of the same things involved in exporting domestically successful products. A recipient country considering adopting a strategy employed by another and foreign government must comprehend the conditions that brought that approach to the forefront, the national objectives of that other government or country, the idiosyncratic characteristics shaping the application of the control system, and so forth.

From his survey of what has been accomplished, Janiszewski, however, is prepared to conclude that the instituting of control systems on the importing of technology, whether by the legislative or administrative approach, does not appear to influence the level of foreign direct investment. Yet we cannot very well know, of course, what would have been the case if controls had not been introduced, nor whether different types of control systems might have had different effects on investment decisions, nor – switching now from quantitative to qualitative concerns – do we know how these controls might have influenced the character of foreign direct investment. Janiszewski does not address the latter issue very much, except to say that it appears to be the case that the existence of control regulation does not hinder the transfer of sophisticated, latest-generation technology. Yet it should be clear that this is much too important an issue to be dealt with in passing. Economic orthodoxy suggests that any such government intervention into the market (as regulation of technology transfer obviously is) is bound to yield inefficiencies and suboptimal outcomes that could be manifested in higher costs, diminished technology creation, and the like. Although these prospects would frighten any self-respecting traditional economist, it definitely is not likely to have the same effect on importing-country leaders if they feel (as frequently they do) that such intervention is the necessary and appropriate path for the achievement of national objectives, a pursuit that dominates the so-called economic costs and inefficiencies that may need to be absorbed. In fact, they may even argue that unhindered technology transfer is not at all cost-effective from their viewpoint, and that the benefits that economic orthodoxy speaks about are strictly in the eye of the beholder. My aim is not to take sides in this dispute, but simply to indicate that the assessment of "applied systems of regulation of technology transfer" is far more complicated than is indicated in the paper.

Also more complicated than presented is the negotiation process typically found in regard to technology transfer. Too clear a distinction is made when the protagonists in the negotiaiton are described as supplier transnational enterprises and their home governments on the one side and host governments (and perhaps some firms) on the other. This is very misleading and obscures the degree to which the recipient side is marked by cross-cutting cleavages and loyalties that serve to undermine any clean two-part division. In other words, typically the nature of the alignments in negotiation processes are likely to be quite complex. For example, even if there is present on the technology-receiving side substantial agreement on general development goals, there may at the same time be substantial disagreement on goal priorities and/or the specific public policies and practices that are likely to actualize these goals, and under the most efficient and least turbulent conditions. In point of fact, we know on the basis of extensive experience that such disagreement is very likely to be the case, if only because technology transfer in particular and international trade and investment in general have differential effects regarding classes, ethnic groups, geographical regions, sectors, industries, companies, and so forth within recipient developing countries. The generation of "backward linkages" deriving from the import of technology is sufficient alone to complicate greatly the simple "supplier/recipient" adversary dichotomy. One of the most difficult tasks many national governments in technology-importing developing countries face is dealing with domestic interests aligned with foreign transnational enterprises as a result of "backward linkage" effects, as well as other indigenous groups that for a variety of philosophical or other reasons manifest similar orientations. Consequently, one should be very careful not to imply that a "national perspective" representing strong consensus confronts supplier firms in negotiation processes. The shape of the negotiations is usually far more messy than suggested, and for understandable reasons. And it will be difficult to comprehend negotiation positions or outcomes if this messiness is not taken into account.

The paper ends on an interesting note as Janiszewski attempts to articulate "expected developments in the regulations of [the] technology transfer process." He anticipates that although developing countries will tend more in the future "to curb the uncontrolled flow of technology," negotiations will become more orderly and less rancorous as "controlling policies and measures gradually evolve into complex technological policies." I find the argument he offers to support these expectations persuasive, although at times it takes on an improbably optimistic tone. Janiszewski chooses to avoid discussing events or situations that may upset or retard the developments he anticipates. Instead he prefers to assert that among major suppliers and importers the "aims and rules of both sides of the 'technology game' are by and large similar." Perhaps, but again the actual situation is far more complicated than is represented. It is difficult to generalize, since certainly the recipient countries do not all have the same interest or approach on technology, nor the same ability to finance the acquisition of technology, nor the same development aims or security imperatives that shape demand structures. And, moreover, it is not clear just how the present energy problem will shape priorities and policies on all sides. Thus the general trend Janiszewski speaks about, even if it is the case, may obscure as much as it reveals about what is likely to occur in specific negotiations between supplier firms and recipient countries in regard to the trading of technology.

11 Technology Exports and Technology Transfer Controls

John H. Dunning

In the past two decades a good deal of attention has been focused on the effects of the international transfer of technology (and other resources) on recipient or host countries.(1) It has only been in the last few years that consideration has been given to some of the implications of this transfer for the sending or home countries. Such consideration has been prompted primarily by a growing sense of concern, articulated by politicians, businessmen, and academics alike, that the industrialized developed world – as represented mainly by the OECD countries – is either benefiting insufficiently from the export of technology or technological capacity,(2) or being adversely affected by it. That these concerns are being expressed at a time when host countries, particularly developing countries, are taking action to control the amount and form of technological imports and tilt the terms of trade in their favor has led some observers to be pessimistic about the prospects for the transfer of technology, at least between North and South. In the mid-1950s the exchange of technology was generally thought to be a positive-sum game, with both exporting and importing nations benefiting, but the mid-1970s produced situations in which both parties perceived themselves to be worse off as a result of the transfer. Though prima facie implausible, such a result is theoretically possible, particularly where the parties to the exchange aim to maximize relative rather than absolute economic gains, and/or have different political or cultural perceptions.

Apart from intergovernment grants and loans, the main institution for the transmission of resources in the postwar period has been the transnational corporation, operating through a network of wholly majority-owned foreign affiliates.(3) In such cases, the transfer of technology is internalized, in the sense that there is no change in its ownership and the transferrer continues to exercise de jure direction over the use made of the resources. It is the nature and effects of this control, and the fact that technology is transmitted as part of a package of resources, that distinguishes this vehicle of transference from that of "arm's length" exchange between independent sellers and buyers. It is the identification and evaluation of the consequences that arise from the transfer and dissemination of technology by TNCs that is the subject of this paper.

The presentation will proceed in the following way. First, I will describe the main causes of concern expressed by home countries about technology exports from home countries. Second, I will summarize the extent and pattern of these exports over the last two decades and the role of TNCs in promoting them. Third, I will consider the effects this has had on the economic goals of home countries. Fourth, I will discuss some of the policies pursued by home governments and those which might be pursued toward technology transfer by their own TNCs.

In proceeding in this way, I will adopt an "if . . . then" as well as a "because . . . then" approach, mainly because data limitations allow only partial answers to these questions. Nevertheless, the evidence suggests that, all too frequently, policymakers are persuaded to adopt a particular stance about technology exports by TNCs without an adequate appreciation of the assumptions they are making in the attribution of their effects, or without a proper evaluation of the alternative policy options available.

CONCERNS OF HOME COUNTRIES

The immediate interest of most policymakers in early industrialized countries in the effects of technology exports by their own TNCs stems from a slowing down of international economic growth, a downswing in technological innovation, rising domestic unemployment, a marked slowing down of the increase in industrial productivity, and widespread inflation. All of these are viewed as symptoms, in part at least, of a decline in international competitiveness. Such concern has been most often voiced in Sweden, the United States, and the United Kingdom, industrial countries that have experienced the most dramatic fall in their share of world trade in manufactured goods and relative industrial status, vis-a-vis both other industrialized countries, and the rest of the world.(4)

Partly related to these trends is the increasing effort of developing countries to shift the balance of economic and industrial power from the North to the South. This is illustrated by the philosophy and intent of the New International Economic Order and the Lima declaration of UNIDO in 1975, which aims to increase the share of world industrial production by developing countries to 25 percent by the year 2000. Attempts by such countries to pursue a policy of rapid industrialization are taken as signals by the existing industrial countries that the postwar order of economic interdependence, fashioned at Bretton Woods and Havana, is no longer acceptable, and that some developing countries, particularly so-called newly industrialized countries, are a competitive force to be reckoned with. Taken together with the expansion of alternative sources of technology, for example Eastern Europe, these phenomena have made the established industrial powers fearful not only of an erosion of their industrial hegemony but an undermining of their economic structure.

This might not matter so much within a framework of nations at similar stages of economic development and a common political ideology, such as broadly exists within the OECD. But for the most part this is not so. The political perceptions and economic philosophy of many developing countries anxious to acquire technology are sufficiently different from those of most technology-exporting countries as to make the latter worried lest the added economic leverage that the technology brings might be used to the detriment of their strategic and other aspirations. Here, as has been pointed by Hawkins and Gladwin in Chapter 8 of this volume, a conflict may arise between the humanitarian, economic, and political goals of the home countries. The matter is further complicated by the effects that some kinds of noncommercial technology transfers or sharing may have on the home country's commercial technological capacity. The export of weapons by the United States may help support innovation in the military aircraft and missile industries, with consequential spillover effects on the domestic aerospace and electronics industries. Just as the cutback in the U.S. space research program has reduced the flow of commercially useful technology, so any retrenchment of defense spending, including arms exports to "friendly" nations, could have similar effects.

The Hawkins and Gladwin paper has dealt with some of these issues from the angle of the United States. Having acknowledged that these are some of the most important components of the political economy of technological transfer, in this chapter I will limit my attention to the transfer of technology undertaken through normal business channels.

While no commentator would go so far as to attribute all of the current economic difficulties of the established industrial countries to the emergence of some of the developing countries as industrialized powers, and still less to the transfer of technology by TNCs to these countries, some (most noticeably Baranson, 1979; Gilpin, 1975; and some of the labor unions) assign it a major role, at least from a U.S. perspective. There are two main thrusts to this allegation. The first is that TNCs, by engaging in foreign direct investment, have diverted their energies away from technology-innovating activities in the home country. Thus it is pointed out that between 1968 and 1976 the percentage of money spent on research and development (R & D) in the United States slipped from 2.8 percent of gross national product (GNP) to 2.25 percent, and that whereas in the early 1960s the United States accounted for 75 percent of the world's industrial R & D, by the mid-1970s this proportion had fallen to about 55 percent.

The second and main charge against technological exports is that they improve the international competitiveness of firms in the recipient country at the expense of firms in the sending country. Thus while the United States has led the world in the export of technology in semiconductor electronics over the last twenty years, between 1965 and

1977 the U.S. trade balance for electronics and communications equipment (excluding computers) has continually worsened (Meyer, 1978). This in turn (so the argument goes) erodes the market base of the technology-exporting firms and makes the costs of R & D more difficult to recoup. In the case of labor-intensive industries, the effect is more dramatically and immediately felt on domestic employment. In the context of North-South technology transfers, the debate has had less to do with falling export shares in the advanced industrial countries and more with increasing import competition faced by them. Nonetheless, it is a variant of the "we want to sell more milk and fewer cows" type of argument (Meyer, 1978).

These, then, are some of the main concerns currently felt to a greater or lesser degree by all developed countries. How far are they justified? How far can they be attributed to TNCs? What, if anything, can be done about them?

SOME THEORETICAL POINTS

I take as my starting point the proposition that any transfer of proprietary technology is undertaken by firms in the belief that the transfer will benefit the enterprise as a whole. Where a firm sells or leases technology to other firms, it is presumed that the terms and conditions of the exchange will fully compensate it for the opportunity costs of the supplying of the technology (Graham, 1978; see also Chapter 5 of this volume). In some cases a transfer will only take place subject to the fulfillment of restrictive covenants on the part of the transferee. Thus a license to exploit a patent may only be granted on the understanding that the licensee will only market the product embodying the technology in his own country; or if certain production methods and specifications are adhered to; or if a satisfactory standard of maintenance and after-sales servicing of the product is guaranteed; or if certain tie-in arrangements are accepted. Where a firm is transferring the technology to one of its own affiliates, then subject to the regulations and policies of the host governments, it has full and continuing control over its use. A priori one may presume that, from the transferring firm's viewpoint, in the short and medium run at least the transfer will be beneficial. In the long run the matter is less certain, since it is possible that, as a result of the competitive stimulus from foreign direct investment, indigenous firms may become technologically more progressive and outcompete the foreign affiliate.(5)

Depending on whether its sale is internalized or externalized, there are various gains to the transferrer from the transfer. To the firm selling under contract, these are primarily the revenue it receives from the sale of technology. As long as the marginal revenue exceeds the marginal opportunity costs(6) (including externalities) of supplying the technology, the firm will find the transfer worthwhile. In practice, for many firms selling technology that would have been produced in any case for use by the firm at home, the costs reduce to the marginal negotiating and transaction costs; in other instances, and for technology-supplying specialists in particular, such as building and construction firms, project engineers, systems analysts, petrochemical consultants, and management consultants, the resource costs of producing the technology are relevant (Graham, 1978; Teece, 1977). To the TNC, transferring (or sharing) technology by the direct investment route is the best means of capturing the full economic rent of the technology and any other ownership-specific advantages transferred with it (Helleiner, 1975; Dunning, 1977). Most technology of this kind is product, material, process, managerial, organizational, marketing, and financial knowledge that gives the TNC a competitive edge over its indigenous competitors. The gains from the transfer are those that accrue to the enterprise as a whole; they include not only the profits earned on the capital invested, but all the other benefits that arise from international production, including the securing of new markets that help spread the R & D and other overhead costs of the parent company, thus helping the TNC to maintain and extend its competitive strength.

But TNCs may also transfer technological capacity and, by so doing, advance the ability of the recipient country to produce technology for itself. This may be done either because it is cheaper to set up new R & D facilities abroad, rather than expand facilities in the home country, or because there are some kinds of R & D best done in the host country, or because differences in factor endowments and/or markets enable it to specialize in particular kinds of R & D in different countries. Examples include R & D into tropical diseases by pharmaceutical companies in tropical locations; new methods of cultivating or blending tea and tobacco in tea- and tobacco-producing regions; and research into labor-intensive production technologies for the production of motor vehicles, machine tools, and processed foods in low-labor-cost countries. Again, quite apart from the efforts of host governments to attract such technological capacity, TNCs will tend to locate their R & D activities in countries that they think will best advance their own interests.

Excepting, then, cases of business misjudgment, to suggest that a home country may not benefit from an export of technology by its firms is to suggest that the social opportunity costs of such exports exceed the private opportunity costs. To see that this may be so one only has to look at the different goals held by TNCs and home countries. TNCs are primarily interested in making profits independently of where these profits are earned. They will, consequently engage in foreign production and transfer the necessary technology with this objective in mind. Investing countries, on the other hand, are interested in the location of production of their firms but from a wider perspective; their goals include the growth of gross national product, maintenance of full employment, control of inflation, buildup of indigenous technological capacity, promition of the most efficient structure of resource allocation, and so on. There is no presumption that, in achieving their own objectives, firms will necessarily be advancing those of their home countries.

At the same time, it would be wrong to judge the macroeconomic consequences of a transfer of technology by a TNC by its microeconomic opportunity costs, _or_ to attribute such costs to the TNC. Let us look at a simple example. Suppose that as a result of the transfer of technology by U.S. TNCs U.S. affiliates in South Korea are able to outcompete domestic producers in the supply of black-and-white televisions for the U.S. market, and that jobs are displaced in the U.S. television industry. Suppose also that those made unemployed do not find other work. Then the immediate gains to the United States of the transfer of technology will be the profits, net of foreign tax, earned by the foreign affiliates, and any reduction in the price of television sets passed on to U.S. consumers, while the opportunity cost will be any loss in the gross domestic product caused by the increase in unemployment. To the TNC, on the other hand, the effect of the transfer may be higher sales and profits than otherwise would have been possible.

It is obvious that this conclusion _may_ be entirely false. It may be quite wrong to attribute any fall in domestic employment to foreign direct investment, simply because if the investment had not been made, another nation's firm might have done it, which would not only have resulted in the same fall in domestic jobs and output, but the U.S. economy would have lost the profits that otherwise would have been earned by its own TNC. Moreover, the unemployment caused may only be temporary and, over time, the labor displaced may be employed in more productive activities, either in the same firm or elsewhere in the economy, thus raising rather than lowering the gross domestic product.

The word "may" has been used throughout the previous paragraph because, whatever the theoretical validity of the argument, in practice the effects may be very different. For a host of reasons, labor displaced by a transfer of technology may not be easily reemployed, and the costs of adjustment assistance may more than outweigh the benefits; moreover, the argument that the recipient economy will obtain the technology it needs in any event and outcompete U.S. producers cuts little ice with those who argue that such a possibility can be counteracted by the imposition of import controls.

All this, of course, approaches the question from a microeconomic standpoint. But, obviously, depending on its effects on the host countries, it is not only the costs and benefits of the investing firm that have to be taken into account. Studies carried out in the United Kingdom and United States (see especially Reddaway et al., 1968, and Hufbauer and Adler, 1968) suggest that the main beneficiaries of outward direct investment are often the suppliers of capital equipment and of intermediary products to the foreign affiliates.(7) All or part of these exports would have been lost without the foreign investment, as investment by firms of other nationalities would have led to these goods being bought from their home countries. On the other hand, the extra output abroad may not only replace domestic output by the investing TNC but that of its domestic competitors as well.(8) Again, the effects depend critically on the assumptions made about what would have happened, in both the home and host countries, in the absence of such a transfer of technology and as a consequence of it.

It is worth observing that according to the neoclassical literature – and assuming full employment is maintained in both home and host countries and that there are no market distortions – a transfer or dissemination of technology between countries will increase world real output as, by raising both allocative and technical efficiency, it pushes out the world's production frontier. Moreover, the neoclassical model suggests that both technology-exporting and technology-importing countries will benefit, although some restructuring of resource usage may be necessary. The difficulty of using this model in the current context is that the assumptions underlying it do not necessarily hold good. TNCs do not normally transfer technology in a competitive market situaiton; neither are governments able to ensure that, come what may, full employment is always achieved. Moreover, the welfare functions of countries embrace other goals than the maximization of output, which could be adversely affected by the transfer of technology. However, even if policymakers are forced to deal with second or third best solutions, the principle of the neoclassical model may still be relevant. It may be that the transfer of technology is not as beneficial to the home country as it could be precisely because of market distortions. In such an event, rather than seeking to control the outflow of technology, a policy might be better directed to removing the distortions.

So far, much of this argument has implied that the technology being transferred by TNCs will be used to produce goods competitive with those produced by the transferring firm or country. With investment designed to substitute for exports – so-called import-substituting investment by the recipient country – a lot of technology will be of this kind. But even here experience has shown that where a foreign affiliate is set up to produce one line of goods, its presence may stimulate the import of other lines of goods from the home firm or country. This has proved to be especially common in the case of firms producing consumer goods – televisions, motor vehicles, man-made textiles, processed food products, cosmetics, pharmaceuticals, and so on.

With other kinds of activity by transnational corporations, technology may be used to produce goods complementary or even quite unrelated to those produced by the transferring firm or country.(9) In such cases output and employment in the home country may be increased and its technology base strengthened. Investment by TNCs in trade and distributive activities is of this kind: while it may improve the marketing competence of recipient firms, it may directly increase the exports of goods from the home country. Investment in building and construction, capital equipment, and energy ventures,(10) and by chemical and engineering consultancies may have the same effect. In other words, a disembodied technology transfer may give rise to the transfer of technology embodied in goods. It is, of course, a moot point how far technology <u>is</u> transferred through goods, but for the purposes of our analysis we shall treat such exports as a benefit to the home firm and country.

The transfer of technology through international vertical backward integration by TNCs in resource-based industries may also result in a spinoff in technological capacity in the home country, depending on the extent to which secondary processing activities are undertaken in host or home countries. Even if there is a transfer of expansion of technological capacity, imports of capital equipment are likely to take place, and these are more likely to be sourced from the home than from other countries. In some service industries, for example banking and international tourism, similar spinoff effects are likely.

Finally, there are likely to be multiplier effects for the income generated from the transfer of technology, whatever its kind. In the nineteenth century these yielded substantial benefits to the U.K. economy. Rising incomes in recipient countries, generated by the export of British capital and expertise, provided markets for U.K. manufactured goods, which helped to finance new investment and, through economies of firm size, reduce the prices of goods supplied to home markets. Although there is no economy currently as dominant as the U.K. once was, depending on the domestic value-added of foreign affiliates and the recipient country's marginal propensity to import from the investing country, gains may accrue to the latter. Work done by Hufbauer and Adler (1968) suggests that, for the U.S. economy, these effects are often significant.

It must be admitted that some of the effects on the home country ascribed to the transfer of technology by TNCs would be better ascribed to foreign direct investment in general. My defense of the present approach is that the unique ingredient of most kinds of foreign direct investment is technology, be it the technology of product, materials production, management, organization, or marketing. Without this component, direct investment would become portfolio investment. The main incentive for a firm to internalize the flow of resources across national boundaries is to capture the full economic rent on the package of technological ingredients that are specific to it. It is no coincidence that TNCs tend to dominate in industries that have a high technological content, and that, because of the benefits from internalization, they are the main producers and transmitters of technology.(11)

Nevertheless, given that the focus of this volume is on the effects of transfer of technology per se, it is reasonable to look at the extent to which TNCs do transfer technology across national boundaries, how this affects the host country, and what the short- and medium-term repercussions on the home country are. The previous paragraphs have suggested certain principles that will determine these effects. These include, first, the market conditions in which the transfer of technology takes place; second, the form of technological transfer, and, in particular whether it is likely to complement or substitute that produced at home; third, the strategies of the investing or technological exporting firms; and fourth, the goals of home and host governments and the policies pursued by them to achieve these goals.

THE EXTENT OF TECHNOLOGICAL TRANSFER BY TNCS TO LESS DEVELOPED COUNTRIES

The Role of Transnational Corporations

What, then, has, been the extent and significance of technological transfer by TNCs to developing countries?

The most widely published indicator of the extent of foreign involvement by TNCs is value of outward direct investment. Latest figures published by the United Nations (1978) show that at the end of 1976 the cumulative value of direct investment by TNCs of all nationalities was $287 billion. Between 1967 and 1976 foreign direct investment stock increased by an average of 11.7 percent per annum and, in the latter year, accounted for 6.9 percent of the GNP of member countries of the Development Assistance Committee of the OECD. In the mid 1970s the value of international production exceeded that in international trade. In the case of the United States, the foreign direct capital stake export of goods ratio, in the early 1970s, was 4:1; for the U.K. was just over 2:1 (United Nations, 1973).

In 1976 about 95 percent of the foreign investment stake was owned by TNCs of developed countries, and about three-quarters of this was located in the developed countries (United Nations, 1978). The latter ratio has steadily increased since 1967, when it was 69 percent. Of the total receipts from technology exports by France, West Germany, Italy, Japan, Netherlands, the United Kingdom, and the United States in 1976, one-fifth originated from the developing countries; this ratio has also steadily declined from a quarter in the late 1960s (OECD, 1977).(12) These data suggest that technology exported by OECD TNCs is primarily to other OECD countries and not to developing countries. It is also worth observing that investment flows, particularly in recent years, have become more symmetrical among the OECD countries. Whereas in the late 1950s the United States was by far and away the dominant capital exporter, with all European countries and Japan importing more capital than they exported, the last decade has seen the rate of foreign direct investment in the United States growing much faster than that of foreign investment by the United States.(13) Since inward and outward direct investment flows, like trade, are not entirely unrelated to each other, the critics of outward technology flows would do well to take account of the possible effects on inward direct investment of controlling such flows.(14)

As recipients of foreign direct investment and technology, developing countries may be divided into a number of fairly distinct groups. First there are the OPEC countries, which in 1975 accounted for 23 percent of the foreign capital stake in developing countries, and whose imports of technology were heavily biased toward petroleum-related activities. Second, tax haven countries accounted for a further 13 percent; here there was little, if any, technological transfer. Third are the mainly industrializing countries, of which eleven accounted for 42 percent of the total foreign capital stake in 1975. These included the larger and newly industrialized countries,(15) whose imports of technology were largely directed to the manufacturing sector.(16) Fourth is a mixed bag of developing countries that mainly attract foreign capital and technology to specialized resource sectors and to fairly standardized import-substituting manufacturing activities. By area, Latin America and the Caribbean accounted for about one-half of the total foreign investment in developing countries in 1972 and Africa and Asia for one-fifth each. For manufacturing industry alone, the respective proportions were 67 percent, 12 percent, and 19 percent (United Nations, 1978).(17)

How important are the affiliates of TNCs to the economies of developing countries? Tables 11.1 and 11.2 set out some details of the shares of foreign affiliates in the manufacturing sector of a sample of developing countries. They reveal that their contribution is, in general, considerably more important in Latin America than in Asia; details on African countries are too scant to be reliable, although in some (for example, Nigeria, Kenya, and Ghana) their participation is known to be significant. The data also show that the degree of transnationals' participation to be negatively associated with the host country's contribution to total manufacturing exports and/or increase in manufacturing exports.(18)

Interesting as these data may be, more relevant to our present theme is the effect that the transfer of technology by TNCs may have on the <u>structure</u> of industry in developing countries. In this connection, a report by the UN Center on Transnational Corporations (UN, 1979) reveals that within manufacturing industry, the pattern of involvement by TNCs in developing countries is very similar to that in developed countries. There are differences <u>between</u> countries within the developing world, and particularly among the newly industrialized countries, but, as table 11.3 sets out, in both groups of countries TNCs play their most important role in those sectors characterized either by above-average technological intensity or by the differentiation of their products, for example the chemical, engineering, motor vehicles, and food, drink, and tobacco industries. By contrast, indigenous firms in both groups of countries dominate in labor-intensive industries producing goods that require fairly standardized technology, such as textiles, wearing apparel, and leather products.

The importance of the affiliates of TNCs in high- to medium-technology industries suggests that the transfer of technology by TNCs has helped to build up these industries in these countries. However, there are some notable country differences. In Singapore, for example, foreign corporations account for 47 percent of the assets in the textiles and clothing industry. The figures, though lower, are also significant in other Southeast Asian countries, notably Taiwan and South Korea. Japanese companies are known to have a particular strength in the textiles industry. In 1975 their share of all foreign investment in this industry was more than three times their average share of foreign investments in developing countries; however, part of the explanation for the Japanese involvement is their strength in the man-made fibers sector,(19) which is more appropriately classified as part of the chemical industry.(20)

Table 11.1. Participation Rates of Transnational Corporations in Manufacturing Sectors of Selected Developing Countries

	% of All Firms				Total Manufacturing Exports	
	Sales	Assets	Employment	Exports	1976 $M	1976 - 1965 value
Latin America						
Argentina	31.0 (1972)		11.0 (1972)	At least 30 (1969)	975.5	12.0
Brazil	49.0 (1974)	29.0(1974)		39.9 (1973)	2,192.3	21.0
Chile	20.0 (1964)				118.7(1974)	7.2[a]
Colombia				27.7 (1973)	383.6	11.4
Mexico	28.0 (1970)			35.0 (1973)	2,089.9	14.0
Peru	46.0 (1969)				51.8(1974)	4.2[b]
Asia						
Hong Kong			10.8 (1971)	10.0 (1973)	7,859.3	7.9
India	13.0 (1973)			6.2 (1973)	2,802.7	3.5
Republic of Korea				15.0 (1972)	6,746.8	65.0
Malaysia	50.0 (1971)			28.1 (1971)	798.5	11.8
Pakistan				8.0 (1973)	677.2	3.6
Singapore			30.0 (1968)	75.0 (1972)	2,920.3	9.7
Taiwan				15.5 (1973)	6,921.0	37.1
Thailand			8.9 (1970)		510.6	42.2
Africa						
Nigeria	70.0 (1968)				18.4	2.2
Ghana	50.0 (1974)				8.8(1975)	1.8[c]
Southern Europe and Middle East						
Greece	35.1 (1973)	41.2(1972)	28.2 (1973)	62.3 (1973)	1,252.3	28.7
Israel				28.0 (1976)	1,854.7	6.7
Spain	11.0 (1974)			32.3 (1973)	6,025.2	15.8
Turkey		41.0(1974)		3.0 (1973)	465.7	42.7

a: 1965-74 b: 1966-74 c: 1965-75

Source: Sales and employment from United Nations, 1978 (table III 54, p. 263; N.B. footnote and sources to table); Exports from Donges and Muller-Ohlsen, 1978; for Argentina, from Nayyar, 1978; for Greece from Ph.D. thesis prepared by V. Papandreou; University of Reading; Value of total manufacturing exports by country and increase since 1965 from Keesing, 1978, annex B.

Table 11.2. Developing Countries by Degree of Transnational Penetration in Manufacturing Industry and Export Performance

		Exports	
		A	B
High	Brazil	4.9	21.0
	Ghana	0.0	1.8
	Greece	2.8	28.7
	Malaysia	1.8	11.8
	Nigeria	0.0	2.2
	Peru	0.1	4.2
	Singapore	6.5	9.7
		16.2	13.3
Medium	Argentina	2.2	12.0
	Chile	0.3	7.2
	Colombia	0.8	11.4
	Mexico	4.7	14.0
	Israel	4.2	6.7
	Spain	13.5	15.8
		25.6	12.2
Low	Hong Kong	17.6	7.9
	India	6.3	3.5
	Pakistan	1.5	3.6
	Korea	15.1	65.0
	Taiwan	15.5	37.1
	Thailand	1.1	42.1
	Turkey	1.0	42.7
		58.2	11.3

A: Share of total manufacturing exports from developing countries listed (1976 or nearest date).

B: 1976 (or nearest date) ÷ 1965 manufacturing exports.

Source: See table 11.1.

Table 11.3. Industries in Developed and Developing Countries by Degree of Transnational Penetration of Industries and Export Performance

	Developed Countries	Developing Countries	Exports A	Exports B
	Petroleum processing	Petroleum processing	ni	ni
	Electrical machinery	Rubber	nsa	nsa
	Transportation equipment	Electrical machinery	15.7	14.7
	Instruments	Transportation equipment	2.3	8.5
High	Chemical products (pharmaceutical products and petrochemicals)	Instruments	nsa	nsa
		Chemical (pharmaceutical products and petrochemicals)	7.5	2.4
	Rubber products	Nonelectrical machinery	4.0	7.8
	Nonelectrical machinery	Food (incl. beverage and tobacco)	ni	ni
			29.5	6.2
	Fabricated metal products	Primary metal	3.6	4.6
	Primary metals	Fabricated metal products	4.9	7.2
Medium	Food (incl. beverage and tobacco)	Construction materials (stone, clay, cement, etc.)	3.2	2.2
	Paper and paper products	Printing and publishing	nsa	nsa
	Printing and publishing	Textile	13.6	3.3
	Construction materials (stone, clay, cement, etc.)	Miscellaneous manufactures	11.0	4.3
			36.3	3.6
	Textile	Wearing apparel	21.8	6.4
Low	Wearing apparel	Wood, wood products, and furniture	5.3	6.5
	Wood, wood products, and furniture	Leather and leather products	6.9	5.9
			34.0	6.3

Source: See table 11.1.

Note: The developed countries are Australia, Austria, Canada, the Federal Republic of Germany, France, Japan, Sweden, and the United Kingdom. The developing countries are Argentina, Brazil, Hong Kong, India, Mexico, Singapore, and Turkey.

A: Share of the total manufacturing imports from developing by developed countries in 1974.

B: 1974 - 1967 exports

ni: = not included

nsa: = not separately available

Types of Foreign Investment and the Nature of Technology Transfer

It would be dangerous to read too much into the kind of data just presented. One important deficiency is that the contribution of the foreign affiliates of TNCs is usually expressed in terms of gross sales rather than local value added. By itself this may tell us nothing about the extent and form of the production undertaken, nor about the imported technology content. The pharmaceutical industry provides an excellent illustration. There is very little research and development in pharmaceuticals done in developing countries by TNCs, and in only a few of the larger newly industrial nations are pharmaceutical chemicals produced in any quantity. By far the greater part of the value added created by foreign affiliates of TNCs consists of labor-intensive formula or dosage preparations and the packaging of the final product. It could well be that within the sectors dominated by TNCs, the operations actually carried out in developing countries are labor intensive, while in the developed countries they may be capital or technology intensive.

Much of course will depend on the type of foreign investment undertaken, its position in the "product cycle," and the policies of host governments. Import-substituting activities normally start with simple finishing operations to a product that is mainly made elsewhere, and gradually work backward to other, technologically more intensive manufacturing stages. By contrast, investment in resource-based industries may start with extractive activities but later extend forward, to secondary processing operations. Investments intended to take advantage of cheap and abundant supplies of semiskilled labor may initially be directed to labor-intensive production processes and later embrace more capital- or technology-intensive processes. Nevertheless, though the particular processes undertaken by TNCs may not be technology intensive, the final output may be classified as the product of an industry normally regarded as such. Technology aids the use of all factor endowments, but it does not necessarily make for a more technology-intensive industrial structure.

As far as the transfer of technological capacity is concerned, data on U.S. TNCs show that of the research and development undertaken by them, only 8 percent was undertaken outside the United States in 1970, and only 0.5 percent in developing countries. Other estimates suggest that the developing countries only account for 2 percent of the world research and development, compared with 12 percent of world manufacturing production and 10 percent of manufacturing exports.

The conclusion suggested by these data is that although TNCs may transfer most of their technology within industries classified as high or medium technology intensive, in some cases, at least in developing countries, the technology will be used to improve the efficiency of labor-intensive processes of production rather than to promote the development of technology-intensive sectors. This is most likely to occur in the so-called high-technology industries in which some form of international specialization of products or processes is practiced, and where intrafirm transactions also tend to be concentrated.

The Nature of Competition Between Developed and Developing Countries

Table 11.4, which is derived from a study done for the International Labor Organization (Sabolo and Trajtenberg, 1976), sets out the breakdown of activities of foreign affiliates of TNCs in developing regions in 1970. Among other things, it reveals that resource-based and local-market-oriented investments were far more important than labor-oriented investments in three of the major regions listed. In Asia (which includes India) labor-oriented activities were of equal importance with the other two groups; the evidence suggests that in the 1970s they have become considerably more significant.

Table 11.4. Foreign Direct Investment by Orientation in the Developing Regions (1970)

Region	Type of Investment(%)			All Investments	
	Raw-material-oriented	Local-market-oriented	Labor-oriented	%	Million $ U.S.
Southern Europe	19	62	19	6.3	2.6
Africa	60	34	6	19.4	7.9
Latin America	33	62	5	51.1	20.8
Middle East	91	9	-	8.8	3.6
Asia	30	36	34	14.4	5.8
Total (million $ U.S.)	17.0	19.6	4.0	-	40.7
Overall distribution(%)	42	48	10	100	100

Source: Sabolo and Trajtenberg, 1976.

For the purposes of our analysis, this distinction is an important one. Not only may the raison d'etre for foreign direct investment be very different, but so might its consequences on the home country. Though there is some concern about the effects of technology exports induced by all kinds of foreign direct investment, most recent attention has been directed to the labor-intensive offshore activities of TNCs. It seems to be accepted that foreign investment designed to serve host-country markets is not often a real substitute for home-country exports. Import controls, non-tariff barriers, availability of local materials, transport costs, the need to be near customers, and the behavior of competitors often combine to make the transfer of technology, through direct investment, a <u>necessary</u> condition for supplying the local market. Moreover, since the type of technology required for this may be available from other TNCs, the cost of not supplying the market may be to lose it to a rival.

With investment in labor-intensive processes or products designed to supply markets outside the host country, and particularly those currently supplied by the home country, the attitude is less accommodating. Here it is asserted that, in the absence of this investment, the competition faced by the investing country would not (or need not) have been so great. This is for two reasons: first, the technology required to mount the investment in the host country is unlikely to be as standardized as in the import-substituting case; and second, because the home country can always protect its own industries from foreign competition generated by another countries' TNCs. But what role have TNCs played in these industries?

In 1973, ten developing countries accounted for nearly four-fifths of the export of manufactured goods imported by developed from developing countries. Five of these countries – Hong Kong, Singapore, South Korea, Brazil, and Mexico – are, along with Greece, Portugal, Spain, Turkey, and Yugoslavia among the newly industrial countries, whose share in the total imports of manufactures to OECD countries increased from 2.5 percent in 1963 to more than 8 percent in 1977. These newly industrial nations have also recorded a faster-than-average rate of growth in industrial production; by 1976 they were responsible for about 9 percent of market-economy production, compared with 5 percent in 1963. But, of the fifty top exports of manufactured products, which in 1976 accounted for about 80 percent of the exports of these countries, TNCs were actively involved only in about one-quarter (see table 11.2). These included those most competitive to those produced in OECD countries, namely such things as textiles, clothing, and leather goods.(21) The one sector in which TNCs do tend to dominate is electrical machinery and consumer electronics, the exports of which rose by 46 percent between 1967 and 1974 (Plesch, 1978).(22) Even so, this does not mean that TNCs have had no influence in the extent and direction of exports. Far from it – transnationals' subcontracting and buying groups, particularly those of Japanese origin, have played a decisive role, even though the production is in the hands of indigenous firms.(23) Overall, it has been estimated that TNCs accounted for around 15 percent of manufacturing exports from the less developed countries in 1974 (Nayyar, 1978).

The U.K. Case

Let us illustrate this general point by relating the percentage of total U.K. imports originating from the newly industrial countries(24) to the extent of TNC involvement in particular industries, as set out in table 11.2. In 1977, imports from the newly industrial countries accounted for 10 percent of all imports of manufactured goods, or 3 percent of total supplies to the U.K. market.(25) Of these imports, those involving products from industries with a low penetration by transnational corporations (U.K.-based or otherwise) accounted for 40 percent; those with a medium penetration, for 35 percent; and those with a high penetration, for 25 percent. A crude estimate of the percentage of these imports actually produced by TNCs would be 10 to 12 percent. The percentage of U.K. imports originating from the newly industrial countries varied from 57.5 percent in the case of clothing, 47.5 percent in travel goods, handbags, etc., and 43.1 percent in footwear to 2.2 percent in transport equipment, 6.8 percent in instruments and photographic equipment, and 8.6 percent in electrical machinery. Some 21.0 percent of imported radio sets and 9.1 percent of television sets came from the newly industrial countries, along with 30.3 percent of watches and 4.5 percent of cameras.(26)

Forms of Technological Transfer

The previous paragraphs have shown three main things:

1. The greater part of the direct investment by TNCs in developing countries has been directed towards import substitution and resource-based activities, with the exception of southern Europe and Southeast Asia, where it has been also directed to production of labor-intensive products and/or labor-intensive processes of high- or medium-technology products.

2. Much of this investment has been prompted by a shift in the locational advantages of production from the developed to the developing countries, such that if TNCs of developed countries had (individually) responded by not investing they would have lost all or most of these markets.

3. Most of the growth of manufacturing exports from newer industrial countries has been in countries and/or industries in which TNCs do not generally play a dominant role. Moreover, the revealed comparative advantage of developing countries has improved most significantly in sectors in which the degree of transnational penetration is low.(27) Exceptions include electrical machinery and photographic supplies. In other words, a substantial part of the increasing import competition from these countries has little or nothing to do with technology transfer by TNCs.(28)

We now turn to some of the implications of these facts for the advanced industrial countries.

First, from the viewpoint of international resource allocation, the transfer of technology by TNCs has had mixed effects. Where it has been prompted by import controls imposed by host countries, apart from those justified by the optimum tariff and infant industry arguments, it is likely to have had less beneficial effects than would exports. Where this has been coupled with TNCs acting as oligopolists and undertaking investment to advance product differentiation or to protect market shares, the result has almost certainly been a misuse of resource allocation. Professor Kojima (1978) believes that much of U.S. foreign investment in technology-intensive industries has been of an antitrade kind, and operates to the disadvantage of both host and home countries. This is not a fully persuasive argument; insofar as there is a case in favor of intraindustry trade, there is presumably one for intraindustry investment as well. But the main point is that the Kojima criticism would be better directed to the activities of TNCs within the advanced countries, since a much smaller percentage of such activities in the developing countries is undertaken for defensive oligopolistic reasons. On the other hand, investment by transnational corporations to take advantage of resource or labor cost differentials has probably benefited the allocation of resources, since it has freed the flow of technology that might otherwise not have taken place, or taken place to the same extent. Professor Kojima regards the technology exported by TNCs to promote this kind of activity as trade-promoting, and in accord with the principle of comparative advantage.

From the viewpoint of individual home countries, the crucial questions are: What would have happened had not the technology been exported? What effects can be attributed to it? What is the reaction of investing firms and home countries to these effects?

Let us now briefly look at the three types of activities by TNCs.

Investment to supply the local market. For firms of one nationality to supply foreign markets, they must have some competitive or ownership advantage over local or other foreign firms. (For a discussion of these, see Hirsch, 1976; Dunning, 1977; Lall and Streeten, 1977.) If they choose to supply that market from a local production base rather than by exports, then it is assumed this is because location advantages favor the host rather than the home country. The product cycle theory of investment suggests there is a natural progression from exports to foreign production. Thomas Horst (1974) has estimated that in the early stages of foreign production by U.S. firms, U.S. exports increase along with the production; eventually, however, as production becomes established, exports decline. His calculations, however, make no assumption about what would have happened had the production not taken place. Other research suggests that, in those cases where it is realistic to assume that foreign investment substitutes for domestic (or other foreign) investment in the host country, it may permit exports to be higher than they otherwise would have been (Reddaway, Potter, and Taylor, 1968; Hufbauer and Adler, 1968).

Earlier I described some of the ways that the exports of the home country might be directly increased or reduced as a result of an internalized transfer of technology. There are other less obvious but no less important effects on the home country. For example, an expansion of the market for a firm's products may bring with it economies of increased size, and the spreading of organization and administrative overheads and R & D expenditure. Where there are pressures on the firm's domestic operations, production abroad may release capacity for more productive use at home. There may be technological feedback from the affiliate to the parent company, especially where investment is made in a more advanced industrial country (Reddaway, Potter, and Taylor, 1968). Firms may be better able to take advantage of the geographical diversification of their activities in their purchasing arrangements, fund raising, and management recruitment. By having their assets spread over many countries and denominated in different currencies, they can better cover their risks; they can take advantage of leads and lags, protect themselves against exchange rate changes, engage in transfer price manipulation, and so on. As a consequence of the transfer (or sharing) of the technology of production, the investing enterprise almost always strengthens its technology of information and choice, and in some cases its technological capacity as well. Companies such as Philips of Eindhoven provide an excellent example of how domestic capacity released as a result of investment in developing countries may be used to produce new or upgraded lines and/or be given over to technologically more advanced or complex production processes (Graham, 1978).

Resource-based investment. The purpose of resource-based investment is usually to supply the home firm or country and other countries with resources. The technology required for exploration, extraction, and processing, together with the large amount of capital required, gives TNCs ownership advantages over domestic competitors in a number of industries (United Nations, 1979). Here there is no substitution between foreign and domestic investment. The transfer is of technology to increase or protect the existing supply of minerals or materials, hopefully on improved terms; moreover, it may also strengthen the technological capacity of home companies in engineering, chemical design, and consultancy and maintenance work.

In some cases, technology transfer to promote secondary processing activities may be at the expense of secondary processing in domestic markets. Again, it may be that the firm has little choice but to locate or relocate the processing operations in host countries in response to the policies pursued by the latters' governments. But the basic equation is the relationship between social and private opportunity costs. The amount or kind of technology transferred may be too high because foreign direct investment confers lower net benefits from a social rather than a private viewpoint.

Export-platform investment. I have suggested that it is the transfer of technology associated with export-platform investment that is likely to have the most effect on domestic employment, as it is concentrated in labor-intensive activities. Unlike import-substituting investment, where if a TNC wishes to supply a foreign market it must produce in that market, firms do have options on the location of their export-platform investments.

There are two forms of this kind of investment. One is to produce complete goods and/or services that require substantial inputs of labor. Here the advantages of specialization and division of labor would seem to dictate that TNCs should concentrate their production of labor- or resource-intensive goods in labor-and resource-rich countries, and capital- or technology-intensive goods in countries rich in capital and technology.

The second kind of export-platform investments is in labor-intensive parts of a production process for sale in world markets with the capital- or technology-intensive part of the production process being performed in capital- or technology-rich countries. Again, this conforms to the principle of the international divison of labor, except insofar as the action is prompted by distorted markets or government policies.

Perhaps the reason why this kind of investment is strongly criticized is that the critics believe that without it the adverse effects on domestic employment could more easily be arrested. The fact that if the home country's TNCs did not invest other firms would do so and compete the home firms out of the market cuts no ice, because it is also argued that this can be prevented by import controls. The fact that consumers may have to pay a higher price for their products and that domestic TNCS will earn lower profits is regarded as an acceptable price to pay for minimizing unemployment. But, to repeat an earlier point, it is not the transfer of technology by TNCs per se that is the root of the difficulty, but the general dissemination of technology from both developed and socialist countries to developing countries, coupled with free trade.

REACTIONS OF HOME COUNTRIES

One of the main conclusions of an earlier section of this paper was that in the manufacturing sectors that are most likely to generate adverse effects on the established industrial countries from the transfer of technology to developing countries – and particularly to the newly industrial nations – the role of TNCs was not generally a significant one. Such advantages as they did possess were mainly in the marketing of goods in the industrialized countries; hence the presence of multinational buying groups, trading companies, and joint ventures with local firms. The major exception seemed to be where the technology was ownership-specific, and where it could only be advantageously exploited within the same firm, rather than (for example) by licensing or subcontracting. In such cases the TNCs remained the chief vehicle for transfer of technology and the main cause of concern to the home economies.

Yet even in the industries that are the most highly penetrated by TNCs, their role is a decreasing one in the more rapidly developig countries in Southeast Asia. Moreover, unlike the dominance of U.S. transnationals in Latin America and U.K. transnationals in many of the Commonwealth developing countries, the geographical parentage of TNCs in Southeast Asia is more widely diffused. This means that not only is the impact of any one home country's TNCs likely to be less significant, but the opportunity cost to any one country not investing in these areas is correspondingly greater.

But it is not only through equity investments that TNCs may transfer technology. Following the example of Japan, many newly industrial Southeast Asian countries are using a variety of nonequity routes to attract foreign technology. These include not only straightforward licensing agreements and management contracts, but a whole variety of other forms, involving various institutions, for example local firms, host governments, international agencies, and consortia of TNCs (United Nations, 1979; Baranson, 1979). As a consequence, not only are an increasing number of manufacturing TNCs setting up international project, consultancy, and advisory divisions to deal with external technology sales, especially to the newer-industrial, Middle East, and Eastern European countries, but consultancy firms in project design and evaluation, industrial and chemical engineering, and managerial systems are mushrooming. Indeed, it seems likely that the growth of some new industrial countries, notably South Korea and Taiwan, will rest more on their ability to induce externalized technology flows and technological capacity than to attract foreign direct investment. To the established industrial countries the relative benefits of consultancy contracts are less clearcut. A firm like Dunlop-Pirelli argues that had its projects division not sold know-how kits on tire factory design, construction, and operation to various less developed countries in Eastern Europe and elsewhere, other tire companies would have done so. But it admits that in the long run the advanced industrial countries may not benefit as much as they would had their TNCs been directly involved.(29)

In the case of the transfer of petrochemical know-how, the possible clash of interests is more serious. Contractors like Foster Wheeler and Lummus are primarily concerned with selling knowledge of how to construct and operate a chemical plant and market its end products. Their interests are served by an increasing output of petrochemicals, and they will do what they can to help potential buyers obtain the necessary financing from sources like the World Bank. The petrochemical producers, on the other hand, will only invest in developing countries if they believe they can make a reasonable profit from selling the output they produce, and normally they will only wish to transfer knowledge of plant construction and operation if they believe that their own ability to sell petrochemicals in third markets is not adversely affected. Many petrochemical producers consider that there is currently a surplus of petrochemical capacity in developing countries, and that this, aided and abetted by the transfer of technology and capital from developed countries, is forcing down the price of petrochemical products to uncompetitive levels. The situation is further aggravated wherever host governments subsidize production and/or dump petrochemicals on the world market.

The problem just described is not confined to the petrochemicals industry; it has shown itself in the mining industry, noticeably copper (Dunning, 1979) and several other branches of manufacturing industry. It results in a genuine conflict of interests between the suppliers and the existing producers of end products using that technology, and is often exacerbated by the policies of host governments wishing to develop their own industries and break into the international market. It is the export of technology through specialized consultants that home governments need to be more perturbed about than the transfer of technology through their manufacturing TNCs, whose interests, at least in this instance, coincide with those of their governments. On the other hand, this is not something that an individual advanced industrial country can attempt to curb without incurring even greater loss to itself. Some kind of international stabilization agreement is needed to prevent excessive technology flows from causing instabilities in the output and prices of essential commodities.

I have said that, at a micro level, any adverse effects that a transfer of technology may have on the investing firm's capacity may provoke various responses, ranging from a cutback in domestic capacity to a restructuring of production into new product and process lines. Where the latter is successfully achieved, the firm has borne its own adjustment costs. Where it is not able to do this, then unemployment results. I have argued that this is most likely to be the case in declining labor-intensive industries, although, even in expanding industries, restructuring is likely to take place toward more technology- or capital-intensive products or processes. Here it is worth observing that most of the industries in which TNCs have been active in the newly industrial countries – particularly of the export-platform kind – are those in which technological change has been the most rapid and the effects on production processes, products, and labor requirements have been the most pronounced.(30) This has been the most obvious in the computer and consumer electronics sectors. Take a company like Philips of Eindhoven. Employment in the complex of plants located at Eindhoven hardly changed between 1965 and 1978, yet real output doubled. Over the same period Philips's employment in the developing countries increased by 125 percent. The company claims that technical development in the field of audio and video equipment, and particularly in the technology of microelectronics and computerization, has caused a major restructuring of its worldwide operations from which both the company and the Netherlands have benefited. Be that as it may, it is certainly true that today's leaders in the consumer electronics industries are also those that have been active in foreign activities. The failure of U.K. companies to engage in such investment may well have weakened rather than strengthened the position of the U.K. electronics industry in world markets.

The most obvious problems of technology exports occur in industries whose domestic markets are static or declining, where there is little product or process innovation, where output is specialized, and where production methods are labor intensive. The textile and clothing industry is the one most typically cited. In Western Europe it is estimated that 400,000 people, one-sixth of the labor force, have lost their jobs because of import competition from Asian and North African factories, where wages range from one-half to one-twentieth of those in West Germany or Belgium.(31) Between 1970 and 1976 exports from the newly industrial countries rose sixfold, and it has only been as a result of the multifiber agreement that imports since 1977 have slowed. At the same time the antiprotectionist lobby argues that the developed countries have more to lose from protectionism than the developing countries do, because the rate of growth of exports from the developed to the developing countries is greater than that from the developing to the developed countries.

Very often the problem of these declining industries – like those of the basic industries in the United Kingdom after the First World War – is highlighted by the fact that they tend to be located in the less prosperous regions of the home economies and use the type of labor that cannot easily find alternative employment or be retrained. The new jobs being created require different skills than those they replace. In such cases structural unemployment persists. The EEC Commission estimates that more than 1.6 million jobs in Europe are at risk as a result of import competition from developing countries. In the United Kingdom it is reckoned by the Foreign and Commonwealth Office that, between 1970 and 1975, some 47,500 jobs were lost as a result of import competition from developing countries, three-quarters of which are in the textiles and clothing industry, and that over the next ten years the gross displacement of labor could be four times that amount. Though even this latter figure is only around 10 percent of the total amount of U.K. unemployment in 1978, it is of concern because it is likely to affect particular regions and types of workers very severely indeed.

That this is an age-old problem, and is the inevitable result of the emergence of newly industrializing economies, is of little comfort to those adversely affected. Neither is the fact that TNCs are not often actively involved in the industries in question. But the fear is that the import competition from a comparatively few industries today will spread to many others tomorrow. As the newer industrial countries climb up the industrial ladder, will they not attract more markets away from the developed countries? And if they do, will not the TNCs become increasingly involved? In this respect, the participation of TNCs in audio and video equipment and in synthetic fibers and auto components in some countries in Asia and Latin America is cited as a foretaste of the pattern of the future. But here the experience of Japan is salutary. Japan increased its share of world trade in manufactured goods from 6.9 percent in 1960 to 16.0 percent in 1978 with a minimum contribution from foreign TNCs. Yet to achieve this resurgence of industrial strength Japan imported a very substantial amount of technology while building up indigenous ability to absorb, alter, and adapt that technology. For example, between 1950 and 1970 the Japanese government approved 8,324 contracts made by Japanese concerns involving purchases of technology from Western enterprises (Ozawa, 1972).

It is a moot point whether Europe and the United States would have been better off without the Japanese economic miracle. The point at issue is that it would have been difficult, if not impossible, for these countries to have used normal commercial channels to arrest the miracle. At the same time, it is at least arguable that the United States and Europe would have been better off had they been allowed a stake in this miracle. In contrast to Japan are Brazil and Mexico, whose rates of growth of industrial production since the late 1960s have outpaced that of Japan, but in whose prosperity foreign-based, and particularly U.S., TNCs have fully participated. The dictum "if you can't beat them, join them" seems very apt in this case.

The World Bank estimates that exports of manufactured goods from developing countries will rise by about 12 percent per annum between 1975 and 1985 (Keesing, 1978a). Assuming that this prediction is accurate, established industrial countries will be affected differently. Low-wage-cost developed countries might be expected to be penetrated less than high-wage countries, for example, the United States and Sweden; on the other hand, low wages are often a symptom of low productivity and lack of industrial competitiveness, in which case the weaker developed countries might suffer more than the stronger. The Foreign and Commonwealth Office study estimates that between 1978 and 1988 the increase of U.K. imports from the newer industrial countries could be on the order of 8 percent of the present output of manufactured goods. It suggested that the gross labor displacement might be more than this proportion of the manufacturing labor force because the industries likely to be most affected are those of higher-than-average labor intensity, and proposed that firms react to this additional competition by adopting more capital- or technology-intensive methods of production. To some extent this labor displacement will be offset by increased employment in the industries serving the newer industrial nations, but as the study argues, it is the gross labor displacement that is the relevant indication of the adjustment task.

Different industries will be differently affected, but the study asserts that in spite of the inroads into the United Kingdom's (revealed) comparative advantage by the changing comparative advantage of the developing countries (Donges and Riedel, 1977), there are many sectors in which the United Kingdom's competitive position should (or perhaps one should say could) improve.

CONCLUSION: SOME POLICY IMPLICATIONS

It should be clear from the foregoing that the impact of the transfer of technology carried out by TNCs on the economies of home countries is but a tip of the iceberg of the adjustment problems now facing and likely to face the established industrial countries in the foreseeable future. The fact is that although TNCs continue to account for the bulk of the commercial research and development carried out in the Western world, only a small amount of this is directly or immediatley relevant to the present industrialization program of the developing countries. Most of the product, process, materials, production, marketing, managerial marketing, and organizational know-how currently required for much of this industrialization is not proprietary to individual firms and is generally available in the marketplace.

However, in the next decade the picture may change quite dramatically, as the newly industrial countries become more sophisticated in their product structure and production techniques. The first question advanced industrial nations have to ask themselves – provided they are allowed to – is, How far do they wish to be involved in this industrialization program? This chapter has no more than hinted that the net social benefits of being involved, through technology transfer by their own TNCs, may be higher than those any of alternative action they may pursue.

In response to increasing and more pervasive competition from the newer industrial countries there are basically three policies open to the governments of the advanced industrial nations. First, they can resort to some kind of protectionism. Second, they can enter into agreement with their newly industrialized competitors to control the latter's increases in manufacturing exports or to import from the advanced industrial countries as much as they export to them (the assurance of fair trading comes into this category). And third, they can seek to identify their likely revealed comparative advantage in resource allocation in the future and introduce policies to encourage the necessary restructuring. In the long run, assuming it is successful, the last is obviously the first, best solution; among other things, it would imply a lot more attention by governments to fostering research and development and innovative activities, more than any of the OECD countries appear to have aid up to now. (For other policy prescriptions see Hogan, 1978.) But in the interim the problem is how to minimize the adjustment costs of market disruptions to maximize the smoothness of the industrial restructuring, and where appropriate, to encourage firms affected by competition from newly industrial countries to be more efficient. To solve these problems, which because of market rigidities may well mean controlling the speed of the restructuring process, some form of import control should not be ruled out.(32)

It is, however, difficult to understand how controlling the activities of manufacturing transnationals in these newly industrial countries can meet these goals, any more than the controls exerted by the United Kingdom on the export of technology in the nineteenth century realized its objective (Rosenberg, 1978).(33) If attention is needed in this direction it should be to removing any distorting influences that encourage TNCs to transfer more or less or a different type of technology and as a result may interfere with their returning the maximum benefits to the home country. Again, action may involve changes in international patent, monetary, or tax system measures, some of which at least would operate against the interests of the technology-receiving countries.

The area where home governments need to exert most attention, apart from those mentioned in Chapter 8, is where their firms, acting in a mainly consultative capacity, are transferring technology that cannot be obtained elsewhere. Here the social opportunity costs of transfer may be considerably greater than the private opportunity costs, and some way of redressing this difference may be necessary.

I have argued that advanced industrial countries are likely to be affected differently by the industrialization of the developing world. I have also asserted that one reason why a particular firm or country may be reluctant to control technology exports is that its competitors may not do so. This suggests that the main area of international competition will continue to be between countries of the developed world. Only if they adopted a common policy toward the actions of their TNCs would it be possible to control the outflow of technology, which is specific to TNCs, without any developed country feeling it was worse off as a result. While such a policy, at least on economic grounds, seems very unlikely, the efforts of the developing countries, including the newly industrial ones, to strengthen their bargaining power vis-a-vis TNCs from the developed world may eventually provoke a countervailing reaction to redress the balance which the advanced industrial countries believe to be against their own interest (Bergsten, 1974). A less controversial route to follow might be the search for harmonized policies between North and South to smooth the adjustment process, which might be of benefit to both older and newer industrialized countries.

One final observation. This chapter began with the expression of concern by developed countries about the possible adverse effects of exporting technology to developing countries through their own TNCs. It ends by asking the question, Can developed countries not afford to export technology to developing countries through their own TNCS? For if the activities of TNCs are looked at not as a threat to domestic investment, jobs, and technological capacity, but as a way of (1) exploiting international markets, (2) ensuring a stake in the prosperity of developing countries, (3) financing technological capacity, and (4) protecting or advancing the international competitive position of one advanced industrial country relative to another, the question of whether North-South technology transfer is a "good" or a "bad" thing takes on completely new meaning. My argument is not that this latter proposition is in any way proven, but that it deserves at least as much attention as the alternative thesis, which is currently more in vogue and being more vigorously researched.

NOTES

(1) For excellent summaries of the main issues involved see Helleiner (1975) and Moxon (1979).

(2) Technology is defined as a flow of all kinds of commercial applicable knowledge: technological capacity is defined as a stock of assets, both physical and human, capable of generating technology flows.

(3) For example, in 1975 about 80 percent of all receipts of royalties and fees by the United States from the sale of technology originated from the foreign affiliates of U.S. transnational corporations (United Nations, 1978).

(4) These countries' share of world manufacturing exports fell from 36.5 percent in 1967 to 28.8 percent in 1977. Of the other industrialized nations, Japan has increased its share from 9.8 percent to 14.5 percent while France and West Germany have maintained theirs at around 9 percent and 20 percent respectively.

(5) As has happened in the U.K. pharmaceutical industry, where the share of output of indigenous firms has risen in the 1970s.

(6) That is to say, a firm must expect to earn at least the opportunity cost of the capital invested in technology-creating activities.

(7) A recent study by Kawaguchi (1979) suggests that Japanese joint ventures and subsidiaries in Southeast Asia buy between 30 and 100 percent of their inputs (in value terms) from Japan (see especially table V.3, p. 31).

(8) For example, IBM has been one of the main losers of the arrangement made by the U.S. Amdahl Corporation to supply the Japanese computer firm Fujitsu with advanced computer technology; see Baranson, 1976.

(9) In the macro sense it is difficult to conceive of unrelated goods, as all technology may, indirectly, affect the international competitive position of the recipient country.

(10) Such as the French-Mexican nuclear energy project.

(11) In 1970 it was estimated that U.S. TNCs accounted for 55 percent of all the R & D undertaken in U.S. manufacturing industry.

(12) In a symposium organized by Business International on the "Transfer of Technology" in May 1978, one speaker estimated that as little as 3 percent of the value of technology transfers took place between the developed and developing countries.

(13) In 1976 the outflow of direct investment from the United States was 2.5 times greater than in 1967-69; the inflow of direct investment was 5 times greater. The respective figures for West Germany were 4.6 and 2.4, for France 4.3 and 3.0, for the United Kingdom 3.3 and 2.2, and for Japan 10.3 and 1.9.

(14) In 1972 the United States accounted for 60 percent of the combined positive balance of technological payments of the OECD countries listed earlier; by 1976 this proportion had fallen to 45 percent (OECD, 1977).

(15) The coverage of this category varies according to source. The OECD and UNCTAD list Greece, Portugal, Spain and Turkey, Yugo-

slavia, Hong Kong, Singapore, South Korea, Taiwan, Brazil, and Mexico. However, the U.K. Foreign and Commonwealth Office in a recent report also includes India, Malaysia, Pakistan, Philippines, Thailand, Argentina, Israel, Malta, Pland, Romania, and Hungary. The eleven mentioned in the text are Brazil, Mexico, India, Malaysia, Argentina, Singapore, Hong Kong, South Korea, and the Philippines.

(16) Three-quarters of the inward capital stake of these eleven countries was in manufacturing industry.

(17) Since this date the percentage of manufacturing investment directed to Asia has increased significantly.

(18) In 1972 the ratio between the stock of investment by TNCs in the manufacturing sector and manufacturing exports of Latin American and Caribbean countries was 2:35; the corresponding figure for African developing countries was 2:80; and for Asian and Oceanian developing countries, 0:28.

(19) Since this date the percentage of manufacturing investment directed to Asia has increased significantly.

(20) It is noteworthy that the rate of increase in exports from developing countries has been greatest in sectors where transnational corporate involvement is small. Had the exports of food, drink, and tobacco been included in Group One, the increase in exports for that group would have fallen from 6.2 to 4.2.

(21) For example, Keesing (1978b) estimated that transnational manufacturing affiliates only account for about 5 percent of the exports of textiles and clothing from developing countries; and the percentage is even lower for leather goods. Again, the proportions vary between TNCs of different nationalities, with the Japanese being the most actively involved in these sectors.

(22) Moreover, Plesch estimates that in this sector between 72 and 100 percent of the exports are intrafirm. As far as individual newly industrial countries are concerned, the role of TNCs varies considerably. In Hong Kong foreign investment provided work for 68 percent of the workforce engaged in electrical engineering in 1972; in Singapore the corresponding figure in 1968 was 67 percent. In South Korea in the early 1970s foreign firms were supplying 25 percent of the output and 50 percent of electronic exports. On the other hand, in Yugoslavia, which among the newly industrial countries is the leading exporter of domestic electrical appliances and electric power machinery, the role of TNCs is very limited indeed.

(23) One estimate by the Netherlands Research Institute is that 70 percent of all Hong Kong's exports of clothing passes through Western retail traders and import merchants.

(24) As defined in the U.K. Foreign and Commonwealth Office (1979) study; see footnote 3, p. 18.

(25) This is a lower proportion than in other OECD countries. For example, using the OECD definition of newly industrial countries, they accounted for 7.0 percent of U.K. manufactured imports, 20 percent of U.K. imports, 8.9 percent of German imports, and 8.1 percent of OECD imports as a whole.

(26) It is worth observing that an earlier newly industrial country – Japan – is now a more important supplier of sophisticated products, e.g., cars, motorcycles, and consumer electronic goods.

(27) Donges and Riedel, 1977. Of the thirty-six product groups in which the revealed comparative advantage of developing countries in 1972-73 was greater than 1, in only six was the degree of transnational penetration high. Of the fifty-five groups in which the revealed comparative advantage of developing countries was less than 1, the corresponding penetration figure was 16.

(28) Excepting the technology of marketing or market areas, as made available by transnational buying and subcontracting groups.

(29) Indeed, one of the costs is now being seen in the form of cheap rubber tire imports from Eastern Europe economies.

(30) For example, it has been estimated that over the period 1962-75 fifty times more jobs were lost through growth of labor productivity than through growth of imports from developing countries (Keesing, 1978b).

(31) The corresponding figure for the United States is 225,000.

(32) Johnson (1974) makes the point that both adjustment assistance and adjustment safeguards should be considered in the general context of economic change, rather than in the context of change mediated through changes due to international technology transfer.

(33) The sluggishness of the United Kingdom to adapt to the structural changes required of its industrial economy in the late nineteenth century is an object lesson for established industrial countries faced with competition from newly industrial rivals.

REFERENCES

Baranson, J. 1976. A new generation of technology exports. Foreign Policy 25 (Winter 1976-77).

Baranson, J. 1979. Technology transfer to developing countries. New York: Praeger.

Bergsten, C.F. 1974. Coming investment wars? Foreign Affairs 53 (October).

Donges, J.B., and Riedel, J. 1977. The expansion of manufactured exports in developing countries: An empirical assessment of supply and demand issues. Weltwirtschaftliches Archiv 113.

Donges, J.B., and L. Mueller-Ohlsen. 1978. Aussenwirtschaftsstrategien und Industrialisierung in Entwicklungslaendern. Tuebingen: J.C.B. Mohr (Kieler Studien no. 157).

Dunning, J.H. 1977. Trade, location of economic activity and the MNE: A search for an eclectic approach. In The international allocation of economic activity, ed. B. Ohlin, P.O. Hesselborn, and P.M. Wijkmon. London: Macmillan.

Dunning, J.H. 1979. Multinational mining companies and governments; A new detente? Multinational Business, no. 1.

Gilpin, R. 1975. U.S. Power and the Multinational Corporation. London: Macmillan.

Graham, E.M. 1978. The terms of transfer on technology to the developing nations: A survey of the major issues." Unpublished paper prepared for the OECD.

Grossman, G.M. 1978. "Import competition from developed and developing countries." Unpublished paper.

Helleiner, G.K. 1975. The role of multinational corporations in the less developed countries. Trade in Technology World Development 3, no. 4 (April).

Hirsch, S. 1976. An international trade and investment theory of the firm. Oxford Economic Papers, July 28.

Hogan, W.G. 1978. Questions on structural adjustment policies. Working papers, Department of Economics, University of Sydney, September.

Horst, T. 1974. American exports and foreign direct investment. Harvard Institute of Economic Research, Discussion paper no. 362, May.

Hufbauer, G.C., and Adler, M. 1968. U.S. manufacturing investment and the balance of payments. U.S. Treasury Department, Tax policy research study No. 1, Washington.

Johnson, H.G. 1974. Technological change and comparative advantage: An advanced country's viewpoint. Journal of World Trade Law.

Kawaguchi, N.B. 1979. The role of Japanese firms in the manufactured exports of developing countries. Unpublished paper, World Bank, November.

Keesing, D.B. 1978a. World trade and output of manufactures: Structural trends and developing countries exports. Washington: World Bank.

Keesing, D.B. 1978b. Developing countries' exports of textiles and clothing: Perspectives and policy changes. Unpublished paper, World Bank, May.

Kojima, K. 1978. Direct foreign investment. London: Croom Helm.

Lall, S., and Streeten, P. 1977. Foreign investment transnationals and developing countries. London: Macmillan.

Meyer, H.E. 1978. Those worrisome technology exports. Fortune, May 22.

Moxon, R.W. 1979. The costs, conditions and adaptation of MNC technology in developing countries. In Economic policy and the multinational enterprise, ed. R. Hawkins. New York: JAU Press.

National Science Foundation. 1974. The effects of international technology transfers on the U.S. economy. Washington: National Science Foundation.

Nayyar, D. 1978. Transnational corporations and manufactured exports from poor countries. Economic Journal 88 (March): 349.

OECD. 1977. Data concerning the balance of technological payments in certain OECD member countries: Statistical data and methodological analysis. Unpublished document, DSTI/SPR 77.2 Paris: OECD, November.

Ozawa, T. 1972. Should the United States restrict the technology trade? MSU Business Topics 20 (Autumn).

Plesch, P.A. 1978. Developing countries' exports of electronics and electrical engineering products. Unpublished paper, World Bank, February.

Reddaway, W.B.; Potter, S.T.; and Taylor, C.T. 1968. The Effects of U.K. Direct Investment Overseas. Cambridge: Cambridge University Press.

Rosenberg, N. 1978. The international transfer of industrial technology. Paper presented for OECD, September.

Sabolo, Y., and Trajtenberg, R. 1976. The impact of transnational enterprises on employment in the developing countries. Geneva: International Labor Organization, January.

Teece, D.J. 1977. Technology transfer by multinational firms: The resource cost of transferring technological know-how. Economic Journal 87 (December).

U.K. Foreign and Commonwealth Office. 1979. The Newly Industrialized Countries and the Adjustment Problem. Government Economic Service Working Paper No. 18, January.

United Nations. 1973. Multinational corporations in world development. Department of Economic and Social Affairs, E73 II A11.

United Nations. 1978. Transnational corporations in world development: A Re-examination. Department of Economic and Social Affairs, E78 II A5.

United Nations. 1979. Transnational corporations and the long term objectives of the developing countries. Unpublished document, Center on Transnational Corporations.

Wilkins, M. 1974. The Role of Private Business in the International Diffusion of Technology. Journal of Economic History 34 (March).

Comments

Eugene Eschbach

Is controlling transnational corporations a realistic mechanism for controlling technology transfer? That is, are TNCs so important and so unique to technology transfer that controlling their actions will indeed control technology transfer? I think not, even though TNCs can, and do, provide complete packaging of technology that helps projects go ahead once they are authorized by all concerned. And as I interpret Professor Dunning's remarks, neither does he believe that the TNCs enjoy a unique position of leverage.

I suspect that perhaps the action of TNCs is in response to technology transfer opportunities, and is even just a symptom, and at the most a valuable agent, especially in instances where outside capital is needed. I suspect that while a measure of technology transfer is indeed brought about through TNCs, in the absence of these entitites, technology transfer would take place eventually, and possibly the transfer would be of greater substance, in the eyes of some observers. In particular, real (substantive) transfer occurs only when enclaves in the host company are "ready," are not until then. And unless we measure "readiness" in our studies of technology transfer, such as the action of TNCs, it may not be properly recognized how true it is that they can only be a part of real transfer. However, action by TNCs may provide a modicum of technology transfer sooner, even though it may be imperfect by some measures. And if it is sooner, long-run concerns may be better served, especially in regard to establishing and holding market positions.

Professor Dunning suggests the attempt at some control by way of protectionism or comfortable accommodations through fair trade. A system based upon relative comparative advantages is acknowledged to serve an ultimate role. He also speculates that to regulate technology transfer, advanced industrial countries' governments might control firms principally involved in a consulting capacity. Short of complete isolation among nations, I do not believe technology transfer will be controlled in the long run, for reasons having to do with the very nature of technology and the resourcefulness of man. However, short of control, technology transfer could be "adjusted" by mechanisms being discussed in forums such as this one.

As we proceed, I believe we should keep in mind a distinction between the transfer of general technological substance and the transfer of specific structure. Transfer of specific structure, I suspect, is far easier, and is often the major transfer when a well-developed and well-demonstrated chemical or physical manufacturing operation is put in place as a routine operation without the necessity of establishing the fundamentals of the process. That this situation is often the case is, I believe, captured in descriptions of the "product cycle."

It would be helpful if we could measure the degree of excellence (and thereby the general substance) with which technology transfer is achieved, rather than having to work with macromeasurements, particularly financial flows, as important as they are. In most cases, at the outset at least, only the apparent structure of technology is transferred, and the substance of technology remains to be transferred. However, this may be too harsh and too preliminary an assessment, because later, unless technical obsolescence occurs, once a structure is in place the flesh can be filled in. Even if technical obsolescence does occur in the product or the process, the "flesh" may be filled in with yet another product or process. For example, sophisticated light bulb production followed the demise of vacuum tube production in some developing countries.

The study of technology transfer is in part confounded because many technologies are changing very rapidly. However, technology involved in a transfer process is most likely to be static, in keeping with the production cycle considerations. Thus the perception may be that the transfer process in inadequate, when in fact technology was transferred but only to become out of date. The speed with which technology changes is not appreciated by many, particularly those in developing countries and their well-meaning supporters elsewhere. Accommodating change is, of course, a mark of technological sophistication, and to achieve it, the substance of technology must be in place. Thus one measure that I do not see being discussed in these deliberations and that I am concerned about is whether or not the wand of substance was successfully passed or not.

Also, I would like to further refine some technical points involving the apparent efficiency of technology transfer. In the instance of some developing countries, a criticism is often leveled that the mass production process is for them inappropriate technology. While this may be a legitimate observation for automated mechanical processes (semi-automatic or fully manual processes may be better in some countries), it is often not true for many chemical processes. By the nature of the reversibility of some chemical reactions and the exchanges of heat energy involved, the reactions must be executed as continuous rather than batch processes. Continuous chemical processes require considerable instrumentation and automatic adjustment, and thus little operating manpower. Incidentally, it is far more difficult to transfer the technical substance of continuous chemical processes than that of batch processes, and certainly than that of the processes used to manufacture mechanical devices.

However, before continuing I must ask myself, In what sense should technology transfer be controlled? Should it be controlled to slow it down or should it be controlled to speed it up without bringing on adverse effects? Perhaps it should be controlled to provide more equity and justice to the home as well as the host country. I personally view efforts to slow down and particularly to stop technology transfer as a step in the wrong direction for mankind, and I feel this view is shared by this conference.

Thus I sense that you believe that there are some adjustments to be considered in the technology transfer process that are important to both the home and the host countries other than assuring the ultimate transfer of substance in an appropriate time interval, neither purposely delayed nor naively premature. For the home countries, I am particularly concerned about an appropriate financial return that will provide funding for continued R & D. For the host countries, I am concerned about the many social and economic adjustments that a significant technology transfer can bring about. I am also concerned that it is dysfunctional to not provide technology transfer of substance, which leads to the development of significant responsibility for those indigenous to the host country – this applies to both managerial and scientific professions. Of course, in view of relative home and host country salary schedules and opportunities, host country personnel have a competitive advantage. Home country firms argue that they use qualified host country personnel whenever available, and only wish that there were more.

I recognize that many of these attributes take time to develop and bring about; particularly if the level of training in the host country has not yet brought key enclaves in the country to "readiness." Very possibly we have a tendency these days to think of the time interval for technical transfer in terms that are too short; maybe substantive transfer is actually occurring as fast as is possible.

It is also not clear to me whether or not we should distinguish between transfer of technology from one developed country to another, and transfer of technology from a developed to a developing country. Somehow I should think the measurements (if not the judgment, at least) should be different. For the most part, I believe we are emphasizing the latter, and I shall continue to do so.

Also, with regard to control to enhance transfer, some home countries are concerned about stability in the host countries; this is particularly true of those who make home country investment decisions. Similarly, host countries may be concerned about ensuring continued supplies of key factors from the home countries. Possibly one of the best examples of the latter is the uncertainty with which host countries acquiring nuclear facilities are viewing the United States as a continued supplier of specialized nuclear materials.

In my concern for technology transfer, I find it helpful to distinguish between various characteristics of technology. Depending on who is defining it, technology includes a multiple of attributes, all the way from a product or concept of science through development, to financing, manufacturing, and sales and services. I don't believe that these are all of equal weight, and I think we implicitly recognize this. I find it helpful to think of these attributes as being on three levels of importance: those that are crucial to technology transfer, those that enable technology transfer to take place, and those that facilitate technology transfer. And certainly these attributes have some overlap. I believe that having a process or product is crucial; that having a host government allow a home country to participate in a host country is enabling; and that provision of capital, management, and so on is facilitating. This is a "chicken and egg" situation. Obviously one could argue that having capital is crucial. However, I think of the crucial and the enabling factors as having essentially no substitutes or alternative mechanisms, whereas the facilitating steps, in principle at least, have alternatives. Thus I feel that except for the short run, because of alternatives available for capital and management, TNCs cannot be used as effective control agents for technology transfer unless the specific TNC also controls the crucial factors. And I shall point out that no one can monopolize the crucial factors for long except for highly specialized instances, such as the essence of Coca-Cola or of a perfume. Capital can be raised without home country TNCs; this was particularly true in the development of Japan. While the management, report system, and organizational constructs of a multinational may facilitate operations in a host country, those operations may go on without using the multinational's approach, even though a TNC is more labor-efficient. In fact, some argue that indeed the multinational's methods aren't necessary and that indigenous approaches to management and employment are better. This may well be the case in the eyes of developing countries whose redistribution of the wealth is intrinsically mixed with the operation of the supply sector. In such instances, creation of jobs to fill people's time is perceived to be so important that it can take precedence over generating fewer but higher-productivity jobs that could more rapidly lead to transfer of substance. And yet specialization of labor is a key aspect of technology. Perhaps such countries mean something other than moving a complete production system into a developing country when they speak of technology transfer.

Study may also indicate that to bring about substantive technology transfer in even two or three generations, attributes other than the physical characteristics of technology may be involved. The case of Japan gives cause to ponder this point, unless one dismisses Japan as wholly atypical. Japan has essentially no natural resources as compared with other countries, including most developing countries. And yet Japan developed very rapidly, after an incubation period of perhaps fifty years that included many events usually perceived as setbacks. This suggests to me that, at least within developing countries, there may be attributes (for example, attitudes whose measures we are not taking in these deliberations) that are crucial to technology transfer of substance.

Given an open education and communication system throughout the world, I believe that technology transfer will occur in spite of controls, by virtue of the many avenues of "leakage." I suspect that substantial technology transfer occurs only when infrastructure within host countries is ready. The host country can then make us of implicit and explicit leakage avenues outside of its borders. For example, having students from developing countries do their advanced work in the home countries is a powerful avenue for achieving substantive technical transfer. This represents what I choose to think of as an implicit avenue, particularly so because education is important, not just for specialized training in the sciences or the managerial arts, but for such avenues as continued participation in an "old boy" circuit that develops through the university and continues in professional societies. In addition, the professional and trade literature is an extraordinary agent of transfer, especially so if ever larger numbers of host country students take advanced training in home country universities.

Nor should the patent system be overlooked as an explicit avenue of knowledge for transfer, whether the patents are licensed in the host country or not. By virtue of its scope and its indexing, the patent system is a great library of applied science throughout the Western world. Issued patents are open literature, and unless a patent is secured from, and is enforceable in, a host country, it is not effective except for importation of the goods into a country where the patent is enforced. This latter provision is continually weakened by trade agreements and, of course, is in any case limited to the seventeen years or whatever a patent may allow. It is interesting that a significant portion of that seventeen years has often gone by before manufacture in a host country is undertaken. In addition, in many fields (particularly pharmaceuticals), a goodly portion of the seventeen years elapses in testing and demonstration that the product or process in question is safe.

That published patents are a significant source of knowledge is often not appreciated because observers confuse patents with trade secrets. Basically, a patent consists of two parts: the allowed claims, which is what really constitutes the franchised activity (or, in some eyes, the monopolistic part of the patent), and the teaching portion, which precedes the allowed claims. As a source of knowledge about technical transfer, the teaching portion is authoritative, because to receive the privilege of exercising the allowed claims the inventor must lucidly and completely "teach" his invention. As a result, anyone familiar with the art can use the teaching portion of the patent as a "recipe" to reproduce the results presented in the claims. If this is not true and can be so demonstrated by one skilled in the art, the patent can be declared invalid. Thus the teaching portion of a patent is a technology transfer mechanism from inventors to others skilled in the art, that is, to those who are "ready." Of course, patents may purposely be secured for only part of a system, and a trade secret, unpublished and presumably unknown to anyone else, can be another part of the system that is vital to its overall functions as a commercial entity. Such a strategy, of course, is seldom used successfully because this merely invites, and allows, others to focus R & D on the missing link or to make a substitute.

Among communications and education mechanisms that can lead to technology transfer are the specific approaches, such as consultants and commissioned studies of all sorts ("brain picking"). Home countries could, in principle, curb these through export licensing procedures, but I think this would involve at most only short delays unless open educational and professional communication systems were also done away with. In effect, the TNCs provide consulting in-house, and this may appear more palatable to the home country because economic returns result over a continued time period. One could speculate that consulting should be paid on the basis of long-term value rather than direct, one-shot fees disguised as costs. And, of course, the same argument could be made for education. However, the competition among the many potential suppliers cast doubt on the efficacy of this approach over the long term.

If we are concerned about the control of technology transfer, particularly about slowing it down, we should recognize that industrial spying (for lack of a more gentle but equally accurate term) is a fact of life for all enterprise. However, even without outright specific spying, knowledge and understanding of a specific field allow development of a product or process, particularly once it is well known that it is possible and when samples can be secured. Examples are manifold. Once the transistor was announced, other semiconductors were made to function as an electric valve without specific knowledge of the transistor as it was articulated by Bell Telephone Company scientists. Similar or even the same discoveries are often made in two or three places about the world within a year's time or so, by apparently independent workers – independent, with the possible exception of exchanges having to do with the nature, but not necessarily the specifics, of their work, through formal and informal professional channels.

Also, technology transfer leakage can readily occur by "reverse engineering" for assembled products, particularly of a physical sort. Obviously, an automobile can be copied directly, with the possible exception of exotic alloys and plastics that require a little longer to figure out or to develop an alternative for. However, reverse engineering may not work on some chemical, pharmaceutical, or other exotic products; there may be no way to determine how to economically compound the material by examining the product.

Of course, applied science is only one side of technology, and other factors must be in place, or be brought to bear, if technology is indeed to be relevant. However, in my mind, the existence of many "leaking parts" casts considerable doubt on the contention that scientific and managerial technology transfer can be slowed or eliminated without total isolation, and thus much of our discussion here may not be particularly relevant except for short-term effects. However, through such study, presumably aspects can be identified that could allow technology transfer to be enhanced in such a fashion as to satisfy more fully both the host and the home countries.

It is important when assessing the extent to which technology transfer has occurred, that we distinguish between technology transfer and the provision for differentiated products for host markets. Mention was made of technology transfer among the OECD nations by Dr. Dunning. While I do not have access to that data, from my experience I would suspect that this is a method of providing product differentiation for slightly different markets, and that technology transfer in the substantive sense has scarcely occurred nor is required. Certainly Volkswagens manufactured in Brazil for Brazilians are only somewhat different machines than the Volkswagens manufactured in the United States for the U.S. market. I suspect that we would find that much of the technology transfer that we associate with multinationals is the provision of differentiated products, especially transfer among the OECD countries. And while some substantive technology transfer may occur under these circumstances, it is not significant as compared with the changes in perspective for a host country that transfer of new technology for new products makes possible.

The extent to which technology is transferred is, I think, measured variously by observers depending on their perception of the interest of the home country versus the interest of the host country. Dr. Dunning observed that the multinationals do not sponsor significant amounts of research and development in the host country, implying perhaps that this may be a willful holdback, perhaps in an effort to control technology transfer. I would like to suggest that this effect of carrying on a process or manufacturing activity in one area and retaining R & D in another is not unique to the international scene. There is considerable evidence that within a given country there are significant regional preferences in the location of centers for R & D and planning and managing activities. For 100 years the center of the R & D and much of the planning done in the United States was in the Northeast, and only recently is it shifting to California. Many companies have attempted to establish divisional R & D laboratories, perhaps with a product orientation, only to end up reconsolidating the more basic R & D at an original site and making the regional R & D laboratory a works lab mainly concerned with testing to support quality control. The Northwest United States is now becoming a site for assembly operations for firms such as Hewlett-Packard, but Hewlett-Packard's major R & D remains in the San Francisco area.

A minimum number and size of technically and otherwise sophisticated activities appear to be a requirement for maintaining a stable and creative community, even though the members of that community will be working in laboratories that compete with each other. Few scientists want to go out to the "boonies," so to speak. Those of us who have been involved with recruiting young scientists for sites such as Hanford, Washington, where I have been for thirty years, have encountered this reluctance again and again. Only recently, with the environmental movement being what it is, have the "boonies" become attractive to a growing number of people. In the instance of all aspects of technology that I am familiar with (managerial, financial, technological, promotional) there appears to be a tendency for personnel to want to be in a peer-group-acceptable enclave. Moreover, there may well be a tendency on the part of management to want R & D laboratories and planning activities in home territory and perhaps under surveillance. This peer group regionalism even prevails with operations units. A manager in an outlying plant is uneasy about his chances of being promoted to the executive suite in corporate headquarters simply by virtue of his physical isolation. Thus outlying plants are adopted with some reluctance. I see no reason for holding back technology transfer by TNCs merely because they don't establish laboratories in host countries.

Also, I am not comfortable with justifying control of transfer of selected technology so as to slow it just because it is observed that a slackening in some of the major industrialized countries (namely, the United States, United Kingdom, and Sweden) occurs in the same time frame with technology transfer. IN recent years innovations have slacked off in the United States. However, at the same time, many other factors could contribute to the slowdown: the value of money has increased substantially, and the enforcement of moving targets with respect to environmental constraints and consumer's warranties has increased. Moreover, it is now more attractive for corporations to make acquisitions than to originate new activities. Thus, whatever has caused it, I suspect that the reluctance to make long-term investments, especially on new inventions, is far more important to slowing the productivity of developed nations than is the extent to which technology transfer has weakened home country positions or directly discouraged long-term investments in a home country. In the instance of the United States, of course, I am thinking of technology transfer to Japan, and the apparent growing dominance of the Japanese in the television and hi-fi markets in the United States. U.S. dominance in computers still prevails. It is noteworthy that while R & D and innovation in computers in the United States has increased since 1960, attention to television R & D and innovation, in real terms, has dropped off since 1956, according to my sources. In short, we were not necessarily "put out" of television development by technology transfer – we gave up.

Then, aside from our agriculture, is the uniqueness of "Yankee know-how" a myth? When I find myself in discussions about the efficiency of production, I am quite uncomfortable with the tendency to give the Americans unqualified leadership in "Yankee know-how." I suspect that the apparent U.S. efficiency in manufacturing and production generally is associated with our extraordinary scope and depth of natural resources, coupled with our historical shortage of available labor and our ability to draw on technology from elsewhere. In view of these two factors, we found it economic to utilize continuous, open-ended process concepts, extensively exploiting our resources, and to employ automated machines, especially those from Europe. The automatic screw machine or lathe, the four-slide machine, the centerless grinder, and so on were direct lifts from Europe generally, and the continuous chemical process is a direct lift from Germany. Textile machinery from England is possibly even a more convincing example, with the minor exception of Eli Whitney's cotton gin. The manufacture in the United States of products with interchangeable parts was brought about by the country's labor shortage, and in essence depended for success totally upon the utilization of European machine tools and meteorological science. Therefore, it is most difficult to sort out the uniqueness of industrial "Yankee know-how," and thereby to view technology transfer as occurring with the United States as a net loser home country. I also really wonder whether any of the developed nations are net losers. Certainly some basic products such as the transistor and the computer belong to the United States. However, even the computer was first formulated and operated in England. Very possibly the United States' superiority is in "Yankee marketing," and not know-how in general. The automobile, for example, was invented in Europe, but Henry Ford made it available to the mass consumer. And marketing is, of course, part of the technology that can be transferred.

I would like to emphasize what I believe is a fundamental characteristic, or possible flaw, in our technological system that may be common to many nations of the world. This characteristic, or possible flaw, and not technology transfer, is what is limiting our relative competitiveness in the world. What I sense is basically this: that after achieving substantial technological progress that improves living standards, organizational constructs become so enthralled with the virtues of their technological system that they wish to institutionalize it in its present form, and then only rationalize, rather than even ameliorate, perceived defects. Eventually, technology, including the universities, begins to walk in lockstep with industries and the government, whatever the government may be. This, I think, leads to an inflexibility that stulifies change. (Of course, this characteristic may be inevitable, leading to a maximum level of general welfare through technology.)

The results appear in several ways: a diminution of adventuresome capitalism; an increase of rampant inflation not associated with increases or decreases in costs due to scarcities; and an increase in governmental restraints. Not wishing to explicitly recognize these factors can lead to blaming "technology transfer" for weakening a host or home country.

The development of technology throughout the world is not understood even by those involved. For example, many in the United States do not realize that the world as a whole, not just the United States, develops the crucial parts of technology and that the United States has no monopoly on it. In fact, many of the inventions associated with U.S. industry were made abroad and brought to the United States. Only recently (the last forty years) has the United States been emminent in science per se in terms of learning (if that is important to invention and innovation). The laundry list of examples is far too long to dwell on here. Aside from nuclear energy, major ideas introduced into the United States include electronic television (the iconoscope), cellophane, penicillin, motor cars, electric lights, stainless steel, high-speed tool steel, and tungsten carbide. Recently the United States has also adopted fully engineered technology such as the triple-deck railway car for transporting automobiles (from West Germany), the horizontal climbing crane (from France), the flotation process for plate glass (from Britain), the basic oxygen furnace (from Austria), the electrochemical milling machine (from Russia), the hovercraft (from Britain), and the fundamentals of the digital computer (from Britain).

And I cannot leave this discussion of technology without discussing further its crucial spark – invention and discovery – in relation to the role of large companies. Try as we may, we have not been able to identify simple predictors of inventive activity or propensity. Many believe that inventions could be increased by massive government or company investments in basic R & D. Several studies of twentieth-century inventions of prime importance all point to the conclusion that most innovations of great economic impact are not causally related to the sophisticated scientific discoveries – or to the R & D expenditures of large companies. We may have one counterexample in the emergence of the microelectronics industry as a multibusiness, billion-dollar, international enterprise, following hard on the heels of the discovery of the transistor; but many small companies were, and are, involved. Moreover, most technological innovations are based on well-known, not too recent scientific and engineering knowledge. I am going to argue further that the TNCs are not the primary source of good ideas. Of course, TNCs can develop alternatives and make decisions and have sources of money, each important to implementing sophisticated technical activity, although that activity may be based on someone else's idea. Nonetheless, this role of TNCs is an important role, the worth of which should be recognized and not diminished because of their perceived lacks in other regards.

In the United States, over 80 percent of private R & D dollars is spent by 200 large companies. But more than two-thirds of the crucial ideas that resulted in important technologies came from independent inventors or small firms. And just as it is impossible to establish a correlation between total size of an effort and results, it is also impossible to establish any correlation between geography and inventive results. The basic inventions that have proven most important in this century have come from throughout the world. Patented invention per laboratory worker has been decreasing in the United States since 1925.

I am personally quite concerned that a significant cause of low inventive output may be the overwhelming dominance of scientific and technological thought by so-called logical, deductive empiricism, enhanced by institutional constraints – including those that occur in large companies. Much of day-to-day classical science is based upon the building-block approach – one step at a time. Thus, it orients thought to accept things as they are and limits advances to only incremental, apparently logical, steps (an attitude that is deadly to creativity). Much of applied science is defensive (concern for the environment and so on). This, too, is deadly to creativity.

It is, however, very important that we recognize that there may be irreconcilable differences between the organizational constructs (or lack of constructs) that might enhance invention and those organizational constructs that enhance technological refinement, especially in today's regime of constraints. While lamenting the dominance of large organizations, including the TNCs since they appear not to encourage invention, I wish to emphasize the apparently excellent role they play in refinement, which is extremely important to technology transfer. This may be a paradox with respect to substantive technology transfer. Certainly an unrefined concept would have difficulty being transferred in an economic fashion. And at the same time, a highly refined process may be transferred in structural form but may not appear to directly contribute to the transfer of substance. And yet refinement is key to achieving world market shares, especially for consumer goods that are warranted with respect to service, and for consumer goods that must meet changing environmental standards. Thus it may be that we must be content to transfer only parts of technology, at a time wherein we are interpreting the term "technology" in the largest sense, embracing invention through production, administration, financing, and social impact. Moreover, perhaps we should recognize that the propensity for invention belongs to no one country and to no one organizational construct and, therefore, is not available for "transfer."

By its very characteristics and by virtue of inherent leakages through a variety of communications mechanisms, then, the transfer of technology is difficult to control in the long run (especially to speed up, particularly the transfer of substance, since the host country must be "ready," if technology transfer is to "stick"). Technology is not static and is not the domain of a single nation for long. Once a new product or process is introduced, even of a fundamental sort, then others knowing that it can be done confirm it rapidly or discover a viable alternative. In the long run there are no secrets in technology. Thus, technologically oriented products and processes cannot be frozen to provide a world monopoly. Rather, one must proceed with development, refinement, and an occasional new product to maintain one's overall competitiveness. No nation can be secure in world markets by merely keeping its technology secret. Aside from the export of some labor-intensive industries, it is doubtful that technology transfer in itself causes a home country to lose significant long-term market positions.

To conclude ever so briefly: In view of the inherent nature of technology per se, and of internal considerations that are badgering the environmental impact statements, I do not believe that, short of isolation, technology transfer can be controlled beyond the short run, given our existing open education and communications systems and given the resourcefulness of man. I feel that in meetings such as this, much of the discussion is involved with measurement and analysis of the wrong factors. At best, a shaky beginning has been made; at worst, the beginning is on the wrong foundations and we are wasting our time.

Comments

George Moldelski

In his paper Professor Dunning asks the question; What is the justification for the fears aroused in certain "home" countries about the effects of transnational corporate activities, as formulated by such writers as Gilpin?(1) Is there a meaningful relationship between technology transfers to the less developed countries by transnational corporations, and the industrial deterioration shown by certain developed countries (including the US) on the one hand, and the upsurge experienced in particular by the newly industrializing countries in recent years on the other?

Professor Dunning's answer is, briefly, no, there is not; hence (except for some adjustment planning) there is largely no need for home country controls over transnational technology transfers to less developed countries. The reasons for this conclusion are founded both on the classical theory of the international division of labor and on what we know about transnational activities from recent reports and studies. Such data suggest that all transnational assets in less developed countries are now about one-quarter of the total, and that technology transfers to less developed countries may account for 20 percent or even less of the total of such transfers. About one-half of these transfers go to the dozen or so newly industrializing countries (including Brazil and Mexico in South America, and India, South Korea, Taiwan and Singapore in Asia). The evidence tends to show that in fields in which the transnational stake is high (high-technology and product-differentiation idustries), export participation, and hence potential damage to the markets of the "home" countries, is low. Similarly, transnational participation in labor-intensive, export-platform industries is relatively light (on the order of 10 percent), the only specific exceptions being the electronics and photo industries. Thus the substantial part of the competition from newer industrial countries has nothing to do with TNCs, although some conflicts of interest may have arisen in regard to consultancy arrangements for establishing new industries (such as petrochemicals), hence in regard to unbundled technology.

Professor Dunning's case is well argued and supported, but it is also unsurprising. It is widely recognized that transnational interests "are well served by conventional modes of technology transfer," but what about unbundled technology? Given also the relatively small size of the TNC stake in less developed countries, maybe there is no ground for concern about it. And yet it is transnational investment in the oil industry and the attendant intracompany technology transfers that made OPEC possible and wrought a dramatic change in the terms of exchange, in regard to oil and other fields (including technology transfer).

But it is only fair to recall, too, that Gilpin's arguments and concerns were directed not primarily at the LDCs. His concept of "core" referred only to the United States as a world power and, for him, the periphery meant in the first place Western Europe and Japan, and only secondarily the Third World. Hence the diffusion process he was analyzing was one that was tending to shift the locus of world power away from the United States toward some new center in the periphery, but not really the LDCs or even the newly industrialized countries. Hence in terms of his political-economic analysis the questions to ask concerned that 75 percent of transnational activity located in the developed countries and the 80 percent of technology transfers occurring there, some of it in unbundled form (for example, in Japan). It could be that the relationship here is the same as that with LDCs, but then again it also might not be, and the effect could be broader. Gilpin might have overestimated the impact of technology diffusion on the decline of U.S. power, and generally might have followed what might be described as a "diffusionist" model of power deconcentration,(2) therefore attaching undue importance to the mechanisms of diffusion involved in technology transfers. It could well be that the contemporary economic strength of Western Europe, Japan, and the new industriali countries has more to do with national evolution than with technology diffusion, but surely we need to keep in mind both sources of change.

The Dunning paper lumps together the interests of all the developed countries (presumably as expressed through the OECD or at economic summits). Undoubtedly an important common set of interests does exist in regard to TNCs and TT, especially since their relations have become increasingly symmetrical in these matters in recent years (in part as the result of earlier technology diffusion). These common interests might also extend to coopting the newly industrialized nations into their ranks, or even to establishing a common front in relation to the rest of the LDCs.

But we also need to bear in mind, in certain contexts, the special position of the United States as a world power, and the implications of changes in that role. For even the analysis that concentrates on the effects of TT through "normal business channels" needs to bear in mind the political framework within which such transactions occur.

The post-1945 world order was based on the leadership of the United States, and that resided in part in military predominance (including a monopoly on nuclear strategic weapons systems and technology) and in part in a high-growth, high-technology economy. In turn, this leadership formed the general political framework that also set the foundations and the conditions for a variety of specialized markets, including those for technology transfers, and those too were dominated by the United States (three-quarters of world R & D, a major part of TT).

Against this background we might formulate three propositions relating technology transfer to recent changes in the structure of world policies:

1. The process of TT has contributed to power deconcentration. Nuclear proliferation may be one instance of that process, exports of military technologies another, the general marketing of unbundled technologies yet another. If that is so, then the process serves to alter the world political structure, and affects the role of the United States in the world.

2. A change in the world political structure toward deconcentration may be expected to alter the terms in which technology transfer occurs, by reducing monopoly rents previously earned and so tending to lower the price of technology, and raise the demand for it, and reduce its supply.

3. Based on past experience, such changes in the world political structure, while creating conditions of greater freedom, have also been generally destabilizing and, in particular, have also led to instability in markets for technology transfers (as in the late nineteenth century). Hence advocates of freer technology transfer also need to be proponents of positive changes in world politics that lend greater stability to its structure.

Viewed, therefore, from the perspective of U.S. concerns for world order and in terms of world system analysis, we cannot conclude that TTs through TNCs or other channels must necessarily be of positive significance for the United States or for the world as a whole. Such transactions need to be watched closely and with great care, and especially with an eye on long-term trends.

NOTES

(1) U.S. Power and the Multinational Corporation (1975).

(2) The "diffusionist" model assumes that the mechanism of world power deconcentration is "diffusion," a spread of strategic innovations and practices from the center to the periphery, as against an "evolutionist" model, which postulates that the capacity for spontaneous development is widespread and is activated in suitable conditions. This is a replay of the classic debate in nineteenth century anthropology about the nature of the process of cultural diffusion.

12 Technology Transfer and Control Systems: The Japanese Experience

Terutomo Ozawa

When we use the term "technology transfers," we obviously cover a very broad set of activities that include not only market-based (or commercial) but also non-market-based (or noncommercial) transactions.(1) This chapter, however, is concerned solely with market-based transfers; hence it is equally concerned only with market-related control systems (that is, control systems such as official technical aid programs, for example, are outside its scope). Moreover, a control system as the term is used in this paper, means a suprafirm, official control system that is designed to affect the firm's behavior as both a supplier and a recipient of technology. Individual firms do exercise control over their technology transactions by using different modes of transfers. In this sense, they have their own control systems. Here, nevertheless, such control systems are treated as market behaviors, monopolistic or monoponistic market behaviors as the case may be. For the very exercise of their profit-motivated control is the core of many problems created at the national and international levels. It is basically these activities that necessitate suprafirm controls.

Suprafirm, official control systems can be established at three different levels: national, regional, and global. A national control system is naturally intended to maximize the country's own gains from technology transfers. As the system moves above the national level, it aims at coordinating or restricting individual national control systems as well as individual transactors' behaviors in the marketplace. This means that the individual unit, whether a nation or a firm, may be required to sacrifice its own gains, to some extent if not totally, in the interest of a region to which it belongs or in the interest of global harmony. The effectiveness of control also tends to diminish at the international level, since national interest is less directly represented and enforcement becomes difficult. No doubt both the scope and the effectiveness of a control system are greatest at the national level. This chapter focuses on Japan's national control system, as it is the only one relevant to the postwar Japanese experience.

Why, then, do we in general need market-related control systems at the national level? Control systems of this sort are justified under two circumstances: (1) when the incidence of market failure necessitates measures to improve the market mechanism; and (2) when the incidence of market insufficiency for achieving a given national objective calls for measures to supplement the market mechanism. In the first situation, the market is malfunctioning as a result of its internal deficiencies;(2) hence some sort of control is required to eliminate or counteract those deficiencies. In the second case, however, the market is functioning as it is expected to function, but market forces are considered not sufficient to bring about a given desired result; hence some sort of control is needed to reinforce or back up the market mechanism. The former situation may be summarized as the "market does not work so let's improve it" situation, the second as the "market works but more is desired" situation. In either case, therefore, the control system introduced is market-oriented in its intention and execution.

The market-related control system, if its intent is to make up for market failure on the part of technology recipients, implies that once sufficient market forces are restored the system ought to be dismantled. In other words, what may be called the "infant technology" theory of control – akin to the "infant industry" theory of trade protection – has economic validity if and only if such control is intended to improve the recipient economy's technological ability and bargaining strength in technology trade and only if such control is eventually lifted.

In this chapter, technology transfer is also viewed essentially as an economic activity observable in the sequential process of invention (conception of technical feasibility), innovation (commercialization), and diffusion (widespread use following demonstration). Technology transfer occurs most frequently in the diffusion phase after technology has been successfully developed, but the transfer of rudimentary research technology may also take place at the beginning of, or in the midst of, the innovation stage. In the latter instance, a great deal of adaptive and creative research and development (R & D) is required for commercialization on the part of the technology recipient.(3) Given the absorptive capacity of recipients, the gains that may accrue to recipients are likely to be much greater in the innovation-phase transfers than in the diffusion-phase transfers. Conversely, the loss, if any, to transferors is likely to be less in the latter phase.

The Japanese experience in the post-World War-II period is perhaps an ideal laboratory in which we can examine the relationship between market forces and a national control system over technology transfers interacting to achieve national economic objectives. It is a case where we can observe the experience of Japan, initially that of being onesidedly an active technology recipient and then, more recently, an increasingly active technology supplier. In fact, as will be discussed below, these "opposing" roles are not unrelated. On the contrary, the very active role of Japan as a recipient has led to its recently expanding role as a technology supplier.

EXPERIENCE AS A TECHNOLOGY IMPORTER

Technology Import Control System

Japan's successful absorption of advanced Western technology in the postwar period and the favorable impact of that technology on its industrialization are well known. In 1966, for example, William W. Lockwood observed matter-of-factly:

> Hailed sometimes as a "miracle," there is nothing mysterious about Japan's high-pitched growth. The factors immediately at work are well known. . . . The principal gain . . . has come from a sustained rise in the productivity of labor. . . . A key impulse of this productivity gain has come from the great wave of technological borrowing from abroad since the occupation. The isolation and destruction of the war had caused Japan to lag well behind the West in the industrial arts. As soon as possible her businessmen and public agencies moved energetically to repair the gap, importing the latest technologies over a wide front to rebuild their industrial plant.(4)

Thus the importation of advanced Western technology was clearly a major driving force behind Japan's economic "miracle" throughout the 1950s and 1960s. Since detailed studies of this phenomenon have been made elsewhere,(5) I will not go into all of the many interesting and wide-ranging elements of this Japanese experience; instead I will concern myself solely with the particular relationship between market forces and the control system as it applies to technology imports.

Japan's successful adoption and adaptation of foreign technology are often considered superlative examples for the developing countries to emulate (particularly Japan's success in "unbundling" technology from foreign capital ownership and control). But the following fact seems not so widely known: the Japanese experience was founded on massive importations of what might be called "inappropriate" technology – at least in the early stages of postwar technology borrowing, the crucial stages in which the forces of technological momentum were released, creating both forward and backward technological linkages throughout Japanese industry. For, as is implied in Lockwood's reference to "the latest technologies over a wide front," a great many of the technologies Japan imported at the outset appeared "inappropriate" because they were highly capital intensive and resource-consuming, features incongruous with the factor endowments of early postwar Japan.

Chalmers Johnson makes a more direct observation of this experience:

> In postwar Japan, the government consciously led a massive, popular, difficult campaign to develop a capital-intensive, technology-intensive industrial structure in the face of the theory of comparative costs, which would have recommended labor-inten-

> sive types of industries for a country with a large population, few resources, and little accumulated capital. . . .
>
> If Japan had pursued a comparative-costs, light-industry economy policy in the postwar world – more or less like that envisaged by the Allied Occupation authorities – it is extremely unlikely that it could have raised per capita income above that prevailing in comparably organized economies (such as in several Latin American countries).(6)

In fact, a detailed statistical analysis of this development strategy was made by Tuvia Blumenthal and Morris Teubal, who concluded:

> One feature of Japanese economic development has been the introduction of "future oriented technologies" – namely, technologies whose use was unjustified by existing conditions, but which were efficient in view of future circumstances. We believe that this phenomenon can be regarded as an important reason behind the rapid development of the Japanese economy.(7)

Thus Japan's postwar technology borrowing was "unorthodox" in the sense that it ran counter to the theory of comparative costs, since it concentrated on the latest technologies necessary to establish exactly those industries that were highly intensive in the use of capital and natural resources, the very factors of production Japan badly lacked. Yet by 1973, for example, Japan, as compared to the United States and West Germany, came to exhibit "the highest concentration of industry in the sectors consuming the most resources and the lowest concentration in the sectors consuming the least resources."(8)

This "unorthodox" approach obviously required intervention by the government in the market mechanism, its guiding hand. Besides, the market system itself was still in disarray in the immediate postwar chaos, with shortages of many vital industrial resources. The only abundant factor was labor – labor relatively well trained and highly motivated to work, thanks in part to war-time production experience. With the market clearly so defective and so weak in its capacity to absorb foreign technology, some governmental measures were required. Moreover, even if the market mechanism had been normal, it would not by itself have guided Japan into building the capital-intensive, resource-intensive sectors of the heavy and chemical industries, the sectors that postwar Japan desired to develop. In this sense, early postwar Japan exhibited an acute case of market failure, or the "market does not work so let's improve it" situation.

Japan's postwar system of controlling technology imports was introduced by the Foreign Exchange and Foreign Trade Control Law of 1949 – generally known as the Foreign Exchange Control Law – and by the Law Concerning Foreign Investment of 1950 – commonly called the Foreign Investment Law. Under these two laws governing external transactions, technology import contracts are classified into two categories, types A and B. Type A covers those contracts that have a life of more than one year, with the payment of royalties and fees guaranteed to be made in foreign currency, and falls under the Foreign Investment Law. Type B deals with those contracts that have a life of less than one year, and is subject to the Foreign Exchange Control Law. The latter normally includes incidental arrangements, such as an invitation to foreign engineers or the acquisition of drawings. The use of a patent or a brand name or the transfer of major know-how necessitates a contract in type A because of the long-term nature of the transaction.(9)

All applications for type A contracts must be submitted, via the Bank of Japan, to the Minister of Finance, who in turn has them deliberated on at the Foreign Investment Council, a screening organ set up under the Foreign Investment Law. Applications for type B contracts pertaining to industrial property rights are subject to approval by the Minister of International Trade and Industry; all others are approved by the Minister of Finance.(10)

When the Foreign Investment Law was legislated, it took the positive approach of attempting to enhance the economy's capacity to earn foreign exchange by fostering both import-competing and export industries at home, whereas the Foreign Exchange Control Law was passive, in the sense that it was designed to ration available foreign exchange reserves and to optimize their use. The positive tone of the former is clearly set forth in its preamble:

> The purpose of this Law is to create a sound basis for foreign investments in Japan, by limiting the induction of foreign capital to that which will contribute to the self-support and sound development of the Japanese economy as well as to the improvement of the international balance of payments, by securing remittances arising from foreign investments, and by providing for adequate protection of such investments.

In this law, technology assets are treated as capital. Two phrases are worth special attention. The phrase "limiting the induction of foreign investment to that which will contribute to the self-support and sound development of the Japanese economy" is obviously based on the supposition that without appropriate government controls the acquisition of foreign capital and technology would lead to the domination of Japanese industry by foreign interests. This was the basic assumption adopted by the government in formulating its policy for the acquisition of foreign capital and technology – the same basic policy assumption, in fact, that Japan has held ever since the industrialization initiated after the Meiji Restoration of 1868. The other phrase, "providing for remittances arising from foreign investment," related to the government's intention to provide a full guarantee of the repatriation of earnings once the investment project was approved. Although the fundamental objective of the law was to develop domestic industries with the help of foreign capital and technology – but without being dominated by foreign interests – the guaranteed remittances of investment earnings were emphasized as a primary justification of the controls.(11)

One significant role played by the government was that of a "countervailing power" protecting – in the government's view – Japanese firms that were desperately in need of modern technology and, as a consequence, had weak bargaining power vis-a-vis foreign suppliers of technology. Administrative guidance discouraged "undesirable" competition among Japanese firms for the purchase of a given type of technology as much as possible. As a result, competition for a given type of technology offered by a particular supplier was eliminated in most cases, and certainly in those involving major industries, whose development was under close supervision by the government. This no doubt strengthened the bargaining position of Japanese industry. After carefully studying the trend of royalty rates agreed upon by the Japanese firms over the period of 1963 to 1971, Merton Peck and Shuji Tamura observe:

> The mid-1960s, with the lowest royalty rates, represented the period of most intensive MITI intervention. The requirement of MITI approval served to reduce competition among Japanese firms, especially since there was sometimes informal designation by the MITI of a particular Japanese firm to negotiate with a specific foreign company. Beyond this, the MITI would delay its approval or make it conditional upon revisions that would lower rates. . . .
>
> In 1966, the MITI announced that it would limit its revisions to agreements with royalty rates of less than 3 percent, a figure well below the prevailing average rate and one that allowed the MITI to continue playing an active role in most agreements.(12)

The guidelines also included a fixed fee not exceeding $50,000 and a term of less than seven years.(13) But there were many exceptions.

In this connection, it is interesting to note a different conclusion arrived at by Franklin R. Root in his study:

> The most important implication of our examination of international technology transfer pricing is that host governments can do very little directly to lower the price of technology imports. . . .
>
> The imposition by a host government of constraints on the licensor (such as percentage royalty ceilings, duration of the agreement, trademark rights, etc.) can be nullified by adjustments in technology packages (for example, less technical service); adjustments in compensation arrangements (for example, greater lump-sum payments); or by the refusal of foreign firms to enter licensing agreements. In sum, the effectiveness of regulatory measures to lower technology transfer prices (apart

> from their administrative effectiveness) is highly questionable. Even if they were effective, the opportunity cost in lower availability of desired technology can be high.(14)

How can we reconcile these two contrasting analyses? My speculative response is that perhaps the Japanese government's intervention had more a "passive" effect than an "active" effect on royalty rates and other terms of contract, in the sense that it prevented licensors from charging excessively high prices rather than forcing them to lower prices drastically.(15) In fact, the passive approach might have been the true modus operandi of such control: Japan was interested in purchasing foreign technology at prices that were reasonable and acceptable for the contracting parties.

Foreign licensors did complain about the government's intervention, but kept supplying technology – a clear sign that the prices they received were acceptable to them. Yet if they were selling at prices lower than they normally charged, what motivated them? One conceivable factor was the growing Japanese market; even if the price of technology was low, the total revenue from such a growing market would certainly be greater than they could expect anywhere else. In this sense, the Japanese market was a price-elastic market, and some licensors might have decided to supply technology at even lower prices instead of refusing to enter licensing agreements. (Total revenue consideration often seems to be left out of discussions on the pricing of technology. There is good reason for licensors to tend to demand higher prices in the developing countries that offer only a small price-inelastic market.)

What has been perhaps more frustrating for the foreign firms is that government regulations concerning foreign business operations in Japan, such as the Foreign Investment Law, are purposely left undefined and unspecific to give greater discretionary power to the administrative bureaucracy. In fact, this characteristic applies practically to any legislation in Japan (hence it is not necessarily intended to discriminate against foreign firms). This situation is succinctly explained by Dan F. Henderson:

> Legislation originating in the ministries is characteristically drafted in very abstract terms that leave, at least in regulatory statutes, many important distributive decisions, as well as detailed application, to be worked out by the regulations, the making of which is again delegated to the self-same bureaucrats. Though the phenomenon of delegating to the administration important rule-making powers is not itself peculiarly Japanese, the net effect of the practice confers upon its bureaucrats law-making powers so enormous in degree that it is almost different in kind from the results, striven for at least, in western countries more experienced with administration-under-law, buttressed by ideas of equal protection and procedural due process and implemented by a tradition of judicial review of administrative

acts. These latter institutions are as yet little developed and of little practical relief to foreign business in Japan.(16)

The net result of bureaucratic power is persistence and long-term consistency in Japan's policy programs, since they are not hampered by changes in administration. Yet it also means to business that it must deal with the bureaucrats rather than with the formal written regulations.

Worse yet for the outsiders are the internal ministerial rules and regulations – known as naiki – that govern administrative procedures such as those involving so-called case-by-case screening. Because they are internal, naiki are not made public; outsiders can only get a feeling of prevailing internal rules through close consultations with officials. Thus government decisions and instructions pertaining to the screening of foreign investment or technology import proposals are practically always communicated verbally rather than in writing. Left largely in the dark, the business sector has no recourse but to consult the officials in person; this sort of practice naturally gives the bureaucracy a powerful basis for administrative guidance.

The frustration of foreign firms caused by such bureaucratic controls is well expressed in a statement by the Committee for Invisible Transactions of the OECD in 1968:

> All [applications] are carefully examined and the general approach is restrictive. Whatever criteria are applied are not embodied in any formal regulations, although some guidelines to Japanese importers of foreign technology are laid down in semi-official publications. In accordance with general Japanese practice official instructions for amendments to the original contract are only given verbally, never in writing, and foreign representations in the responsible quarters are thus made extremely difficult. Apart from the application of general principles of "exceptional adjustment," it happens that, after the Japanese and the foreign partner have agreed on the terms of a contract, the authorities make approval conditional on amendments, usually in favour of the former. The scope of the technology is frequently changed; the royalties and initial payments are reduced; minimum royalties and back royalties are eliminated; arrangements must be made for the Japanese partner to get privileged access to certain foreign markets; provisions are disallowed under which the Japanese partner renounces manufacture after the expiry of the contract or to make certain competitive products; sub-licensing is made subject to further government approval; undertakings are deleted under which the Japanese partner would hand over a list of his customers at the end of the contract; the duration of the contract is reduced; automatic renewal is excluded, etc. All these changes of the terms of his contract are imposed upon the resident (as well as the non-resident) whether he wishes it or not, although, as long

> as they are in his favour, he would be unlikely to object to this form of official guardianship over his interests. Government action is generally supposed to be for the good of its nationals, but in this case it takes an unusual direct form of interference with bona fide private contracts.
>
> It is recalled, moreover, that any arrangements require the prior approval of the Ministry of Finance or of other Ministries concerned under which subsidiaries, joint ventures or branches use unregistered technical information or services, designs, specifications, drawings etc. free of payment.(17)

The government did have broad guidelines for technology imports. They are stipulated in the Foreign Investment Law and the Foreign Exchange Control Law. As is shown in table 12.1, the former law set forth two sets of approval standards, "positive" and "negative," that needed to be satisfied before applications were validated, whereas the latter law stipulated only the negative criteria. But these positive and negative criteria left a great deal of room for administrative discretion. In short, Japan's technology import control system has resided in the discretionary power of the bureaucrats rather than in the formally written regulations.

It is also important to stress here that Japan's technology import control system was not a mere control system per se; it constituted an integal part of Japan's comprehensive scheme of industrialization, the thrust of which went far beyond what could have been predicted from its initial economic conditions in the early postwar period. This is clearly reflected in the fact that the Foreign Investment Law required the government to "make public a list of the kinds of desired technological assistance from foreign investors" (Article 7). The first such list, issued in 1950, included thirty-four desired technologies considered essential to develop the heavy and chemical industries. The second list, that of 1959, indicated a further effort to introduce more sophisticated new industries, such as jet aircraft and electronics products, and to raise productivity by introducing automation technology.(18) In the early days of the control program, Japanese firms seeking approval were also required to give a statement of the economic benefits to themselves of imported technology and the anticipated effect upon the Japanese economy. They were thus compelled to conduct a sort of what is now known as "technology assessment."

Throughout the postwar period Japan endeavored to reorganize its industrial structure from labor-intensive light manufacturing toward more capital-intensive heavy and chemical sectors. This structural transformation had to be assisted by technology imports as well as by various fiscal measures, including tax benefits, accelerated depreciation, and special allocations of foreign exchange for the importation of machinery, equipment, and raw materials – all "control" measures designed to back up the market mechanism. The technology import control system was an essential "joint" instrument for facilitating this national scheme of industrial transformation.

Table 12.1. Positive and Negative Approval Standards for Technology Imports and Inward Foreign Investment

	The Foreign Investment Law	The Foreign Exchange Control Law
Positive criteria	1. Contracts contribute to improvement of the balance of payments 2. Contracts contribute to development of important industries or public works 3. Contracts are necessary for contination, renewal, and/or associated changes of technical contracts for the important industries and public works	
Negative criteria	1. Contracts are unfair and illegal 2. Contracts involve fraud, coercion, or undue pressure 3. Foreign investments will have an adverse effect on the recovery of the Japanese economy 4. The acquired foreign currencies lack sufficient convertibility	1. The balance of payments will be adversely affected 2. Recovery of the national economy will be hampered 3. Capital movements in question are believed to be for the purpose of evading the existing laws and regulations 4. Capital movements in question are deemed undesirable in light of the condition of the foreign exchange reserves

Source: Compiled from Robert S. Ozaki, The Control of Imports and Foreign Capital in Japan (New York: Praeger, 1972), pp. 80-81.

The role of technology imports in this scheme is best exemplified by the government's use of a so-called staggered entry strategy, combined with a "one set" principle; under this strategy each major industrial group was encouraged to set up the same new key industries – hence "one set" – by acquiring advanced foreign technology at staggered intervals.(19) This process increased the bargaining position of Japanese enterprises in negotiating with foreign technology suppliers because only one (or, at most, a selected few) was allowed to enter a new industry at a time. Moreover, it also ensured the establishment of plants of efficient economic scale by restricting competition. For example, when the polyethylene industry was introduced into Japan, the first entrant allowed to "monopolize" the industry was the Sumitomo group, which in 1955 established Sumitomo Chemical Company with technology purchased from Imperial Chemical Industry of Britain. Then in 1957 the Mitsubishi group was chosen to follow, with technology acquired from BASF of West Germany. Three years later it was the Mitsui group's turn, with technology obtained from DuPont of the United States it set up Misui Polychemical. And still later on, other groups and other companies within the same industrial group were gradually allowed to enter the industry – all with the help of foreign technology.(20) A similar pattern was also apparent during the course of the development of Japan's synthetic fiber industry. The first entrant permitted to enter the production of nylon in 1951 was Toyo Rayon (the Mitsui group). The second entrant, approved in 1954, was Nippon Rayon (the Sanwa group), which secured technical assistance from Inventa A.G. of Switzerland. It was, however, nearly ten years before the third group of entrants were allowed to move into the nylon market, which by then had proved extremely lucrative because of the monopolistic position given to the earlier entrants.(21) Such "orderly" entry did not occur in every industry, of course, but these examples illustrate how the technology import control system was used to develop key domestic industries under the close guidance of the government. The system was thus closely coordinated with other industrialization promotion programs.

Viewed from the above perspective, Japan's technology import control has served not only as a device to improve the bargaining position of Japanese licensees and Japanese joint-venture partners but also – probably even more important – as a device to direct the direction of Japan's industrialization. In the words of Merton Peck, "More important than any stated rationale for controls is their impact on the price of technology, the composition of the imported technology, and the industrial structure – all areas in which the evidence suggests that government controls did make a different."(22) The impact on the industrial structure was to produce a number of Japanese enterprises that were as monopolistic in terms of size – and hence of bargaining power – as the foreign suppliers of technology.(23)

The control system used in Japan was essentially an administrative, ad hoc process rather than a legislative, rule-governed system. In other words, the effectiveness of control resided in the administrators' capacity to informally use the formal regulations, which were in fact left ambiguous in many instances.

Liberalization of Controls

As might have been expected, the need for market intervention by the government diminished with the rapid growth of domestic industries, many of which soon developed trade competitiveness in the world market. A study made by MITI in 1962, for example, revealed that imported technologies had contributed significantly to Japan's trade expansion. The ratio of the exports of licensed manufactures to total exports in the electrical machinery industry, for instance, climbed from about 10 percent in 1952 to more than 50 percent in 1960; for iron and steel it rose from less than 1 percent to more than 35 percent over the same period; for chemicals, from less than 1 percent to about 24 percent.(24) These ratios may not be precise because of inevitable errors in estimating and sampling, yet there is no doubt about the favorable effect of imported technology on the quality and variety of Japanese exports.

The Japanese recipients' capacity to absorb technologies, increasingly more sophisticated ones in particular, also grew considerably in the early 1960s, as their rising expenditures on R & D suggested. One major Japanese corporation after another set up a corporate research center, many of them modeled on such centers in the United States. This phenomenon reflected a shift in the nature of the technologies imported, a shift from diffusion-phase transfers toward innovation-phase transfers, as Japan began to narrow its technology gap vis-a-vis other advanced countries. Another survey made by MITI in 1963 showed that about one-third of Japanese firms' R & D expenditures over the 1957-62 period was devoted to "processing" imported technology. (The processing consisted of two basic activities: 1) R & D to modify and perfect imported "raw" technology; and 2) production design and engineering and the laying out of plant facilities to install new equipment and machinery.)(25)

With a noticeable improvement in Japanese industry's technological capacity and competitiveness, liberalization of controls over technology imports became justified. Moreover, additional pressures for decontrol came from outside, as Japan began to join international economic organizations as a full-fledged member: in 1964 Japan became an Article 8 country of the International Monetary Fund and a member of the Organization for Economic Cooperation and Development (OECD). This meant that Japan had to remove restrictions on capital movements and invisible trade transactions, including technology transfers. Actually, the first liberalization step for invisible trade transactions was taken in 1959, followed by the second and third steps in 1960. The first step was to expand the scope of approval to allow the acquisition of less significant technologies, including those related to the production of locally oriented consumer goods. Guaranteed remittances for these "new" contracts, however, were left subject to the proviso of a temporary suspension in the event of an adverse balance-of-payments condition. The 1960 liberalization changed the approval standard from the "positive" to the "negative" screening standard. In 1966 a so-called one-month rule, whereby each application was as a rule to be processed within one month, was introduced, and about one-third of the items on the previous checklist for approval were deleted. It was not until 1968, however, that a substantial technology import liberalization took place. In that year, all technology import contracts with compensation less than $50,000 in each case, for sectors other than those specifically restricted by the government (that is, aircraft, weapons, explosives, nuclear energy, space exploration, computers, and petrochemicals), began to be approved automatically by the Bank of Japan unless the government filed an injunction no later than one month after application. In 1972 further relaxation was introduced: automatic approval was extended to all the seven previously restricted fields except computers and petrochemicals. And finally, on July 1, 1974, the remaining exceptions were removed.(26) Yet so long as the screening mechanism, the Foreign Investment Council, exists, "the government has reserved its authority to apply controls, and it is unclear how active the MITI remains, on an informal basis, in the review process."(27)

In this context it is worthwhile to note an exchange of views between the Japanese government and the OECD's Committee for Invisible Transactions that took place in 1963, prior to Japan's joining the OECD. The Japanese government presented the following view:

> Since technical assistance largely contributes to the development of the economy, it is the policy of the Government of Japan to authorize contracts as simply and as speedily as possible. The Government of Japan moreover envisages as the ultimate goal complete liberalization, taking into consideration problems special to Japan, viz., the widespread existence of excessive competition and the actual situation of small and medium enterprises.
>
> However, while the above difficulties remain, cases might arise requiring minimum measures of adjustment thereon. The object of the screening system is to provide for such exceptional adjustments. The Government of Japan, therefore, will study forthwith a substantial reduction in the period required for examination.(28)

The OECD's reaction was as follows:

> At the time, the Committee for Invisible Transactions expressed its surprise "that a country with high growth targets and great technical skill should find it necessary to have any restrictions at all." The Committee recalled that, as a rule, technological exchanges lead to greater productivity and growth and that

> foreign exchange spent on foreign technical assistance and intellectual property should normally be more than offset by savings on imports or increased exports. No other Member State of the Organization has a reservation on either of the above items. The Committee did not endorse the Japanese claim that restrictions on foreign technology are an appropriate way of dealing with such problems as arise from the existence of numerous small and medium-sized enterprises. It added that it would seem that all industrialized countries have had to deal with similar difficulties at one time or another.(29)

No doubt Japan had by then emerged as "a country with high growth targets and great technical skill," and the Japanese government's trepidation, as far as "the actual situation of small and medium enterprises" was concerned, was not quite justified – viewed with hindsight, at least, since there were "few reported cases of notable disruptions among small businesses in Japan on account of the induction of foreign technology, and practically all applications for technical assistance to the small business sectors have been validated."(30) On the other hand, "the widespread existence of excessive competition" has, indeed, been a major problem rather "special to Japan" because of her unique industrial structure dominated by so-called keiretsu or industrial groups. Excessive competition has been responsible, partially yet significantly, for the prevalence of the importation of duplicate technology – defined as "agreements covering products and processes already brought into Japan by other Japanese firms through previous technology agreements."(31) Such import duplications rose from 36 percent of all the type A agreements in 1963 to 74 percent in 1970.(32)

But the true motive of the Japanese government, particularly that of the MITI, might have been its eagerness to retain control not so much over technology imports themselves as over private industry. As has been pointed out by Chalmers Johnson, the MITI exerts a most pervasive and potent influence on industry, despite the fact that it is the smallest of Japan's five major ministries in personnel size and budget; its power rests on the laws whose enforcement it overseas.

> The essence of MITI's power derives from its legally sanctioned "license and approval authority" (kyoninkaken), something that it has exercised in varying degrees and over changing activities since the Meiji period. All analysts agree that the ministry's increased reliance on administrative guidance during the 1960s was a direct result of the loss of significant portions of its license and approval authority due to economic liberalization.(33)

One interesting episode throws much light on the MITI's struggle for continued economic "guardianship."

> The incident most symbolic of the cracks developing in the Japanese economy occurred in the realm of Japan's changing international economic relations rather than on the domestic front. This was the aborted attempt by Shigeru Sabashi in 1963 to have the Diet pass a law promoting the development of specified industries. Sabashi was then director of MITI's Industrial Policy Bureau. Yielding to persistent U.S. demands for liberalization as early as 1960, the Japanese government announced a broad program for the liberalization of trade and foreign exchange transactions, and this stripped MITI officials of their power to restrict imports – their most effective tool in directing the progress of economic expansion. Sabashi and his men launched a counteroffensive that was aimed at reconstructing government-industrial unity and establishing a new industrial order for Japan, proclaiming that the bill they were proposing would protect Japanese industry against the pressure of foreign competition.
>
> Leaders of private industry, however, were thinking differently. They preferred to ride out the shock of liberalization by making voluntary adjustments through a relaxation of constraints under the Antimonopoly Law and the formation of cartels. They adamantly opposed Sabashi's bill for fear it would lead to a revival of bureaucratic control of the economy, and ultimately succeeded in having the bill rejected.
>
> This episode indicates how Japanese industry had by that time grown into a giant too powerful to be controlled by MITI bureaucrats. In fact it marked the watershed in MITI's postwar leadership over private industry.(34)

Thus there was strong indication that by the mid-1960s Japan's private industry had grown into a powerful economic entity that no longer really required the government's visible hand as a countervailing power in its bargaining with foreign technology suppliers. The effective guidance of, and the "third party" role played by, the government had succeeded in assisting Japanese industry to become noticeably monopolistic, nearly sufficiently so to cope with foreign monopolistic elements – if further relaxation of the antimonopoly regulations were to be made. Ironically, the very success of industrial policy came to undermine the economic authority of the government, as the work of coordinating economic activities was performed more and more by restored market forces rather than by official controls.

The effect of liberalization is clearly mirrored in the trend of Japan's technology-import contracts. The annual number of such contracts, both types A and B, is shown in figure 12.1. The number of type A contracts zoomed after the liberalization of 1968, strong evidence that substantial decontrol took place. Also, interestingly enough, although up until 1967 type B contracts had increased *pari passu* with type A contracts, since 1968 type B contracts have shown a declining trend. Perhaps this development may indicate in part that Japan's

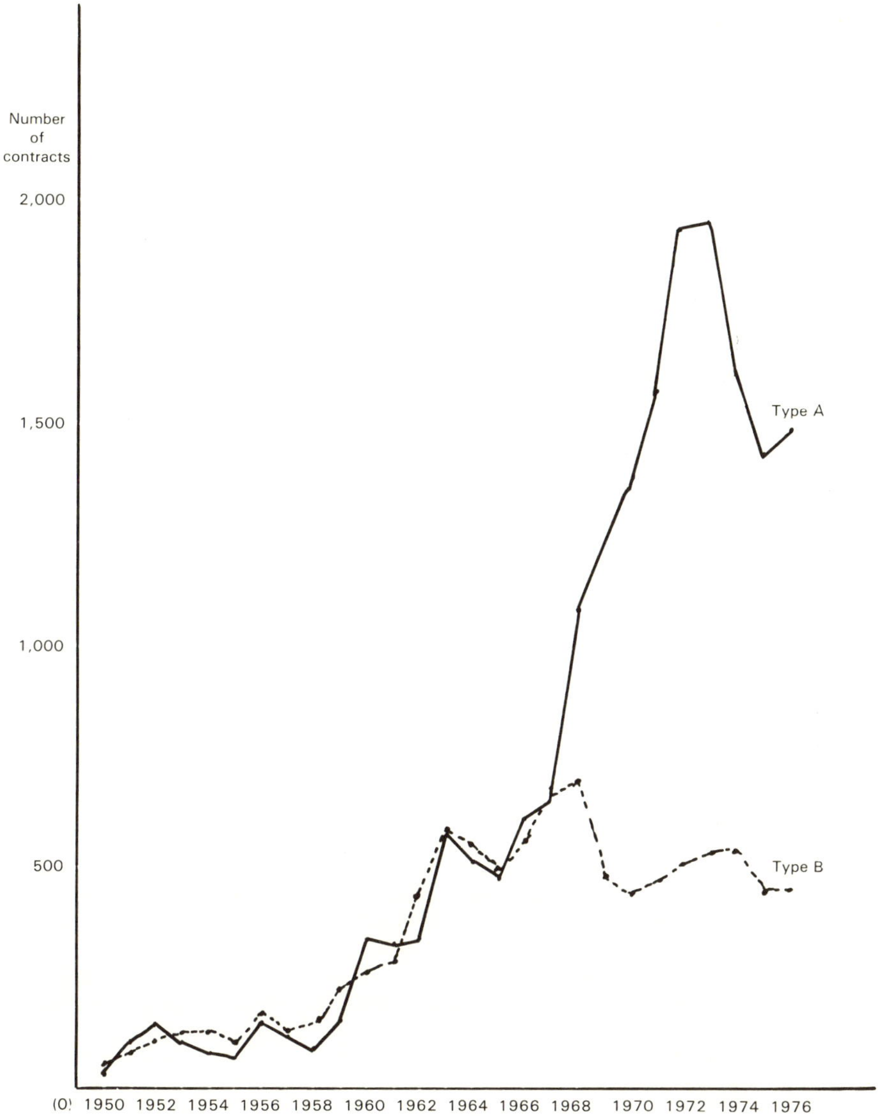

Fig. 12.1. Technology import contracts, types A and B, 1950-76.
Source: Science and Technology Agency, Kagaku Gijutsu Hakusho (White paper on science and technology), annual issues.

overall technological capacity had, by then, advanced to such an extent that industry had become less and less dependent on incidental or auxiliary technical assistance purchased under type B contracts. This conjecture may not be wide of the mark, since in the computer industry, an advanced, high-technology industry, type B contracts are even now increasing, along with type A contracts.(35) Yet one might also argue that because until 1968 the Japanese firms were simply not allowed to acquire foreign technology as freely under type A contracts as under type B contracts, they had to make as much use of the latter as possible. In other words, after 1968 those contracts that might well have been of type B came to appear as type A, moving, that is, from short-term to long-term contracts. It is equall noteworthy that the total number of technology import contracts began to decline in 1974, the year the hypergrowth of Japan's economy came to a sudden halt. This decline in technology imports was obviously a result of market forces, evidence that technology imports became essentially a function of the market mechanism.

Another set of statistics reveals an interesting change in one of the most controversial terms of contracts, export market restrictions. As figure 12.2 shows, those contracts that prohibited exports from Japan were only 10 percent of the total number of contracts in 1968, but they came to account for as much as 39 percent in 1976. Japan's Science and Technology Agency interprets this development as a sign of the toughening stance of, and more stringent supply conditions demanded by, foreign technology suppliers.(36) This does not necessarily mean, however, that the bargaining power of Japanese technology buyers has been deteriorating; it may even denote the emergence of more monopolistic behaviors on the part of Japanese industry in colluding with foreign technology suppliers to divide their markets. Put differently, they seem to have, indeed, been successful in "riding out the shock of liberalization by making voluntary adjustments through a relaxation of constraints under the Antimonopoly Law and the formation of cartels" since the mid-1960s.

In sum, the Japanese experience as an active technology importer demonstrates two important relationships: (1) a relationship between transnational diffusion of technology and the recipient's capacity to absorb and adopt outside-originated technology as an integral part of its self-sustaining knowledge system; and (2) another crucial relationship between the national control system and the official administrators' capacity to use the system most effectively in guiding the inflows of technology. In the first relationship, Japanese industry was able to shift very quickly from the diffusion phase to the innovation phase of technology acquisition by stepping up its own R & D activity, both adaptive and original.(37) In the second relationship, the Japanese government, that is, the ministries in charge of controlling technology imports, developed the control system on a both formal and informal basis, in close coordination with other key industrial policies, to achieve the national goals of economic growth and trade expansion. What the Japanese experience points out is (1) that the effectiveness of tech-

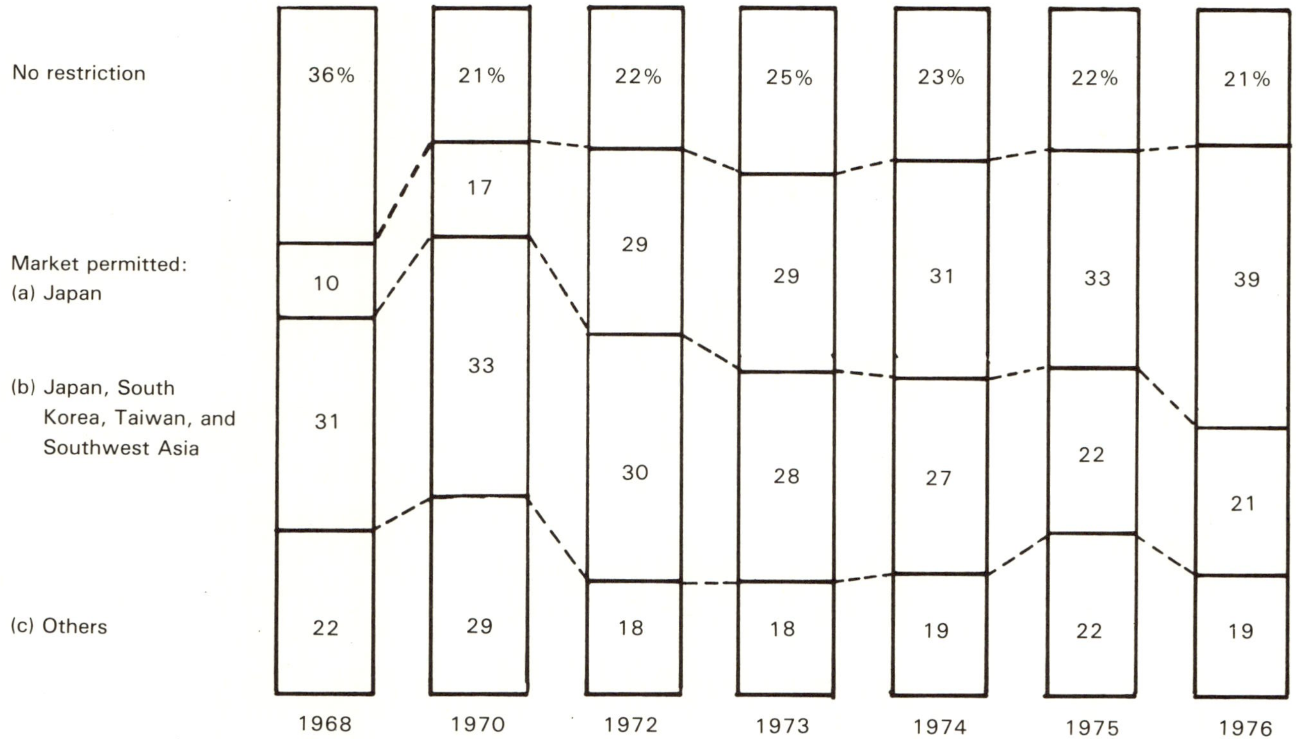

Fig. 12.2. Restrictions imposed on Japanese licensees.
Source: Science and Technology Agency, Kagaku Gijutsu Hakusho (White paper on science and technology), (Tokyo, 1977), p. 44.

nology transfer is basically a function of the recipient's absorptive capacity, and (2) that, similarly, the effectiveness of any control system is determined by the official administrators' capacity to run the system so as to achieve the goals established – two parallel sets of functions we tend to forget, since they are so obvious.

Another distinct characteristic of Japan's control system was its strong market-supplementary approach. It was designed to foster market forces on the part of private industry. And when private industry attained sufficient bargaining power vis-a-vis technology suppliers, the control system was largely dismantled. It was thus a successful case of the "infant technology" approach to controls.

EXPERIENCE AS A TECHNOLOGY SUPPLIER

Eager Disseminator of Industrialization

Although Japan is well known as an adroit user of foreign industrial arts, it has lately demonstrated its newly acquired capacity to supply industrial knowledge and skills very competitively, particularly to the developing countries. Indeed, Japan has emerged as an active disseminator of industrial knowledge, assisting the Third World to industrialize in such sectors as synthetic textiles, shipbuilding, iron and steel, petrochemicals, and aluminum smelting. This activity on the part of Japan is one that is likely to have significant implications for the rest of the industrialized world, because these are the very industries in which the industrialized countries are confronted with so-called adjustment problems in the form of glutted markets and excess production capacities.

The exporting of technology also often entails "externality" problems, such as those already considered serious in the United States, namely an adverse effect on trade balance and domestic employment, the latter being the most serious problem in the eyes of U.S. labor unions.(38) The Burke-Hartke bill is only a recent reminder. Technology outflows, particularly where the latest innovations are at issue, are looked at by some as a self-destructive mechanism of the Pax Americana, and some form of neomercantilistic controls is even suggested.(39) Perhaps surprisingly, Japan at the moment is but little concerned about these issues. On the contrary, government, industry, and even labor in Japan, far from restricting the overseas sale of technology, are on the whole favorably disposed toward encouraging it. One may be tempted to dismiss this question quickly by saying that Japan is still an insignificant supplier of technology in terms of both volume and the nature of its exportable technology, and that Japan has therefore not reached the stage at which the problems mentioned above would inevitably assume serious proportions. There is some truth in this off-the-cuff reaction, but the Japanese experience is certainly worth exploring in some depth, since it reveals a number of interesting "anomalous" features, features important for the formulation of technology policies on the part of other countries that export technology as well as those that import it.

Salient Features

In the first place, it may be a surprise to many to hear that since 1972 Japan has become a net exporter of technology so far as the balance of trade in new contracts is concerned,(40) although Japan's overall balance sheet, if existing old contracts are included, is still in a huge deficit, with the payments of royalties and fees about five times as large as the receipts. But because of a much faster growth in the receipts than in the payments, Japan's trade balance in technology has been showing steady improvement (table 12.2). This development is not surprising if one realizes how the Japanese themselves worked on imported technology to come up with many improvements and new technological breakthroughs.

Table 12.2. Japan's Balance of Trade in Technology
(in million dollars)

Year	A. Receipt	B. Payment	A ÷ B (%)
1966	19	192	9.9
1967	27	239	11.3
1968	34	314	10.8
1969	46	368	12.5
1970	59	433	13.6
1971	60	488	12.3
1972	74	572	12.9
1973	88	715	12.3
1974	113	718	15.7
1975	161	712	22.6
1976	173	846	20.4

Source: Science and Technology Agency, Kagaku Gijutsu Yoran [Indicators of science and technology] (Tokyo: 1978), p. 140.

Note: Based on balance-of-payments statistics prepared by the Bank of Japan.

In fact, the receipts from technology exports are not insignificant even by international standards. In 1976, for example, Japan received $173 million (converted from the yen figure at the then-prevailing rate of $296.6 to a dollar), an amount about one third of what Britain received and slightly more than half what West Germany received.(41) The annual receipts are equivalent to as much as 46 percent of R & D funds in the construction industry, 31 percent in the nonferrous metals industry, 28 percent in the textile industry, and 15 percent in the iron and steel industry.(42) These industries are exactly those which now produce surpluses in technology trade even when existing old contracts are included.(43)

As one might expect, knowing Japan's sphere of influence, both economic and political, the best customers of Japanese technology are Asian countries, especially neighboring countries such as South Korea and Taiwan; together they account for close to 40 percent of Japan's earnings from technology exports. The next best customer is Europe, with a 26 percent share; the third is the United States, accounting for about 16 percent; and Latin America and other regions take the balance of about 18 percent.(44) Japan is thus exporting more technology to the developing countries than to other industrialized countries.

Are there any major differences in the types of technology Japan exports to these different regions? How about the modes of transfer? Several years ago I wrote a report on technology transfers from Japan for the United Nations Institute for Training and Research (UNITAR) that pointed out the following characteristics:(45)

1. In general, Japanese technology exported to the developing countries is quite different in nature from that directed to advanced countries. The latter consists largely of patented, high-level technology that is usually transferred through licensing agreements. In contrast, the technology transferred to the developing countries is not so specific in its technical nature. It is mostly in the form of know-how or even in the form of general industrial experience, covering a wide spectrum of production activities (such as assembly techniques, material selection, combination, and treatment techniques, machine operation and maintenance techniques, provision of blueprints and technical data, training of engineers and operators, plant layout, selection and installation of machinery and equipment, quality and cost controls, and inventory management).

2. Although there are some cases of licensing arrangement, ownership (if not total) and management of overseas ventures are the predominant transfer mechanism of Japan's technology exports to the developing countries. This is due partly to the fact that the Japanese technology exported to the developing countries is mostly embodied in the know-how and skills pointed out above and cannot be "disembodied" from the management. Besides, a typical contract of technology transfer to the developing countries tends to be comprehensive, extending from production to management, and even to marketing. Its effectiveness necessarily requires the actual participation of Japanese firms in the operation of local industrial ventures. And this usually results in the formation of joint ventures.

3. Japanese firms have an additional motive for opting for direct foreign investment. Since they cannot expect profits from the sale of standardized technology, the Japanese transferors try to compensate for technology transfer by securing or monopolizing the supply of intermediate goods and raw materials. To this end, capital ownership and management participation become strategically important.

4. Japan operates as an intermediary in technology transfers between industrialized and developing economies. This role derives from the intermediate status of the Japanese economy and the incessant catching-up process of its industrial modernization in the postwar period. A rapid assimilation of Western technology has contributed to the unprecedented pace of industrial transformation in Japan, and its export of industrial know-how to the developing countries has essentially been the spillover of its own technological transformation.

5. Only recently did Japan itself acquire industrial skills on a massive scale from advanced Western countries. For this reason, Japan has a unique, valuable experience ("modernization experience") to share with the developing countries.

6. Because of the narrowness of their markets, the developing countries unavoidably face what may be called the diseconomies of scaling down. Yet the goods produced with Japanese technology there are mostly standardized and technologically unsophisticated ones; they are of the same general kind as those produced by small- and medium-sized firms in Japan. There is not much problem of producing on a small scale. Besides, the Japanese themselves have learned how to cope with the problems of small-scale operations at home. They can apply the lessons of their own experience. All in all, Japanese technology has a comparative advantage in labor-intensive small-scale ventures.

These features are still predominant characteristics of the nature and manner of Japan's technology transfers. But the past eight years have seen Japan's role as a technology supplier further expand and have produced many additional features we must recognize; and the passage of time has also given me a wider perspective from which to view the phenomenon and more time in which to analyze it.

Despite the noticeable expansion of technology outflows and some serious structural problems at home, Japan appears to still be committed to furthering the overseas sale of technology, without much concern about fostering competition overseas. In fact, a "catching-up competition" against Japanese products is strongly felt not only in third country markets (notably in the United States) but also in the home market itself. Yet there is no move yet on the part of Japanese economic planners or even on the part of labor unions to hinder technology outflow. In what follows, we will examine the forces underlying Japan's exports of technology to the developing countries and discuss why they are largley left to the marketplace – without much direct interference by the government.

Developing Countries As Customers

Japan's eagerness and capacity to export technology to the developing countries, particularly in Asia, are closely related to its success as a technology importer. As noted earlier, the persistent acquisition of front-edge technology has resulted in a continuous transformation and upgrading of industry, initially away from traditional light manufacturing sectors and toward modern heavy and chemical sectors, and then most recently toward knowledge-intensive sectors, notably electronics. In this process, industry has experienced the appearance and shedding of quickly outmoded technology and production activities on the bottom rung of the industrial ladder.

This does not mean, however, that in postwar Japan the labor-intensive light manufacturing sectors immediately declined in absolute terms. On the contrary, these industries expanded also. According to Hugh Patrick and Henry Rosovsky:

> The share of labor-intensive light industries, notably processed foods and beverages and textiles, declined sharply. In contrast, electrical machinery (especially consumer electronics) and transport equipment (especially ships and automobiles) grew particularly rapidly. It must be kept in mind that these changes in the relative importance of industries occurred within the context of overall rapid growth. Even industries in relative decline grew absolutely, in fact rapidly in comparison with other industrial nations.(46)

As a matter of fact, early postwar Japan relied heavily on the export of labor-intensive light manufactures as a way of earning precious foreign exchange. In 1950, for example, Japan's top export item was textiles (cotton, rayon, and silk products), which accounted for more than 36 percent of the total value of exports. But as the industrial transformation proceeded, Japan's export pattern gradually shifted to one of more sophisticated, higher-value-added types. By 1970 the top export items had come to be steel, ships, automobiles, electronics products, and synthetic textiles – a change that closely mirrored a continuous industrial transformation at home.(47)

This rapid structural transformation caused many strains as the capital-intensive modern sector dominated by large-scale enterprises expanded along with the labor-intensive sector composed mostly of small-scale enterprises, all against the backdrop of the structural duality or the "unevenness" of Japanese industry.

One such strain appeared in the labor market. In the mid-1960s Japan began to experience an acute shortage of labor, particularly of young high-school graduates willing to work as blue-collar workers. The shortage of workers occurred in all age groups except the 50-year-old-and-over group, but the most serious shortage was in the 19-year-old-and-under category. The ratio of job offers to job seekers for this age group rose from about 2 to 1 in 1965 to no less than 5 to 1 in 1970. As

James C. Abegglen put it, "When Japanese companies complain of a 'shortage of labor' they usually are in fact complaining that middle school graduates – the least costly in terms of wages and most accommodating in terms of starting jobs – are diminishing sharply in numbers as the educational level of the whole society moves sharply upward."(48)

This tight labor market for young workers, and their rising wages, affected small- and medium-sized firms more adversely than large ones. The latter were, by and large, more successful in holding down the rise of total corporate wage payments by narrowing the wage differentials between age groups based on the traditional seniority system. In contrast, small- and medium-sized firms had wage differentials that were too narrow to manipulate. Furthermore, to attract new graduates they often had to pay higher wages relative to the workers' productivity than large firms, which were in a position to instead offer better fringe benefits and greater job security.

Despite their gleaming modern facilities, even big corporations experienced difficulty in recruiting young factory workers for assembly-type operations, which were considered unbearably monotonous and tedious by youth. The worst squeeze, however, was on small- and medium-sized manufacturing firms in the labor-intensive sector, where working facilities were the least attractive. The White Paper on Small and Medium Enterprises for 1968 made the following report on the situation:

> At home, the accentuation of labor shortages eliminated low wages, the most important factor on which the existence of small enterprises depended. On the international scene the small enterprises have been directly or indirectly hit by increases in competition coming from the developing countries, the progress in the liberalization of entry of foreign capital to Japan and the stronger internationalization of the Japanese economy.(49)

As a consequence, an intensive search started for labor "the least costly in terms of wages and most accommodating in starting jobs" – a search going beyond national boundaries and extending to labor-abundant countries. Thus the strained labor market and the catching-up competition gave a strong push effect to the overseas migration of Japanese firms, notably small-scale ones operating at the labor-intensive bottom level of Japan's industrial structure.

There were also equally strong pull effects on the part of these host countries. In the first place, these countries had a plentiful supply of young workers, since more than half of their population was under 19 years old, the very category of labor in which Japan had begun to experience a serious shortage. Moreover, Taiwan opened a tax-free export processing zone in 1955, and South Korea followed suit in 1971, these being zones designed to attract foreign investment in labor-intensive manufacturing operations under favorable tax incentives. Geographical proximity clearly favored Japanese firms, because the processing activities in the zones involved the shipping and reshipping of parts and components and their assembled products; transportation costs were obviously an important consideration. In addition, wages in Asian countries were in many cases purposely kept low, while labor unionism was suppressed by the local governments.

The net result has been an intra-Asian spread of Japan's dual industrial structure; small-scale, labor-intensive manufacturing operations that had to depend on low-wage labor started to migrate to neighboring countries.(50)

This structural adjustment and its accompanying technology transfer can be placed in a theoretical perspective. According to Kiyoshi Kojima:

> Let us now introduce foreign direct investments which are undertaken by a firm in X-industry of country A [Japan] so as to improve technology of the same X-industry in country B [the developing country]. Such direct investment is stimulated by the fact that the production of X-goods . . . under the international commodity price ratio . . . gives lower rewards both to labour and capital in that industry as compared with the other industry Y, and labour and capital must shift from the less profitable X-industry to the more profitable Y-industry (at home). . . . This is an internal structural adjustment. But there is another possibility for a firm in X-industry to use its accumulated technology and managerial skills: that is in foreign direct investment.(51)

Kojima's model can be elaborated on in the following way: When rapid change in conditions of factor endowments results in loss of competitive edge, the contracting industrial sector must release economic resources that then can be absorbed either internally (through intracountry, intersector transfers) or externally (through intercountry transfers). Here it is important to distinguish between those resources that are transferable to the expanding sector at home and those that are nontransferable at home. The former include non-sector-specific resources such as labor, industrial sites, money capital, and part of physical capital (equipment, plant, office buildings, and the like), while the latter consist of sector-specific resources such as technology. Interestingly enough, however, those intraeconomy nontrasnferable resources actually become transferable internationally – by way of various transfer mechanisms – to the same industrial sector that is about to develop in the developing countries because factor endowments conditions there are more congenial for that particular sector.(52)

This type of resource transfer has an important implication for the welfare of the world as a whole (setting aside, for the moment, the inevitable question of the distribution of such gains between the two countries involved), since such resources, if kept at home, will be simply wasted. In other words, so far as the sector-specific technology of the declining industry is concerned, the opportunity cost of technology transferred internationally is zero (or near zero because technology released by those firms that have abandoned domestic production may be used by other surviving firms in the same declining sector at home). Hence this kind of technology transfer is definitely a positive-sum game, a characteristic that underlies a generally strong enthusiasm displayed by both Japanese industry and developing host countries for this economic activity.

Foreign direct investment is now conceptually treated as a phenomenon of knowledge transfer rather than as one of capital transfer in the conventional sense. This approach is, for example, emphasized by Richard E. Caves:

> For the possession of some special asset to lead the firm to invest abroad, two conditions must be satisfied. First, the asset must partake of the character of a public good within the firm, such as knowledge fundamental to the production of a profitably saleable commodity. Any advantage embodied in knowledge, information or technique that yields a positive return over direct costs in the market where it is first proven can potentially do the same in other markets without need to incur again the sunk costs associated with its initial discovery. Knowledge would seem to be the prototypical asset displaying the character of a public good proprietary to the firm, but it is not the only one. The essential feature of an asset conducive to foreign investment is not that its opportunity cost should be zero, but that it should be low relative to the return attainable via foreign investment.(53)

Caves is referring to some substantially unique asset (such as patents, differentiated products, or marketing skills) specific to a large, oligopolistic firm in some technology-based, capital-intensive industry. Yet it is interesting to observe that the technology transferred by Japan's relatively small-scale firms operating in more competitive, labor-intensive industries also displays the character of a public good – not only within the firm but also within the industry. (In this sense, this type of Japanese technology is more truly a public good than that possessed by the oligopolistic firm envisaged by Caves.) There are additional differences: (1) the type of Japanese technology involved is more industry-specific (that is, specific to a declining industry) than firm-specific; (2) Japanese firms' willingness to transfer technology stems from the structural adjustment of their economy rather than from their own technological progress; and (3) the condition of the near-zero-opportunity cost of technology to be transferred prevails because without international transfers such technology would simply be discarded at home (that is, the return attainable via foreign investment – insignificant though it may be – is always positive relative to the opportunity cost).

This way of looking at the transfer of Japanese technology to the developing countries goes a long way toward explaining why "the

Japanese transferors try to compensate for technology transfer by securing or monopolizing the supply of intermediate goods and raw materials" and why "capital ownership and management participation become strategically important" for their technology transfers. Here the mode of transfer itself (rather than the technology transferred) becomes a source of quasi-rent for the Japanese firms.

This rather unusual situation surrounding technology transfers from Japan's declining industries has another important theoretical implication as well. Japanese firms in such a contracting sector, struggling to survive, are finding it much easier to cross national borders and reestablish themselves in more labor-abundant countries than to cross industrial boundaries by entering the new expanding industrial sector at home. This "cross the national rather than industrial border" behavior (or what Caves calls "high international coupled with low inter-industry mobility of capital")(54) is considered typical of highly oligopolistic firms. Yet we can observe the same pattern of behavior in Japan's relatively competitive industries.

All this means that the urge to go abroad is, as was emphasized before, greatest among the small and medium-sized firms whose operation is relatively intensive in the use of labor. They usually produce light manufactures such as textiles, simple electrical appliances, sundries, and simple electronics parts and components. Many of them are subcontractors, producing goods on a subcontracting or consignment basis, or "captive firms" of large dominant firms, including trading companies. It is not unusual for these small firms' overseas investment to have been encouraged and both financially and managerially supported by trading companies serving as overseas investment partners. These ventures are usually joined in by local partners (formerly, in many instances, the local distributors of Japanese exports). As a result, the usual pattern is that Japan's trading companies take 20 to 30 percent of ownership, Japanese manufacturers another 20 to 30 percent, and the local partners take the rest.(55)

Normally, the investing small firms that lack capital "capitalize" their technology and good will as equity capital in their overseas ventures, and in addition contribute what little capital they can raise and invest (or lend as loans). The close tie-in between Japanese trading companies and manufacturers explains, at least in part, why as much as 80 percent of Japan's technology exports is said to be intermediated by trading companies.(56)

The motive of trading companies in setting up overseas ventures was initially somewhat suspect; they were more interested in creating their own business opportunities to export capital equipment, sell input materials, and market whatever their overseas ventures produced than in running successful ventures as joint partners. This sort of attitude on the part of trading companies led to unprofitable operation of these small-scale overseas ventures, since their manufacturer partners, inexperienced as they were in international operations, could hardly be expected to succeed in running their overseas ventures without managerial and financial help from their trading companies. But the latter eventually realized the seriousness of the situation and reorganized their corporate structures in order to play a more active role as investors than as traders.(57)

In sum, the rapid process of industrial transformation that occurred under the stimulus of imported technology rendered obsolete a great deal of existing or even recently acquired industrial knowledge and skills and made them useless at home – unless they were transferred to other countries that could make use of them. A massive infusion of new technology exerted the Schumpeterian force of "creative destruction." That is, the quick transfer to the expanding sector at home of many economic resources released from the contracting sector was made all the more difficult because of the capital-intensive sector's dramatic technological sophistication and more nearly monopolistic structure. In other words, the degree of the monopolistic elements of the economy as a whole must have increased as the deepening of capital and technology occurred in the expanding sector with the introduction of advanced foreign technologies. And the resulting "structural gap" raised barriers to entry into the expanding sector by the firms operating in the contracting sector. Thus labor shortages and the equalizing tendency of wages compelled the firms in the contracting sector to transfer production (hence technology) to labor-abundant neighboring countries. The ultimate consequences were the "cross the national rather than industrial boundary" behavior and the "intra-Asian spread of Japan's dual industrial structure."

Looked at from this viewpoint, the transfer of technology through Japan's overseas ventures in labor-abundant countries has been essentially a by-product of Japan's industrial transformation, and has been perfectly compatible with its national interest. It is not surprising, then, to hear a high-ranking Japanese government official advocate such transfers straightforwardly:

> While certain segments of the industrial structure are being encouraged, there must be modification of those industries where productivity is low, where technology is stagnant and where there is reliance on simplistic intensified use of labour. . . .
>
> The solution of this problem is to be found according to economic logic, in progressively giving away industries to other countries, much as a big brother gives his out-grown clothes to his younger brother. In this way, a country's own industries become more sophisticated.(58)

There is a catch, however, in this strategy: What will happen if the younger brother starts growing faster than his big brother and demands the same clothes he is now wearing? Japan is about to be confronted (in fact is already confronted to some extent) with this dilemma, now that Japan has shifted from an era of fast growth to one of slower growth at the very time when neighboring countries are accelerating their economic growth.

In short, so far as Japan's transfer of technology to developing countries is concerned, it has been at least up to now, a perfectly rational economic behavior from both a private and a public point of view, a behavior governed by the macroeconomic forces of both domestic and international markets. There has been no need for Japan as a technology supplier to institute controls so as to intervene in the market. The government did, in fact, introduce some additional incentive measures to help small and medium-sized firms migrate overseas, since most of them are incapable of doing so on their own. These measures will be discussed later. The whole situation so far can be summed up as the "market does work but more is desired" situation.

Labor Acquiescence

Even though such technology transfer has been largely in Japan's national interest, one may still be puzzled why labor has acquiesced in management's decision to "export" jobs to other countries. Here we must understand some institutional characteristics of Japan's labor market beyond the fact that there have been serious shortages of young factory workers.

As is well known, Japanese workers are "organized" around their own companies. All the workers, both blue collar and white collar, employed by a single company comprise a company-wide labor union. There is no industrial craft union organized on an intercompany, economy-wide basis along the occupational lines of crafts. Most individual enterprise unions in turn are affiliated with nationally federated organizations, Sohyo (predominantly white-collar workers, notably in the government sector) and Domei (mostly blue-collar workers), the former twice as large in membership as the latter.(59)

Sohyo, with its strongest constituency in government workers (national railway and postal workers and the like), has little worry about the job-draining impact of overseas investment and technology outflows.(60) Neither do its private-service-sector members, because the jobs transferred are basically factory-related employment. Even the blue-collar workers belonging to the Domei-affiliated unions have so far not been much disturbed by the shift of production overseas. The main reason has been largely the practice of lifetime employment, and the special care and efforts taken by management to minimize the "laying-off" costs of output curtailment and plant shutdowns on their workers, even when such actions were necessary in the past. More important, because of their strong sense of belonging to their own individual companies, the economic interests of union members are invariably identified with those of their own companies.(61) The company's economic destiny is identical with the workers' own; if it is necessary for the interests of their company to sell technology overseas, so be it. If such action causes a curtailment of the company's operation affecting its workers, the company normally tries to transfer them to other production units within the company or to other companies within the company's affiliated industrial group before it dismisses excess workers.

Actually, those workers who are most likely to be affected directly by the overseas shift of labor-intensive production are not union members at all. There are millions of nonunionized workers who actually serve as shock-absorbers, so to speak. This institutional arrangement is referred to in a recent remark made by a leading businessman, Ichiro Hattori, senior managing director of the Seiko watch company:

> There are problems which arise from the management's effort to maintain stable employment for regular employees even during bad times while trying to reduce slack capacity at the same time. In such circumstances, the management often chooses to dismiss seasonal or part-time workers, who are not considered an integral part of the corporate community, and to sharply reduce the purchase from subcontracting firms by transferring the work to the internal employees. In my opinion, this obvious imbalance between big corporations and small firms, and the inequality among their employees, are likely to raise a serious issue in the Japanese employment practice, which provides another evidence that the employees together with their labor union are concerned primarily with the well-being of their own community and clearly set a boundary around them against the outside world.(62)

Thus the seasonal or temporary workers within large enterprises and subcontracting firms (typically small-scale firms, including cottage-type, family-owned enterprises) bear the brunt of whatever economic dislocations Japan's technology transfer or overseas investment may cause by shifting labor-intensive production to neighboring countries. The seasonal or temporary workers are mostly either middle-school or female workers. As pointed out earlier, the supply of middle-school as labor has been dwindling in absolute terms. Female workers, on the other hand, have traditionally been employed as temporary workers while they are young, for a few years prior to their marriage. True, the participation of married females in the labor force is on the rise; the rate of their participation increased from 24 percent in 1970 to 44 percent in 1977. But as much as 83 percent of them prefer temporary employment, since they consider their jobs secondary to their household.(63) Consequently they can be hired on a temporary basis when need arises and can be dismissed easily without creating much social strain. Thus the enlarged female labor force seems actually to be providing more resiliency and flexibility to Japan's labor force – as long as the majority of them remain temporary workers.

Another important reason for the subsector workers' acquiescence to the export of jobs is that while low-paying manufacturing jobs in factories are exported, new employment opportunities have been created in Japan's swiftly expanding service industries. For example, manufacturing employment declined from 11,440,000 in 1970 to 11,260,000 in 1977, whereas service-related employment expanded from 12,890,000 to 16,510,000 over the same period, a gain of 28 percent.(64)

And there still seems to be room for the Japanese economy to expand the tertiary sector, since the sector is relatively small compared to its counterparts in other advanced countries (in 1975, for example, it was 52 percent of the GNP in Japan, as compared to 65 percent in the United States and 58 percent in Sweden).(65)

For the foregoing reasons, despite a sharp rise in unemployment rate (from 1.4 percent in 1973 to 2.3 percent in 1978), neither technology transfer nor overseas investment has been brought up as a direct cause – or even as a scapegoat – for unemployment. True, there are passing remarks in official documents to the effect that the "chasing-up" competition from the developing countries has been actually assisted by Japan's own drive for plant exports, and for capital and technology tie-ups with local interests, but there is no serious expression of concern about this issue.

The Balance-of-Payments "Surplus" Problem

Although Japan runs a huge trade deficit with other Asian countries as a group because of Japan's dependence on importing natural resources, it constantly maintains a large trade surplus with resource-poor Asian countries, notably South Korea, Taiwan, Hong Kong, and Singapore. One major cause of Japan's trade imbalance with these countries is their heavy reliance on Japanese supplies of industrial goods, including input materials to be used by Japanese manufacturing ventures located in these countries.

It is important to note here that Japan's overseas ventures do not necessarily use Japanese technology. One very comprehensive survey taken by MITI in 1975 showed that 5.3 percent of the manufacturing ventures surveyed (that is, 47 firms out of 886 firms surveyed on a worldwide basis) were based principally on production technology not supplied by the Japanese partners. Nevertheless, so far as Japanese manufacturing ventures in Asia and Latin America are concerned, nearly all of them (98 percent) were using as their principal production methods either the Japanese partners' technology or some combination between Japanese and local or third-country technology.(66)

The developing countries' demand for Japanese technology and capital was initially created as a result of their policies of import-substituting industrialization. Particularly in the late 1950s and the early 1960s, many Japanese manufacturers were compelled to set up small-scale local production in the developing countries, notably Brazil, Taiwan, and Thailand. These Japanese ventures were primarily "tariff factories" or assembly operations that were heavily dependent on raw materials, parts and components, and other input materials supplied from Japan. In many instances, the investing Japanese firms themselves asked the host countries to raise tariffs so as to make their operation economically feasible. This practice was widely known and in fact often a target of criticism. But it was perhaps necessary for the investors, since the local markets of the Asian host countries were so small that the investors were inevitably confronted with a problem of scaling down that raised the local unit cost of production.(67) Without sufficiently high protective tariffs, the investing Japanese firms could not have competed either with their own exports or with exports from their Japanese competitors.

As the local markets expanded, so did the imports of industrial goods from Japan. In an effort to reduce local production costs, many of these overseas manufacturing ventures gradually shifted from local to export markets. Local governments began to emphasize export-oriented industrialization (or what economists call "an outward-looking strategy") instead of import-substituting industrialization (or "an inward-looking strategy"), and provided various incentive programs. This policy change, which occurred in the early 1960s, also coincided with Japan's need to transfer labor-intensive manufacturing activities overseas.

The swift expansion of local industries meant an ever-rising need to acquire input materials from Japan, since the developing countries, particularly in Asia, followed close on the heels of Japanese industry, taking up exactly those industrial activities that had become incompatible with Japan's rapidly changed factor endowments (hence the activities in which Japan was quickly losing trade competitiveness) but that were ideally suited to the relatively labor-abundant developing countries (namely, activities that could be developed into competitive ones there with the use of Japanese capital and technology). In fact, South Korea, for example, officially designated a dozen industrial sectors as ones the country considered desirable to develop – precisely those so-called twilight industries that Japan had begun to lose competitiveness in and was therefore eager to transplant.

All these developments led to an irony: the developing countries' initial effort to substitute local production for imports from Japan actually resulted in a much greater import dependency on Japan. In a sense, they unwittingly came to be locked into such trade relationships. For the initial imports of finished products from Japan were stoppable if, for example, the developing countries wanted to cut back for balance-of-payments reasons. Besides, similar imports could easily be purchased from somewhere else without much difficulty. But the industrial materials that newly established local manufacturing industries had to import from Japan were so vital that they could not be cut off even temporarily without causing serious economic disruptions. What is more, this "locked in" effect has been intensified, particularly in those countries that have huge trade deficits with Japan, since they usually agree to balance a trade deficit with capital inflows – that is, with more direct foreign investment from Japan.(68)

The host countries are, however, not totally irrational in this behavior. For in the first place Japanese ventures, being relatively labor-intensive in operation, can contribute considerably to local employment. The standardized and unsophisticated technologies transplanted by these ventures are, in a sense, "appropriate" ones for the developing countries (although criticism has been voiced from some quarters of the developing host countries that Japan exports "obsolete," not "appropriate," technology). More important, by moving into those industrial activities in which Japan is losing trade competitiveness and by taking over Japan's previous export markets, the host countries are able to develop export industries.(69) In fact, they have been successful in capturing a growing share of the Western market for such light manufactures as textiles, toys, plywood, black-and-white TV sets, radios, and the like, while Japan's market share in these exports has declined considerably. In other words, their trade imbalance with Japan has been translated into their export drive to (and trade surplus with) third country markets, particularly the United States and Europe.

Japan's transfer of technology and capital has so far definitely contributed to its balance of trade. Although Japan has lost the bottom levels of these advanced countries' markets, it has captured the upper market segments very successfully with its own new exports, such as color TV sets, home video tape recorders (VTRs), microwave ovens, digital watches, and calculators. A study made by MITI on this export-creating effect of Japan's overseas manufacturing investment estimated that in 1973 the value of investment-induced exports was as high as \$3,090 million, whereas the value of exports replaced by overseas production was \$1,650 million. That is, overseas production encouraged, on the whole, about \$2 in new exports for every \$ exports replaced.(70)

Transfers to Other Industrialized Countries

So far we have focused on the dissemination of rather standardized, soon-to-be obsolete or even already obsolete technology from Japan to the developing countries in the form of Japan's overseas manufacturing ventures. In other words, this type of technology transfer is occurring at the very end of the invention-innovation-diffusion sequence. (This does not mean, however, that Japan always sells only those bottom-end technologies. When Japan exports modern plant equipment, such as that required for petrochemicals and steel mills, the latest technology and operational know-how are sold as supplementary services via technological assistance contracts.)

In this section I will examine Japan's transfer of the latest technology to other advanced economies. Given the fact that technology sold to these countries is normally Japan's front-edge innovation, can such innovation-phase transfers still be regarded as an activity in Japan's national interest and hence something to be encouraged, as it is in the case of technology transfers to the developing countries? Don't such transfers jeopardize Japan's newly attained technological strength? Don't they cause unemployment in skill-intensive sectors?

The technology Japan transfers to other industrialized countries is basically of three types:

1. technology transferred under cross-licensing agreements;
2. technology transferred unilaterally under licensing agreements; and
3. technology "transplanted" through overseas ventures.

We will take a look at each of the three.

Cross-Licensing

With increasing recognition of Japan's technological capability – and as a way for foreign technology suppliers to secure her latest industrial knowledge – the number of cross-licensing agreements and joint R & D contracts has been showing a rising trend ever since the mid-1960s (see table 12.3).

This route of technology transfer involves the instant outflow of Japan's very latest technological accomplishments. And it always assures two-way, mutually beneficial transactions. Even so, this type of exchange may not always be beneficial from either a national or a global welfare point of view. For it may intensify monopolistic rivalry between dominant firms across national borders; under a cross-licensing agreement, for example, two dominant firms located in different countries can mutually reinforce monopolistic position vis-a-vis their compatriot firms. In other words, internal diffusion of technology is hampered, while international diffusion is accomplished only between the dominant firms. But at the moment, perhaps because the number of cross-licensing agreements has not yet reached a significant enough proportion to raise the question, this type of transfer arrangement appears to be left out of the policy considerations of the Japanese government.

There is, however, one exception, involving the case of computer-related VLSI (very large-scale integration) technology separately developed since 1976 by two research groups under the leadership of the Japanese government. (One group, led by MITI's Agency of Industrial Science and Technology, comprises Nippon Electric, Hitachi, Fujitsu, Toshiba, and Mitsubishi Electric; the other, joined by the MITI's first three companies, is centered on the governmental Nippon Telegraph & Telephone Public Corporation.) The two groups initially showed no intention of licensing any outside firms, but later decided to partially open their respective technology to both domestic and foreign firms on a case-by-case basis. This decision was reportedly prompted by the strong demand for licensing by U.S. computer makers, notably IBM, which maintain cross-licensing contracts with the Japanese companies involved in the government-sponsored research groups.(71) Thus cross-licensing has become a legal constraint on an attempt by the Japanese government to withhold a certain type of the latest innovations. As the Japanese government's role in R & D on frontier industrial knowledge expands in the form of research collaboration with the private sector, this kind of problem is expected to arise more frequently.

Table 12.3. Cross-Licensing Agreements and Joint R & D Contracts

Industry	1970	1971	1972	1973	1974	1975	1976
Chemicals	4	12 (3)	15 (2)	7	11 (2)	5	12
Metals	2	4	5	3	1 (1)	4 (1)	3
Nonelectric machinery	3	11 (2)	4 (2)	14 (1)	14 (5)	12 (2)	34 (6)
Electric machinery	10	25	8	32	15 (1)	26	19
Others	7	2	5	12	9 (1)	2	1
Total	26	54 (5)	37 (4)	68 (1)	50 (10)	49 (3)	73 (6)

Sources: Compiled from Science and Technology Agency, Gaikoku Gijutsu Donyu Nenji Hokoku [Annual Report on the Induction of Foreign Technology], annual issues.

Note: Number of cross licensing agreements, both type A and type B, including joint R & D contracts shown in parentheses.

Unilateral Licensing

Since the buyers in other industrialized countries have high absorptive capacities, and since until only recently production costs there had been generally higher than in Japan, Japan's technology has usually been sold under licensing agreements instead of through direct foreign investment. If a particular piece of technology sold is only a small component of the total system of the technological capacity of a licensee, the sale of such technology contributes only incrementally to the competitive strength of the foreign buyer, while such a sale may constitute a substantial source of income for a Japanese licensor. Furthermore, Japan's technology, especially in an instance where it has only recently been patented, may still be in rudimentary stages of development, requiring expensive R & D for commercialization, a task that may be beyond the means of the licensor.

Even if the technology exported is significant in nature, assisting the foreign licensees to improve their productivity and competitiveness substantially (as is often the case with the sale of Japan's steel-making technology), Japanese possessors of technology may be quite willing to make such an export, partly because there is no way of exploiting technology in the internatioal market other than by licensing; the Japanese may either have already captured a sizable part of the foreign market and may even be operating under "voluntary" export controls, or the alternative of setting up direct manufacturing ventures abroad using such technology may be prohibitively expensive and risky. Given these constraints on both exporting and direct local production, licensing is the only remaining avenue through which to exploit the firm's technical advantage. In fact, this type of licensing is referred to as a "soft" export in Japan's industrial community. Corporate emphasis on soft exports began in the early 1970s, when Japan began to experience huge trade surpluses. Many firms started to set up special sales forces to increase revenues from technology exports through licensing.

Direct Foreign Investment

Since the industrialized Western countries normally impose no restriction on foreign ownership except in certain industries, Japanese ventures there are usually wholly owned subsidiaries, particularly if they have some firm-specific technology or if their advantage resides in managerial skills. Hence this type of venture constitutes a technological transplant in a foreign enclave rather than direct technology transfer to the local economy. The venture is much more of an enclave when most of the parts and components are imported from Japan and local production remains basically an assembly operation.

A recent sharp rise in the value of the yen has made Japanese exports less competitive and overseas ventures more attractive, notably in the United States. Lured also by the favorable investment climate of the United States, Japanese manufacturers, such as those of miniature bearings, color TV sets, and chemicals, have moved into local production. For example, practically all the major Japanese manufacturers of color TV sets are now in the United States, some of them already successful as "local" producers and others about to begin operation. It is noteworthy that Hitachi's plan to tie up with General Electric to set up a joint venture in the United States was disapproved by the U.S. Justice Department on the ground that such a tie-up would eliminate "significant existing and potential competition."(72) The deal would have merged "Hitachi's advanced technological capabilities and 2.2 percent of the U.S. market share with GE's U.S. TV production facilities and more than 6 % share of the market."(73) Technology transfer was thus prevented by the U.S. Justice Department for antitrust reasons. In fact, in 1978 there was a similar incident: Japan's Toray Industries, the top producer of carbon fibers, was negotiating to form a joint venture with Union Carbide in the United States. However, "the plan was abandoned because of their anticipation that the Toray-UCC joint venture might infringe upon U.S. antitrust laws."(74) Toray instead licensed its technology for carbon-fiber production to Union Carbide so that the latter could build "the world's largest plant in the United States."(75)

In connection with the decision of Japan's color TV industry to shift production to the United States rather than to export from home, one may ask whether such move creates a serious unemployment problem for Japanese manufacturers of color TV sets. The answer is largely in the negative. They are highly diversified manufacturers of consumer electronics; their exports of color TV sets are increasingly being replaced, for example, by those of home VTRs, a hot export item. (The Wall Street Journal reported: "Through the first 10 months of 1978, VTR exports to the U.S. totaled 439,257 units with a value of $273 million – gains of 98 percent in volume and 50 percent in value from a year earlier. Color TV exports to the U.S. during the period were valued at $353 million.")(76) Secondly, there are newly emerged markets to which they can divert their exporting activities. For example, Hitachi, which now has decided to set up shop on the outskirts of Los Angeles all by itself, is reportedly pinning its hopes on China as a new export market, from which "it has received a large order for 100,000 color TV sets."(77) The Middle East has also become a significant "bazaar" for consumer goods produced in Japan, one catering to both locals and foreign migrant workers. The rigors of unemployment are further cushioned by Japan's traditional style of personnel management, discussed earlier.

Promotion of Technology Exports

Although Japan once instituted a rather elaborate and effective system of controls designed to restrict the importation of foreign technology, it has, asymmetrically, left the exportation of her own technology largely to the dictates of international market forces. In general, the overseas sale of technology has been, and still is, looked upon just like any other sale of exportable goods and services; hence such sale has in fact been encouraged in principle, not restricted, as an additional source of earning foreign exchange. In many instances, moreover, technology exports are a by-product or even a necessary "complementary" activity of Japan's export efforts, particularly when plant exports are involved: the foreign buyers, notably in the developing countries, need to learn how to operate and maintain their newly acquired plants and therefore technology assistance contracts are a usual feature accompanying such exports.

Moreover, Japan's considerable dependence upon imported technology results in a substantial deficit on its technology-trade account. A nation's technology-trade balance serves as an important indicator of that nation's scientific and technological standing in the international community (in fact, this balance is now officially used as such an indicator by the Japanese government – along with other indicators such as R & D expenditures as a ratio of GNP and the number of patents acquired overseas – in its computation of the comparative level of Japan's general technological strength). Consequently, there is an implicit policy prescription that Japan should narrow this technology-trade gap by encouraging more exports of its own technology (even though such a deficit itself, regardless of its size, may not necessarily signify a loss of welfare from an economic point of view).

The fact that Japan has long regarded technology exports as a foreign-exchange earning activity is reflected in the only restriction placed by the government on such exports, in terms of the Foreign Exchange and Foreign Trade Control Law of 1949. A single clause in that law introduced a restriction:

> Article 43. Unless authorized under the provisions of Cabinet order, no resident shall render services to a nonresident without an adequate payment provided in accordance with this Law.

This simply means that no resident is permitted to sell technology free of charge, but he is not restricted from such sale so long as he receives an adequate compensation.(78)

True, later on a cabinet order was issued to the effect that no technology export is permitted if it is deemed (1) to directly or indirectly exert an adverse effect on the balance of payments; (2) to deter the growth of the national economy; or (3) to be taken to circumvent legal restrictions (for example, for purposes of capital flight or other illegal purposes). This cabinet order, by setting the "negative" standards, does point to the possible conflicts between technology exports and the national interest, in the form of an adverse impact on the balance of payments under item (1) and "job exports" under item (2). Yet Japan has not so far found it necessary to invoke these regulations.(79)

At the moment, in fact, technology exports are actively encouraged under various programs. For example, Japan has a special tax-exemption program for receipts from technology exports. As a rule, a maximum of 70 percent of such receipts by a given firm can be exempted from the corporate tax on the firm's overseas taxable income, as long as the exemption is within 50 percent of the firm's taxable income from technology sales. A 20 percent deduction also applies to the income from incidental technical assistance services (such as feasibility studies, preparation of drawings and blueprints, technical advice, supervision, and inspection). This program is actually a remnant of the special tax-exemption program given to all Japanese exports of goods and services that was suspended at the end of March 1964 on the occasion of Japan's becoming a regular member of the IMF and the GATT in that year. Japan retained the tax-exemption privilege for technology exports, reasoning that such a promotional measure for "invisible" trade would not violate the rules established by the GATT. It was initially continued on a seven-year provisional basis but has now been extended.(80)

There is also a special loan program for technology exports administered by Japan's Export-Import Bank. The funds necessary to the preparation for and implementation of the overseas sale of technology services can be borrowed up to 100 percent – usually in the form of a combination loan from the Export-Import Bank (with a maximum of 80 percent) and a city bank (with a minimum of 20 percent). The loan carries a maximum maturity of fifteen years (normally from three months to five years), with an annual rate of interest ranging from 4 to 7 percent.(81) In addition, technology exports are indirectly fostered in connection with Japan's promotional measures for overseas investments, since industrial properties are in many cases capitalized as equity capital in overseas ventures.

As noted earlier, technology assets possessed by Japan's small and medium-sized firms are transferred mostly through their overseas ventures. In order to assist these firms specifically – and also the developing countries interested in hosting joint ventures with these Japanese firms – the Japan Overseas Development Corporation, a government agency, conducts preliminary surveys on the practicality of proposed ventures in the host countries and extends interest-free loans to those small Japanese firms participating in its approved ventures; the loans have a maximum term of twenty years and cover a maximum of four-fifths (normally, however, two-thirds) of the required capital. The government-affiliated financial institutions established to assist small firms (such as the Small Business Finance Corporation, the People's Finance Corporation, and the Central Bank for Commercial and Industrial Cooperatives) provide advice and commercial loans for overseas investment.(82)

SUMMARY AND CONCLUSIONS

Japan's experience as an active technology importer and an equally active exporter within a relatively short span of time in the postwar period provides an ideal laboratory in which to study interactions between market forces and control systems. If we agree that Japan's control program over technology inflows has been successfully executed as a vital instrument to achieve the national goals of economic growth and trade competitiveness(83) – if not so much as a tool to force foreign technology suppliers to sell licenses at bargain prices – one major lesson we can learn is that the control system was effectively designed to foster the strength of the domestic market mechanism. It was intended to make domestic industry structurally as strong as the foreign industry that supplies technology. Put simply, it was intended to make Japanese industry as monopolistic as that in advanced Western countries, particularly in the United States, by fostering the development of large-scale enterprises in high-technology sectors by the arrangement for staggered entry, by relaxation of the Antimonopoly Law, by the formation of rationalization cartels, and by other industrial policies. The dynamic economic growth that resulted made foreign firms more willing to supply technology and also induced the newly established Japanese firms in high-technology sectors to conduct their own R & D.

The importation of advanced foreign technology was used as a powerful stimulant to transform Japan's industrial structure, and the exports of technology to the developing countries have been an instrument for shedding labor-intensive, low-value-added industries. For transfer control mechanisms to be effective, two crucial factors must be satisfied: the recipient firm's capacity to absorb and adapt foreign technology and the official administrator's capacity to integrate such mechanisms with other industrial policies. Without these the control system will be ineffective, no matter how elaborate and formal it is. In other words, the "infant technology" argument will not be valid.

The market conditions with which Japan was confronted as a buyer of technology in the early postwar period was the "market does not work" situation that justified extensive governmental controls, whereas the more recent experience of Japan as a seller of technology corresponds fundamentally with the "market does work" situation. In the former experience, the theory of comparative costs was ignored, at least initially, in Japan's efforts to introduce new industries whose factor requirements were incompatible with Japan's resource endowments; in the latter experience, however, it has been upheld in Japan's technology exports to the developing countries. Japan has moved to liberalize technology imports, relying more and more on the market mechanism. And there is no prospect of reversal of this policy. So far as technology exports are concerned, Japan has been relying on the market mechanism from the beginning – with some additional promotional measures. And this general attitude is likely to persist as long as Japan continues to upgrade its industrial structure (although there may be some isolated cases of attempts to retain the latest technology within Japan).

In this chapter I have looked at only proprietary technology transfers. But no doubt the Japanese acquired valuable industrial knowledge from overseas through nonproprietary channels as well. For example, the Japan Productivity Center (JPC), established in 1955 under U.S. sponsorship, organized and sent overseas hundreds of study tour groups (each usually consisting of about twenty executives and technical experts, with the tour lasting four to six weeks). They visited corporate headquarters, factories, research centers, trade associations, academic institutions, and wherever else they could learn firsthand the operations of modern industries. Gary Saxonhouse's observation is pertinent:

> Although various sources attribute as much as 30 to 60 percent of the postwar Japanese economic growth to technological improvement and assimilation, as late as 1970 Japanese firms were paying no more than $500 million in licensing fees and royalties. It is impossible to rationalize this small figure by assuming massive mistakes by foreign licensors and/or overwhelming monopsony power by Japanese licensees acting in concert with the Japanese government. It seems more likely that most technology and technological information either reached Japan through nonproprietary channels or was produced by the Japanese themselves.(84)

Indeed, we cannot assess total technological progress in Japan in terms of proprietary technology transfers alone; nonproprietary transfers and indigenous R & D also are crucial components. Controls and promotional measures on nonproprietary channels must have played roles as significant as those on proprietary channels. Similarly, a large part of Japan's dissemination of technology to the developing countries, notably in Asia, are carried out through nonproprietary channels. In fact, the JPC has been transformed into the Asian Productivity Center, which is now active in promoting this mode of transfers.

Furthermore, this chapter addresses itself only to Japan's postwar experience. Yet it is not the first time that Japan has committed itself to an all-out assimilation of foreign technologies. In fact, its experience as a successful technology borrower goes as far back as the sixth century, when Japan initiated a determined effort to pattern its society after that of the T'ang dynasty in China. The entire endeavor was more than an incident of technology transfer in the modern sense of the term. For it was the religious, social, and political aspects of T'ang China that Japan strove to adopt into the existing Japanese society. In the process, nonetheless, a variety of significant Chinese industrial techniques in such fields as architecture, weaving, paper making, pottery, lacquer ware, farming, mining, and metallurgy were learned by the Japanese. More important, as a result of this ancient experience of massive cultural adoption and adaptation that lasted, though with diminishing vigor, up to the middle of the ninth century, Japan developed a distinctive tradition of assimilating alien cultures, however and whenver practical. It is a valuable national heritage that has proved instrumental for Japan's modernization in later periods.

The second extensive effort to adopt foreign technology took place after the Meiji Restoration of 1868. It was the menacing visit of the steam-powered "black ships" and their overpowering cannons that brought to an end Japan's seclusion policy of two centuries. The Japanese quickly realized how industrially backward they were, and embarked on a nationalized program to modernize their society. Fearful of the Western colonialism then rampant, the Meiji government made it a policy to prohibit the participation of foreign equity capital in Japanese industries, despite the immense need for capital. "Learn Western knowledge but refrain from depending on foreign capital and management" became a slogan. A large number of foreign engineers and technical experts were hired by the government as industrial tutors who gave the Japanese on-the-job training mostly at government-established "model" plants. In addition, an equally large number of carefully selected Japanese officials and students were sent to Europe and the United States for study. Upon their return home, they soon replaced foreigners at factories and colleges. Japan was thus able to acquire Western technologies in "unbundled" form, and mostly through non-proprietary channels.

These previous successful experiences are obviously not unrelated to the more recent experience this chapter has examined. In fact, we should view the latter essentially as the continuation of Japan's effort to catch up with the West in industrial endeavors: the historical momentum has clearly long existed. After World War II the Japanese felt no qualms about adopting the Western way of industrial production and consumption. Their past experience provided psychological security. Japan went all out for Western technologies.

NOTES

(1) This distinction corresponds to the one between proprietary and nonproprietary technology transfers defined in John H. Dunning's "Alternative Channels and Modes of International Resource Transmission," Chapter 1 in this volume. The term "market-based" instead of "market" is used here, since the latter connotes only licensing or "arm's length" transactions, excluding direct foreign investment; the direct investment route is also included in market-based transactions.

(2) In the context of international technology transfer, market deficiency includes – in addition to a breakdown of the normal market mechanism – a lack of sufficient bargaining strength on the part of domestic firms vis-a-vis foreign firms. In other words, the market that is considered normal in a "closed" economy may be found deficient in an "open" economy from a national point of view.

(3) This analytical framework is presented in Terutomo Ozawa, "Technology Transfer," in The Encyclopedia of Japan (New York: Kodansha International, forthcoming).

(4) William W. Lockwood, "Political Economy," in The United States and Japan ed. Herbert Passin (Englewood Cliffs, N.J.: Prentice-Hall, 1966), p. 103.

(5) See Merton J. Peck (with the collaboration of Shuji Tamura), "Technology," in Asia's New Giant: How the Japanese Economy Works ed. Hugh Patrick and Henry Rosovsky (Washington: Brookings Institution, 1976), chap. 8, pp. 525-85; and Terutomo Ozawa, Japan's Technological Challenge to the West, 1950-1974: Motivation and Accomplishment (Cambridge, Mass.: M.I.T. Press, 1974).

(6) Chalmers Johnson, "MITI and Japanese International Economic Policy," in The Foreign Policy of Modern Japan, ed. Robert A. Scalapino (Berkeley: University of California Press, 1977), p. 276.

(7) Tuvia Blumenthal and Morris Teubal, "Factor Proportions and future Oriented Technology: Theory and an Application to Japan," mimeographed, 1975.

(8) Terutomo Ozawa, "Japan's Resource Dependency and Overseas Investment," Journal of World Trade Law 11, no. 1 (January-February 1977): 60.

(9) Robert S. Ozaki, The Control of Imports and Foreign Capital in Japan (New York: Praeger, 1972), p. 93; and Ozawa, Japan's Technological Challenge, pp. 16-18. A detailed judicial analysis of the Foreign Exchange Control Law and the Foreign Investment Law, as they apply to technological assistance agreements, is given in Dan F. Henderson, Foreign Enterprise in Japan: Laws and Policies (Chapel Hill: University of North Carolina Press, 1973), chap. 6, esp. pp. 229-32.

(10) Ozaki, Control of Imports, pp. 93-94.

(11) Ozawa, Japan's Technological Challenge, p. 17. See also Henderson, Foreign Enterprise in Japan, pp. 216-18.

(12) Peck and Tamura, "Technology," p. 548.

(13) Ozaki, Control of Imports, p. 95.

(14) Franklin R. Root, "The Pricing of International Technology Transfers via Nonaffiliate Licensing Arrangements," Chapter 5 in the present volume.

(15) By "price" is meant a total compensation, involving not only the royalty rate but also lump-sum fees and other terms of contract, both pecuniary and nonpecuniary.

(16) Henderson, Foreign Enterprise in Japan, p. 200. A similar point is emphasized in Edwin O. Reischauer, The Japanese (Cambridge: Harvard University Press, 1977), pp. 288-89.

(17) OECD, Liberalization of International Capital Movements: Japan (Paris: OECD, 1968), pp. 57-58.

(18) These lists are reproduced in Ozawa, Japan's Technological Challenge, p. 21.

(19) The first theoretical analysis of the "one set" principle was presented in Yoshikazu Miyazaki, Sengo Nihon no Keizai Kiko [Economic structure of postwar Japan] (Tokyo: Shin Hyoronsha, 1966).

(20) This example of staggered entry is cited in Peck and Tamura, "Technology," pp. 554-55.

(21) Hoshimi Uchida, Gosei Sen'i Kogyo [Synthetic fiber industry] (Tokyo: Toyokeizai Shinposha, 1970), pp. 200-05.

(22) Peck and Tamura, "Technology," pp. 545-46.

(23) For the postwar trend of the increasing concentration of market power, see Kozo Yamamura, Economic Policy in Postwar Japan: Growth versus Economic Democracy (Berkeley: University of California Press, 1967). For example, he states: "The role of the government was and is to encourage the process of concentration with all the power at its command. Especially after 1958 the government provided the leadership which took varied forms ranging from expressions of desire for increased rationalization by the Prime Minister to suggesting and then actively participating in the negotiation of a merger between two leading automobile firms. . . . The government was and is providing added pressure to mold the market structure which was and is also following its own momentum generated by the economic forces of postwar Japan (increasing "lumpiness" of technology, needs for rapid capital accumulation, and increased international competitive ability). As these two forces worked together in shaping the market structure, one cannot hope to isolate the government's contribution to the increase in concentration of the market structure." (pp. 106-107.)

(24) MITI, Gaikoku Gijutsu Donyu no Genjo to Mondaiten [Current status and problems of foreign technology absorption] (Tokyo: MITI, 1962).

(25) MITI, Gijutsu Doko Chosa Hokokusho [Report on the trend of technology] (Tokyo: MITI, 1963).

(26) These liberalization steps are constructed from information contained in Ozaki, Control of Imports, pp. 94-95; and Japan's Science and

Technology Agency, Kagaku Gijutsu Hakusho [White paper on science and technology] (Tokyo, 1978), p. 166. See also Henderson, Foreign Enterprise in Japan, chap. 7, pp. 237-90.

(27) Peck and Tamura, "Technology," p. 546.

(28) OECD, Liberalization, p. 57.

(29) Ibid., p. 57.

(30) Ozaki, Control of Imports, p. 94, quoted in Peck and Tamura, "Technology," p. 545.

(31) Peck and Tamura, "Technology," p. 540.

(32) Ibid., pp. 539-40.

(33) Johnson, "MITI," pp. 246-47.

(34) Seiichiro Saito, "Tsusho Sangyosho Ton," in Shokun, July 1978; translated into "The Trouble with MITI," Japan Echo 5, no. 3 (1978): 59-60. Emphasis added.

(35) Science and Technology Agency, Kagaku Gijutsu Hakusho, p. 30.

(36) Ibid., p. 43.

(37) For more evidence on this effort, see Ozawa, Japan's Technological Challenge, chap. 5.

(38) These issues are discussed, for example, in Robert Hawkins and Thomas Gladwin, "Conflicts in the International Transfer of Technology: A U.S. Home Country View," Chapter 8 in the present volume.

(39) See Robert Gilpin, U.S. Power and the Multinational Corporation (New York: Basic Books, 1975); and Jack Baranson, "Technology Exports Can Hurt Us," Foreign Policy 7, no. 25 (Winter 1977).

(40) Science and Technology Agency, Gaikoku Gijutsu Donyu Nenji Hokoku [Report on the induction of foreign technology] (Tokyo, 1975), p. 27.

(41) Computed from data in Science and Technology Agency, Kagaku Gijutsu Hakusho, pp. 164-65.

(42) Computed from data in Science and Technology Agency, Kagaku Gijutsu Yoran [Indicators of Science and Technology] (Tokyo: 1978).

(43) Ibid., pp. 146-47.

(44) Ibid., p. 149.

(45) Terutomo Ozawa, Transfer of Technology from Japan to Developing Countries, UNITAR Research Report No. 7, 1971.

(46) Hugh Patrick and Henry Rosovsky, "Japan's Economic Performance: An Overview," in Patrick and Rosovsky, Asia's New Giant, pp. 20-21. Emphasis added.

(47) MITI, Tsusho Hakusho [White paper on international trade] (Tokyo: 1971).

(48) James C. Abegglen, Management and Worker: The Japanese Solution (Tokyo: Kodansha International, 1973), p. 182.

(49) Quoted in K. Bieda, The Structure and Operation of the Japanese Economy (Sydney: John Wiley, 1970), p. 196.

(50) The analysis of labor shortages in Japan and the intra-Asian spread of Japan's dual industrial structure presented in this section is based on Terutomo Ozawa, "Multinationalism, Japanese Style," Columbia Journal of World Business 7, no. 6 (November-December 1972): 33-42; and Multinationalism, Japanese Style: The Political Economy of Outward Dependency (Princeton: Princeton University Press, 1979), chaps. 2 and 3.

(51) Kiyoshi Kojima, "International Trade and Foreign Investment: Substitutes or Complements," Hitotsubashi Journal of Economics 16, no. 1 (June 1975): 8.

(52) For a theoretical elaboration, see Terutomo Ozawa, "International Investment and Industrial Structure: New Theoretical Implications from the Japanese Experience," Oxford Economic Papers 31, no. 1 (March 1979): 72-92. See also Ozawa, Multinationalism, chap. 2.

(53) Richard E. Caves, "International Corporations: The Industrial Economics of Foreign Investment," Economica 38 (1971).

(54) Ibid., p. 2. This and other theoretical implications are discussed in Ozawa, Multinationalism, chap. 2.

(55) Ozawa, Transfer of Technology, p. 13.

(56) This percentage is pointed out in Noboru Tsuda, Gijutsu Boheki no Chishiki [Knowledge of technology trade] (Tokyo: Nippon Keizai Shimbun, 1969), p. 115.

(57) An excellent discussion of this internal reorganization of trading companies is presented in M.Y. Yoshino, Japan's Multinational Enterprises (Cambridge: Harvard University Press, 1976), chap. 4.

(58) Quoted in William V. Rapp, "Japan's Industrial Policy," in The Japanese Economy in International Perspective, ed. Isaiah Frank (Baltimore: Johns Hopkins University Press, 1975), p. 55.

(59) For Japan's labor unions, see, for example, Reischauer, The Japanese, pp. 186-87; and Walter Galenson (with the collaboration of Konosuke Odaka), "The Japanese Labor Market," in Patrick and Rosovsky, Asia's New Giant, pp. 587-672.

(60) Yoshi Tsurumi, The Japanese Are Coming: A Multinational Interaction of Firms and Politics (Cambridge, Mass.: Ballinger, 1976), p. 299.

(61) For the Japanese blue-collar workers, see Robert E. Cole, Japanese Blue Collar: The Changing Tradition (Berkeley: University of California Press, 1971); for the white-collar workers, see Thomas P. Rohlen, For Harmony and Strength: Japanese White-Collar Organization in Anthropological Perspective (Berkeley: University of California Press, 1974).

(62) Ichiro Hattori, "A Proposition on Efficient Decision-Making in the Japanese Corporation," Columbia Journal of World Business 13 (Summer 1978): 10. Emphasis added.

(63) Economic Planning Agency, "Nenji Keizai Hokoku," ESP, no. 76, August 1978, p. 159.

(64) Ibid., p. 159.

(65) Ibid., p. 159.

(66) MITI, 1977-Wagakunikigyo no Kaigaijigyo Katsudo [Overseas business activities of our national enterprises] (Tokyo, 1977), p. 98.

(67) For the scaling-down phenomenon, see Ozawa, Transfer of Technology, pp. 21-25.

(68) This "locked in" effect is discussed in Ozawa, Multinationalism, chap. 6.

(69) This trade creation effect is examined in Kiyoshi Kojima, "Transfer of Technology to Developing Countries – Japanese Type versus American Type," Hitotsubashi Journal of Economics 17 (February 1977): 1-14; also Direct Foreign Investment: A Japanese Model of Multinational Business Operations (London: Croom Helm, 1978).

(70) MITI, 1975-Tsusho Hakusho [White paper on international trade] , p. 434.

(71) "NTT is going to partially open VLSI technology," Japan Economic Journal, December 12, 1978, p. 13.

(72) "U.S. Challenges Planned Venture of GE, Hitachi," Wall Street Journal, November 29, 1978, p. 2.

(73) Ibid., p. 2.

(74) "Toray Licenses know-how on carbon fiber to UCC," Japan Economic Journal, January 16, 1979, p. 9.

(75) Ibid., p. 9.

(76) Masayoshi Kanabayashi, "Videotape Recorders are Becoming Japan's Top Electronic Export to the U.S.," Wall Street Journal, January 4, 1979, p. 1.

(77) "Hitachi Gives Up for Time Being Plan to Enter U.S.," Japan Economic Journal, January 30, 1979, p. 9.

(78) Tsuda, Gijutsu Boheki no Chishiki, p. 96.

(79) Ibid., p. 99.

(80) Industrial Technology Institution, Kenkyu Kaihatsu Josei Seido [Promotional measures for R & D] (Tokyo, 1975), p. 103.

(81) Tsuda, Gijutsu Boheki no Chishiki, p. 118.

(82) Small and Medium Enterprise Agency, Chusho Kigyo Hakusho [White paper on small and medium enterprises] (Tokyo, 1976), pp. 345 and 370.

(83) Merton Peck agrees very strongly on this view. See Peck and Tamura, op. cit., p. 581.

(84) Gary R. Saxonhouse, "The World Economy and Japanese Foreign Economic Policy," in Robert A. Scalapino, ed., The Foreign Policy of Modern Japan (Berkeley: University of California Press, 1977), p. 289.

REFERENCES

Abegglen, James C. Management and Worker: The Japanese Solution. Tokyo: Kodansha International, 1973.

Baranson, Jack. Technology Exports Can Hurt Us. Foreign Policy, 7, no. 25 (Winter 1977).

Bieda, Ken. The Structure and Operation of the Japanese Economy. Sydney: John Wiley, 1970.

Blumenthal, Tuvia, and Teuba, Morris. Factor Proportions and Future Oriented Technology: Theory and an Application to Japan. Mimeographed, 1975.

Caves, Richard E. International Corporations: The Industrial Economics of Foreign Investment. Economica 38 (1971).

Cole, Robert E. Japanese Blue Collar: The Changing Tradition. Berkeley: University of California Press, 1971.

Economic Planning Agency, Japanese Government. "Nenji Keizai Hokoku." [Annual economic report] ESP, no. 76, August 1978.

Galenson, Walter (with the collaboration of Konosuke Odaka). The Japanese Labor Market. In Asia's New Giant: How the Japanese Economy Works, ed. Hugh Patrick and Henry Rosovsky. Washington: Brookings Institution, 1976.

Gilpin, Robert. U.S. Power and the Multinational Corporation. New York: Basic Books, 1975.

Hattori, Ichiro. A Proposition on Efficient Decision-Making in the Japanese Corporation. Columbia Journal of World Business 13 (Summer 1978).

Henderson, Dan F. Foreign Enterprise in Japan: Laws and Policies. Chapel Hill: University of North Carolina Press, 1973.

Hitachi Gives Up For Time Being To Enter U.S. Japan Economic Journal, January 30, 1979, p. 9.

Industrial Technology Institution. Kenkyu Kaihatsu Josei Seido [Promotional measures for R & D]. Tokyo, 1975.

Johnson, Chalmers. "MITI and Japanese International Economic Policy." In The Foreign Policy of Modern Japan, ed. Robert A. Scalapino. Berkeley: University of California Press, 1977.

Kanabayashi, Masayoshi. "Videotape Recorders Are Becoming Japan's Top Electronic Export to the U.S." Wall Street Journal, January 4, 1979, p. 1.

Kojima, Kiyoshi. Direct Foreign Investment: A Japanese Model of Multinational Business Operations. London: Croom Helm, 1978.

______. International Trade and Foreign Investment: Substitutes or Complements. Hitotsubashi Journal of Economics 16, no. 1 (June 1975).

______. Transfer of Technology to Developing Countries – Japanese Type versus American Type. Hitotsubashi Journal of Economics 17, (February 1977).

Lockwood, William W. Political Economy. In The United States and Japan ed. Herbert Passin. Englewood, N.J.: Prentice-Hall, 1966.

Ministry of International Trade and Industry (MITI). Gaikoku Gijutsu Donyu no Genjo to Mondaiten [Current status and problems of foreign technology absorption]. Tokyo, 1962.

______. Gijutsu Doko Chosa Hokokusho [Report on the trend of technology]. Tokyo, 1963.

______. Tsusho Hakusho [White paper on international trade]. Tokyo, 1971 and 1975.

______. Wagakuni Kigyo no Kaigaijigyo Katsudo [Overseas business activities of our national enterprises]. Tokyo, 1977.

Miyazaki, Yoshikazu. Sengo Nihon no Keizai Kiko [Economic structure of postwar Japan]. Tokyo: Shin Hyoronsha, 1966.

NTT is going to partially open VLSI technology. Japan Economic Journal, December 12, 1978, p. 9.
OECD. Liberalization of International Capital Movements: Japan. Paris: OECD, 1968.
Ozaki, Robert S. The Control of Imports and Foreign Capital in Japan. New York: Praeger, 1972.
Ozawa, Terutomo. International Investment and Industrial Structure: New Theoretical Implications from the Japanese Experience. Oxford Economic Papers 31, no. 1 (March 1979).
______. Japan's Resource Dependency and Overseas Investment. Journal of World Trade Law, 11, no. 1 (January-February 1977).
______. Japan's Technological Challenge to the West, 1950-1974: Motivation and Accomplishment. (Cambridge: M.I.T. Press, 1974.
______. Multinationalism, Japanese Style. Columbia Journal of World Business 7, no. 6 (November-December 1972).
______. Multinationalism, Japanese Style: The Political Economy of Outward Dependency. Princeton: Princeton University Press, 1979.
______. Technology Transfer. In The Encyclopedia of Japan. New York: Kodansha International, forthcoming.
______. Transfer of Technology from Japan to Developing Countries. New York: UNITAR, 1971.
Patrick, Hugh, and Rosovsky, Henry. Japan's Economic Performance: An Overview. In Asia's New Giant: How the Japanese Economy Works, ed. Hugh Patrick and Henry Rosovsky. Washington: Brookings Institution, 1976.
Peck, Merton J. (with the collaboration of Shuji Tamura). Technology. In Asia's New Giant: How the Japanese Economy Works, ed. Hugh Patrick and Henry Rosovsky. (Washington: Brookings Institution, 1976).
Rapp, William V. Japan's Industrial Policy. In The Japanese Economy in International Perspective, ed. Isaiah Frank. Baltimore: Johns Hopkins University Press, 1975.
Reischauer, Edwin O. The Japanese. Cambridge: Harvard University Press, 1977.
Rohlen, Thomas P. For Harmony and Strength: Japanese White-Collar Organization in Anthropological Perspective. Berkeley: University of California Press, 1974.
Saito, Seiichiro. Tsusho Sangyosho Ron. Shokun, July 1978. Translated into The Trouble with MITI. Japan Echo 5, no. 3 (1978).
Saxonhouse, Gary R. The World Economy and Japanese Foreign Economic Policy. In The Foreign Policy of Modern Japan, ed. Robert A. Scalapino. Berkeley: University of California Press, 1977.
Science and Technology Agency, Gaikoku Gijutsu Donyu Nenji Hokoku [Annual report on the induction of foreign technology]. Tokyo, 1975.
______. Kagaku Gijutsu Hakusho [White paper on science and technology]. Tokyo, 1978.
______. Kagaku Gijutsu Yoran [Indicators of science and technology]. Tokyo, 1978.

Small and Medium Enterprise Agency. Chuso Kigyo Hakusho [White paper on small and medium enterprises]. Tokyo, 1976.

Toray Licenses Know-How on Caron Fiber to UCC. Japan Economic Journal, January 16, 1979.

Tsuda, Noboru. Gijutsu Boheki no Chishiki [Knowledge of technology trade]. Tokyo: Nippon Keizai Shimbun, 1969.

Tsurumi, Yoshi. The Japanese Are Coming: A Multinational Interaction of Firms and Politics. Cambridge, Mass.: Ballinger, 1976.

Uchida, Hoshimi. Gosei Sen'i Kogyo [Synthetic fiber industry]. Tokyo: Toyo Keizai Shinposha, 1970.

U.S. Challenges Planned Venture of GE, Hitachi. Wall Street Journal, November 29, 1978.

Yamamura, Kozo. Economic Policy in Postwar Japan: Growth versus Economic Democracy. Berkeley: University of California Press, 1967.

Yoshino, M.Y. Japan's Multinational Enterprise. Cambridge: Harvard University Press, 1976.

Comments

Thomas Roehl

Professor Ozawa's thorough coverage of the Japanese postwar experience with technology transfer gives us a good case study for considering what the prerequisites for a successful control system are. It seems to me that Professor Ozawa's paper shows that for the Japanese, and perhaps for many other nations, rules and codes are not enough to make a successful technology transfer control system. If we are to generalize at all from the Japanese experience, we must conclude that the system succeeded because it was not ad hoc. Rather, Japan's technology transfer control system was a part of a larger system that not only sought a widely shared policy goal of rapid infusion of technology into heavy industry, but that clearly set, again with wide agreement, the role of foreign resources in that process.

Technology transfer control systems are only successful if the firms acquiring technology can utilize the knowledge. Japan's preferential finance system guaranteed resources to technology-acquiring firms in what the country considered key industries. Foreign exchange was also channeled according to these priorities. These decisions were generally agreed to by all parties, so that there was little quarreling with the control system's decisions. When most actors are in agreement with the decision rules, any system will work better, and the concern over inefficiencies in administration of the control system is less cogent. The fact that MITI waited almost fifteen years after the war to issue a list of desired technologies indicates that most people knew what the priorities were.

Japan moved into what Professor Ozawa calls "inappropriate" technology under these control system conditions. Perhaps a better term might be "new technology," since Japan had progressed through stages where it adopted secondhand and then up-to-date technology in the textile and other light industries prior to the war, and the change was more that of taking the next step rather than of skipping a stage, which the word "inappropriate" might imply. Rather than being handled case by case, the administration of technology transfer had what Ozawa called a general legislative mandate. That is key, and does not require national or international codes for effective implementation.

Ozawa's comments on the bargaining role of the government also suggest the significance of a government ally for a firm negotiating for technology. This is another area not easily codified. As a businessman, you would like to feel that you could call on government to increase your bargaining position in technology transfer, but you would not want to limit your flexibility in negotiating contracts. With the basic priorities set and agreed to by business and government, Japan could set up a flexible program, where government pressure need only be applied when too large a percentage of the contract gains went to the foreign firm. By evaluating the whole agreement, rather than each part, firms did not sacrifice much negotiating flexibility under the approval system.

Firms would not want a government to assign the right to negotiate for foreign technology to one firm. While Ozawa suggests a designation of one firm at each point in time, he makes clear that other firms could license competing technologies, subject to capacity limits. Unless there is faith in the government decision maker's willingness to avoid favoritism, however, other countries' firms might hesitate to adopt such flexible rules.

Two additional points make any technology transfer control system in Japan work better. First, the size of the market influences the effect of any given rules, and Japan's large and rapidly growing economy lessened the supply-reducing effects of any rules. That condition will not always transfer to other countries using the same system. Second, as Ozawa states, Japan set up clear property rights on the royalties to a foreign firm. This again makes it clear that returns are available if a foreign firm chooses to follow the system. This type of a property right delineation might make a national code more attractive to a foreign firm.

When Professor Ozawa moves to a discussion of outward technology, he again offers us information useful in evaluating the effects of national technology transfer control systems. Since Japan has given its firms a free hand in outward technology transfer, less developed countries have a larger number of alternative sources as a result. This is true not only in the addition of Japanese sophisticated technology, but also of less sophisticated technology from smaller firms that, as Ozawa documents, have chosen to cross national borders rather than cross industry borders to reassign resources no longer valued at market rates in their industry in Japan. To the extent that this has improved LDC company bargaining positions, the formal codes' effects are changed. If inadequate bargaining power is one reason for a formal code, a widening of source for technology transfer makes the economies of a code less favorable. If we can assume that more and more countries will offer transfers of technology as they follow Japan through the transition Ozawa describes, then Ozawa's "fadeout" of Japan's control system also becomes important for code builders in other countries. A code that will be useful in the next few years may find lower benefit streams in the future if bargaining units continue to increase.

Finally, some comments on Japan's experience as a host country for technology transfer. Japan's experience, as Professor Ozawa sketches it, tells us how much that policy depends on the domestic and international constraints the country faces. Even though alternative sources may push toward more supply of technology, country-specific factors are important too. To Ozawa's list I might add the inability of such Japanese technology-intensive products as steel to further penetrate foreign markets, and the current weakening in the domestic demand for managerial talent due to lower growth. If managerial talent is rented when technology is sold, these factors may well be a more significant force than rules in enabling LDC firms to drive good bargains for Japanese technology.

Professor Ozawa states there has been very little Japanese government pressure to stop these transfers. I think he need not take quite so extreme a position. Japan is certainly concerned that the shortening of the product cycle is decreasing potential returns on the R & D effort, due to increased LDC bargaining power. And unions, at least recently, have become worried about the "boomerang effect" of technology flows on domestic employment. I think it would be safer to couch his conclusion, with which I agree, in terms of those trade and external constraints that, along with Dunning's argument, make a "hands off" policy the best one for Japan.

13 Technology Transfer Control Systems: The Case of East and Southeast Asian Developing Countries

R. Hal Mason

The purpose of this chapter is to examine the methods used by the developing countries of East and Southeast Asia to control and monitor the acquisition and transfer of technology.

The countries examined are South Korea, Taiwan, the Philippines, Thailand, Malaysia, Singapore and Indonesia. Technology transfer has become a somewhat contentious subject. Developing countries often feel that historically they have been badly used, first by colonialists and subsequently by multinational firms from advanced countries. All of the countries examined here, with the exception of Thailand, have been colonized at one time or another by an industrialized power: South Korea and Taiwan by the Japanese, the Philippines by the Spanish and Americans, Malaysia and Singapore by the British, and Indonesia by the Dutch. Moreover, there continue to be feelings of hostility toward the Japanese stemming from Japan's role in World War II. Other cultural elements also play a part in focusing feelings of nationalism throughout the area. Much of this nationalistic bent, in turn, tempers individual countries' stances toward multinational enterprises and the transfer of technology.

These countries recognize that, if they are to acquire technology, they must deal with multinational firms and, in some respects at least, are dependent upon them as suppliers of technology. Prior to World War II these countries were colonies, and large national firms of the colonizers received preferential treatment in exploiting resource concessions, establishing agricultural plantations, and serving the domestic market. Indeed, many of the largest enterprises in some of these developing countries were established during the colonial period and in

*Financial support for the research underlying this chapter was provided by the International Business Machines Program on International Business Studies during the 1977-78 academic year, when I was an IBM Postdoctoral Fellow.

some instances remain under foreign ownership control today. But the colonial period was different from the post-World-War-II period. The colonial period was largely characterized by the independent foreign enterprise. It was seldom a part of a much larger entity such as the multinational firm we have seen emerge in the postwar period. Moreover, in terms of technology transfers the foreign-owned enterprise was little different from any other enterprise, that is, it was generally not tied into a parent firm from which much of its technology and knowhow sprang.

Under colonialism, it was the mother country (the colonizer), through its ability to govern, that could grant favors and concessions and in turn influence the mix and level of economic activity. Usually the colony was left in a dependent position and was cast into the role of being a producer of primary commodities and semifinished goods. The mother country specialized in the production of finished goods, which were then sent back to the colonies to be marketed.(1) The early performance of multinational firms seems to hold some analogies with colonialism. Indeed, the term "neocolonialism" is used to refer to the role of the multinational firm. Developing countries see an analogy between multinational firms and colonizing countries because they bring about certain kinds of dependency relationships, and the roles parcelled out to subsidiaries appear to be very similar to the roles that colonial firms played in an earlier era, that is, specialization on primary production, intermediate and low technology industry.

While nationalism may have been at least partially an outgrowth of colonialism, that alone cannot be the full explanation. The countries of Latin America are a long distance in time from their colonial days. Yet the most virulent forms of nationalism appear there. Undoubtedly the reaction to colonialism contributed to nationalism, but so too have imperialism, foreign investment, and technological dependence. Regardless of the proximate cause, nationalism has become a reality, and out of nationalism has grown the desire to control or at least strongly influence the acquisition and application of technology. The developing countries see that one of the major contributions multinational firms can make to economic development and industrialization(2) is to supply technology and knowhow. Yet they also see that multinational firms, through their technological expertise, can create conditions that place the country in a continuing position of dependency. Thus there are widespread attempts within the developing countries to alter this relationship and to decouple the technology, wherever possible, from foreign control – especially control exerted via foreign ownership.

This characterization, if given no further explication, would be misleading because it is overly simplified. The acquisition of technology is not being totally controlled, nor is control of foreign investment the total or even necessarily the primary avenue being used to influence the flows of technology and their composition. Indeed, it should not be suggested that countries wield great control. They can and do influence technology flows, but they hardly control them. The tools they have devised are imperfect and at best provide only a loose rein by which to guide economic activity, industrial composition, and the activities of both foreign and domestic firms. Moreover, nationalism often yields to economic pragmatism when certain technologies are considered to be of great value. The rules of the game are seldom tightly defined, and even when they are exceptions are permitted and even encouraged. Within this context, I will attempt in this chapter to describe and analyze the technology acquisition strategies and tools devised to implement these strategies in the developing countries of East and Southeast Asia.

As a prologue to that effort it should prove useful to briefly highlight some of the methods being used to influence technology transfers. I can then describe how each country has fashioned its policies to utilize these tools in pursuit of its own perceived objectives. Finally, I shall examine the Association of Southeast Asian Nations (ASEAN) and explore the efforts of this group to find unity wherever feasible. Certainly, it must be noted that there is not unanimity among this group of nations, which are culturally dissimilar. They also are economically diverse, in terms of resource availability and levels of economic development, and politically diverse in terms of attitudes toward democratic processes, nationalism, and foreign participation in economic affairs. Finally, as a group, they look to the United States and Japan as primary sources of technology, know-how, capital, and financial aid, while individually still maintaining strong ties, in several instances, with their former colonizers and/or other countries with which they hold a strong cultural or religious affinity. Thus it is difficult as best to generalize. Yet there do seem to be some common threads that can be explored. This I try to do.

A GENERALIZED MODEL OF TECHNOLOGY TRANSFER CONTROL

All of the countries to be explored do have some things in common in terms of the apparatuses they have installed to review, monitor, and control technology transfers through direct foreign investment and certain contractual arrangements. The countries differ in terms of which industries are included. This is what one would expect, given that their resources, objectives, and priorities differ. All of the countries apply some or all of the following paraphernalia to new investments.

1. A set of industrial priorities
2. Incentive programs tied to the industrial priorities
3. A Board of Investments (or equivalent agency) to screen new investments
4. An interagency screening process that usually covers incentives (Board of Investments), foreign exchange implications (Ministry of Finance or Central Bank), capital requirements and ownership structure (Board of Investments and Ministry of Finance), investment priorities and technology acquisition (Ministry or Board of Science and Technology, Ministry of Industry and Commerce),

import and export implications (Ministry of Trade), and licensing agreements and/or registration of patents and trademarks (Board of Science and Technology and Board of Patents and Commercial Registration)
5. A list of overcrowded industries
6. An incentive program to disperse new industry away from congested urban centers
7. A set of rules on ownership structure of the enterprise and the application of patented technology
8. A list of industries open to nationals only.

In addition, some of the countries have plans under law by which foreign entities must reduce their share of ownership to some specified level by spinning off share (equity) capital to nationals within some specified time horizon. Also, there may be controls over how many expatriates may be employed as a proportion of the workforce – especially as a proportion to those positions involving technical and managerial skills. Through all of these devices there are several objectives being pursued, but it is apparent that the control over technology transfer by foreigners is at least an implicit objective of considerable importance.

It is largely typical that these countries, with a few exceptions, do exert some control over:

1. The industries open to foreign investment
2. The ownership and financial structure of enterprises in which foreigners participate
3. The nature and terms of licensing and franchising agreements
4. The composition of the workforce and the amounts and types of training undertaken
5. The location of new or expanding foreign enterprises
6. The level of repatriated profits and service fees.

Through these controls they can and do manipulate the owners of technology in terms of the types of technology transferred, the degree to which these are diffused or internalized by nationals and the price paid for the use of these technologies. At the same time, however, there may be adverse side effects, making certain desirable technologies unavailable to the country. But the empirical evidence on which we could make a judgment on this aspect of technology transfer controls is virtually nonexistent.

Another element of national controls and the pricing and acquisition of technology is that countries vary in their length of experience and their maturity in dealing with controls. Taiwan and Singapore, for example, have considerably more experience than do the other countries, although South Korea is rapidly catching up. There is considerable emulation of the Taiwanese approach, which appears to be adopted from the earlier Japanese experience. The Philippines have a system that has been adapted from the Mexican experience. To my mind, four major tendencies are in evidence. First, there is considerable exchange of information among Boards of Investment regarding methods of control over foreign participation and technology transfer as well as the bargains struck in making technology acquisitions. Second, there is a clear recognition that industries differ with respect to the value of their technologies and the terms under which it will be made available. Third, countries differ in terms of the degree to which they see the multinational firm as a vehicle through which national policies can be or should be implemented. And fourth, the countries, at least along some dimensions, see themselves as competing with one another in the acquisition of technology because they are attempting to attract or stimulate many of the same industries. Thus while there are some areas of cooperation in the exchange of information among countries, there are also areas of competitiveness in their efforts to obtain what are deemed to be particularly attractive industries and technologies.

Despite having many similarities in their technology acquisition and control strategies, there are major differences, too – especially in terms of how liberal the policies are. Or, stated differently, countries differ in their degree of reliance on the market mechanism and freedom of trade. A rank ordering from most to least liberal might be as follows:(3)

1. Singapore
2. Taiwan
3. South Korea
4. Malaysia
5. Thailand
6. Philippines
7. Indonesia

At this point it should prove useful to present a generalized case as to how a Board of Investments functions and how a foreign investor obtains approval for an investment proposal. Also, we can review the process of obtaining approval for a licensing or technical aid agreement. The following section will then move to the more specific, and indicate how individual countries deviate from this norm. The case presented is somewhat of a composite of what the various countries do. Let us begin with a description of how a Board of Investments is structured.

The typical Board of Investment is a working group with expertise in examining feasibility studies and, in some cases, in making such studies. The board itself is made up of members who are for the most part at the ministry level of government, and includes all ministries having a vital interest in interpreting and implementing investment and licensing regulations. Thus the typical board will have ten to fifteen members, including the following:

1. Prime minister or his designee, as chairman of the board
2. Secretary general of the Board of Invesments

3. Secretary general of the Economic Planning Board
4. Minister of industry
5. Minister of agriculture
6. Minister of comerce
7. Minister of finance
8. Secretary general of the Agency for Science and Technology
9. Governor of deputy governor of the Central Bank

In addition, some countries have a foreign contract review board, and the head of this board may also sit on the Board of Investments. Also, there may be a sprinkling of prominent citizens, not members of governmental bodies, who are usually selected for their particular expertise.

While the board does screen proposals and makes final decisions, it is supported by a staff under the direction of the secretary general of the board. This staff is composed of technocrats who have industry or sector specific knowledge. For example, the technical staff may be subdivided into four or five analytical groups, such as:

1. Metals and mining
2. Electrical and nonelectrical machinery
3. Agriculture and agribusiness
4. Chemicals and petrochemicals
5. Service industries
6. Miscellaneous industries (construction materials; textiles, clothing, and footwear; woodworking).

When a foreign firm submits an investment proposal, it is submitted on forms prescribed by the Board of Investments and must indicate:

1. Total capitalization and capital structure
2. Local partner(s), if any
3. Production capacity and projected annual production
4. Sales quantity to domestic market, unit prices, total sales, and major customers
5. Sales quantity to foreign market, etc. (as above)
6. Sources and quantities of raw materials
7. Types and brands of machinery to be purchased, unit price and sources.
8. Expected time to start up
9. Projected profits
10. Foreign exchange requirements
11. Number of expatriates to be employed
12. Total employment
13. Training plan
14. Plan for bringing nationals into ownership and management (if required).

The feasibility study and project plan is submitted to the board by the prospective investors. This study is then passed along to an analyst in the appropriate industry group for review and analysis. At this point the analyst is primarily interested in whether the study is complete and all of the necessary information has been submitted, and can also screen and flag specific pieces of information and suggest changes. For example, the proposal may specify certain types of raw materials, components, or equipment that are expected to be imported even though equivalent or usable counterparts are available locally. The analyst can ask for clarification as to why locally unavailable equivalents have not been specified. After this technical review, the proposal may be approved and, if so, it goes to the board for review. However, it may also be returned to the prospective investor for revision or clarification. Once it is approved by the technocrats there may still be a lengthy process ahead. Indeed, in some countries the proposal may be shunted from agency to agency until it has been reviewed by all of those involved. In the past this has created considerable red tape for foreign investors. To get around this problem, countries have established "one-stop shopping," whereby either the Board of Investments takes responsibility for delivering the proposal to all responsible agencies and expediting its movement from agency to agency or the responsible reviewing officers in these agencies are periodically loaned to the Board of Investments so proposals can be reviewed expeditiously.

Once the review process is completed, the board examines each proposal in light of the analyses of all agencies. The analyses lay out any real or potential conflicts the proposal has with existing regulations. The board has the option of accepting the proposal, rejecting it, returning the proposal to the investor for further clarification, or returning it to technical analysts for additional examination. Even if the proposal is rejected, the investors can appeal the decision.

If a license or technical aid agreement is involved, it will be reviewed by a technology review board or a specialized agency, such as an Agency of Science and Technology. There it will be checked to see whether the agreement is valid in terms of the reasonableness of the royalty rate, the duration of the contract, the desirability or need for the specified technology to the country, and any restrictive clauses in the contract. Restrictive clauses may be illegal. However, even if they are not, the application to license technology may be denied if there are:

1. Restrictions that prohibit the licensee from contesting the validity of patents to be used or any other patents of the licensor
2. Restrictions as to the use of the subject matter of the patent or any unpatented know-how
3. Restrictions on exports
4. Charges or royalties after expiration of the patent
5. Prohibitions, with some exceptions, on the use of know-how after expiration of the contract

6. Requirements, with some exceptions, to use only certain imported inputs from the licensor so as to prohibit imports from third parties
7. Requirements to use particular trademarks as a condition of acquiring the knowhow
8. Requirements for the licensee to acquire other patents not desired by him
9. Exclusive arrangements that forbid the licensee to acquire knowhow from other parties
10. Requirements that the licensee use only the distribution channels of the licensor.

All of the countries except Thailand do have patent protection laws, and patents can be registered – that is, there is reciprocal recognition of patents. However, as we shall see in the following section on the practices of individual countries, there is considerable variation from this generalized model. The important point to be established here is that the countries examined do have control systems in place. They vary one from another in terms of how sophisticated they are in applying these systems. They also differ in terms of the technical resources available to undertake the implied studies and reviews.

Singapore relies heavily on the market mechanism and considers most licensing agreements as matters to be settled between private parties. Singapore does provide incentives and accordingly does review investment proposals. Of those countries having complex review procedures, Taiwan, South Korea, and the Philippines are considerably more sophisticated than are Malaysia, Thailand, and Indonesia. Largely this is a matter of administrative experience rather than the degree to which there are regulatory laws and decrees; all countries have the latter.

A REVIEW OF INDIVIDUAL COUNTRY REGULATIONS AND PRACTICES

It would be impossible to catalogue all of the laws and regulations regarding foreign investment and technology transfer of each country examined here. In this section, I merely attempt to highlight those aspects of country policies that should have a considerable bearing on the transfer of technology, in terms of which industries are open to foreigners, which industries receive incentives, and how licensing arrangement are handled.

South Korea

South Korea does not foreclose foreigners from any industrial sector. However, in the military goods, power, railways, telecommunications, coal mining, tungsten, tobacco, ginseng, and salt industries foreign capital is permitted only on a negotiated basis, since the government does exert control over these industries. There are no restrictions on the proportion of foreign equity in any venture, but there is official preference for joint ventures and particularly for a minority position by foreigners. Wholly foreign-owned enterprises are allowed in the export zones, in fields where the enterprise introduces new and sophisticated technologies or management systems, and in instances where the enterprise would diversify the sources of foreign capital inflow, that is, enterprises from countries having little or no investment in Korea before. Industries or projects provided incentives are those which:

1. will develop the machine and chemical industries – especially heavy industry
2. will use little in the way of imported inputs
3. will provide new technology or innovations in the Korean economy
4. will be export oriented and labor intensive.

The incentives offered are general to all foreign invested enterprises. South Korea does not have a published schedule of priorities. However, it periodically publishes a list of industrial activities in which heavy foreign participation is welcome. Tax incentives include exemption from the corporation tax, income tax, property tax, and acquisition tax in direct proportion to the foreigner's share of equity capital. Exemptions amount to 100 percent for the first five years and 50 percent for the following three years. Customs duties on imported capital goods required by the enterprises's operations are not applied. Also, the tax on dividends is waived for five years and applied at a 50 percent rate for an additional three years. The repatriation of capital is at the governments' discretion and may be limited to 20 percent of the initial capital after the enterprise has been in operation for two years. Such limits would only be applied in case of a foreign exchange shortage. Repatriation of licensing royalties and technical fees is subject to authorization. Contracts on technology licenses are reviewed by the Korean Institute of Science and Technology. Korea does recognize foreign patents. Also, the rate of royalty and duration of contracts is largely considered a matter for decision by private parties. As an additional incentive to employ labor and upgrade skills, the South Korean government bears a part of the cost of on-the-job training of new workers. And finally, foreign enterprise has all of the same guarantees as local enterprise in all aspects of conducting business.

South Korea's Economic Planning Board has a Bureau of Foreign Investment Promotion, an organizational entity which reviews investment and technology contract proposals. It is assisted in its technical analyses by the Korean Institute of Science and Technology and the Korea Development Institute. In terms of the Economic Planning Board's guidelines, there are some refinements to the broad statements made above. The guidelines used by the board on ownership and capitalization are as follows.

1. A foreign enterprise must invest at least $200,000, except in the electronics and machinery industries where the minimum is $100,000. Export industries that use domestic raw materials and are recognized to be inducing the application of new technology are allowed a minimum scale of $50,000.

2. Joint ventures of 50 percent local and 50 percent foreign capital are preferred. However, in simple labor-intensive operations the foreigner should have less than a 50 percent share in ownership unless the enterprise is bonded and located in an export-processing zone where it uses only labor and all raw materials are imported. Under these circumstances the enterprise can be 100 percent foreign owned, but does not have to be. Any venture serving the domestic market must be less than 50 percent foreign owned unless the enterprise is recognized to be diffusing technology to other industries in which case the foreigner can have more than 50 percent ownership. It is also permissible for the foreigner to own 100 percent of the enterprise if there are multiple technologies involved and the enterprise provides access to international markets.

3. Any technology licensing agreement of less than three years' duration having a royalty rate of 3 percent to 4 percent of net sales and a down payment of less than $100,000 will be automatically approved. Anything exceeding these figures must be submitted to the Economic Planning Board's Office of Foreign Contract Review. All submissions are worked on a case-by-case basis.

South Korea's long-range goal is to place foreign and domestic investors on an equal footing, dismantle the existing review apparatus, and move to complete liberalization of investment, that is, rely upon the market mechanism almost entirely. There is also a shift away from Japanese technology. Priorities, while not stated formally, do favor the acquisition of American and European technology because Korea now has the ability to adapt the more sophisticated technologies supplied by U.S.- and European-based firms. Also, Korea has become a supplier of simple technologies to other developing countries, especially in light industry such as tires, shoes, textiles, and ceramics. Some of these exported technologies are modified versions of Japanese technologies acquired earlier. The longer-range plan calls for the acquisition of technology in heavy chemicals, petrochemicals, machine tools, and hotel management and tourism. Korea has also become involved in the export of joint projects with U.S. firms in the heavy construction field, including refineries, cement plants, and the like. Korea supplies many of the steel shapes and the bulk of the construction labor; U.S.-based firms supply the computerized systems and much of the engineering and architectural know-how.

Examination of South Korea's development suggests that, like Japan before it, South Korea has become a prodigious consumer of technology. However, unlike Japan, its policies toward foreign ownership are much more liberal. Moreover, it is liberalizing further or expects to on a much shorter time scale than has Japan. Korea has a well-developed and sophisticated system by which it can review the technology acquisition process. And through a system of flexible priorities it can use the incentive system to induce changes in the industrial structure. This system is applied to foreign investments and technical aid or licensing agreements. However, the same system applies to domestic investors as well, so in other words there is no discrimination against either, except insofar as foreigners may be asked to avoid 100 percent or even majority ownership in certain instances. There is room for negotiation, and the willingness to bend on the part of the Economic Planning Board is a matter of the perceived benefits emanating from the new enterprise – especially as this relates to the technology to be transferred.

Taiwan

Taiwan's system of foreign investment review and incentives is very similar to South Korea's. However, there are some sectors that are closed to foreign investors: military goods, power, railways, petroleum refining, telecommunications, sugar, fertilizers, and alcoholic beverages. As to the proportion of foreign equity, there are no specific restrictions outside of the sectors just mentioned, although joint ventures are preferred. Also, Taiwan specifically encourages investment by overseas Chinese. The overseas Chinese are treated as nationals if they are willing to forego foreign exchange privileges extended to foreigners but not to nationals. Otherwise they are treated like any other foreigner. There are what are considered to be overcrowded industries in Taiwan. However, there is no published list because Taiwan does not wish to discourage investment proposals. Proposals are examined on a case-by-case basis by the Industrial Development and Investment Center. Yet it should be noted that nationals and overseas Chinese would receive preference in these industries unless the foreign investor intends to export 100 percent of the output. Some of these overcrowded industries are plywood production; mushroom, asparagus and pineapple production; and canned food processing.

Some 157 manufacturing industries are eligible for incentives. In addition, certain categories of handicrafts, mining, agriculture, forestry, fisheries, animal husbandry, transportation, warehousing, public housing construction, tourism, technical services, and heavy equipment construction are also encouraged through incentives. The incentive system includes:

1. A five-year tax holiday on income tax beginning on the date when the enterprise first markets its products
2. A one- to four-year deferment of the tax holiday for capital-intensive or technology-intensive investments requiring long lead times
3. A tax rate ceiling of 22 percent on new enterprises or expansions of old ones in encouraged industries

4. Export financing for packages of plant equipment and machinery
5. Development financing for scientists and engineers to establish technology-intensive enterprises (R & D costs can also be expensed as incurred)
6. Exemption from business tax on exports
7. Deferment of duties on imported capital goods where suitable local equipment is not available
8. Exemption from import duties on instruments and equipment used in research and development activities

There are other incentives, but these are the major ones. Whether a foreign investor is eligible to invest depends upon whether or not the investment or the services to be offered can be supplied by nationals. If so, there would be no need for foreign investors including overseas Chinese. The distinction is that even if a foreigner were willing to give up foreign exchange privileges (the privilege of repatriating capital), as long as a national can undertake the investment the foreigner's proposal would be rejected in favor of the national.

While there are numerous industrial categories being encouraged, there nevertheless is a set of guidelines used by the Investment Commission, an interagency group including members from the Industrial Development and Investment Center which screens investment proposals. Those new enterprises that meet one or more of the following criteria are eligible to apply for the incentives.

1. Produces goods regarded as daily necessities, but not currently produced domestically – minimum required investment is approximately $1,325,000 U.S.
2. Produces import substitutes for goods similar to those currently being imported – minimum investment as above
3. Adopts new technology or introduces new equipment so as to improve quality or lower costs of production
4. Develops indigenous resources or utilizes by-products so as to promote improved productivity
5. Produces goods of which a minimum of 50 percent is supplied to defense authorities
6. Establishes or expands operations that benefit development of the economy with 100 percent of new output for export.

While in principle foreign nationals can own 100 percent of the equity capital in any approved enterprise, the Investment Commission prefers local participation. Even long-standing foreign investments in some fields have been urged to spin off a part of their equity to nationals, especially those heavily involved in serving the domestic market. However, if the foreign investor feels strongly that it would be difficult to cooperate with nationals, the Investment Commission and the Industrial Development and Investment Center (IDIC) tend not to impose. Also, Taiwan does not ordinarily allow foreigners to buy up locally owned firms. Foreigners are allowed to come in as partners if there is an increase in capitalization and production facilities are expanded. However, there are no particular incentives for the creation of joint ventures other than avoiding the displeasure of the reviewing agencies of government.

There are two screening boards, the Investment Commission and an Export Processing Zone Board. The latter is of lower rank and is authorized by the Investment Commission; it screens investments destined for the export processing zones. The criteria used in screening are:

1. The entire output must be exported
2. Local value added should not be less than 25 percent of total value added
3. The minimum paid-up capital must be no less than about $160,000 U.S.

The Investment Commission is the only countrywide screening board. Investment proposals are submitted to this commission. If it is a manufacturing investment it will be reviewed by the Industrial Development Bureau and the IDIC. If a foreign loan is involved the proposal will also be reviewed by the Bureau of Foreign Exchange. The Board of Directors of the Investment Commission is composed of:

1. Vice-minister of Economic Affairs (chairman of board)
2. Vice-minister of Finance
3. Vice-minister of Foreign Affairs
4. Vice-minister of Communications
5. Governor of the Central Board
6. Vice-minister of Interior
7. Director of the Department of Reconstruction of Taiwan
8. Director of the Department of Reconstruction of Taipei City
9. Director of Health

Also attending are the director and deputy director of the IDIC and the Director of the Industrial Development Bureau (IDB).

This board meets every two weeks and deals with approximately forty cases at each meeting. These cases include licensing and technical aid agreements. While there is no guideline on royalty rates to be charged on licensing agreements, a rule of thumb of 3 to 5 percent of sales is used. However, if a licensor can show a signed agreement with an independent company that has agreed to a rate of over 5 percent, the proposal will most likely be accepted. The average duration of licensing agreements is five years, but a longer period is permissible if the technology is particularly desirable and is complex or sophisticated. Licenses can be renewed by both the licensor and licensee reapplying for extension. The board uses past examples of industry practice within particular industries as a guideline in making decisions. If a proposal appears to be unreasonable, the board may ask the IDB or IDIC to confer with the investor or licensor and ask him to restudy the proposal.

Both groups have wide experience in and knowledge of industrial practices. If an investor's proposal is considered too optimistic they may recommend rejection, because they do not wish to have unsuccessful ventures.

Taiwan has patent laws that are similar to those of the United States. Patents on new inventions run for fifteen years. Models can be patented for ten years, industrial designs for five years. Trademarks are recognized and can be registered for ten years and renewed for periods of ten years.

Once an investment has been approved the Investment Commission does follow up, examining products being produced, capital remitted, and marketing in relation to the proposed sales plan. Import licenses and other documents are used to check on equipment and other items brought in by the investor. And if the investment is eligible for incentives, the operation will be checked by the investment authorities.

From the above description it can be seen that Taiwan has a very extensive review system by which it can control the acquisition and application of technology. It uses an extensive system of incentives and post-investment reviews to shape and monitor the contribution of foreign technology to industrialization.

The Philippines

Of all of the countries under study here the Philippines has perhaps the most stringent technology control system. It is in many ways the most legalistic, detailed, and formalized of all.

The Philippines forecloses foreigners from participating in several industries, including rural banking, mass media, retail trade, rice and corn production (except in some instances), and certain military goods. There is also temporary foreclosure in what are considered overcrowded industries. The list of these industries changes from time to time, but currently there are some forty industries on the list, including galvanized iron sheeting, nails, pencils, matches, radios, and room air conditioners. However, if the entire output is to be exported and/or the new investment would contribute to dispersal of industry away from congested urban Manila, investment in an overcrowded industry may be allowed.

While the Philippines excludes foreigners from participation in only a few industries, there are very strict rules regarding the proportion of equity capital a foreigner can hold. These rules are a part of the Investment Priorities Plan. The proportion of allowable foreign equity varies by industry groupings, which are:

1. Restricted industries
2. Pioneer preferred industries
3. Nonpioneer preferred industries
4. Tourism
5. Other industries.

In addition there is an Export Priorities Plan, which allows registration even in nonpioneer industries.

In the restricted industries, Filipinos must own at least 60 percent of the equity, and in some instances the foreigner's share must be as little as 25 percent. The restricted industries include:

1. Banking, except for new consolidations of branches or agencies of foreign banks
2. Savings and loan associations and finance companies
3. Public utilities
4. Domestic air transport
5. Fishing and fishing vessels
6. Concessions to utilize geothermal energy, natural gas, and methane gas
7. Natural resources and land
8. Coastal trade

In the pioneer preferred areas foreigners may own 100 percent of the equity, but this must be reduced to 40 percent within thirty years. In the nonpioneer preferred areas foreigners may own no more than 40 percent of the equity, except when measured capacity is not achieved within three years, at which point the enterprise may be "liberalized" and treated under the rules of a pioneer preferred enterprise, as above. Under the Export Priorities Plan, even a nonpioneer preferred enterprise may register and be treated as a pioneer preferred enterprise if it exports at least 70 percent of its output. Also 100 percent ownership is allowed in the export processing zone where 100 percent of output is exported, and in industrial estates designed to disperse industry. If an enterprise qualifies under the Tourism Priorities Plan, the foreigner may hold up to 100 percent of the equity but is then treated as a pioneer preferred enterprise. However, the thirty-year period may be extended by another ten years in certain instances under the Tourism Priorities Plan. All other areas of investment may allow 100 percent ownership, but when foreign equity exceeds 30 percent the proposal will be reviewed by the Board of Investments. Permission for such an investment may be granted if:

1. There is no conflict with existing laws limiting the degree of foreign ownership
2. There is no clear danger of promoting a monopoly
3. The field is not already adequately served by Filipinos
4. The investment does not conflict with the Investment Priorities Plan in force at the time
5. The investment contributes to sound and balanced development of the national economy.

The list of industries encouraged or welcomed under incentives can change from time to time. However, in the Tenth Investment Priorities Plan, published in June 1977, some 100 industries were listed. At the same time, the Export Priorities Plan listed nearly 300 industries being promoted. Thus there is considerable detail available to foreign investors as to the investment priorities and areas where foreign participation is welcomed.

The Philippines has several incentive programs, including those under the Investment Incentives Act, the Export Incentives Act, the Tourism Industry Investment Program, and the Agricultural Investment Incentives Act. The objectives of the program emphasize:

1. Labor-intensive industry
2. Industry dispersal
3. Small and medium scale industries in the metro-Manila area
4. Intensification of exports, especially labor-intensive manufactures
5. Processing of materials destined for export markets.

Firms eligible for incentives are those with 40 percent or less foreign participation in listed nonpioneer areas or those with 100 percent or less foreign participation in pioneer areas, nonpioneer liberalized areas, export oriented enterprises, and promoted enterprises in export zones or industrial estates. For them the following incentives are offered:

1. Exemption from income taxes:
   a. 100 percent for first five years
   b. 75 percent for years six through eight
   c. 50 percent for years nine and ten
   d. 20 percent for years eleven and twelve
   e. 10 percent for years thirteen through fifteen.
2. Foreign nationals may be employed as up to 5 percent of total employment for five years, but after five years only the president, treasurer, and general manager, if foreign nationals, may be retained
3. Tariff protection not to exceed 50 percent of the value of importable items similar to those being produced
4. Reduction of taxable income in the amount of expenditures on direct labor costs plus costs of locally procured raw materials for export enterprises
5. Tax credit for locally procured machinery of 100 percent of customs duties that would have been paid if equivalent machinery had been imported
6. Tax exemption on imported capital equipment not manufactured and available locally
7. Exemption from capital gains tax on the sale of capital assets after a five-year period of operation of an initial investment.

There are other incentives, but the above appear to be the major ones. In addition, however, the Philippines also protects patents and other proprietary rights, such as trademarks and copyrights, that have been registered with the government.

The important distinction between the Philippines and Taiwan and South Korea is that foreigners do not have the same access to the incentive programs that nationals do. Whether they have access depends, in large degree, upon their willingness to share ownership with nations. Moreover, while there are some sectors in which 100 percent ownership is allowed, and these tend to fall within the pioneer areas, in the long run the foreigner must accept a minority position in ownership.

It should be noted that the Philippines, while affording patent protection, does closely review licensing agreements for restrictive clauses. Thus unlike most of the countries being examined, licensing agreements (or other agreements to use proprietary rights) are not matters to be settled between private parties. In January 1978 President Marcos established, through Presidential Decree 1263, a more restrictive set of rules on licensing agreements, which limit royalty rates to 5 percent per annum of net sales and challenges restrictive clauses in licensing agreements. Unlike Central Bank Circular 93, which provided the guidelines prior to the decree, there now appears to be little flexibility for the Board of Investments to exempt certain agreements.

Any investment exceeding 30 percent foreign participation or technical aid proposal begins with a submission to the Central Bank of the Philippines, because there are likely to be remittances of profits or royalties.(4) The proposal is in the form of a feasibility study that describes the project the investor wishes to register for incentives. The proposal is then referred to the Board of Investments. On the board's staff there are approximately seventy-five specialists who evaluate investment proposals and technical aid agreements. There are three industry groups that cover metals and mining (this group is responsible for electronics, engineering industries, capital equipment, metals, and minerals), chemicals, and agriculture and agribusiness. In addition, there is a Research Information and Statistics Division that looks after all other product groupings not covered by those industry groups noted above. If the investment is eligible for and request incentives, the review process takes about two to three months. Moreover, if the reviewing subgroup decides that the investment would result in "overcrowding" the industry it would be rejected for incentives unless it had other redeeming virtues. A legal staff also reviews the proposal to determine the desirability of the investment. If it meets the criteria, a license to do business will be issued.

Should there be a technical agreement involving a patent recognized by the Philippines, that agreement is reviewed by a newly established Technical Resources Center (TRC). (Any agreement not involving a patent registered in the Philippines goes directly from the Central Bank to the Board of Investments ad hoc review committee.) With the TRC is a Technical Assessment and Registry Program (TARP). A study group reviews the agreement and makes a recommendation to the director of the TARP, who then renders a decision. During the evaluation process the TARP can consult with any government agency, including the Board of Investments. Once a technical agreement is accepted, it is returned to the Central Bank where the decision on royalty rate is made. If the royalty rate violates the 5 percent guideline, or if the duration of the contract is considered overly long, there is an interagency consultation with the TRC to determine whether the cost and/or duration are reasonable for the degree of technology being transferred.(5) Once it is approved, administration of any technical aid agreement is turned over to the Board of Investments.

There is an appeals procedure for rejected proposals and the Board of Investment does make exceptions in certain cases, so there is some room for negotiation.

To summarize, the Philippines has an elaborate and sophisticated system for reviewing investments and technical agreements. There is a system of incentives designed to encourage investors in labor-intensive, export-oriented, nonurban, and high-technology investments. The Philippines' laws are very nationalistic, but there is room for negotiation and liberalization on certain types of investments or technology transfers, especially where the technology being acquired is considered particularly attractive.

Thailand

Thailand's technology control system is very much a composite of the programs already described. However, it seems to be somewhat simpler and less legalistic than that of the Philippines and not so concerned about high-technology industries as any of those described to this point. Also, it appears that Thailand has not had as much experience as the Philippines, South Korea or Taiwan.

There are some sectors foreclosed to foreigners (public utilities, rural banking, insurance, savings banks, and certain military goods) and these are industrial groupings that require at least 50 percent Thai ownership. Thailand has three categories. Category A includes industries reserved for Thai majority ownership and control, and includes rice and salt farming, numerous personal services, dealing in local sale of agricultural products, and building construction. Category B requires all new enterprises to be under Thai control in several types of farming, fishing, industrial handicrafts, printing and newspaper publishing, plywood, retail trade, tour agencies, and several service industries. Existing foreign-owned enterprises established before November 26, 1972, can continue under foreign control in Category B. Category C encompasses all of those industries in which foreign ownership and control are allowed for both newly established and older enterprises (those established prior to November 26, 1972). Older enterprises in Category C are assigned an annual growth rate that cannot be exceeded, with the results that older enterprises may shrink as a proportion in Category C. The Category C list is less lengthy than that of Category B and includes wholesaling, exporting, machinery retailing, sales of food and beverages to promote tourism, light manufacturing, and businesses not included in Categories A or B.

All of those industries not included in the three lists can receive incentives under appropriate conditions. However, joint ventures are encouraged in all fields but not required as in those noted above.

There are so-called overcrowded industries in Thailand, but whether an industry is overcrowded is a decision for the Board of Investments. Like Taiwan, Thailand does not publish a list of overcrowded industries. Moreover, an industry that is on the promoted list may be removed by the board if it becomes overcrowded in the board's judgment.

The incentive system is similar to those in other countries and is targeted on:

1. Agricultural industries
2. Labor-intensive industries
3. Export-oriented industries
4. Industrial dispersal from the Bangkok area.

However, it would appear that Thailand continues to rely substantially on import substitution and a relatively high tariff structure as part of its development strategy. The Board of Investments can offer special tariff protection to new industries by placing a levy of up to 50 percent on the C.I.F. value of the imported good. This is in addition to any already existing levies.(6) In addition to tariff protection, Thailand offers the following incentives in promoted industries:

1. Exemption from import duties on imported capital equipment providing local equivalents are not available
2. Exemption from income tax for between three and eight years depending upon the percentage of local content and/or the number of workers employed.

If the enterprise is located in an Investment Promotion Zone (a part of the dispersal of industry program), there are additional incentives including:

1. Reduction of import duties and business taxes by up to 50 percent on raw materials for up to five years, provided comparable local materials are unavailable
2. Reduction of up to 90 percent of business taxes on the sale of products for up to five years
3. Income tax deductions of up to 25 percent of the value of buildings and equipment
4. Reduction of up to 50 percent of income tax for the sixth through tenth years of operation
5. On goods produced for export, exemption of export duties, exemption from import duties on raw materials, income tax deduction of 2 percent on increases in year-to-year income, and exemption from business tax for local suppliers of raw materials to the exporter.

There are value-added requirements and incentives for firms exporting less than 20 percent of output in industries promoted into import substituting activities.

While Thailand protects trademarks and copyrights, it does not have a patent law and there is little recourse in case of patent infringement. The conditions of technical agreements are largely ignored in Thailand; however, there may be an attempt to regulate technology agreements in the future. Through two recently established groups, the Department of Commercial Registration and the Applied Scientific Research Corporation, there is an attempt to examine the adaptation of technology and the technology being used by enterprises having technology transfer agreements. The Bank of Thailand has the authority to regulate royalties, but there is no actual control at the moment. In the main the bank merely records remittances.

Thailand's Board of Investments appears to have broader authority than do similar agencies in the other countries studied. Yet the functions carried out and the methods used are quite similar; the submission of proposals and methods used to screen them are similar. Thailand has moved to "one-stop" shopping." The Board of Investment helps to guide the proposal through the review process, and if the proposal is approved, it then is expedited to the Ministry of Industry, which is the granting agency for business licenses, and the Ministry of Commerce, which registers all new business projects. The review process takes from four to six weeks.

Some 225 investment proposals are reviewed annually by about ten or so project analysts, who use a form of benefit-cost analysis to examine proposals for their economic and engineering soundness in relation to the objectives of the investment promotion program. The project analyst on any particular proposal will make a recommendation as to what type of incentive package (selected from among some twenty-six incentives) the investor should receive. Projects can vary from one another in this respect. A set of conditions may also be attached by the analyst; that is, he specifies what incentives the applicant should receive and under what conditions. There is some effort on the part of the board to examine for alternative technologies by using assistance from the United Nations Conference on Trade and Development (UNCTAD).

In terms of the final selection of projects, there is an interagency committee composed of thirteen members and five advisors. Theoretically the prime minister is chairman of this board, but in fact the minister of Industry (vice-chairman) chairs meetings.

Another role played by the board's staff is that of follow-up. There is a continuing effort to see that Thai nationals are gradually moved into positions of responsibility. Alien work permits, which must be renewed every year, may be denied to certain expatriates so as to force foreign enterprises to employ nationals in key positions. However, this is a subject for negotiation between the enterprise and the board. So far it appears that the pressure to reduce the proportion of aliens in high-level jobs has not been great.

To summarize, Thailand's technology transfer control system is not as well developed as those of South Korea, Taiwan, and the Philippines. However, the structure of incentives and ownership requirements are similar to those of the Philippines. The Thai Board of Investment appears to have considerably broader powers and authority than do other boards, yet final decisions remain in the hands of an interagency committee.

Malaysia

Malaysia's technology transfer control system is similar to that of the Philippines in many respects, but does not include some of the strictures of the latter. Foreigners are only excluded from ownership participation in public utilities, domestic air transport, and production of some military goods. Also no areas are specifically restricted in terms of the proportion of foreign equity. However, there is strong pressure to have at least 51 percent Malaysian ownership in enterprises exclusively devoted to serving the domestic market. This is not an absolute requirement. Ownership proportions are negotiable, and if a particularly desirable investment is proposed and it serves only the domestic market, majority foreign ownership may be allowed. In "export only" enterprises (80 percent of output exported and all raw materials imported), foreigners may hold 100 percent of the equity. It is nevertheless true that the Malaysian government has a preference for joint ventures – particularly in those enterprises that depend heavily upon the domestic market and/or use Malaysian resources, especially depletable resources. Between these two extremes there is room for negotiation.

A wide variety of industries is encouraged via incentives including:

1. Manufactured food products
2. Growing and processing of agricultural crops
3. Fishing
4. Cattle farming, processing, and by-products
5. Dairying
6. Silkworm culture
7. Manufacture of wood products
8. Natural rubber
9. Chemicals
10. Agricultural implements
11. Handtools
12. Machinery
13. Transport equipment
14. Motor vehicle parts
15. Iron and steel
16. Electronics
17. Instruments and precision products.

Malaysia provides incentives for pioneer industries, export-oriented enterprises, and enterprises that are labor-intensive or at least use a large number of employees. The incentive program can be summarized as follows:

1. Pioneer status industries
   Exemption from income tax:
   a. two years if fixed capital is less than $250,000 Malaysian dollars
   b. three years if fixed capital is between $250,000 to $500,000 Malaysian dollars
   c. four years if fixed capital is between $500,000 to $1,000,000 Malaysian dollars
   d. five years if fixed capital is more than $1,000,000 Malaysian dollars

   A further exemption of one year for each of the following conditions met:

   e. sited in a development area (dispersal)
   f. product a priority product (based on priority list)
   g. Malaysian content of 50 percent or more

2. Labor utilization relief (similar to pioneer status industries, based on employment
   Exemption of income tax:
   a. 50 to 100 employees – two years
   b. 101 to 200 employees – three years
   c. 201 to 350 employees – four years
   d. 351 and above employees – five years

   Conditions e, f, and g are same as above

3. An investment tax credit of 25 percent is offered in nonpioneer industries and is increased by an additional 5 percent for each of the conditions (e, f, and g above) met.

There are additional incentives for hotels, export growth of an enterprise, and dispersal of industry.

The maximum tax holiday is ten years when the enterprise locates in an incentive geographic area and when both the amount of fixed capital and the amount of employment are coupled with product priority and Malaysian content incentives.

While ownership proportions are negotiable, there is an implicit guideline on the part of the Federal Industrial Development Authority (FIDA), the agency that is the competent authority, similar in function to the Board of Investments in the Philippines or Thailand. Higher technology and labor-intensive investments are desired, but there a major part of the output must be exported if the foreigner is to hold a majority share in ownership. In all cases where the enterprise is heavily dependent on the domestic market, regardless of its level of labor intensity or technological sophistication, there must be a Malaysian majority. (When the term "Malaysian" is used in this context, what is meant is that the local partner should be of Malay extraction rather than Malaysian Chinese or Indian.) In a joint venture in which locals hold more than 30 percent of ownership, at least 30 percent of the ownership capital must be reversed for Malaysians of Malay ancestry (Bumiputra). If less than 30 percent of ownership is local, the entirety of the local share then must be held by Malays. The objective is to shift the ownership structure of trade and manufacturing away from foreign to local ownership. By 1990 Malaysia is targeting on the following shares:

	Current ownership shares	Desired shares for 1990
Malays (Bumiputra)	8.0%	30%
Other Malaysians	37.3	40
Foreigners	54.9	30

It is only current transactions that are being screened. Old ownership relationships are honored – there are no rules on disinvestment.

There is a short list of overcrowded industries that includes steel and cement.

The screening process by the FIDA is similar to those discussed above. There is a Project Processing Division, within which there are three deputy directors, each of whom specializes on industry groupings, as follows:

Group #1	Group #2	Group #3
Transportation equipment	Industrial chemicals	Rubber and plastics
Electrical and electronics	Nonmetallic Products	Agro industries
Textiles	Wood products	Palm oil
Hotels	Paper products	Fabricated metals
Miscellaneous		Miscellaneous metal products

FIDA examines proposals with respect to plans for bringing Malays into key positions. The foreign investor must justify his use of expatriates in terms of the skills needed, and there is some pressure to see that Malays are trained for certain of these positions. Moreover, the investor is under pressure to match the composition of his workforce to the average composition of the workforce in the external labor pool. The goal nationally is to have 55 percent Malays, 30 percent Malaysian Chinese, 10 percent Malaysian Indians, and 5 percent others.

Malaysia does recognize licensing agreements on patented technology. However, there is a guideline on royalties of 2 percent on sales and contract duration of five years (the contract can be renewed for five-year intervals). Technical aid agreements generally are not accepted. Licenses are screened by the Industries Division of the Ministry of Trade and Industry, which follows up on investments and licensing agreements and enforces the conditions negotiated with the foreign investor by the FIDA.

To summarize, Malaysia, like other countries examined so far, does have a set of priority or pioneer industries with an incentive scheme to entice investments into these industries. Emphasis is on employment creation (not necessarily labor intensity) and export development. Ownership proportions are substantially negotiable, providing there are exports and local content other than depletable resources. Also, the investor is under considerable pressure to employ and train Malays in preference to domestic Chinese, Indians, or others or non-Malay ancestry.(7)

Singapore

Singapore has the fewest rules and regulations on foreign investment of any of the countries in the Far East. There is heavy reliance on the market mechanism, with few import controls. Some sectors are closed to foreign investors: public utilities, domestic air transport, and some military goods. Also, Singapore promotes joint ventures in some fields, whereby the government owns a share along with domestic or foreign partners. Examples are shipbuilding; construction projects for public housing, industrial estates, and parks; steel, ocean shipping; and petroleum refining. Other than these instances, there are no restrictions on the areas in which foreign investors can participate. Although Singapore does prefer to have local participation in foreign ventures, the government is very flexible in this respect.

Investment incentives are provided in pioneer industries and to firms producing for export. Virtually any industry can be classified as a pioneer industry if developmental prospects are considered favorable. The decision is up to the minister of Finance, who can designate an industry as having pioneer status. Incentives may also be offered to existing firms that are expanding their operations, are exporting, are high-technology, or are providing desired types of training to upgrade skills.

Skill-intensive, export-oriented industries using high levels of technology are preferred. Singapore is extremely concerned about continuously upgrading the skills of its labor force. As an island economy with little in the way of natural resources, Singapore recognizes that its continued growth, development, and economic health depend largely upon continued improvement of the labor force and its productivity.

Incentives offered include the following.

1. Pioneer firms are exempt from income taxes for a period of from five years up to ten years if they meet the technological and employment standards. The length of the tax holiday depends on the industry amount of investment, number of jobs created, and level of technology.

2. Expansions qualify for the tax holiday if the expenditure on plant and equipment exceeds \$10 million Singapore dollars.

3. In export industries the investor can be exempted from 90 percent of business taxes on the incremental growth in export earnings for five years.

4. Export enterprises having pioneer status receive a tax holiday of up to ten years, plus a three-year concession of incremental growth in export earnings.

5. Firms under number (4) above may have this privilege extended to fifteen years if invested capital exceeds \$1 Singapore dollars in a majority foreign enterprise or \$150 million Singapore dollars in an enterprise that has majority local capital and minority foreign capital.

The minister of Finance can extend the export tax privileges to any enterprise beyond fifteen years if, in his judgment, the enterprise is particularly deserving.

Singapore recognizes and abides by the patent laws of the United Kingdom. Only patents registered in the U.K. are recognized, but they must be registered in Singapore within three years of their approval in the U.K. Singapore has its own trademark law. Upon registration, a trademark is valid for seven years but it can be renewed for additional successive periods of fourteen years. There are no restrictions on royalty payments or the duration licensing contracts – negotiations are a matter to be agreed upon between private parties. However most foreign firms, because of the smallness of the Singaporean market, prefer to invest rather than license to locally controlled firms.

Singapore's screening process is similar to but much less protracted and legalistic than that of any of the countries already described. The Economic Development Board is the competent authority. It administers the tax incentives, financial assistance, manpower training, and export assistance programs. One important difference emerges. The board itself is much less involved in the screening and selection of foreign investments than in other countries. In Singapore the board is more concerned with the design of new strategies for Singapore's overall development. It is supported by a staff that is functionally structured into divisions including:

1. International operations
2. Investment services
3. Projects
4. Manpower
5. Administration.

These divisions are under the direction of an executive committee. Through its International Operations staff the board is active in foreign countries in the identification of desired investments and particular enterprises to be solicited to locate investments in Singapore. This division is supported by the Projects Division, which conducts industry studies and attempts to identify industries and technologies that should represent desirable opportunities for Singapore's economic and technological development. It also evaluates applications for pioneer status and other incentives. Once an investment is attracted to Singapore, the Investment Services Division works with the investor to assist with implementation of the investment plan. The Manpower Division supports all of the above efforts by conducting manpower studies and administering training programs, some of which are in partnership with the private sector. Singapore also finances the training of selected individuals who are sent overseas to acquire specialized technological skills.

Singapore's program for attracting foreign technology is designed around the notion of creating skills in advance of the firm's arrival. Singaporean officials do not say that their intent is to share in the economic rents, but that is largely the outcome an economist would expect. Singapore, like Japan before it, is sending its best graduates abroad to work in foreign firms on the equivalent of apprenticeship programs in high-technology industries. Singapore sees that they are trained in the language of their host country prior to leaving Singapore; it also pays for their educational and other expenses while abroad. The industries selected for study are targeted by the Singapore Economic Development Board. These generally are industries that are at or near the technological forefront in electronics, avionics, machine tools, and the like. Once the skills are available, the leading firms in these industries are approached and the board sells the advantages of locating in Singapore. The program has been effective; the educational and industrialization programs tend to be mutually supporting. Some of the firms that have helped train in Singapore, and in the United States, Europe, and elsewhere, have subsequently become investors because of their high regard for the technical capabilities of their Singaporean students.

To summarize, Singapore has most of the same incentives offered by other countries. However, it is much more flexible in terms of ownership structure and the classification of pioneer industries. Also, it has a well-developed system to assist investors in the implementation of new investment. Training is extremely important to the technology acquisition program and Singapore attempts to train ahead of the need for new skills. Yet there is a close matching between the skills being created and the new investments being solicited to locate in Singapore.

Indonesia

The Indonesian system of technology and investment control is perhaps the most nationalistic of those examined.(8) Joint ventures are required throughout; that is, Indonesia ownership capital must be included. Foreigners are foreclosed from investing in the following activities:

1. Arms and ammunition, explosives, and armaments
2. Coal mining
3. Trade and distribution, except importing and exporting connected with domestic manufacturing.

In the exploitation of natural depletable resources, foreigners are permitted to use work contracts or concessions.

Indonesians are required to have majority control in electric power, harbors, telecommunications and mass media, shipping, domestic aviation, public utilities, atomic energy, and automobile production. Foreigners are largely foreclosed from activities involving livestock and food production and processing, logging, and numerous other such items. The list extends to more than 100 products. In virtually all other instances, foreigners can own the majority of equity capital but must reduce their share to a minority position within ten years. There are a few exceptions. If the investment is highly labor-intensive and/or produces exclusively for export, the government may be willing to allow 100 percent foreign ownership.

There is a sizeable list of industries that have been classified into three groupings: priority, preferred, and nonpreferred. Priority industries are mainly those that utilize natural resources and hence are agro-based, forest-based, petroleum-based, or mineral-based industries. However, there are some not fitting this description, such as pharmaceuticals, toolmaking, electrical and nonelectrical machinery, automotive parts, shipbuilding, hotels and motels, and low-cost apartments and houses.

Priority and preferred projects receive incentives; tax holidays are extended only to the priority classification. Nonpreferred industries are those in which foreign participation is allowed, but no special effort is made to attract them.

For the priority industries, current law permits the granting of fiscal incentives to include exemption from corporate tax and dividend tax for two years and an additional period of up to four or more years on a negotiated basis. As an alternative to this tax holiday, the investor can opt for an investment tax credit of 20 percent of the approved project cost spread over a four-year period, that is, 5 percent per year. The typical exemption of import duties on imported capital equipment is also in place. Expansions of existing enterprises in priority industries also are afforded these fiscal incentives.

To summarize, Indonesia's technology control system is not only more restrictive with respect to foreign ownership than are those of other countries, but also the incentives offered look to be considerably less generous. However, it should also be noted that Indonesia is well endowed with natural resources and is relying considerably more heavily on the development of these than it is upon manufacturing industry in its development strategy. Indonesia has adopted a "one-step" investment service through the Capital Investment Coordinating board. From published reports by the board one can conclude that the screening process is quite similar to the process used in Malaysia or Thailand.

TECHNOLOGY TRANSFER CONTROL SYSTEMS AND ASEAN

It is evident from a review of the technology transfer control systems of the member countries of the Association of Southeast Asian Nations (the Philippines, Thialand, Malaysia, Singapore, and Indonesia) that there are some similarities among these systems. But there are also some vast differences. A serious question that appears not to have been addressed is, How will these countries coordinate their incentive programs and foreign investment and licensing regulations so as to foster the rationalization of industry across the total market composed of the five countries? As a group the ASEAN countries do seem to be desirous of achieving the efficiencies that could be gained through complementation programs. Yet in many respects the countries are competitive. They individually seek to attract many of the same industries and investors. However, they differ in terms of the acquisition strategies pursued in this endeavor – particularly since these strategies also pursue nationalistic goals. At the most nationalistic end of the spectrum are Indonesia and the Philippines, but the Philippines seems to offer a more generous package of incentives. In midspectrum are Thailand and Malaysia, which are somewhat less nationalistic and also offer fairly generous incentive programs. And at the other extreme is Singapore, which lacks natural resources, offers attractive incentives, and has a very open system that is less compromised by nationalistic tendencies.

Perhaps incentives are needed in today's world because one country must compete with all others to attract foreign investors and transferors of technology. Yet incentive systems are akin to the prisoner's dilemma: if none offered incentives none would need to, but if one country offers incentives, they all must do so. Perhaps in the end competitive forces merely neutralize the effects of incentive systems. The result is that everybody is worse off to the extent that all must incur the expense of maintaining an administrative apparatus to implement incentive programs.

Insofar as the ASEAN countries have differentiated their programs in terms of the industries receiving priority treatment, and insofar as these industries build upon individual countries' comparative advantages, incentive programs may be useful in the same way that advertising is – incentives may gain the attention of those who would not have invested otherwise. But then again they may not. And even if they do, the impact is dulled once all of the countries concerned are playing the same game. Moreover, from an ASEAN perspective the technology control systems of individual countries may be nothing but a roadblock to useful cooperation.

There are several conclusions we can draw from this review. First, it is quite evident that the developing countries of East and Southeast Asia have constructed some rather sophisticated systems through which they can screen investments and technological agreements. Second, an examination of the evidence suggests that many of the claims emanating from developing countries about multinational firms' superior bargaining power vis-a-vis the host countries' bargaining power are much overdrawn. Certainly in the countries studied here there are highly knowledgeable people involved in the screening of foreign investments and licensing agreements. There is knowledge of industry practice. I should say that, if anything, there is overregulation and overcontrol in many instances, which, rather than reducing the cost of technology, either elevates the cost or denies particular countries access to the technologies they most desire. Only Singapore seems to have a sound understanding of what is required to attract technologies appropriate to its needs and stage of development. Third, the incentive systems are targeted on priority industries rather than on industry generally. Or stated differently, these countries have backed away from generalized approaches to attracting industry and foreign technology purely for the sake of industrialization, job creation, and generation of foreign exchange. There is more attention being given to the notion of comparative advantage and available resources. However, it should also be noted that the incentives very well might not be needed in many instances for this very reason. Finally, we can conclude that there is considerable diversity among the programs examined. And without a very detailed study of exactly how each incentive program functions in particular cases, one cannot say whether or not one country has a superior system when compared with another. But one question that needs to be raised is the extent to which these incentive systems may merely neutralize one another. Certainly this is a question the ASEAN countries must grapple with as they attempt to develop complementation programs and rationalize several industries across the five markets.

The Aims of ASEAN

The Association of Southeast Asian Nations is now some 13 years old – it was formed in August 1967. The Bangkok or ASEAN Declaration committed the five countries to regional cooperation with the following aims:

1. Accelerate economic growth and social progress
2. Promote regional peace and stability
3. Promote collaboration on common interests in economic, social, cultural, technical, and administrative fields
4. Provide assistance to one another in training and research
5. Collaborate for greater utilization of agriculture and industry, trade expansion, transportation, and communications
6. Promote Southeast Asian studies
7. Maintain close cooperation with international and regional groupings having similar aims.

Through the ASEAN Concord of February 1976 the five countries, in support of the economic cooperation objectives, agreed that they would cooperate in the establishment of large-scale ASEAN-wide projects, give priority to projects that utilize the available resources of the member countries, and contribute to increased food production. Also, they agreed to cooperate in the development of intra-ASEAN trade and to work together to improve trading opportunities between individual ASEAN members and between individual ASEAN members and the international trading community. These efforts were seen as leading to cooperation in the field of technology and production so as to improve quality of export products, increase production of existing products, and develop new products that would diversify the export base.

At the second meeting of ASEAN economic ministers in March 1976, specific project recommendations were made, including an examination of the feasibility of immediately establishing ASEAN-wide projects, as follows:

Project	Country
Urea	Indonesia
Urea	Malaysia
Superphosphates	Philippines
Diesel engines	Singapore
Soda ash	Thailand

These projects were to be designed to supply ASEAN-wide needs. Other projects to be investigated included newsprint, potash, and integrated steel and petrochemical industries. It was also agreed that information would be exchanged on various sectors with a view to establishing complementation programs. Feasibility studies were proposed as well for the metalworking and machine tools, fisheries, tin-plating, heavy-duty rubber tire, and electronic components industries. Finally, it was agreed that ASEAN industrial projects should receive ASEAN-wide preferential trading arrangements, in other words a reduction of import tariffs by all members on the products emanating from the projects and from complementation programs.

The amount and type of trading preference depended upon whether the products involved showed a good possibility for import substitution in ASEAN. Excluded would be products already in surplus in ASEAN or products that are particularly sensitive to one or more member countries.

By August 1977, at the time of the tenth anniversary meeting of ASEAN heads of government, only the urea project study for Indonesia had been completed. The other feasibility studies for key projects were nearing completion at that time. It was hoped that the Indonesian urea project could be launched by mid-1978. In September 1977 the ASEAN economic ministers agreed that each country would present a list of some fifty items on which it was willing to grant trade concessions in each future trade negotiation round. However, ASEAN is still a long way from becoming an economic union or even a customs union. Nevertheless, the five countries have made considerable progress in their efforts to move toward economic and trade liberalization. The main implications of this for foreign investors and technology transfer resides in the ASEAN-wide industrial projects and the complementation programs. Most of the items so far volunteered for tariff reduction in a straightforward trade liberalization are relatively low-technology products. Moreover, the ASEAN preference rate provided on these items is not overly generous.

ASEAN and Technology Transfers

It seems apparent that the ASEAN-wide industrial projects will not involve foreign private investors. Rules specify that these projects will be jointly owned among the five countries, with the host country owning 60 percent of the equity. Each project will have only five shareholders, that is, the five countries. Each country will nominate a shareholder agency or company, but one-third of the shareholder agency should be owned by the nominating government. Thus there is room for private participation but it would seem unlikely that any enterprise not headquartered within ASEAN and not majority-controlled by nationals would be nominated as the shareholding agency, because of the one-third government ownership rule.

From the perspective of ASEAN-wide developments as they impinge on technology transfers from non-ASEAN sources, it would seem that the industrial complementation programs are of greatest importance. Indeed, the complementation programs have been more or less delegated to the private sector. The object of complementation is to take advantage of the longer production runs and economies of scale offered by the large ASEAN market as opposed to the historical situation of isolated, individual country markets.

Currently the complementation program includes eleven industry clubs, each of which deals with a particular set of products. These eleven industry clubs are:

1. Automotive Industry Club
2. Electrical and Electronic Federation
3. Ford Processing Industries Club
4. Agricultural Machinery Industry Club
5. Ceramics Club
6. Iron and Steel Club
7. Rubber-Based Industry Club
8. Sheet Glass, Glass Containers, and Soda Ash Clubs
9. Pulp and Paper Industry Club
10. Cement Manufacturers Federation
11. Chemical Industries Club.

So far only the Automotive Industry Club has been highly active. The other groups have held plenary sessions but few had formalized into clubs. Most have established some type of study group or groups to examine the possibilities for complementation. Foreign investors are participating in these clubs.

In general, there seems not to be any major effort to examine or attract "appropriate" technologies. The effort is largely predicated upon market expansion and rationalization instead. In large part the ASEAN effort aims to ameliorate past mistakes created by indiscriminate import substitution programs. Only Singapore among the five countries has moved swiftly toward free trade. This in and of itself makes ASEAN less valuable to Singapore than it might be to the other four countries. Moreover, even to date there have been many sticking points in the attempt to establish ASEAN-wide projects.

Some Conclusions on ASEAN

As was noted in the opening comments of this section, ASEAN has only begun to grapple with many of the differences among its members. And so far it does not appear that much has been done to examine intercountry differences in technology transfer control systems. While private-sector industry clubs can make recommendations regarding "best locations" for particular types of factories, these clubs cannot deal with the vast differences among the five countries in terms of the incentives they offer, their attitudes toward and control over foreign participation, and their requirements that nationals be included in ownership and/or be trained to deal with modern technology and managerial systems. What is best for ASEAN may not satisfy the foreign investor's needs, in terms of his preferences to locate in a particular country. The latter may rest more on language, culture, attitudes, and past business dealings than it does on which country has the apparent comparative advantage.

SOME CONCLUDING REMARKS

It was not the purpose of this chapter to place the technology transfer control systems of the East and Southeast Asian countries in a global perspective. However, the material seems worthy of some additional comments in light of the United Nations' proposed Code of Conduct on Technology Transfers.

It is evident from the materials reviewed here that these countries not only have sophisticated technology transfer control systems in place, but also have an awareness of the cost and value of technology. They seem to be shifting away from import substitution toward labor-intensive, export-oriented outputs. The technologies most desired and the priorities established support these aims. No longer are they in the position of having to rely upon the multinational firms to assess their technological needs. They are largely able to do this for themselves. Perhaps they have demonstrated that there is little need for international regulation of technology transfers. Indeed, one could hypothesize that an international agreement would do little if anything to enhance their position in bargaining for or attracting technology.

The systems in place are, for the most part, functioning well. Information is exchanged. Most Board of Investment are in a position to evaluate the technology and its costs and to evaluate these in an international context without a global agreement. I suspect that much the same could be said of the Latin American countries and India. There are able people in a position to evaluate technology packages. The administrative apparatus is also in place to guard against excessive gains that might be achieved were one party to a transaction to be grossly ignorant of industry practice or the cost of technology under competitive conditions.

Several developing countries have complained that international firms do not fully transfer their technology. They cite patent laws and patent ownership as being a major hinderance.(9) They also cite the reluctance of international firms to transfer more than the final assembly stages of production. But most of these countries are missing the point because they fail to understand what the key ingredients of technology transfer are. Even if patents were freely given and all stages of production were transferred, there would still be little improvement because, so far as a new enterprise is concerned, it is the disembodied technology owned by the transferor that is the critical ingredient. This kind of technology, generally, is not for sale. If the developing countries are to achieve a level comparable to the advanced countries, they are going to have to follow the example of Japan. They must create the complementary factors that go with the physical technology; they must acquire the ability to create the disembodied technology so that it can eventually become embodied within productive enterprises. Until they do, they cannot capture the economic rents. Various institutional arrangements will differ in terms of their ability to accomplish this in light of the competitive strengths that yield technology suppliers their economic rents. Those rents will be reduced in some direct proportion to the developing countries' abilities to create the complementary factors (know-how) that yield those rents. A good example of a country which is attempting to do precisely this is Singapore.

Singapore, because it is an island economy, recognizes that the only resource it has to sell is the skills of its people, and hence it seeks to continuously upgrade those skills. It also recognizes that it needs the technology of international firms and that skills are the complementary factor required if one is to be a major recipient of technology transfers in the sense discussed here. Ultimately, of course, industrialization and the acquisition and creation of technology depends upon human skills. This is a basic truth not sufficiently appreciated by some developing countries. Too many look to multinational firms to create these skills. And while multinationals do a great deal of training, they cannot be expected to locate their technology in or bring more sophisticated technologies to countries where the complementary factors do not exist at all. The cost to them in relation to the benefits are just too great.

On balance I detect a growing understanding between multinational firms and the countries receiving flows of technology. My own evaluation is that the formation of Board of Investment and the creation of investment regulations have brought with them a great deal of good. One of the most important benefits has been to bring to the developing countries a recognition that they are not powerless. There is a new confidence based on a decade or so of intensive learning. Investment boards are now endowed with knowledgeable young technocrats who understand the value of technology transfers, who can assess the expected contribution of the technology, and who are aware of what prices are paid for technology around the world. What seems to be lacking in all of this is a full appreciation and understanding of the need for appropriate technologies.

NOTES

(1) This is a much oversimplified stereotype; I merely wish to draw an analogy. It is also true that one could debate the economics of colonialism, especially as to whether or not outcomes would have been much different had there been free international trade instead. It may simply be that there are penalties to bring a late arrival to the industrial age. Free trade might have yielded the same outcome in terms of degree and types of specialization.

(2) In many quarters industrialization has become almost synonymous with economic development. This is a view, however, that has been largely discredited in recent years. Industrialization is increasingly recognized as being only a part of the total process of economic development. Unfortunately, preoccupation with industrialization has often resulted in the neglect of agriculture and small-scale industry.

(3) These rankings are made without pretense of exactness. Indeed, some countries are more liberal than others along certain policy dimensions, less liberal on others. These rankings are based on my judgment, which in turn is based upon a review of investment and other regulations and interviews with Boards of Investment.

(4) All investment proposals having less than 30 percent foreign participation that are not aimed at the overcrowded industries are registered with the Securities and Exchange Commission, which is the reviewing body in these cases. However, the investor must also file a copy of the proposal with the Board of Investments.

(5) Presidential Decree 1263 had muddied the waters at the time of my visit to the Philippines. There was doubt whether this kind of flexibility

was still open to the interagency committee, even if it felt that a particularly meritorious case should be allowed a royalty rate exceeding the guideline. All cases not involving a Philippine patent still come under the Central Bank's discretion and are not affected by Decree 1263.

(6) While I was visiting Thailand in 1978, some 141 industrial items were subjected to an increase in tariff levies.

(7) It should be noted that Malays, on average, are not as well educated as the Malaysian Chinese and Indians. Also, because Malays have historically been employed in agricultural or handicraft industries they generally are less well trained in industrial skills.

(8) Because of scheduling problems, I was unable to arrange an interview with the Capital Investment Coordinating Board. The material presented here has been obtained from Indonesian documents on foreign investment laws and regulations. Accordingly, I am unable to report on the Indonesian screening process.

(9) For a Latin American viewpoint see Constantine Vaitsos, "Patents Revisited: Their Function in Developing Countries," in Science Technology and Development: The Political Economy of Technical Advance in Underdeveloped Countries, ed. Charles Cooper (London: Frank Cass, 1973).

References

General References

Allen, Thomas W. Industrial Development Strategies and Foreign Investment Policies of Southeast Asian and South Pacific Developing Countries. In International Business in the Pacific Basin, ed. R. Hal Mason. Lexington, Mass.: Lexinton Books, D.C. Heath, 1978, pp. 51-91.

______. The ASEAN Report, vols. 1 and 2. Hong Kong: Book Division, Asian Wall Street Journal, 1979.

Kohlhagen, Steven W. Host Country Policies and the Flow of Direct Investment in the ASEAN Countries. In Mason, International Business in the Pacific Basin, pp. 93-115.

Mason, R. Hal. Technology Acquisition in the Pacific Basin: Direct Foreign Investment versus Unpackaged Technology. In Mason, International Business in the Pacific Basin, pp. 117-29.

______. Technology Transfers: A Comparison of American and Japanese Practices in Developing Countries. Working paper presented before the Japan Society Business Educational Program Workshop, UCLA, May 30-June 4, 1978.

Korea

Economic Planning Board. Foreign Capital Inducement Act. Seoul, Korea, no date.

______. Guide to Investment in Korea. Seoul, Korea, 1977.

Taiwan

Industrial Development and Investment Center, Statute for Encouragement of Investment, Taipei, Taiwan, no date.

______. Statute for Investment by Foreign Nationals. Taipei, Taiwan, January 1978.

______. Statute for Investment by Overseas Chinese. Taipei, Taiwan, October 1977.

______. Categories and Criteria of Productive Enterprises Eligible for Encouragement. Taipei, Taiwan, no date.

______. Regulations Governing Customs Bonded Factories. Taipei, Taiwan, no date.

______. Protection of Industrial Property and Technology in the Republic of China. Taipei, Taiwan, September 1977.

Philippines

Board of Investments. The Investment and Export Incentives Acts (with Corresponding Amendments). Manila: Government Printing Office, November 1974.

______. Questions and Answers on Foreign Investments in the Philippines. Manila: Board of Investment, August 1977.

______. Investment Opportunities in the Philippines. Manila: Board of Investment, August 1977.

______. Guidelines in Evaluating Licensing Agreements. Manila: Board of Investment, mimeo, no date.

Bautista, Lilia R., "Policies on Transfer of Technology in the Philippines." Special study. Manila: Board of Investments, January 1978.

Thailand

Office of the Board of Investment. Investment Promotion Act B.E. 2520. Bangkok: Government Gazette, 1977.

______. Collection of Laws Pertaining to Investment Promotion. Bangkok, May 1973.

______. Procedures for the Implementation of Promoted Projects. Bangkok, no date.

______. Business Opportunities: Labor Intensive Industries in Thailand. Bangkok, no date.

International Legal Counsellors of Thailand, Thailand Business Legal Handbook. Bangkok: Board of Investments and Bangkok Branch of Chase Manhattan Bank, N.A., January 1976.

Malaysia

Federal Industrial Development Authority. A Basic Guidebook for Potential Investors. Kuala Lumpur, June 1973.

______. Supplement to a Basic Guidebook for Potential Investors, Kuala Lumpur, no date but current to existing law.

Singapore

Business International Corp. Investing, Licensing and Trading Conditions Abroad: Singapore. New York: Business International, April 1977.

Singapore Economic Development Board. Annual Report: 1976-77. Singapore, 1977.

Indonesia

Indonesian Investment Coordinating Board. Investment Opportunities; Taxation; New Investment Procedures; Incentives for Investors; Investment Priorities. All Djakarta, 1978.

Comments

Louis T. Wells, Jr.

Professor Mason's paper is largely descriptive of what certain developing countries are now doing with regard to controlling foreign investment flows. The paper hardly attempts to answer John Dunning's questions about whether the countries are making themselves better off or worse off by controlling technology flows. On my first reading of the paper I wondered whether I would be able to contribute to a discussion, since the paper is largely descriptive and my first thought was that it would be difficult to disagree with a description of administrative systems. However, there are a number of elements that do raise questions in my mind.

The paper really explored two themes: first, what the policies of the countries are and second, what the processes are through which the policies are administered. I am going to talk about each of these. But it seems to me that the most important contribution I can make is on the second theme, the description of the processes, since very little has been done in the literature to describe how foreign investment policies are actually administered.

The policies themselves fall into three areas: first, attempts to influence the cost of technology; second, attempts to influence the quality or the kind of technology that comes in; and third, attempts to influence the quantity of technology that flows in. I think the paper covered the first and third of those three areas rather well. In the second area the paper failed to mention some important policies. There is a good reason for the omission: few policies in the region Professor Mason covered are explicitly designed to influence the kind of technology. Nevertheless, certain policies in the region have significant effects on the kind of technology imported, regardless of the original aims of the policymakers.

Four kinds of policies were mentioned for all the countries in the group. First, there are policies that prohibit certain practices on the part of a foreign firm coming into the country. Second, there are policies that require certain practices; for example, there are requirements for joint ownership, which are designed to decrease the cost of technology. (Whether the policies do or do not actually reduce cost is an empirical question.) The paper touches on a third type of policy designed to decrease costs: efforts to promote regional cooperation. I wasn't quite sure where Professor Mason came out on the effectiveness of policies of this type. My own experience in the region is that there isn't any real cooperation among the countries when it comes to foreign investment.

There was a fourth policy area that the paper barely touched, but a few illustrations suggest to me that policies in that area are a little more than the paper indicated. The policies are those through which governments attempt to increase the bargaining power of their own firms in dealing with foreigners. Let me just illustrate with one example from the Philippines, for the pharmaceuticals industry. The country has a forced working provision in its patent laws: if a patent holder does not manufacture under the patent in the country, he loses his patent protection. At the same time, the country imposes a limit of 5 percent on royalty payments in the pharmaceuticals industry. The result is that foreign firms are quite willing to license to local firms that show an ability to copy the foreign innovation. The policy appears to have been quite effective in bringing in pharmaceutical technology at a relatively low cost. In fact, one result is a firm in the Philippines that is now beginning to invest in other countries in the region, taking overseas the technology that it previously imported. At home, that same firm is selling at rather low prices pharmaceutical products based on foreign know-how.

There were some potential policy areas that were unexplored in the paper, because the countries Professor Mason covered have not instituted policies of these types. In some of the areas, Japan has taken important steps. Since Japan has served as something of a model in the region, it is tempting to examine Japanese policies that Southeast Asian countries have chosen not to follow. For example, the Japanese government has attempted to keep different firms in the country from purchasing similar technology. If they do so, the country might have to pay two or three times for the same technology. There seems to be relatively little effort in the countries Professor Mason covered to do similar things.

A number of policies used in the Southeast Asian region affect the quantity of technology imported, but they were not mentioned as such. One policy, mentioned in a different context, is followed in almost all the countries – that of closing certain industries to further investment when the industry is considered to be "overcrowded." Although such policies are not designed to affect the quality of technology, they do have an important impact on it, for they keep old technology in place, and reduce competition in the particular industry. As a result, there is a reduction of the pressure on firms to choose lowest-cost ways of manufacturing.

Another policy mentioned in the paper was that of governments toward tariffs on imported capital equipment. Almost all of the countries in the region give rather liberal exemptions for the import of capital equipment. This policy itself affects the kind of technology that is imported.

There are other policies that also affect the quality of technology in fact. A number of the countries in the region ban the import of second-hand equipment. This policy has a significant effect on the capital intensity of technology that is imported, although the ban is widely ignored. At least one country in the region, the one from which I will draw many of my illustrations, imposes an excise tax on products produced by certain kinds of technology in an effort to discourage the use of particular processes. Another kind of policy that, regardless of its purpose, influences the kind of technology is tax incentive arrangements. Typically, the region's incentives give greater rewards for large projects. Several countries in the region give extra tax holidays if the capital invested is over a certain amount. Professor Mason pointed out that one country tried to offset the undesired incentive from such programs by also giving a tax holiday for projects that create a large amount of employment.

In some of the countries there is an ethnic bias in the approval process that certainly affects the quality of technology imported: Chinese investors simply are not particularly welcomed. But Chinese investors from Hong Kong and Taiwan tend to bring in technology that is quite different from what the multinationals import.

Finally, the policies of state banking institutions influence the quality of technology in the region. In the country I'm most familiar with it is almost impossible to obtain a loan from a bank for anything other than fixed assets. As a result a businessman is likely to choose a capital-intensive technology, because he can obtain a loan for the equipment but would find it difficult to get a loan for the larger amount of working capital required for a typical labor-intensive project. Second, the financial institutions in that particular country are very poor in collecting on their loans. It is tempting to borrow for capital equipment, especially if one might not have to repay the loan if the project turns out to be a poor one. Third, interest rates are usually subsidized in the area, and interest rates for preferred projects are far below what the market rate would be. Thus capital-intensive projects are encouraged.

The bureaucratic procedures followed in the region also influence the quality of technology imported. They strongly discourage investments by small firms because the small firms tend not to be able to afford the long waiting time or the many trips that are involved in getting a project approved. The result is a disproportionate amount of foreign investment from large firms and an increase in the kind of technology that accompanies that kind of project.

The paper seemed to cover virtually all policies in the region that are designed to influence the quantity of technology imported. The countries are rather consistent in that they are all trying to increase the import of technology. In another area of the world, Professor Mason might have found quite different policies. In some countries governments are not unambiguously trying to increase the flow of technology flow abroad.

The most fascinating part of Professor Mason's paper is the effort to describe how the control processes work in practice. When I first read the paper I believed its descriptions. However, when I reached the descriptions of the country where I have had the most experience my doubts grew. Professor Mason did admit that he had not gone through the same interview process in Indonesia that he had followed in other countries. But I tried to guess what the Indonesians would have told him if he had conducted the interviews there. When I tried to fill in those blanks about Indonesia, I concluded that I did not believe the description that emerged. Mason would have been told that the Indonesians have set up a "one-stop" approval process; they have a board of investments with authority, according to the law, to make the final decision with regard to an investment application. On the board, there is high-level representation from the relevant ministries. But that description is simply not the way the process works. In the first place, when the board has a meeting, attendance from the ministries is not complete and is not high-level. The decisions of the board are rarely final. Wise foreign investors have learned that. They file their application with the foreign investment board and then go to individual ministries to do much of their negotiation. They have learned from the experience of earlier investors, who were granted tax holidays, exemption from income taxes, exemption on import duties, and so on by the board, but later discovered that the Ministry of Finance did not necessarily feel bound by the commitment of the board. My experience in one country makes me suspect that there may be similar stories in some of the other countries. I must conclude that brief interviews with the representatives of investment boards are likley to lead to idealized descriptions of the process that are far from the practice one actually encounters.

In fact, the differences in theory and practice are still greater than what I have described. In Indonesia, a large number of the transfers of technology are completely outside of the foreign investment board, despite what the law says. I am sure that the same is true of other countries in the region.

In addition, there are many "informal" investment arrangements that are not registered at all as foreign investments – they skip the process completely. Some Hong Kong firms go into Indonesia with joint ventures with no documentation whatsoever: they simply set up a foreign investment project, but never register as foreign investors. And finally, in Indonesia arrangements such as management contracts are exempted from screening by the board of investments, if there is no foreign equity participation.

Professor Mason would be told in his interview that the Board of Investment staff uses very good analytical techniques, and that they look at the economic impact of a proposal on the country. In fact, they rarely do. They have on occasion attempted social cost-benefit analysis, but they have not used the technique extensively. They do not review the proposed technology and compare it with alternative technologies. They simply do not have the required information and skills.

In most countries the boards do not know what terms have been agreed to in similar circumstances in other countries. At best they have press reports. I recently was involved in negotiations in which the host government was trying to press a foreign company to agree to terms that the company had agreed to in another country in the region. All the government officials had to go on were the press reports about the other agreement. Since the negotiators did not have the actual agreement, they did not know the critical definitions that had been accepted in the other country. I was also involved in one country in the region where the government was trying to negotiate a mineral agreement but the negotiators had never seen another mineral agreement for that particular mineral from anywhere in the world. I think that we can overestimate how much information is available to government officials, even though the situation is much better now than it was five years ago.

In most cases the boards do not make a written report on their analysis to accompany their recommendations about a particular proposal. As a result, the underlying analysis is usually lost by the time the recommendation gets to higher decision makers.

Moreover, there are criteria for decisions that are stated in interviews, but when it gets down to the actual approval process there are some criteria that do not ever appear publicly. For example, there is a concern about the ethnic origin of the investor and about the effect of the proposed project on existing enterprises. The decision makers may ask about the effect of the project on the regional balance inside the country. Did a particular region get a project last year? If not, perhaps it should receive one now, regardless of the economics.

The stated rules often require local ownership in a project. In practice the rules may be widely ignored. The foreign investment agreement authorized by the board is unlikely to say that local ownership is not required, but in fact there may be a verbal understanding that some time far in the future the foreign investor will seek a local partner. That understanding is, in many cases, unlikely to be enforced.

Mason, perhaps wisely, does not raise the difficult question of how important corruption is to the decision process. And the paper is silent on the relevance of the legal process in a number of these countries. The trademark laws and the patent laws are, in many cases, no more than laws on the books. They are not laws that can be enforced in the courts, because the tradition in many countries is not that of settling disputes through the courts. Quite different ways of settlement are followed for business disputes.

Finally, there is no formal follow-up by most foreign investment boards. Thus there is no punishment for an investor who makes errors or even lies on the application form. No one ever goes to check to see whether the firm has employed the number of people that it said it was going to employ, or whether it has invested the amount it said it was going to invest.

My final point is that one ought to keep in mind that the investment board approach used in Southeast Asia is almost unique to that region. I remember a few years ago when a U.N. agency set out to teach a course in investment for government officials who negotiate with foreign investors. It was quite easy in Southeast Asia to identify candidates for the course. In most countries in the region there is an equivalent to the board of foreign investment to which the foreign investor goes to apply for permission to invest. But in most of Latin America such an institution simply does not exist. There it was not easy to identify the officials who should take the course, because they were widely scattered in various government agencies and few had positions that involved full-time handling of foreign investment matters.

Professor Mason does not try to give us the bottom line: whether a country that wants to administer controls is better off controlling on an ad hoc basis, with a lot of administrative discretion, or whether it is better off using rather standardized rules of general application. The benefit of the ad hoc method is that the country can, theoretically, optimize terms for a particular project. Where there are big projects that have a major effect on the economy, such a policy may make sense. But there is a real cost: the use of skilled personnel is very high. Moreover, the ad hoc approach increases greatly the opportunities for corruption and it discourages small investors.

In the selection of an approach, the traditions of the host country are important, of course. It would be absurd to try to impose the same administrative system on France that one has in the United Kingdom, for example. The French administrative tradition is very, very different from the British one, and the French are inevitably going to depend a great deal more on ad-hoc-ism, on administrative leeway. The English are inevitably going to depend much more on more generally applicable rules. Likewise, trying to do in Egypt what one does in Indonesia would be very difficult because the Egyptian government structure is organized around what looks like an Eastern European model; there one finds on ministry responsible for all external economic contacts. That structure does not exist in, say, Indonesia. The results will be very different organizational approaches to dealing with foreign investors.

Perhaps the choice of structure for negotiating with foreign investors also depends on the level of development of the country. In looking at the organizations of multinational firms, we have learned something about the role of the international divisions that exist in the very early international stages of an enterprise's expansion abroad. That particular organizational structure seems to make sense at that time, but when the firm has more experience and more expertise, the international division usually disappears. It would not surprise me to discover that developing countries face an analogous situation. Many of the Latin American countries have sufficient expertise in dealing with foreigners that they do not have to concentrate skilled people in a single board, and can instead scatter the expertise in various ministries.

Since the level of a country's development is clearly related to the country's goals, it is also likely to influence approaches to investment policy. Some countries are eager to attract foreigners; they need an organizational structure designed to attract them. Some much more developed countries are less eager to attract certain kinds of firms. For them a quite different organizational structure might be appropriate. And finally, the complexity of problems encountered differs by the level of a country's industrialization. The problem of dealing with IBM in India was quite different from what is faced in the screening process of a country with little industry that is obtaining its first foreign investors. The best solution must depend on the task to be accomplished.

Comments

Joseph McCann

My personal perspective, in reading Hal Mason's paper, is that of an organization theorist concerned with the functional design and effectiveness of the organizational arrangements created to manage technology transfers. Simply put, the hypothesis I advance is that organization does make a difference – one may have good policy, but poor implementation. The organizational design and resulting effectiveness of the investment boards described by Mason become an important moderating variable in the technology transfer and control process.

Using this perspective, the basic issues appear to be:

1. What administrative arrangements and organizational forms currently exist to regulate technology transfers?
2. What strategies and mechanisms have been adopted to do so within these forms?
3. How do these forms and strategies compare across ASEAN members?
4. How effective are these and, in fact, what measures exist which allow any assessment of effectiveness?

Hal Mason's paper is unfortunately too descriptive and broad-brushed in its approach to allow serious organizational assessment. Importantly, taking such an approach builds in an implicit bias for "espoused theories of action" rather than actual "theories in use"; in other words, we are asked to have faith in statements rather than hard data. From my own experience, the split between ideal and real is quite significant.

A major criticism of the national investment board system has been its rationale that economic policies and guidelines shape social outcomes – a common position adopted by central planners. Given higher turbulent environments, policies quickly become irrelevant and inappropriate. Given also the highly social impact of most technologies, such a position or rationale becomes questionable. More explicitly, by creating a borad whose primary mission is to screen investment, the implementation issues surrounding a transferred technology may become undervalued and undermanaged. The boards generally serve as gatekeepers rather than effective coordinating mechanisms for managing technology diffusion. Few investment boards have the capacity, expertise, and political mandate to perform such an interventionist role. Singapore, more than any other country described by Mason, recognizes the second, third, and "nth" order diffusion effects of a transferred technology (for example, their skills-building effort). It must be asked whether investment per se should be the focus of these boards, or whether maximizing and shaping the impact of the technologies transferred within the investment package should be their legitimate concern.

The comparative dimensions offered by Mason are too few for me to feel at ease with. I would like to know more. He focuses on the formalization of the screening criteria (market-determined versus legalistic), the number and kinds of incentives provided, and the general flow of the decision-making process. I would like to offer four more:

1. The relative power and actual authority of a board in creating and shaping development policy vis-a-vis other national agencies
2. The representativeness and accountability of these boards for their decisions:
   a. How are key actors or stakeholders included or excluded in the process?
   b. How early in the process are these stakeholders included?
   c. What evaluation activities take place?
3. The preferred role of a board: .
   a. gatekeeper with control and sanctions emphasized, or
   b. advocate/facilitator with cooperation and "aftercare" emphasized – a proactive role
4. The variety and power of the control strategies actually used (for instance, rewards and sanctions), with emphasis on the richness and variety of such mechanisms.

IMPLICATIONS

Overall I would have to say that the richness of boards as a control strategy at the national level is suspect, given a historical record of policy implementation difficulties and an overreliance on bureaucratic solutions. Screening and incentives programs, while certainly desirable, must be integrally linked to implementation interventions that attempt to optimize the multiplier-linkage effects of a technology transfer. An expanded, more influential role for boards is advocated. Experimentation in this area should be valued.

At a regional level the emergence of clearly negotiated and viable organizational arrangements remains problematic. Regional forms like

ASEAN are important first steps toward the management of externalities and dysfunctional conditions (such as adverse competition and dependency) not resolvable by unilateral national action. The functional roles and responsibilities of some stakeholders in the regionalization process remain too ambiguous (multinational enterprises, international groups, and financial institutions are examples). Nationalism and scarce financial resources are formidable forces that tend to reinforce autonomous rather than collective action.

Ideally a regional entity should perform at least the following functions to be perceived as legitimate:

1. The reduction and management of environmental uncertainty and complexity (for example, as an information disseminator and communication facilitator)
2. The generation of economies of scale for members in attracting and allocating desired resources
3. The creation and maintenance of a coherent, legitimate identity vis-a-vis other types of organizations (such as multinational enterprises)
4. The enforcement, or at least arbitration, of issues concerning the contribution/rewards ratios of each member (such as the equity issue)
5. The capacity-building of each member and of the group collectively (such as skills and expertise levels of staffs).

In conclusion, Hal Mason's paper is helpful in framing many of the organizational issues present within the technology transfer control process. Clearly a serious research agenda needs to be created in this area if we are to better understand how interorganizational arrangements can be improved.

REFERENCES

Galbraith, J.R. Organization Design. Reading, Mass.: Addison-Wesley, 1977.

Litwak, Eugene, and Hylton, Lydia F. Interorganizational Analysis: A Hypothesis on Coordinating Agencies. Administrative Science Quarterly 6 (1962): 395-420.

McCann, Joseph E. Developing Interorganizational Domains. Doctoral dissertation proposal, June 1978.

______. Organizational Variables in the Technology Transfer and Development (TTD) Process: A Conceptual Research Design. Working paper, April 1976.

______. Technology Transfers: The People's Republic of China and Japan with Some Implications for Developing Countries. Working paper, March 1975.

McCann, Joseph E. and Ferry, Diane. Assessing and Managing Inter-unit Interdependence. Academy of Management Review 4, no. 1: 113-19.

Perlmutter, Howard V. Towards a Theory and Practice of Social Architecture. (London: Tavistock, 1965).

Saghafi-nejad, Taghi. Transfer of Technology and Development: A Framework and Some Observations from Iran. In Social Sciences and Problems of Development, ed. K. Farmanfarmaian, Princeton: Princeton University Program in Near Eastern Studies, 1976), pp. 141-170.

Thompson, J.D. Organizations in Action. New York: McGraw-Hill, 1967.

Trist, Eric. A Concept of Organizational Ecology. Speech and concept paper. Management and Behavioral Science Center, Wharton School, University of Pennsylvania, July 1976.

Williamson, Oliver E. Markets and Hierarchies: Analysis and Antitrust Implications. New York: Free Press, 1975.

14 International Controls of Technology Transfer

A.A. Fatouros

Unlike Pascal's truth,(1) modern technology is the same on both sides of the mountain. Automobiles are all alike everywhere on the globe, regardless of political system, culture, or level of development. Whatever the differences among them, actual or desired, the similarities, nay, the identities, remain overwhelming. The "mechanical donkey" Ivan Illich proposes as a substitute for modern automobiles and tractors to fit the needs of the Third World's farmers(2) is closer to the automobile than to the oxcart. It represents a variation within, not outside, the universe of existing technology. The fundamental uniformity of modern technology is a significant facet of its international role and impact.

A second fundamental point has to do with the general acceptance and desirability of technology, as a matter not of normative judgment but of undeniable fact. Acceptance of the value and importance of technology underlies the search for better methods of, or stricter controls over, its transfer. Concern for "appropriate" or "intermediate" technology,(3) or about the resulting value conflicts,(4) however inherently significant and justified, is but a gloss, a *temperamentum* on the acceptance of technology as a valued asset for any nation.

Yet modern technology is much more than a convenient, if complex, tool in the service of man. It is part and parcel of a whole way of perceiving and understanding the world. It has been shaped by, and in its turn has shaped, that approach to life and nature that is characteristic of modern Western culture.(5) Technology is a principal feature of what we prudently call today "modernization" – what in less inhibited times was called "Westernization," "Europeanization," or "civilization" *tout*

*I have greatly profited from comments on an earlier draft of this article from participants in the Seattle session of the conference on Technology Transfer Control Systems. While I have served occasionally as consultant to the Secretariat of United Nations agencies on matters related to the preparation of international codes of conduct, this article reflects solely my own perceptions, positions, and opinions.

court. This, of course, is not the only way to approach reality; it is only relatively recently that it has become dominant. Transfer of technology involves, in the aggregate and apart from specific, narrow, and almost accidental exceptions, an implied undertaking to deal within the bounds of that way of thinking. Along with the elements of the technology, other core elements of the Western approach to reality are directly or indirectly transmitted.

Technology, in its modern form, is not merely one manifestation, among others, of the way of thinking that has given birth to it. It is the leading, most influential manifestation. It plays a clearly hegemonic role(6) and largely determines the shape of the social environment. As Langdon Winner has noted, "technology is itself a political phenomenon . . . modern technics . . . now legislates the conditions of human existence. New technologies are institutional structures within an evolving constitution that gives shape to a new polity. . . ."(7)

Awareness of the points sketchily made in the preceding paragraphs should inform our attempts at an exploration of actual or possible international controls on technology transfers.

SCOPE OF THIS STUDY

Contrasting the scope of the study here undertaken to a possible ("ideal") range of relevant issues leads to certain useful conclusions. It is true that the "definitions" provided are the broadest possible; legitimate disputes could arise as to their exact scope. The point of this discussion, however, is not to offer authoritative definition of terms, but to suggest the broad background against which our limited discussion should be seen.

Controls over international technology transfers are measures and institutions calculated to influence such transfers to serve particular purposes. Such controls in their most inclusive sense may be located at several levels and make take many forms. They may consist of any or all of the following:

1. Unilateral measures by private enterprises with a view to maximizing their benefits from technology transfers or manifesting their benefits from technology transfers or manifesting adherence to certain moral or political values. In a more or less perfect market, such measures would be totally ineffective and would not therefore deserve the name of "controls." However, the more a particular enterprise is in a monopoly position, the more it is likely to be able to influence (control) transfers.

2. Collective measures jointly taken by private enterprises, either (a) to maximize benefits to each of them or (b) to regulate transfers in accordance with certain agreed principles of public policy or social morality.

3. Unilateral governmental (national) controls, designed either (a) to provide incentives or otherwise indirectly influence technology transfers without direct use of public power (imperium) or (b) to regulate such transfers through the direct use of governmental power.

4. Collective governmental (international) measures, on a regional or universal basis, calculated either (a) to influence indirectly national policies concerning technology transfers, or (b) to regulate technology transfers, through international legal action or through national legal action taken in fulfillment of international legal obligations.

This chapter deals primarily with collective private measures and with international controls on a universal or would-be universal basis. The particular types and methods involved will be elaborated upon later. However, the interaction and interdependence of the various types of controls – private with governmental controls, controls at the national, regional, and universal levels – should be kept in mind.

There are many definitions of technology and it is not our intention to repeat them here. The one provided by Denis Goulet is adequate for our purposes: "the systematic application of collective human rationality to the solution of problems by asserting control over nature and over human processes of all kinds.(8)

This definition covers, quite properly, all technology, at whatever level of industrial and scientific development. There is, however, a profound difference between old and new technology – the types of technology developed before and since the industrial revolution. Indeed, modern technology is so different from earlier technologies as to be properly considered something new.(9) Here we shall deal only with the transfer of modern technology; even in the absence of the adjective, the term "technology" will refer to post-Industrial-Revolution technology.

For our purposes, it is the manner in which technology is transmitted across borders that is of particular interest. A comprehensive listing of these would include:

1. The flow of books, journals, newspapers and other published information (as well as personal communications among scientists and other concerned individuals)
2. Population movements between countries, including immigration, return of emigrants, study visits, and other travel
3. Direct foreign investment and the associated transfer of know-how and equipment
4. Import of machinery and equipment
5. Technical assistance programs, multilateral and bilateral, official and private
6. licensing, patent, and know-how agreements.(10)

Whether what is covered by this listing can be better called "supply of technical information," retaining the term "transfer of technology" for a less inclusive list, is primarily a matter of terminological convenience.(11) It is evident that some of these forms and manners of transfer are less susceptible than others of empirical and detailed study; whatever controls may exist or be desirable in their case, they differ in significant ways from the controls relevant to other forms. I shall deal therefore, primarily if not solely, with the last four categories of technology transfers. They have at least two important common features: they are normally, although not exclusively, conducted through commercial (profit-making) channels, and they cover those "elements of technical knowledge which are needed to set up particular new production facilities," as contrasted to new production in a general sense.(12)

The main point of these successive classifications is to make evident that we are dealing with only a part of the entire universe of controls over international technology transfers. We discuss some of the controls over some of the transfers of some kinds of technology. Many types of technology transfers, lato sensu, and of controls on technology transfers remain necessarily outside the area of our current concerns.

WHY INTERNATIONAL CONTROLS?

Before trying to analyze the various kinds of existing, emerging, and proposed controls it is necessary to consider why controls should be required, or perceived as required. In so doing we must look at the current situation, legal and economic, and attempt to see what controls at the international level might add or change.

In discussing the creation of controls, we usually treat the established legal (as well as political and economic) framework of international relations as if it were nonexistent, or at best neutral. We seem to start, as it were, from a tabula rasa on which controls are to be constructed. (My own discussion of controls in the preceding section may be read in this manner.) Yet in reality we are not starting from zero. We start from an already highly developed system of legal (and political and economic) relationships, a system that, far from being neutral, functions (like all legal frameworks) to promote and encourage certain policy ends and discourage others.

That a system of international controls over technology transfers is already in existence becomes evident upon the merest reflection. That system, moreover, is relatively recent: it developed primarily in the late nineteenth century and the first half of the twentieth.(13) By the end of the Second World War it was in place; the legal framework of world trade that developed after the war helped further to strengthen and lubricate it. This system is founded in part on multilateral and bilateral treaty arrangements and the activities of related international agencies, but relies principally on unilateral action by states: enactment of industrial property legislation, acceptance of principles of private international law (conflict of laws) concerning recognition of industrial property rights and enforcement of contractual arrangements across national borders, and so forth.

It is easy to forget the existence of such a control system, because it is not intrusive. As we shall see, the fundamental pattern has been deference to private decisions and actions. As with other laissez faire systems it may seem almost unfair to speak of a set of "controls." That term seems more appropriate to regulatory systems that involve continuous deliberate exercise of decision-making authority by a given body or group. Nonetheless, policy decisions by governments were certainly required for the creation of the existing system – it did not come into being accidentally. If no continuous governmental decisions are now needed (in the main) for its operation, it is because its functioning, by being responsive to the conditions of the international technology market, serves adequately the needs of the countries that by and large dominate that market, as well as the existing technology transfer control system.

The governmental controls currently in effect are founded on (1) formal deference to unilateral national decisions concerning regulation of technology transfers (imports and exports); and (2) actual deference to private decisions, made primarily by means of commercial contracts.

According to the public international law rules currently in effect, states are free to legislate regarding technology, to provide protection for industrial property in their territory, and to regulate such property and technology transfers to and from their territory. The established legal framework is thus formally neutral, based on recognition of the plenary authority of national decision makers. However, its actual operation has been historically far from neutral and impartial. The politically most powerful nations (which have been, in the main, the principal suppliers of modern technology) have utilized the freedom of action allowed by the system in order to insist on "proper protection" for industrial property rights in other countries; on occasion they have exerted strong pressures on less advanced countries to enact "modern" legislation on the subject and to adhere to existing international conventions. In this as in other areas, classical international law provided little protection to weaker countries against open pressures by the Powers. With few exceptions (for example, Japan), it is only very recently that the theoretically available option of enacting regulatory national legislation on technology transfers became relatively real for the technology-importing countries.

The actual fundamental pattern in international technology transfer controls has therefore been that of deference to private decision making. Private individuals (and enterprises) are left free to reach agreement on the terms and modalities of particular technology transfers. That is to say, existing market forces are allowed to determine the terms of the contracts between suppliers and importers of technology. Once the contract is duly concluded, it has behind it the sanction of the state (indeed, of both states involved); it is enforced through use of the governmental machinery and ultimately through the threat of force, whose monopoly is in the hands of the state.

The existing international legal framework, therefore, is not passive or neutral. In the first place, it favors the creation of industrial property rights, that is, the grant of legal protection to holders of technology. Second, it allows considerable freedom to the owners of these rights (or the holders of technical skills whose control over technology has not been perfected through ownership devices, such as unpatented technology, (know-how) in their dealings with buyers. And third, it enforces the contracts so concluded. Economic forces are thus given (relatively) free play, within a strict protective framework: both the creation of industrial property rights and the enforcement of technology transfer contracts are assured by the exercise of public power. Only the intervening bargain is left to the private parties. The existing controls tend therefore to faithfully reflect, in their consequences, the market forces that they serve and strengthen. Their advantages and disadvantages, qualities and faults, are primarily those of the international technology market. If the latter could be said to function efficiently and justly, the same would grosso modo be true of the existing controls.

It would make little sense to attempt here an elaborate discussion of the characteristics, and especially the imperfections of the international technology market.(14) The topic is touched on only in order to provide the basis for a listing of the purposes of the various proposed international controls. While the differences between technology transfers among developed countries and technology transfers between developed and developing countries may be primarily differences in degree, they seem significant enough to make common discussion of the two topics difficult and sometimes misleading. Moreover, most of the recent efforts to establish new international controls have been directed at technology imports by developing countries. The discussion here will share this emphasis, while trying to keep in sight other significant aspects of the problem.

At the center of the issue is the unequal bargaining position of technology-supplying and technology-receiving enterprises and countries. This inequality is, in part at least, inherent in the relationship examined: technology is, after all, largely knowledge, tangible and intangible. The supplier's control over technology implies a superior knowledge of its costs, capabilities, and qualities; the recipient's need for the technology entails a corresponding position of ignorance and uncertainty. In addition, many technology suppliers are otherwise economically powerful, as well. The ultimate consequences of this sort of unequal bargaining position have been often listed: high costs for the recipient, directly – either explicitly or hidden in highly "packaged" technology – or indirectly, by means of restrictive conditions accompanying the technology transfer; limited capability of the recipient to choose the technology most appropriate for local conditions; long-run dependence of the individual enterprise, and of the developing country's economy as a whole, on imported technology; and a concomitant inability to develop indigenous technological capabilities.

At this point, however, a distinction should be stressed that has been slighted in the preceding paragraph. The position of a technology-receiving firm in a technology-recipient country vis-a-vis a technology-supplying firm from a technology-supplying country is in many respects similar to the position of the technology-recipient country vis-a-vis the technology-supplying country, but it is not identical. The two positions do not fully coincide. While the situation of the firm and of the country may be largely similar with respect to explicit direct costs (such as royalties), in the simple sense that it would normally be in the interest of both to pay less for the technology imported, their interests and situations may diverge widely with respect to indirect costs, choices of technology or, even more, technological dependence. In many of these contexts, what may be most profitable for the enterprise may be detrimental to the recipient country's national economy and vice-versa.

The point is rather obvious, but it has important repercussions on the substance of possible government controls on technology transfer. Such controls may aim at assisting technology-recipient enterprises, in order to assist the economies of technology-receiving nations. They may also, however, try to limit the options available to these firms, and allow only those options that are advantageous to the recipient economy as a whole, whether or not they are the most profitable to the individual firm. This may lead to certain peculiar, if not paradoxical, situations, as will be pointed out a little later.

The objectives of most current proposals for international controls over technology transfers consist essentially in the reversal of the traits listed above. The principal aim is the redressing of the bargaining equilibrium between technology-supplying and technology-recipient firms and countries. To bring this about, measures are encouraged that would increase the information at the recipient's disposition, would make legally impossible some of the (more extreme) manners in which the supplier can take advantage of its bargaining position, and would increase the options and possibilities legally and practically available to the recipient. An independent objective (and, at the same time, a partial outcome of the measures mentioned) would be the strengthening of the recipient country's indigenous technological capabilities and avoidance of long-run technological dependence.

To the extent these objectives are, at least in part, independent of one another rather than related as means to ends, difficult choices may have to be made. As already noted, the technology most profitable to a recipient firm may not be the most appropriate in terms of the needs of the national economy of the recipient country. Even the most obvious of the objectives, the preference for lower direct (and total) costs, is not immune to challenge. To the extent that the high costs of foreign technology make its importation difficult or impossible, incentives are thereby created for the development of indigenous technology (which may involve all kinds of "infant" costs). If foreign technology is very cheap and easily available, there may be little economic inducement for the costly and painful efforts needed to develop indigenous technological capabilities.

How far the issues raised in this section are relevant to technology transfers between developed countries is not clear. Many of the developed countries are net importers of technology; their interests may coincide in part with those of the developing countries. It cannot be assumed, however, that the same would be true as far as particular enterprises in the developed countries are concerned, many would probably be technologically sophisticated and economically powerful in their own right. Although developed and developing technology-importing countries may have comparable, if not similar, positions regarding the problem of technology cost, in respect to the degree of availability of effective choice of technologies or the problem of long-range technological dependence, they may not. Be that as it may, increased or exclusive concern for the developing countries' role as technology recipients, may be justified on grounds of their greater need for better terms for the acquisition of technology and their much higher vulnerability to economic pressure and mishaps.

NONGOVERNMENTAL CONTROLS

If we put on one side those private controls that have profit-making (however broadly understood) as their sole and direct aim, we are left with a relatively limited number of actual and possible collective controls, mainly by associations of private businessmen. Enhancement of profit-making opportunities may be an indirect aim of such undertakings (through increased stability in business relations, improvement of "business climate," creation of a benevolent image, and so on), but it is not the only or the immediate one.

The associations we are concerned with here are mostly international (or better transnational); that is, they bring together business groups from various countries. Which type of firm is most influential in such associations varies from group to group. The largest enterprises, for instance, especially multinational (transnational) firms seem to be particularly influential within such groups as the International Chamber of Commerce. National business associations from developing countries may be, and typically are, included, but the international associations generally express the viewpoints of the developed countries' business firms. This in itself is easily explainable; what is more interesting is the extent to which, in most instances, the representatives of Third World business associations appear actually to agree – or not seriously disagree – with the views of their homologues from the developed countries. This is an issue that deserves separate study and cannot be dealt with here.

One significant class of activities by nongovernmental groups covers efforts promoting the standardization of legal instruments and generally increased uniformity in international business transactions.(15) Thus the drafting and dissemination of a standard consultant's contract by the International Federation of Consulting Engineers(16) are a significant control over the type of technology transfer represented by technical consultantships. The purposes of the exercise may be to assist technical consultants in their work, to cut down on legal expenses, to make sure that certain clauses generally thought suitable by experienced consultants are included in all contracts, or to avoid competition between consultants on certain topics. Such efforts are likely to represent the position of moderate and respectable elements within the professional groups concerned, seeking to protect the professionals while avoiding anything that could be seen as taking excessive advantage of the clients. At the same time, they are likely, and indeed are meant, to favor (within respectable limits) the interests of the professionals involved.

A second category of controls, of broader effect, are the "guidelines," "codes," or other such voluntary sets of norms adopted by associations of businessmen or professionals and covering various aspects of their activities.(17) A prime instrument of this kind is the one issued by the International Chamber of Commerce in 1972 under the heading, Guidelines for International Investment.(18) The scope of the instrument, as the title suggests, is broader, but it deals in some detail with technology transfers. It recommends that foreign investors use "reasonable terms and conditions," make "the latest suitable technology available in return for appropriate payment," not require payments for technology "of no real value to the enterprise," and promote the economic development and the growth of technological capacities of the recipient country. A typical instrument of narrower scope, primarily intended to regulate the activities of "professionals" directly engaged in the field under consideration, is the Code of Ethics of the American Institute of Consulting Engineers.(19)

The usefulness of instruments like the ICC guidelines in serving the purposes they purport to promote is limited. Their principal value lies in the fact that they indicate the outside limits of debate on the issues they cover. They establish that certain principles have been accepted as, or conceded to be, legitimate by the leading firms in the fields involved, so that such principles can no longer be treated as subversive or unacceptable. Documents of this type may exert an influence on other firms that own their own might not have been equally "advanced" in their approach to the topic. In other terms, they articulate the views of a hegemonic faction within the business community, the faction whose formulation of the ethics (and economics) of the trade is accepted as governing by the business community as a whole. On these grounds, such documents may have a certain diffuse but real importance.

In more immediate practical terms, their impact is highly doubtful. They usually exhibit three fundamental traits: broad, general, and vaguely qualified language; lack of any claim to coercion or obligation; and absence of any institutional or procedural system of implementation and follow-up. Obviously, which terms are "reasonable" in technology transfer agreements can be a matter of constant dispute between the interested parties; no supplier, however hard-nosed is going to maintain that he intends to exact "unreasonable" remuneration. Even that fuzzy a standard, however, is said to be fully voluntary: presumably a firm that does not follow it is not doing anything wrong, in legal, ethical, or business terms. To make real this lack of any direct or indirect pressure or coercion, no follow-up mechanism, however voluntary, is instituted. The combination of these traits deprives the instruments in question of any real effectiveness. The standard contracts and contractual terms mentioned earlier, while voluntary too, are at least stated in highly specific and concrete language. Guidelines and business codes of conduct, on the other hand, are both wholly voluntary and exceedingly vague.(20) We shall consider a little later whether and how far these faults may be eliminated or tempered when instruments of this type are prepared at the intergovernmental level.

INTERGOVERNMENTAL CONTROLS AND TECHNICAL ASSISTANCE

Several international (intergovernmental) organizations are engaged in activities calculated to influence technology transfers. The range and diversity of these activities are remarkable. Were it not for the very broad definition I am working with here, some might doubt that all such activities could be dubbed "controls." Clearcut classifications and distinctions in this domain are difficult, and the criteria for them often turn out to be tenuous. The rough division that shall be applied here will distinguish the multitude of activities that can be covered by the term "technical assistance" (or "cooperation") from those more abstract and more clearly legally oriented efforts to fashion new legal norms for technology transfers.

Technical assistance activities may assume a large variety of forms. Their basic aim, in this context, is to help recipient developing countries to choose and acquire technology and to improve their longer-range technological capabilities. Three general types of technical assistance may be distinguished.

First, there is direct assistance. In a sense, most international organizations that provide technical assistance may be said to engage in direct transfers of technology. Thus, preinvestment surveys, conducted by or through the U.N. Development Program, involve a gratuitous transfer of technology. The same would be true of FAO programs concerning improvement of land cultivation methods or new varieties of seeds, WHO projects in the field of public health, and so forth. What is being provided usually embodies or involves technology that could have been bought at the open market for a price. Such activities, however, may be considered "controls" only because of their ultimate, sometimes indirect, impact. To the extent they alter the position of the recipient country and give that country an increased opportunity to seek other technology transfers in a slightly better bargaining position, thanks to the assistance, they may be said to be measures influencing international technology transfers.

More relevant to our purpose is the direct assistance provided by the U.N. Industrial Development Organization (UNIDO), an operational international agency charged with promoting industrial development.(21) UNIDO is one of an increasing number of international agencies that have been established not by international convention but by resolution of the U.N. General Assembly. For this reason it is not formally a "specialized agency," although in most respects it functions like one and is indeed on the way to becoming one. A central purpose of UNIDO is the selection and adaptation of apropriate technology for developing countries. It provides technical assistance of various kinds at various levels. Its assistance ranges from providing equipment (incorporating technology) to arranging for surveys and studies and even to constructing entire industrial complexes on a turnkey basis. Although the organization often provides its assistance by contracting with outside private or public agencies, it does not always serve as a mere conduit. Having been committed to helping in the development of alternative technologies, it undertakes not merely to disseminate information concerning available technology but to compare and evaluate alternative technologies in specific contexts.

A second kind of technical assistance may be broadly called "educational" or "informational." It ranges from training individuals in particular technological skills to conducting seminars and meetings devoted to the problems of transfer, acquisition, or assessment of technology, from undertaking or commissioning studies of problems and possibilities in this field to publishing such studies or "guides" and "manuals" of differing degrees of complexity on technology transfer transactions.

Many international agencies engage in this type of activity, within their respective fields of action. Thus, UNIDO as well as UNCTAD, WIPO, the OECD, and the U.N. Center on Transnational Corporations conducts seminars for government officials of developing countries and other meetings of experts devoted, in whole or in part, to problems of technology transfer.(22) The same organizations have published a great number of studies of particular aspects of technology transfer.(23) UNIDO has published a series of "guides" for technology transfer transactions, of increasing comprehensiveness and complexity.(24) WIPO has also published a licensing guide, and the U.N. Economic Commission for Europe is preparing its own guides for international licensing agreements.(25) UNIDO is, moreover, particularly concerned with the longer-range growth of autonomous technological capabilities in the developing countries; its training and educational activities are directed in part to this end.

A last type of technical assistance, related to both previous ones, may be called "institutional." It involves the creation and operation of specific international institutions, on a regional or universal basis, charged with disseminating information, developing appropriate technologies, and otherwise assisting developing countries in technology transfers. UNCTAD has taken the lead in the creation of Centers for the Transfer and Development of Technology, each serving an entire region.(26) UNIDO has established an Industrial and Technological Information Bank covering technology in a limited number of industrial branches.(27)

The technical assistance activities briefly summarized above constitute "controls" on technology transfer in the sense that they are largely directed at enhancing the bargaining position of the technology-importing developing countries. The provision of information and of expert assistance is calculated to allow these countries to take advantage of the available choices and opportunities so as to improve their position, in the short as well as the long run. Such activities thus operate essentially within the established framework of legal rules and economic institutions. In the very long run, this framework may be affected by the changed bargaining position of the developing countries (to the extent the activities in question are successful), but this is not an immediate aim of technical assistance.

ESTABLISHING NEW LEGAL NORMS

A further step is possible: attempting to change the existing national and international rules concerning technology transfer, to restructure the established legal framework for technology transfers. Such an effort, which certain international agencies are now beginning to engage in, also seeks to strengthen the position of the weaker party not only by increasing the legal options at its disposal, but also (perhaps primarily) by restricting the freedom of action of technology suppliers and making it more difficult, or impossible, for the latter to take advantage of their superior skills and economic power.

The moment we begin to deal with legal and therefore potentially binding or coercive action, we must clarify the level at which we are moving. Legal action may be taken at the national or the international (regional or universal) level. We are dealing here with action at the universal level (which may be in fact incompletely universal, when not all states actually participate, but will be termed "universal" here as long as the activities are open to all states, and are not limited to a "region" or group of states). It is not possible, however, to ignore the national level, since international legal action may, and often does, refer to national governmental action.

Many of the activities examined here may involve action that, at the international level, is voluntary – that is to say, does not involve a legal obligation or undertaking – while the corresponding action at the national level is binding on those subject to it. (The distinction between binding and nonbinding international action is not as clear and simple, or as important, as it seems at first glance, as later discussion will show). Consider, for instance, the case of the enactment of a national law regulating technology transfer on the basis of a model law prepared and published but not made legally binding at the international level.

A first category of the activities considered here is in fact the preparation by international agencies of model laws, duly offered for adoption by any country wishing to do so. The World Intellectual Property Organization, in cooperation with various U.N. agencies, has carefully elaborated such texts and has published them with extensive commentary. The very process of elaboration, involving repeated meetings of experts and of national representatives as well as close collaboration with UNCTAD, may be considered an important part of an influential educational process. The earliest such model law was published in 1965 by WIPO's predecessor, BIRPI; it covered the topic of inventions and was particularly directed at developing countries.(28)

This model law is now in the process of radical revision under the auspices of WIPO.(29) In the late sixties two more model laws for developing countries were prepared by WIPO, one on trademarks and unfair competition and one on industrial designs.(30)

In the 1965 version the adaptation of patent law to developing countries took primarily the form of relative simplification and clarification of the established patent system. There was no attempt to propose a systematic scheme for technology transfer regulation. The version currently prepared, however, appears to place particular emphasis on controls over patents and on the possibilities of regulation of technology transfer agreements, often taking its lead from the regulatory legislation enacted in Latin America in the 1970s.(31)

The second main class of activities under consideration involves the preparation and adoption of an international code concerning technology transfers. Provisions on this topic are included in the broader instrument adopted by the OECD to deal with activities of transnational corporations.(32) The OECD guidelines are addressed to the enterprises themselves and are explicitly qualified as voluntary. The section on technology transfer is very brief; it recommends that licensing be done "on reasonable terms and conditions," that technology-supplying enterprises "adopt practices which permit the rapid diffusion of technologies," and "endeavor to ensure" that these plans and activities fit into the recipient countries' plans and "contribute" to the development of local technological capacities. These recommendations are carefully hedged in by qualifications: references to protection of industrial property rights, use of such terms as "to the fullest extent practicable" or "as far as appropriate." Despite the official governmental origin of these provisions, their language and voluntary character appear to greatly diminish their practical effectiveness. It remains to be seen whether the emerging follow-up process may render the provisions more meaningful.

The most important current effort to alter the legal framework of international technology transfers is the initiative of UNCTAD for the preparation and adoption of a code of conduct on the transfer of technology. The balance of this section will be devoted to an examination of the problems arising in connection with this code.

The origins of the proposal for a code of conduct may be found in the early seventies. Concern for technology in developing countries and for technology transfer problems has had a relatively high priority on the agenda of UNCTAD since its creation in 1964. An Intergovernmental Group on Transfer of Technology was created in 1970. A series of studies commissioned or prepared by the UNCTAD secretariat(33) combined with the early stirrings of the movement toward strict national regulation of technology transfers, especially in Latin America (for example, the 1970 Decision 24 of the Andean Pact),(34) to bring about at UNCTAD III a long resolution on technology.(35) Its main emphasis was on modalities of national action by technology-importing and technology-exporting countries; there was only the briefest of references to possible international action of the type that eventually materialized. The very term "international code of conduct" first appeared in early 1973 in a proposal by the developing countries participating in the Intergovernmental Group on Transfer of Technology.(36) The idea took off remarkably fast: by the spring of 1974 it formed part of the General Assembly's Program of Action on the Establishment of a New International Economic Order.(37) Around the same time, in April 1974, the very first draft of a possible code was prepared, under the somewhat incongruous auspices of a Working Group of the Pugwash Conference on Science and World Affairs; the draft was circulated as a UNCTAD document.(38) In the meantime, throughout the period in question, several significant studies by the UNCTAD secretariat were published.(39) By the end of 1975 a new Intergovernmental Group of Experts on a Code of Conduct was set up. While UNCTAD IV was unable to reach any important decision on the major issues still in dispute concerning the form and contents of the code, it gave its blessing to the continuing preparation of a text.(40) Still another group of experts was established and, from November 1976 to July 1978, met in six sessions in which the structure and content of a code of conduct were thoroughly debated and negotiated.(41) A draft composite text was then submitted to the U.N. Conference on an International Code of Conduct on Transfer of Technology,(42) which had a first session in late 1978(43) and a second one in early 1979.(44)

The actual process of negotiation started with the submission by the Group of 77 (the developing countries) of a draft text of a code of conduct, essentially the same as the draft prepared by the Pugwash Conference Working Group.(45) Group B (the developed market-economy countries) and Group D (the socialist countries) later submitted their own drafts.(46) All of the drafts were initially rather hastily put together and were incomplete, with many obscurities and contradictions, not to mention extreme or exaggerated positions. In the course of the discussions each group completed and amended its draft, so that by the end of the first phase of negotiations (July 1978) the competing drafts were complete, fairly comprehensive, and internally reasonably consistent.

Negotiations and debate within the Intergovernmental Group of Experts consisted in part of detailed comparisons of the drafts to identify common positions or language. In a few notable cases the positions of the groups were reconciled by means of new drafts produced by the chairmen of the various sessions, with the "assistance" of the UNCTAD secretariat. In many instances, however, the initial negotiation consisted of merely juxtaposing the disparate language from the various drafts, with an abundance of signs (brackets, underlinings, asterisks) to indicate the origin of each. This process led, in a number of instances, to eventual acceptance of a common text. Still, although the areas of substantial agreement were not inconsiderable, the final draft thus produced (by July 1978) was, in many critical sections, a "composite text" rather than a common draft.

Two significant features of the negotiation process (until the convening of the conference) were the preponderant role of the larger, relatively more advanced, developing countries as leaders and spokesmen for the Group of 77, and the importance of the activities of the UNCTAD secretariat in helping to formulate issues, prepare alternative texts, and generally consult with the states and groups of states involved. The subject matter under negotiation is, of course, more important for the relatively more advanced developing countries than for the poorer and least developed. However, the dominant role of the larger developing countries as representatives of the entire Group of 77 raises the difficult question of the lack of effective protection for the interests of the other less developed countries. The role of the UNCTAD secretariat is equally problematic. It is no secret that it often acts as the principal technical adviser to the Group of 77. At the same time its effectiveness depends on retaining a degree of credibility as a relatively impartial civil service, able to provide background support to the entire operation and to serve on occasion as a mediator between the groups of states. That the secretariat has been able to function to a substantial extent in all these roles is a minor miracle. Difficult questions arise, however, as to its continuing effectiveness, especially as long as the Group of 77 lacks the integrated technical support that a "regional" international secretariat (corresponding to the OECD's role for Group B) would provide.

The draft composite text, as it has emerged from the first session of the conference (November 1978), consists of a preamble (largely agreed upon by the negotiating states) and nine chapters, the very last of which, destined to contain the "final clauses" of the instrument, is at this point an empty shell. The other eight chapters deal with the following topics: definitions and scope of application; objectives and principles; national regulations of transfer-of-technology transactions; regulation of restrictive (and possibly other) practices; guarantees and responsibilities; special treatment for developing countries; international collaboration; and applicable law and settlement of disputes.

Even this bare list makes evident a most important characteristic of the draft code: it contains two main types of provisions, those dealing with actions and measures by governments, and those dealing with transfer-of-technology transactions, their terms, and the rights and obligations of the parties to them. By and large there are by now few profound disagreements as to the provisions of the first type; serious controversies remain as to many of the provisions of the second type.

More specifically, the code accepts the principle that states may regulate transfer of technology transactions subject to a series of prudential considerations and to international law norms – the exact mode of reference to the latter being still hotly debated. It is also generally agreed that special treatment is appropriate for developing countries, but the relevant language is general and permissive rather than mandatory. Various modalities of intergovernmental cooperation, through consultation, exchange of information, and so on, are also agreed upon (again in broad and permissive terms), but proposals for institutional implementation structures are still being acrimoniously debated. No serious effort has been made to find a strictly binding formulation for these provisions, which concern action by governments. Some of these efforts may involve a recognition of the legitimacy of certain types of measures.(47)

As for the provisions regulating technology transfer transactions: although important disagreements persist, the general shape of the provisions is emerging fairly clearly. The code will condemn certain categories of business practices – primarily possible clauses in licensing and similar agreements – starting with restrictive business practices (that is, anticompetitive practices) and possibly also covering certain other practices that impose burdens and obligations deemed excessive and unfair on the technology-receiving enterprise. (The matter is still in dispute.) A series of affirmative obligations of the parties, primarily of the technology suppliers, during the negotiation and performance of technology transfer transactions will also be established. There has been much radical disagreement concerning the problems of applicable law and the methods of dispute settlement, although the shape of an agreed-upon position may now be beginning to emerge.

A checklist of all points of agreement or disagreement would be of limited value, if only because it would be overtaken by events by the time this is published. It is more appropriate to start at a more basic level and refer to the general objectives of the code. Taking the code of conduct as an instrument expressing certain value judgments and perspectives of the international community, we may distinguish four kinds of purposes, of varying degrees of generality and relevance to our present concerns: (1) the code's ultimate, long-range purposes; (2) the objectives served by the preparation of the code; (3) the code's immediate goals; and (4) the desirable features of a reasonably effective code.

As ultimate purposes of the code, we may list international development, the more equitable sharing of the world's resources, and other such broad and very long-range considerations. These purposes influence the general direction of the code and thereby the substance of some particular provisions, but they do not provide practical guidance for the kind of problems we are considering here.

Less distant, yet not immediate, objectives of the preparation and adoption of a code of conduct are those that relate more specifically to the existing international market for technology, seeking to change it in certain manners and directions: improvement of access to technology at reasonable costs, development of the technological capabilities of developing countries, increase of the flow of technology from developed to developing countries, and so on. Such considerations are the criteria against which the success or failure of the entire undertaking will ultimately be judged. But they are still too distant and general to provide practical guidance in the preparation or assessment of the code.

The immediate goals of the preparation of a code of conduct are more important in this context. Without establishing an order of importance (which is variable and debatable), these goals may be listed as follows:

1. The strengthening of the bargaining power of the recipients of technology, thus redressing the present unequal situation in this respect. (Several of the other goals listed here may be seen, at least partly, as means to this end.)

2. The establishment, both in international law and in various national laws, of certain agreed-to principles concerning the methods and modes of transfer of technology, such as the elimination of certain restrictive business practices, the principle of a recipient's freedom of choice between packaged and unpackaged technology, and so on. These principles would protect technology recipients and promote the creation of a fair and more stable technology market.

3. The recognition of the legitimacy ("legitimization") of certain national methods and policies concerning the regulation of international technology transfers, primarily by the governments of technology-importing countries. If a code were adopted by all concerned, this would signify that other countries (such as technology exporters) would accept that the methods and policies in question are lawful and, at least under appropriate circumstances, proper; their application by a technology-recipient country could not then be said to constitute either an unlawful or an unfriendly act. Such considerations are particularly relevant not only to the provisions included in the chapter of the code on "national regulation," but also to those in the chapters on "restrictive practices" and "guarantees and responsibilities."

4. Establishment of a certain degree of harmonization or uniformity in the national laws of various countries concerning transfer of technology. Important elements of such harmonization are: (a) the setting of minimum standards that all countries would undertake to apply, so as to eliminate the possibility of "cutthroat competition" among importing countries through measures offering excessive advantages to technology suppliers, in order to attract them to a particular country; and (b) the elimination, or diminution, of the potential for conflicting regulations among nations.

5. The strengthening and supplementation of national regulation of the importation of technology through international legal rules and measures and through international (intergovernmental) cooperation, thus avoiding gaps in the coverage of relevant laws and rendering their implementation more effective.

6. The formulation of model provisions concerning regulatory policies and methods, which countries that have not yet taken any related national measures could easily adapt to national use. The existence of such models could be seen as an inducement for – or, at any rate, a factor favoring – the adoption of related national measures ("demonstration effect"). Simply listing of possible measures may serve this goal in addition to that of "legitimization."

These goals reflect in the main the point of view of the developing countries, which were the initiators and continue to be the advocates of the code (albeit with markedly diminished enthusiasm). The position of the developed countries is more difficult to spell out clearly. To a large extent it is one of sullen resistance to the idea of an effective code and continuing rearguard action against it. Perhaps an equally significant component is a reluctant realization that something like a code (or, more likely, a set of codes) is unavoidable and that it is important to try to shape it in the manner least harmful to Western interests. Among those decision makers who actually favor the idea of a code (a definite minority in all major developed countries), certain additional goals that parallel (but do not coincide with) those of the Third World may be perceived as important:

1. Creation of an international forum and establishment of procedures for the settlement of disputes, concerning technology transfers to begin with but also concerning investment and other economic issues.

2. Formulation, by international action, of certain legal and quasi-legal standards and rules concerning the conduct and treatment of transnational enterprises. Participation of the developing countries in the process of formulation would presumably preclude them from later challenging these standards.

A proper code of conduct, in addition to being capable of serving the purposes, objectives and goals just listed, should possess certain features that would make it better able to fulfill its functions. These features are:

1. Universality – applicability to transactions involving all (or most) of the countries that participate in world affairs today.
2. Flexibility – allowing for application in circumstances changing over space and over time, under varying conditions in each state, in view of differing development objectives and priorities and of variations in policies, legal systems, and administrative methods. A prime area of such differences relates to the role of private enterprise and the extent of freedom of action allowed private enterpreneurs.
3. Certainly – contributing to stability of expectations on the part of all economic actors concerned, so as to permit long-range planning on both sides.
4. Adaptability – the possibility of future amendments in response to experience from the code's application and to possible changes in the economic environment.

The multiplicity of goals the code must serve and standards that it must meet reflects the fact that each state or group of states seeks to serve many interests at the same time and cannot serve all of them fully to the extent it might otherwise desire. This point can be more concretely illustrated. In all negotiations, each state (or group of states) involved seeks to retain as much freedom of choice (or of action) as it can, while insisting on limitations on the freedom of choice of the other parties. In the context considered here, such a pursuit may contradict other equally important goals and interests. Thus it may be to the advantage of technology-importing states (broadly represented by the group of 77) to limit the choices available to suppliers, while retaining their own freedom of choice – for example, by inserting a national interest exception in all provisions imposing any restrictions on the recipient countries' freedom of action. Such a position may also be justified on the grounds that, for the developing countries, a practical and effective way to exercise control over transnational enterprises and other technology suppliers might be to establish binding restrictive rules of general application and to manipulate the granting of exemptions. (Much of the "tax incentive" legislation in developing countries can be understood in similar terms.) Allowing each recipient country to grant exemptions from the code's rules might lead, however, to possibly serious differences in the actual regulation of technology transfers by importing states and would fail to control "cutthroat competition," that is bidding for technology by promising increased benefits to the suppliers. It would thus tend to undermine one of the principal goals of the code.

One problem that has plagued discussions of the code of conduct on transfers of technology has been that of its legal nature or legal form. Should it be adopted in the form of a recommendation (for example, by U.N. General Assembly resolution or declaration of states), or in legally binding form (as an international convention)? The problem is no doubt important, but its real significance has been exaggerated.(48) To begin with, it is clear that many of the code's provisions are not going to impose any strict obligations on the adopting states, because they are couched in permissive or optional terms, in vague and general language, or in formulations that expressly or by implication allow considerable discretion to the states concerned. Second – and conversely – to the extent that states, by adopting the code, are formally recognizing the lawfulness and propriety of certain otherwise controversial measures and activities, the preclusive effect of such recognition does not depend on the strict legal form of the code. Even if the code is not legally binding, it will no longer be possible for states that have adopted it to argue that measures that are allowed or recommended in it are internationally unlawful.

A legally binding code would enhance effectiveness chiefly in two respects. First, in symbolic terms: states do take binding commitments more seriously and disregard them less (or less often) than they do nonbinding, even if formal, undertakings. Second, and perhaps more important, certain provisions of the code – not many, but important ones – would be more effective if they are part of a binding instrument. This is true, principally, of some of the rules concerning regulation of transfer-of-technology transactions and intergovernmental cooperation in reciprocal recognition and enforcement of measures and decisions.

The first problem may lose some of its importance if proper provision is made for implementation and follow-up of the code. Such provision would signify that proper importance is in fact being paid to the code and that the adopting states do intend to see that it is implemented effectively. The existence and operation of an appropriate implementation machinery do not necessarily depend on adoption of the code in legally binding form. Machinery may be established and various duties imposed on states even with respect to a legally nonbinding instrument. The states adopting the code may be required to submit reports on the progress of their application of it. The relevant services of UNCTAD may continue to study the related issues and to report on the states' application of the code. It may even be possible to establish procedures and institutional machinery for the interpretation of the code and the settlement of related disputes. Measures of this type may significantly enhance the actual effectiveness of the code, regardless of its precise legal form.

The second of the problems just mentioned is closely related to the difficult issue of the interplay of international and national law and policy and decision making. Such interplay raises complex and difficult problems for the preparation and implementation of the code. There is little doubt that much of the impact of the code will depend on the manner and extent of its application by national authorities with respect to a multitude of individual technology transfer transactions. The specific policies to be applied must be elaborated with regard to the features of a particular economy at a particular time. Only national governments can be charged with responsibility for such policies. The same is true for the concrete, individual decisions concerning approval, registration, or disapproval of each transfer of technology transaction. Neither policies nor decision of this type can (or should) be made at the international level. The role of the code will therefore primarily consist in providing a favorable framework for the exercise by states of their policy- and decision-making powers. The code must set certain limits and establish some new international rules concerning transfers of technology. Yet, overstating the case somewhat, one may argue that in most respects the code can only provide the passive background for national action. Only the latter can be adequately active and affirmative.

If the developing states were willing to take advantage of the code, they could simply take it over and adopt a strict version of it as an international convention concluded among developing countries only. Alternatively, each of them could enact as national laws regulations based on the code's chapters on transfer of technology transactions, even if the code does not legally bind them to do so (were the code to be adopted in nonbinding form). Such an attitude on the part of the developing countries would neutralize the dangers of a broad "escape clause" on the grounds of national interest, which, if inserted in the code, might open the way for a regulation-destroying "cutthroat competition." A major problem is that most developing countries are unwilling to take either type of action, partly because they fear dire economic repercussions – the cutting off of the flow of technology that some transnational enterprises and some developed countries are predicting (or threatening) – and partly because their preoccupation with other problems does not permit them to devote to this issue the amount of attention, effort, and resources required.

The difficulties in reaching agreement on a strong enough code, the problems of application, and the doubts as to the attitudes of many of the developing countries are pervasive considerations that suggest that the likely importance and impact of the code should not be overstated.

There exist, moreover, certain inherent limitations to the effectiveness of the instrument chosen, namely the "international code of conduct," the most important of which may be its limited reach over transactions between affiliates of transnational enterprises. Considerable disagreement persists on this issue between developing and developed countries.(49) But even if the proposals of the Group of 77 were accepted, transnational enterprises would still retain a considerable degree of freedom of action. The internal command network of the enterprise is usually sufficient to ensure the compliance of subsidiaries with the parent's policies (more accurately, the policies decided upon by the center of decision, wherever located). As a result, restrictions on the conduct of subsidiaries related to the use of technology may be part of implicit company practices rather than a result of explicit formal arrangements.(50)

In conclusion, the idea of an international code of conduct for transfers of technology reflects certain findings and conclusions as well as certain choices of direction and method for action that were prevalent in the early 1970s among the economists and policymakers of the more advanced among the developing countries. The time lag between the idea and its realization is not too long, by the usual standards of international institutional action, yet it may be far too long in terms of the evolution of the world economy. A certain lack of enthusiasm that may be noticed nowadays among the initial promoters of the idea of a code of conduct (on transfer of technology, transnational enterprises, or restrictive practices) may be in part a consequence of this fact. While still far from being in effect, the code of conduct on transfer of technology hardly seems today an idea geared to the future. It looks rather like a fundamentally moderate attempt to help restructure international economic relations, without unduly disturbing such basic legal and economic institutions as private property, industrial property rights, and sanctity of contracts – merely attempting to regulate them in some of their manifestations, rather than challenge their legitimacy. Such a perspective makes the opposition of the developed countries to various aspects of the code, but primarily to the basic issue of its effectiveness, all the more difficult to defend, if not to understand. One wishes that the matter were closed and the code adopted and turned over to the administrators and technicians for application, so that more attention could be paid to emerging problems and possibilities.

NOTES

(1) "Plaisante justice qu'une rivière borne! Verité au deçà des Pyrénées, erreur au delà." Pascal, Pensées, no. 294 (1670) (L. Brunschvicg, 1904).

(2) See Illich, "Planned Poverty: The End Result of Technical Assistance," in Celebration of Awareness (Garden City, N.Y., 1971), pp. 149, 156.

(3) Cf. E.F. Schumacher, Small is Beautiful: A Study of Economics as if People Mattered (New York, 1973); F. Stewart, Technology and Underdevelopment (Boulder, Colo., 1977).

(4) See D. Goulet, The Uncertain Promise: Value Conflicts in Technology Transfer (Washington, D.C., 1977).

(5) This is not the place for even brief discussion of the history and traits of this "way of perceiving and understanding the world." Special emphasis on the role of "instrumental reason" in modern Western culture has been placed in the work of the Frankfurt School. See, in particular, M. Horkheimer and T. Adorno, Dialectic of Enlightment (J. Cumming transl., New York, 1972 [first German ed. 1944]); and more recently Juergen Habermas, Toward a Rational Society (J.J. Shapiro transl., Boston, 1970); Theory and Practice (J. Viertel transl., Boston, 1973); Legitimation Crisis (T. McCarthy transl., Boston, 1975). And cf. T. McCarthy, The Critical Theory of Juergen Habermas (Cambridge, Mass., 1978) esp. pp. 1-52.

(6) The term "hegemonic" is used here in the special sense associated with the work of Antonio Gramsci; see A. Gramsci, Selections from the Prison Notebooks (Q. Hoare and G. Nowell Smith, eds. and transl., New York, 1971); G.A. Williams, "The Concept of 'Egemonia' in the Thought of Antonio Gramsci: Some Notes on Interpretation," Journal of the History of Ideas 21 (1960): 586; R. Williams, "Base and Superstructure in Marxist Cultural Theory," New Left Review, no. 82 (1973), p. 3.

(7) L. Winner, Autonomous Technology: Technics-out-of-Control as a Theme in Political Thought (Cambridge, Mass., 1977), pp. 323-24. Emphasis omitted.

(8) Goulet, Uncertain Promise, n. 4, p. 6.

(9) For strong affirmation of this novelty and a dramatic contrast between the two technologies, see M. Heidegger, "The Question Concerning Technology," in The Question Concerning Technology and Other Essays (W. Lovitt transl., New York, 1977 [first publ. in Vortraege and Aufsaetze (1954)] , pp. 5, 14-16.

(10) This listing, with minor amendments, follows C.H.G. Oldham, C. Freeman, and E. Turkcan, "The Transfer of Technology to Developing Countries," Study prepared for UNCTAD II (UN Doc. TD/28/Supp. 1, November 10, 1967), p. 8.

(11) For an argument in favor of limiting the concept see C. Cooper and F. Sercovitch, "The Channels and Mechanisms for the Transfer of Technology from Developed to Developing Countries," Study prepared for UNCTAD (UN Doc. TD/B/AC.11/5, April 27, 1971), p. 12. A similarly narrower conception of technology transfer is in fact used in

the present chapter, but it was considered instructive to refer at the outset to the larger processes at work.

(12) Ibid.

(13) A comprehensive study of the growth of the international protection of industrial property, which does not discuss in detail political aspects and relations but still makes evident the extent of deliberate governmental action in the area, is Stephen P. Ladas, Patents, Trademarks and Related Rights: National and International Protection (Cambridge, Mass., 1975), esp. vol. 1. For other critical discussions focused on patents see H. Kronstein and I. Till, "A Reevaluation of the International Patent Convention," Law and Contemporary Problems 12 (1947): 765; and E.T. Penrose, The Economics of the International Patent System (Baltimore, 1951). The ambivalent attitude of developing countries is touched upon in Penrose's book, pp. 220-22; for more recent studies of this issue, see, U. Anderfelt, International Patent-Legislation and Developing Countries (The Hague, 1971); and M. Niance and Y. Plasseraud, Brevets et sous-développement (Paris, 1972).

(14) Among the many related discussions, those found in UNCTAD documents should be specially referred to here, because they largely form the basis for the activities discussed in the last section of the paper. See, in particular, the studies cited in notes 23 and 39. An interesting listing of "obstacles to the acquisition" of technology by developing countries is also found in the World Intellectual Property Organization, Licensing Guide for Developing Countries, (Geneva, 1977), n. 25, pp. 17-22.

(15) Among the many well-known examples of such activities, see the International Chamber of Commerce Incoterms 1953 (international rules for the interpretation of trade terms) and Uniform Customs and Practices for Documentary Credits (1975), both conveniently reproduced in UNIDO, Guidelines for Contracting for Industrial Projects in Developing Countries (UN Doc. ID/149, 1975), pp. 146 and 136 respectively.

(16) "International model form of agreement between client and consulting engineer and International General Rules for agreement between client and consulting engineer (IGRA 1963)," reproduced in UNIDO, Guidelines, n. 15, p. 71.

(17) We do not deal here with the "codes of conduct" adopted by individual (usually transnational) enterprises. To study them properly we would have to undertake a study in depth of business "ethics," especially as they relate to foreign operations (since it seems fairly clear that, traditionally at least, firms apply differing ethical standards to their domestic and foreign operations).

(18) ICC Brochure No. 272 (1972).

(19) Approved 1958, adopted 1959; reproduced in UNIDO, Manual on the Use of Consultants in Developing Countries, (UN Doc. ID/3/Rev. 1, 1972), p. 155. See also the same Institute's Standards of Professional Conduct, ibid., p. 156.

(20) Codes of professional ethics and conduct (like those cited in the previous note) are sometimes enforced by the relevant professional associations. Their policing is generally lax, however, especially with respect to provisions concerning protection of client interests.

(21) For a very useful summary of UNIDO activities in this area, see G. Burdeau and J. Chappez, "Les transferts de technologie dans la pratique de l'ONUDI," in Transfert de technologie et développement ed. P. Judet, P. Kahn, A.C. Kiss and J. Touscoz (Paris, 1977), pp. 227-65. The discussion in this section is based on this study as well as on various reviews and surveys by UNIDO of its own activities.

(22) See, e.g., UNCTAD, "Report on the First UNCTAD/SIDA Training Course on Transfer of Technology," UN Doc. UNCTAD/TT/1, November 1975; OECD Development Center, Transfer of Technology by Multinational Corporations, D. Germidis, ed. (Paris, 1977, 2 vols.); U.N. Centre on Transnational Corporations, "Strengthening the Negotiating Capacity of Developing Countries. Report on a Round Table," UN Doc. E/C.10/50, March 1979.

(23) See the UNCTAD studies cited in note 39. See also the series of case studies prepared by UNCTAD under the general title "Major Issues in Transfer of Technology to Developing Countries": UN Doc. TD/B/AC.11/17, April 1974 (Spain); TD/B/AC.11/18, May 1974 (Hungary); TD/B/AC.11/20, May 1974 (Chile); TD/B/AC.11/21, June 1974 (Ethiopia); TD/B/C.6/6, October 1975 (Sri Lanka); TD/B/C.6/4, October 1975 (pharmaceuticals industry). More recently, see UNCTAD, Handbook on the Acquisition of Technology by Developing Countries (New York, 1978).

Special mention is also due an earlier series of studies published by UNITAR in 1971 to 1974, covering transfer of technology from particular countries (Japan, USSR), into particular countries (the Philippines, Mexico) and with respect to specific industries (automotive, pharmaceutical, petrochemical, electronic, pulp and paper).

(24) See UNIDO, Manual on the Establishment of Industrial Joint-Venture Agreements in Developing Countries UN Doc. ID/68, (1971); Manual on the Use of Consultants; Guidelines for Contracting; Guidelines for the Acquisition of Foreign Technology in Developing Countries, With Special Reference to Technology License Agreements (UN Doc. ID/98, 1973); National Approaches to the Acquisition of Technology (ID/187, 1978).

(25) WIPO, Licensing Guide for Developing Countries. On the preparation by the senior advisers to the Economic Commission of European Governments on Science and Technology of a "Manual on Licensing Procedures and Related Aspects of Technology Transfer," see UN Doc. SC. TECH/R.60, May 13, 1977. Earlier ECE guides on drawing up contracts have covered the international transfer of know-how in the engineering industry (1970), large industrial works (1973), and industrial cooperation (1976).

(26) The secretariat had been asked to "explore the possibility" of such Centers in UNCTAD Resolution 39 (III) (1972). For a comprehensive report and discussion, see UNCTAD, "Transfer of Technology. Action to Strengthen the Technological Capacity of Developing Countries: Policies and Institutions," UN Doc. TD/190/supp. 1, February 1976 (submitted to UNCTAD IV, 1976).

(27) UNIDO, Industrial Development Board, Decision V (XI), June 6, 1977. For an early report on the establishment of the Bank, see UN Doc. A/32/116, June 21, 1977.

(28) United International Bureaux for the Protection of Intellectual Property (BIRPI), Model Law for Developing Countries on Inventions (Geneva, 1965). Around the same time, a U.N. study also addressed the problem; UN, Department of Economic and Social Affairs, The Role of Patents in The Transfer of Technology to Developing Countries (New York, 1965). For more recent related U.N. studies see note 31.

(29) On the recommendation of a WIPO Permanent Committee for the Acquisition by Developing Countries of Technology Related to Industrial Property, a working group on the "Model Law for Developing Countries on Inventions and Know-How" has met several times since late 1974 and has studied successive drafts prepared by the International Bureau of WIPO.

(30) BIRPI, Model Law for Developing Countries on Marks, Trade Names and Unfair Competition (Geneva, 1967); WIPO, Model Law for Developing Countries on Industrial Designs (Geneva, 1971).

(31) As in 1965, a series of recent UNCTAD studies has also dealt with the international patent system. Their contents reflect a substantial change in approach since 1965. See, in particular, UNCTAD, The Role of the Patent System in the Transfer of Technology to Developing Countries (New York, 1975) (in cooperation with the U.N. Department of Economic and Social Affairs and the International Bureau of WIPO); "The International Patent System as an Instrument of Policy for National Development," UN Doc. TD/B/C.6/AC.2/3, July 1975; "Promotion of National Scientific and Technological Capabilities and Revision of the Patent System," UN Doc. TD/B/C.6/AC.2/2, July 1975; "Systems, Including Industrial Property Systems, for Improving the National

Scientific and Technological Infrastructures of the Developing Countries," UN Doc. Td/B/C.6/AC.2/4, August 1975; and "The International Patent System: The Revision of the Paris Convention for the Protection of Industrial Property," UN Doc. TD/B/C.6/AC.3/2, June 1977. A recent study addresses the related topic of trademarks: UNCTAD, "The Impact of Trade Marks on the Development Process of Developing Countries," UN Doc. TD/B/C.6/AC.3/3, June 1977.

(32) Guidelines for Multinational Enterprises, Annex to the Declaration of 21st June, 1976 by Governments of OECD Member Countries on International Investment and Multinational Enterprises, reprinted in International Legal Materials 15 (1976): 969.

(33) See the studies cited in notes 10 and 11. And see also UNCTAD, "Transfer of Technology," UN Doc. TD/105, November 1971 (submitted to UNCTAD III).

(34) Andean Commission, Decision No. 24, Standard Regime for Treatment of Foreign Capitals and for Treatment of Marks, Patents, Licenses and Royalties (December 31, 1970), Int'l Legal Materials, 10 (1971), 1952; repr. as later amended, id. 11 (1972), 126. And see, UNCTAD, "Transfer of Technology. Policies Relating to Technology of the Countries of the Andean Pact: Their Foundations." (A Study by the Junta del Acuerdo de Cartagena prepared at the request of the UNCTAD Secretariat), UN Doc. TD/107, December 1971 (submitted to UNCTAD III).

(35) UNCTAD, Resolution 39 (III), May 16, 1972, in Proceedings of the United Nations Conference on Trade and Development, 3rd sess., Santiago, Chile (New York, 1973), p. 108.

(36) Intergovernmental Group of Experts on Transfer of Technology, Resolution 1 (II), para. 17, in UN Doc. TD/B/424 (Annex I), February 1973. And cf. Resolution 104 (XIII) of the (UNCTAD) Trade and Development Board, September 1973; and Resolution 3 (III) of the Intergovernmental Group of Experts on Transfer of Technology, July 1974, in UN Doc. TD/520, Annex I.

(37) UN GA Resolution 3202 (S-VI), IV (a), May 1, 1974.

(38) See UN doc. TD/B/AC.11/L.12, May 1974.

(39) See, in particular, the following UNCTAD reports: Guidelines for the Study of the Transfer of Technology to Developing Countries (1972): "The Possibility and Feasibility of an International Code of Conduct on Transfer of Technology," UN Doc. TD/B/AC.11/22, June 1974; Major Issues Arising from the Transfer of Technology to Developing Countries (1975); and An International Code of Conduct on Transfer of Technology (1975).

(40) UNCTAD, Resolution 89 (IV), May 30, 1976, in UN Doc. TD/217, May 1976, p. 37. Statements in support of a code of conduct are also contained in the Lima Declaration and Plan of Action on Industrial Development and Cooperation (Second General Conference of UNIDO), Plan of Action, para. 61 (m), March 26, 1975, UN Doc. ID/B/155/Add.1, April 1975; and in the Manila Declaration and Programme of Action (Third Ministerial Meeting of the Group of 77), sec. 5, para. 10-12, February 7, 1976, UN Doc. TD/195, February 1976.

(41) See the successive reports in UN Docs. TD/AC.1/4, November 1976; TD/AC.1/7, April 1977; TD/AC.1/9, August 1977; TD/AC.1/11, November 1977; and TD/AC.1/15, February 1978. The sixth report is the one concerning the text submitted to the conference; see note 42.

(42) UN Doc. TD/CODE TOT/1, September 1978, and TD/CODE TOT/1/Add.1, September 1978. The conference was convoked by virtue of U.N. General Assembly Resolution 32/188, December 17, 1977. For a most useful summary discussion of the draft code at that stage, see the "Background Note" by the UNCTAD secretariat, UN Doc. TD/CODE TOT/4, September 1978, which includes an excellent table listing the "main steps in the work" on the code (Annex I) as well as the text of relevant U.N. resolutions (Annex II).

(43) For the resulting draft text, see UN Doc. TD/CODE TOT/9, November 1978.

(44) For the resulting draft text, see UN Doc. TD/CODE TOT/14, March 1979.

(45) See UNCTAD, "Report of the Intergovernmental Group of Experts on a Code of Conduct on Transfer of Technology," UN Doc. TD/B/C.6/1, May 1975, Annex I and III.

(46) See ibid., Annex II (Group B); UNCTAD, "Report of the Intergovernmental Group of Experts on an International Code of Conduct on Transfer of Technology," UN Doc. TD/AC.1/4, November 1976, Annex III (Group D and Mongolia).

(47) See the discussion of the code's goals below and that of legal form.

(48) Various other considerations may in part account for the differences in the position of the developing and the developed countries on this point – the former insisting on a "legally binding" code, the latter on a "voluntary" one. Many of the developed countries tend to perceive legal rules more rigidly, to "take them more seriously" as coercive and restrictive prescriptions than do many of the developing countries, for whom legal rules are often understood as statements of aspirations rather than strict immediate obligations. To some extent the Third World's emphasis on international rather than national legal

action in this as in other areas reflects its general feeling of dependence on others, whether powerful developed states or the international system in general. International action is thus understood as primarily defensive in character. (It is no less true, of course, that, despite the rhetoric of self-reliance, many developing countries find domestic reform much more difficult, and painful to the ruling elites, than invocations of international solidarity.)

(49) The issue has arisen in connection with the code's scope of application, on the one hand, and the treatment of restrictive practices in transactions between affiliated companies on the other. On the first point, the developed countries limit the code's application to technology transfer transactions crossing national borders; the developing (and the socialist) countries propose to extend its applicability to transactions between a foreign-controlled affiliate and a local firm, both resident in the host country. As to the second question, the developed countries propose that transactions between affiliated enterprises would be subject to the prohibition of restrictive practices only in a limited category of cases; the developing and socialist countries would not allow such special treatment. (As to both questions, compromise positions have been formulated and proosed at the spring 1979 session of the conference. Although they were not agreed upon at that time, they appear likely to be adopted.)

The internal inconsistency of the position of both sides is evident, however justified or unavoidable one may find it. The developing countries are simultaneously affirming that subsidiaries and parents form a single unit (as far as the scope of application is concerned) and that they form discrete units (as far as restrictive practices are concerned). Conversely, the developed countries uphold the separate personality view in the first case and affirm the unity of the firm in the second.

(50) To the extent that the trend toward "disinternalization" noted by John Dunning (see Chapter 1 in this volume) continues to grow in importance, this limitation on the code's effectiveness may lose some of its importance.

Comments

Farouk I. Heiba

The paper by Professor Fatouros presents the background for and the continuous global debate on international regulations and controls for technology transfer.

This paper is basically descriptive, and it is hard to disagree with facts, as Professor Wells of Harvard once mentioned.(1) However, there are some questions to be raised and a number of observations and comments to be discussed. They can be summarized into six basic issues.

First, there are no sufficient justifications or answers given by the author to the following critical questions: Why do we need international regulations and controls of technology transfer? What are the rationale and motivations behind such regulations and controls? For whom should it be designed and in which style? Is it for profit maximization, for monopolistic suppliers, or for global social responsibility, human justice, and world development, or for other goals?

With respect to the statement by Alan Admas that "technology transfer control systems have little rationale,"(2) is it possible to reach a global agreement concerning regulating and controlling technology transfer? If it is, should it be legally binding or voluntary, and in either case, what ought to be done? What can be done? What will be done?

Second, as Professor Kolde observed, one of the loosest and most confusing concepts is that of technology.(3) Thus before designing any model for international regulations and controls of technology transfer, we should understand the meaning, nature, and scope of technology and its implications. With respect to Professor Fatouros's perception of technology, I disagree with the notion that technology is standardized and that it is the same on both sides of the mountain and everywhere on the globe. I do not think the "mechanical donkey" is the same in its purpose or functions or performance in both developed and under-developed societies.

Technology in general varies from one environment to another depending upon its adapted style and function.(4) Also, technology is more than a political phenomenon; it has three major aspects:

1. Materialistic and physical aspects (or hardware technology: machines, capital, computers, and so on)
2. Human aspects (or the software technology, such as know-how)
3. Ideological and mental aspects (the desire and acceptance of modern technology as a social way of life).

Of course certain types of technology are more controllable than others. For example, arms, nuclear power, and energy and space technology are already regulated and controlled by the suppliers or the producers, since they represent a monopolistic behavior in the market of technology. Control of others depends on their styles and marketability. And because we live within a dynamic and turbulent environment, today's technology will be different from tomorrow's technology; thus the adaptation to a continuous control system is required.

Third, Professor Fatouros introduced four types of controls of international technology transfer as measures for serving particular purposes. These four types of controls are unilateral private measures, collective private measures, unilateral governmental (national) controls, and collective governmental (international) measures. I agree to a certain extent with the descriptive purpose of each measure; however, these measures of control lack the mechanism of "make it happen." The author's perception of the "unequal bargaining position" among technology suppliers and technology recipients introduces an important dimension. This is due to the fact that "the technology most profitable to a recipient firm may not be the most appropriate in terms of the needs of the national economy of the recipient country."

Fourth, in discussing the issue of international regulation and control of technology transfer, there is an obvious tendency to discuss symptoms rather than producers, or the core, of the problem. With respect to other factors, one of the core problems regarding the issue of controls is the "trust gap" among the institutional actors,(5) due to the existing xenophobia among technology suppliers and technology recipients. Technology recipients are still suspicious of technology suppliers and producers (who are usually the politically most powerful actors) as a result of the negative impact of the colonial era and the belief that technology suppliers and producers have bad intentions in assisting them technologically toward desirable development. On the other hand, technology suppliers believe that technology recipients, especially in the Third World nations, are unpredictable and represent a political risk, and thus that in order to protect their interest they should control the use of their technology as well as keep the recipients technologically dependent. The more we avoid such xenophobia through acceptable regulations and controls of technology transfer for mutual benefit, the more we close this "trust gap" for the benefit of the world community.

Fifth, the debate over the issue of controlling technology transfer, from the suppliers' side, nonetheless represents monopolistic behavior and expresses the mentality of the "international mafia" of technology transfer. The point of view is, We make technology, thus we have the right to control it, use it, and sell it the way we like at the price we want.

This takes different forms and tones. For example, Don Guertin of Exxon argues that multinational enterprises develop technology basically for their own use; technology is generally developed for proprietary reasons and must be protected by the producer through legalized monopoly rights, that is, patents. He argues that governments cannot force transfers because of the proprietary nature of the technology and also because they view technology as a national asset.(6)

Alan Rapoport of the National Science Foundation and Franklin Root of the Wharton School share a view similar to Don Guertin's in regard to preserving the rights of technology suppliers.(7) The rights of the suppliers are centered on their control over the use of their own technology. If they lose control over the disposition of their technology, they will cease to create technology. Thus, a control system must preserve the rights of the suppliers to sell or not to sell their technology and to bargain for a price they judge to be satisfactory. Also, they demand that technology recipients are to understand the reality: "Don't destroy the whole cake in an attempt to capture a piece."

Unless we eliminate that monopolistic behavior, reaching a global agreement on controlling technology transfer will be unlikely.

Sixth, Professor Fatouros dealt in his paper with only the commercial side of international regulation and control of technology transfer. However, there is another sensitive side of technology transfer, namely, the human side, or what is called the "brain drain." As a part of the process of technology transfer, particularly that of know-how, the "brain drain" represents a serious problem for developing countries, in terms of a negative return on investment in their human resources. I believe that there is an international need for a comprehensive study on this problem to find the best way of regulating and controlling this critical issue, if the technology suppliers, in good faith, want to help the developing countries achieve a desirable future.

In conclusion, I think the author's ideas of intergovernment controls of technology transfer are useful in the field of international controls of technology transfer, but there is a need for more elaboration to see how these ideas could be put into practice. These ideas, with some modifications to avoid the monopolistic behavior of technology suppliers and to design pragmatic measures against future misuse or risk by the recipients, would be helpful in this endeavor. Professor Perlmutter of the Wharton School and Professor Sagafi-nejad of the University of Washington expressed their opinions on this issue as follows: "What appears to be missing are some comprehensive and realistic scenarios for increasing less industrialized countries' self-reliance and national pride. Such scenarios must be pursued with commitment and persistence, with a vision of a better future, and with an understanding that technology must be adapted to the socio-cultural choices each nation prefers. . .(8)

Therefore the author's suggestions as well as other arguments for designing a control system and establishing a code of conducts for technology transfer will not be a solution for the existing problems in that field unless we close the "trust gap" between technology suppliers and recipients and motivate the institutional actors toward achieving common objectives of world development.

NOTES

(1) Louis T. Wells, Jr., in his comment on Chapter 13 of this volume.

(2) Alvin Adams, the U.S. negotiator on codes of conduct, in his discussion during Phase I of the International Conference on Technology Control Systems, Philadelphia, February, 1979.

(3) Endel J. Kolde, in his comment on Chapter 9 of this volume.

(4) See Thomas P. Hughes's "Transfer and Style," Chapter 2 of this volume.

(5) "Institutional actors" usually refers to home country (supplier of technology), host country (recipient of technology), supplier firms, and recipient firms.

(6) Don Guertin, in his comment on Chapter 5.

(7) In discussion during Phase I of the International Conference on Technology Control Systems.

(8) Howard V. Perlmutter and Taghi Saghafi-nejad, "Process or Product? A Social Architectural Perspective of Codes of Conducts for Technology Transfer and Development," Codes of Conduct for the Transfer of Technology: A Critique, Sudan S. Holland, ed. (New York: Council of the Americas, 1976), pp. 95-129.

Comments

Miguel S. Wionczek

I would have been almost ready to agree with most of the substance of Professor Fatouros's paper if it did not contain as its final sentence the following: "One wishes that the matter were closed and the code adopted and turned over to the administrators and technicians for application, so that more attention could be paid to emerging problems and possibilities."

The statement is too categorical for my liking. Nor does it follow the author's brilliant presentation of the code's possibilities and limitations. While the code has both, Professor Fatouros concentrates in the final section of his paper on the code's limitations. But in all fairness, the adoption of the code after over five years of difficult negotiations is one thing; its implementation – both at the international and national levels – is another, even if one accepts that the function of the code (as Professor Fatouros rightly reminds us) is rather modest. Thus to evaluate the potential usefulness of the code one must look not only at the negotiating process and the (largely) agreed-upon text, but at the issue of implementation as well.

It is worth keeping in mind that under no circumstances will the code substitute for the private system of international controls over technology transfer, which happens to be neither liberal nor "neutral." The code will merely create a parallel system of public controls of a theoretically somewhat higher order in the framework of "international law" (whatever that expression means). And, the world being as it is, there are ample reasons to believe that the coexistence of the two international control systems managed by two unequal partners will not decide all conflicts between private versus public interests in favor of the latter, even after proper legal provisions are taken at the national level to carry out the code's general rules. Once the code is approved, the problems will start, not disappear. Their "manageability" will depend not only on national measures but also, to quite an extent, on the use of the code as "a source of law" and a sort of gentle policeman.

The fact that Professor Fatouros did not allow his imagination to wander – at least in the first draft of his paper – on the issue of the possible implementation of the code (or guidelines, for that matter) on the international level is the paper's major shortcoming. It is quite probable that he abstained from such an exercise because the parties directly involved have not yet made up their minds themselves in that respect.

According to some observers, a compromise solution of sorts was in the making at UNCTAD in Geneva in 1980 on this important point, independently – on the surface – of the final decision on the code's legal character. When will the negotiations on the code be finalized? In light of the continued deadlock between LDCs and western industrial countries on two important substantive issues – the scope of restrictive business practices to be banned under the code, and the applicable laws and the settlement of dispute – it is hard to believe that the code will be approved in UNCTAD in 1981. The stagnation of the negotiations reflects, however, more than disagreement on the code itself. It is a part of the overall paralysis of the so-called north-south dialogue related to the deepening worldwide economic crisis. An intergovernmental body is to be established within UNCTAD for the purpose of (1) providing a forum for discussion and exchange of views between states on matters related to the code, in particular its application and the experience gained from its operation; (2) undertaking studies and research for the purpose of furthering the aims of the code; (3) inviting and considering relevant studies, documentation, and reports from within the United Nations system; (4) studying, collecting and disseminating information on matters relating to the code; (5) providing a forum for consultations between the states, and (6) organizing meetings concerning the application of the provisions of the code. Even if the mandate of this intergovernmental body were to state explicitly – as the industrial countries insist – that it is not to act as a tribunal or reach conclusions on the conduct of individual governments or parties, the fact that it will conduct periodic reviews of application of the code and its possible revision suggests strongly that the potential role of the code goes beyond that ascribed to it by Professor Fatouros, namely, of providing "a favorable framework for the exercise by the states of their policy- and decision-making powers." The future of the code depends on its international implementation and not only on its legitimizing function.

The question of implementation is much more important than the legal issue at least to this writer, for very simple reasons. If one accepts that (1) law by itself does not govern social relations unless it is supported by the power of potential sanctions, and (2) in the present transfer-of-technology setup receivers of technology (except in the advanced countries) have much less bargaining (and any other) power than technology producers and holders, then in the absence of sanctions not even an ironclad legally binding convention could help. But national legislations cannot do the trick either in the absence of some sort of bargaining parity.

Thus the issue of a legally binding code versus voluntary guidelines is not the key issue in this case. The insistence by the less developed countries on having a code instead of guidelines only represents the admission by the weaker parties that they are very weak indeed. Hence they would like to see their weakness diminished by any means available, including the "sanctification" of an international regulating instrument.

It is the implementation of the code that can make out of it a real thing. Industrial property owners are quite aware of the importance of that issue. They hoped for the best of two possible worlds: voluntary guidelines, with the implementation mechanism reduced to an informal forum for "observation" of technology transfer practices. As we have heard from one of the participants in the UNCTAD negotiations representing a major Western power, industrial countries continue to consider (almost) unacceptable any "quasi-judiciary" international evaluation of such practices.

There is nothing new in that position. Be reasonable, we are told time and time again; abide by internatonal law as it stands but do not expect us to accept your attempts to equalize it. But from the undeniable fact that the code, considered together with the sort of implementation mechanism described above, will provide some degree of equalization of transfer-of-technology terms and conditions, it follows – at least to me – that the code (or even guidelines) will offer more than just a framework for national action. Of course, no automatic power can be built into the code at the present stage. Moreover, the code may fall apart through the cutthroat competition for technology transfer among the recipient countries, through the drift away from the industrial property system into secrecy by technology holders, or through other, even less expected developments. But after more than five years of negotiations the less developed countries can at least feel that there is going to be no "exploitation without participation" of all parties involved as far as most (albeit not all) international technology transactions are concerned. Whether the stage of "participation without exploitation" is ever reached depends largely on the LDCs.

From the time of the ICC Guidelines for International Investment (1972) and the OECD guidelines for multinational corporations (1976) to those of the UNCTAD code, a lot of distance was covered in the field of technology transfer, just because of the UNCTAD code negotiations. Perhaps without them, for example, the present process of radical revision of the model law on inventions, under the auspices of WIPO, would not even have started. The spinoff or the fallout from the UNCTAD initiative is quite probably larger than it looks at first sight, although it is quite difficult to evaluate it correctly so early in the game.

If one assumes – hopefully – that the code will become a dynamic longer-term exercise, then a lot of ideas related to "emerging problems and possibilities" – to use Professor Fatouros's language – may be incorporated in this ongoing process at a later date. As a framework for national action only the code does not promise too much. But even under such circumstances I would be willing to say that, considering the content of the code, the time and energy dedicated to negotiations on it in Geneva were not spent too badly, particularly in comparison with the generally miserable state of institutional and other changes in North-South relations at UNCTAD and elsewhere.

Finally, a small factual detail. Professor Fatouros notes that the very first draft of a possible code was prepared in the spring of 1974 under the "somewhat incongruous" auspices of the Pugwash Conference. There was nothing incongruous about these auspices – it was just one of those historical accidents, historical in the sense that without that particular very first draft the start of the UNCTAD negotiations would most probably have been delayed for several additional years. The idea of drafting "something" under neutral auspices outside of the U.N. system occurred to me after two years of listening to metaphysical international discussions of the so-called issue of the feasibility of a code in that field. Although they were invited, knowledgeable "private" citizens from two countries did not show up at that 1974 Pugwash ad hoc meeting on the code in Geneva. Guess where they were from: Senegal and Albania? No, from the United States and the Soviet Union. Incongruous? Perhaps. Reflecting the world in which we all live? Certainly.

III

Conclusion

15 Technology Transfer Control Systems: Actors, Issues, and Alternatives

The papers in this volume have raised and discussed from many different viewpoints a wide variety of issues related to alternative technology transfer control systems. The purpose of this concluding chapter is not to summarize these papers, but rather to draw our own conclusions based on the papers and the conference discussions. The questions to be addressed are:

1. What are the prospects for successful implementation of effective control systems at the firm, national, regional, and international levels?
2. What policies by national governments and what other actions are needed for successful implementation?

We begin by describing various possible control systems in terms of their main characteristics. The many alternatives are categorized along the dimensions of origin, goals, and legal status. In order to assess the prospects for these alternatives, the forces favoring and inhibiting successful establishment of control system are then reviewed. Finally, we attempt to spell out what home and host country governments will need to do in order to help bring out the establishment of mutually acceptable technology transfer control systems.

ALTERNATIVE CONTROL SYSTEMS

A technology transfer control system may be any instrument, policy measure, rule, regulation, law, or any other action that influences the flow of technology between supplier and the recipient. In the international flow of technology across boundaries, these instruments can be initiated or located at any point between the supplier and the recipient. In order to better understand these instruments, it is helpful to disaggregate the concept into its component parts.

Technology is defined as organized knowledge for productive purposes. More specifically, however, it refers to commercial technology as the subject of commercial transaction between firms.

Transfer is a spatial term, connoting the movement of technology from one point to another. In the present discussion, the two points are assumed to be separated by national boundaries.

Control refers to influencing or steering in a desired direction, so as to bring a process under some order and regularity.

System is a relative term, and implies any policy, measure, or instrument that is purposeful rather than random. The "degree of systemness," as per systems theory, is left undetermined. By using the term we do not intend to convey any implication as to how integrated, interrelated, or coherent a certain control system may be. Accordingly, in place of technology transfer control "systems," one may use "instruments" or "measures" with impunity.

It is equally important to note what is not meant by the term "technology transfer control system" or the word "control" contained therein. It does not, a priori, mean limiting or restricting the quality, quantity, or terms of the transfer. One must guard especially against the possible misinterpretation that may equate control with restriction. Instead, the term may convey order and regularity.

A more systematic examination of these systems can be made, however, by distinguishing them along three major dimensions: origin, goals, and legal status.

Origin of Control Systems

Control instruments have emanated from subnational, national, regional, and international sources. Statements by individual transnational enterprises in the form of codes of conduct, statements of principles, and standards of social responsibility have proliferated in the last few years. A driving force behind such statements has been the wave of disclosures of unethical or questionable corporate practices. The main objective of such statements by various corporations has been to publicize corporate responsibility in terms of ethical practices, product safety, and responsibility to shareholders, labor, and society. However, these documents invariably contain sections dealing with technology and its transfer and represent one sort of control on international technology transfer.

In addition to these corporate efforts, various industry groups or professional associations have proposed self-policing instruments, for example, statements by the U.S. Chamber of Commerce, the Financial Accounting Standards Board, and the American Institute of Certified Public Accounts. Civic and religious organizations also have intensified their demands for more ethical conduct and corporate responsibility on the part of the business community in general. Such subnational attempts at regulation, of course, have had little more than an educational purpose. Lacking national legal sanctions and enforceability, their impact is limited to moral suasion and increased consciousness.

On the other hand, sovereign governments enjoy the unparalleled ability to permit or prohibit a wide range of activities within their national boundaries. Thus many governments have unilaterally enacted laws and regulations to control the activities of TNEs and the process of technology transfer. These measures vary in their coverage, explicitness, and restrictiveness. Aside from laws governing foreign investment, some recipient countries (notably Mexico, Argentina, Brazil, and Venezuela) have enacted laws pertaining directly to technology transfer. The Mexican law was one of the earliest and most publicized examples of national regulation of technology transfer.

Countries that are primarily sources of technology transfer also have taken measures to regulate the export of technology. The most important of such regulations are those prohibiting or restricting exports on the grounds of national security.

At the reginal level, Decision 24 of the Andean Group is an example of an attempt by a group of nation-states to regulate the flow of technology into the region. The Organization of American States, the Pacific Basin countries, the Organization for Economic Cooperation and Development, the European Economic Community and its associated less industrialized countries (in the Lomé Convention of 1975) have attempted to establish rules, policies, or recommendations on the activities of TNEs and technology transfer.(1)

The most extensive attempts to regulate technology transfer are being made at the international level. Interestingly, the first international agreement on technology transfer was reached nearly a century ago, in 1883, when the Paris Union for the protection of industrial property (patents, trademarks, and inventions) was signed. Current attempts to regulate technology transfer are concentrated within the U.N. Conference on Trade and Development. Since UNCTAD I in 1964, this U.N. body has been interested in improving the terms of trade and accelerating the industrial development of the less industrial countries. UNCTAD III in Santiago (1972) made explicit recommendations regarding technology transfer. And UNCTAD IV in Nairobi (1976) adopted a resolution recommending that work toward a draft international code of conduct for technology transfer be accelerated and a compromise draft be prepared before the end of 1977. Thus UNCTAD has become the pivotal international organization under whose auspices a very extensive exercise began. This has included numerous case studies of countries and industries, methodological papers and monographs, and extensive research on codes of conduct for the transfer of technology.

UNCTAD's Trade and Development Board, its secretariat, and, under their supervision and mandate, an intergovernmental group of experts have been laying the groundwork for a universally applicable code. Throughout these meetings, the three major groupings of countries, namely, the Group of 77 (the LICs), Group B (more industrialized Western countries and Japan) and Group D (the socialist bloc of countries), have been trying to reach a common understanding. Superimposed on these negotiations have been the various UNCTAD resolutions (especially those of UNCTAD III to UNCTAD V) as well as the resolutions of the U.N. General Assembly. In its Sixth Special Session (May 1974), the U.N. General Assembly adopted the Declaration on the Establishment of a New International Economic Order. The Program of Action accompanying that declaration assigned a high priority to a code on technology transfer. That same year a nonpartisan group of scholars, having no official position or mandate, produced the Pugwash draft code of conduct for the transfer of technology, which formed the basis for a draft proposal that the Group of 77 then submitted to the UNCTAD-sponsored intergovernmental group.

The developed countries of Group B, who initially felt no code was necessary, reacted to the Group of 77 proposal instead of developing their own equivalent version. In September 1975, however, the U.S. secretary of state announced that the United States "believes the time has come for the international community to articulate standards of conduct for both enterprises and governments." Some observers believe that this was a turning point in the code formulation process. Subsequent to that policy statement, both Group B and Group D submitted their own drafts for negotiations toward a compromise. Some hoped such a compromise could be achieved during UNCTAD IV in Nairobi in 1976. This did not materialize. Meanwhile, the intergovernmental group of experts charged with negotiating a compromise was able to prepare several consolidated drafts in 1978 and 1979 with numerous controversial issues still in brackets and not yet agreed upon. In June 1978 and in March and November of 1979 (just before and again after UNCTAD V in Manila), different drafts were presented for negotiation.(2) By the end of 1978 the issue had enlarged itself to the point where a series of U.N. conferences were scheduled to reach a compromise. These continued into the April-May 1980 U.N. Conference on an International Code of Conduct on the Transfer of Technology in Geneva. Disagreements still remained, and a draft code was sent to the U.N. General Assembly in April 1981.

Goals of Control Systems

Alternative control systems have varying goals in terms of the functions they are intended to perform, the entities they are designed to control, and the entities that are the intended beneficiaries. Control instruments are directed toward different groups or clusters of entities, depending chiefly on where the instruments originate and who proposes them. In general, they can be classified as: (1) those that aim at controlling the firms (and often only the supplier firm); and (2) those which seek to set out duties and responsibilities for all parties to transfer arrangements. Views are polarized between countries that are primarily importers, and whose goal may be to see stringent rules applied to technology-supplying firms, and the advanced market economies, which insist that rules have to include the duties and responsibilities of both the recipient governments and the supplying firms.

In terms of the objectives underlying the establishment of control systems, the following can be identified:

1. Coordination. Control systems have, on some occasions, simply meant consolidation of several scattered instruments into a single, coherent one. This is true at all levels, be they national, regional, or international, or even the company level. Thus the OECD guidelines were intended primarily as a manifestation of a regional effort at coordination, harmonization, and ensuing risk reduction. Some national measures have also had harmonization in mind. The actual record of success in achieving such intended harmony, however, remains poor.

2. Gatekeeping. Most control systems have involved screening the transfer of technology in order to permit only desired flows and to prevent undesirable behavior by parties to the transfers. Many LDC national systems have this as the major goal, as do supplier systems, regarding the outflow of technology.

3. Facilitation. Other control systems are intended to promote or facilitate desirable forms of technology transfer, often in order to achieve some development objective. This may be accomplished partly by increased coordination and information, but may in other cases require active searching for opportunities rather than reaction to proposals.

4. Increase in bargaining power. Some control systems are designed primarily to enhance the bargaining power of the proponent. International codes being negotiated both at UNCTAD and at the U.N. Centre on Transnational Corporations, as well as the revisions in the Paris Union currently under deliberation, embody this goal.

5. Increase in information flow. Sometimes the purpose of control systems is to increase the flow of information within the organization itself. Thus a unified and integrated national law on technology transfer would enable all ministries, agencies, and decision makers to know the overall policy and framework of the nation. Similarly, a corporation would develop its own code or standard to provide a unified reference point for all executives in its global network.

6. Public relations. There is clearly a "PR advantage" to be had if one gives the outward appearance of "good corporate citizenship" by taking public policy positions in such visible formats as codes or guidelines. In the case of nation-states, laws or other instruments that enhance the country's image as a hospitable low-risk environment, safe haven, or attractive site clearly have their PR advantages.

Legal Status of the Control Systems

An important question raised with respect to any control instrument is its legal status. Should it be voluntary or mandatory, and what policing mechanisms are needed for its implementation? Basically, the legal status of codes or other instruments is determined by whether "courts and cops" are included in its provisions. The two main variants are binding and nonbinding or voluntary. However, there are other variations around these two basic options.(3)

Summary of Alternatives

Given these three basic dimensions of origin, goals, and legal status, many types of control systems can be envisioned, ranging from reliance on the market to a binding international agreement. Their prospects are discussed in the next section.

PROSPECTS FOR ALTERNATIVE CONTROL SYSTEMS

Some control systems have already been implemented, and their results were evaluated to some extent in the conference papers. Others are now being negotiated in various international forums. The prospects for successful functioning of these different systems are discussed below. The first general type of system is that of the free market, supplemented by self-control. The second is national control, either by recipient or supplier country governments. The third is international control, whether regional or global in scope.

The Market as a Control System

Prompted in part by the frustration that results from having to make sense of the interlocking web of control measures and the layer upon layer of instruments, and in part by the transformations in the international technology market, several authors proposed that the market system, possibly supplemented by voluntary self-control or peer control, may indeed provide the best solution for control of technology transfer and thus may be the best strategy.

Dunning reached this observation through his examination of the changing structure and channels of technology flow and the dis-internalization that is occurring. Hawkins and Gladwin, in examining the myriad U.S. instruments for controlling the outward flow of technology, tended to reach recommendations best summarized by Hawkins as "laissez faire with exceptions."

Rather than suggesting an abandonment of efforts at establishing technology transfer control instruments, these authors were pointing out the pitfalls of overregulation and were at the same time suggesting – even if only implicitly – that the underlying motive for control instruments should be to reduce market imperfections and not to add one more layer to the existing confusion.

Driscoll and Wallender, presenting the "multinational's perspective," suggested that the availability of technology from many suppliers has induced developing countries to rely less on legislated or decreed controls on technology transfer and more on bargaining, in order to acquire technology.

The idea that the market can be as good or better a control system as any was also prompted by Root's examination of the pricing issue. He proposed that the final negotiated price for technology must fall within a "bargaining range" that is determined through market conditions and the relative bargaining positions of the two parties. Root concludes that neither host or home governments can do much directly to influence price. At best, a host country may enhance the bargaining position of its technology-acquiring firms by improving the quality of information available to them.

In his examination of technology transfer channels, Dunning proposed a hypothesis that subsequently became known as "the market solution." Dunning's argument, simply put, is that the more perfect the market the better off everyone – both suppliers and acquirers of technology – will be. In examining the alternative channels and modes of technology transfer (and, more generally, international resource transmission) he observes the reversal of the "internalization" process in recent years. Dunning proposed that the propensity of TNCs to internalize technology transfer within their own organization varies directly with the imperfections of the market, and since the international market for technology is moving toward greater numbers of buyers, sellers, and alternative technologies, the suppliers of technology (primarily the TNCs) are disinternalizing. As a result, some decision-making power is shifting to acquiring firms and host governments. This, in turn, is seen as a gradual move toward an improved market mechanism.

Other commentators and discussants, notably Eschbach, pointed out the difficulty, even futility, of controlling the transfer of technology. They point out that transfer of technology through TNCs or through formal contracts represents only a minor part of technology transfer, most of which takes place through trade and an open system of international communications and education.

Gabriel Wilner and Denis Goulet tended to disagree with the notion of the market as the optimum control instrument. Wilner, a lawyer, disputed the adequacy of the international technology market, saying that it is "a seller's market"; and Goulet, a development philosopher, questioned the normative implications of "the market solution," reminding us of the competing normative images of development and alternative models. George Modelski argued that a free market for transfer of technology may lead to economic and political instability, and that international political systems must be improved in order to deal with these problems. Hence in response to the rhetorical question, "Is market a snare and a dillusion?" one tends to say, "It depends on one's 'worldview' as much as it does on the nature of the market."

National Control Systems

Controls by national governments will probably remain important for the indefinite future, but many participants in the conference were skeptical about governments' prospects for achieving their stated objectives.

Home country control programs are established with many objectives in mind, and these are often in conflict. For example, a policy measure prompted by national security or human rights considerations will reduce the flow of technology and thereby potential foreign exchange earnings. At the overall national level, the question must be asked: Are we willing to accept these trade-offs? Only when the importance of the item to be traded is known can one make a proper judgment. Thus coordinating the interplay between these various components can become exceedingly difficult.

We believe, therefore, that at the national level there is a need for more coordination of all the different laws, regulations, and practices that stem from different objectives, and hence result in inconsistencies and contradictions. An overall national policy on technology transfer should emerge as a natural outcome of a careful examination of all such systems and instruments. Such a comprehensive inventory is a necessary requirement of a technology transfer policy.

Even with a clear purpose and policy, however, the ability of any one supplier to achieve its objectives will be limited, given the existence of alternative suppliers and the difficulties of international coordination. International agreements and codes are necessary for effective implementation of many types of technology transfer controls.

Just as the supplier country must find a balance between the various objectives in devising a technology policy, so must the recipient country. Ozawa pointed out that a technology import control system is useful when for some reason markets either do not work or do not achieve given national objectives.

But recipient nations have a variety of objectives pitted against one another – employment versus foreign exchange; capital intensity versus technological absorptive capacity. Nevertheless, in response to various national development requirements, an increasing number of less industrialized countries are adopting unilateral measures. These host country control instruments are intended either to facilitate or to screen the inflow of technology in line with the country's macro- and microlevel objectives. They can take either a legislative or an administrative approach, as described by Janiszewski and by Mason.

Considerable attention in the conference was devoted to the problem of effective implementation of these recipient country control systems. Ozawa pointed out that such systems work best when the country's technical absorption capacities are well developed, as they were in Japan. In that particular case, the control system was used to supplement, rather than replace, the market. In many other countries implementation is far from the espoused policy, often as a result of the lack of coordination, needed skills, or sufficient political power. Decisions are often made outside officially prescribed channels, and often do not address at all some issues, such as the type of technology actually transferred.

International Control Systems

International control systems, both regional and global in scope, have received much attention, and have been the subject of intense negotiations and considerable controversy. The general conclusion of the conference was that regional agreements, such as those of the Andean Group and the OECD, had not achieved what their sponsors had intended, and that progress on an international agreement seems to have reached a standstill. It is important to review the causes of these problems, and what can be done to resume fruitful progress.

Forces favoring international regulation. Although disagreements exist about the desirability and methods of international regulation of technology transfer, some arguments in favor of regulation are widely accepted and provide a powerful rationale for regulation. These and other arguments, some less widely shared, have created political pressures and bureaucratic momentum in the direction of regulation.

The first widely accepted argument for regulation is that, if properly implemented, it improves the functioning of the international technology market. The market for technology is obviously not one of perfect competition; rather, it is a market with significant monopolistic elements, symbolized by patents and trademarks but not limited to them. Such a market invites regulation to prevent the abuses and distortions that can occur when monopolies exist. Of course, agreement on the need for regulation does not mean agreement on the form; but in any case, the interested parties have a basis for negotiation.

The second argument is that there is a need to preempt uncoordinated unilateral policies on the part of individual nation-states. Since many nations are unilaterally implementing legislation concerning technology transfer, and since such legislation can affect other countries directly or indirectly, the possibilities for beggar-thy-neighbor policies are obvious, both between supplier and recipient nations and among recipient countries. For example, some recipient nations now require that imported technology result in an improvement in the country's balance of payments, either by requiring the recipient enterprise to export or by limiting its foreign remittances of profits or fees.

Other nations insist on the creation of local research laboratories, which results in a transfer of employment between nations. In many aspects of the regulation of technology transfer legal jurisdictions overlap, which leads to the possibility of conflicts. Taxation of technology transfers and the regulation of restrictive business practices are examples of policies where conflicts are likely. Although it is possible that such beggar-thy-neighbor policies of individual governments regarding international technology transfer have not been as obvious as they are in international trade, increased intervention by nations in the international technology market argues for international regulation.

These and other arguments have led to considerable political pressure for some form of international regulation in both recipient and supplier countries. The less industrial countries are particularly united on this issue, seeing the possibility of gaining a better bargaining position with the technology suppliers and international assistance in upgrading their technological capabilities. They view an international code on technology transfer as a key part of their program for a New International Economic Order.

The supplier countries now seem to see that it is in their interest to respond positively to this initiative on the part of the developing countries. The negotiation of an international code could be a symbol of the willingness of the rich countries to negotiate with the poor countries on reasonable proposals. It may, without much sacrifice, reduce the mutual hostility that existed in the early 1970s. Since the TNEs have recently come under attack in the rich countries too, there are added political reasons to support efforts to curb their influence. Opposition to such a code has been muted since TNEs have accepted, in large part, the need for some sort of regulation, preferably at the supranational level.

Although there is wide support for a code, it is not clear that all parties want more than a symbolic document. Such a code could meet many of the political objectives without seriously addressing the issues or impinging on each nation's right to do as it wishes.

Forces inhibiting international regulation. As in any international negotiation, differences in national interests exist. In this case there are strong, visible differences between the interests of technology-supplying nations and technology-receiving nations. Also, there are less visible but still powerful differences within the group of technology-receiving nations and within the group of supplying nations. Such conflicts of national interests, as well as a number of technical difficulties involved in formulating international regulations on technology transfer, are significant obstacles to effective regulations.

Supplier and recipient nations see different purposes for international regulations. Although the technical details of their differences are complex, the basic disagreement is simple. Recipient countries want to reduce the monopolistic element in the technology market (their emphasis on free access and no restrictions on use), and to reduce the cost of imported technology. They also want to limit foreign control of technology in the recipient countries (their emphasis on unbundling the technology package and improving national technological capabilities); to put the burden on the supplier for assuring that the technology is appropriate to the country's needs; and to strengthen the ability of individual nations to regulate technology transfer.

Supplier countries have less ambitious goals, seeking only to prevent unreasonable abuses of monopoly power (rather than elimination of monopolies) and to provide a framework of guidelines to facilitate negotiation of technology transfer between interested parties. Therefore they wish to say little about costs, unbundling, or the appropriateness of technology, leaving these aspects to private negotiations. Supplier countries are not interested in strengthening the ability of recipients to regulate technology transfer; they feel that recipients have that ability already, and would rather restrain national regulations. While progress has been made in the UNCTAD negotiations in narrowing these differences, a wide gap remains and will continue to make the implementation of a code very difficult. In addition to real differences in interests between supplier and recipient countries, perception gaps exist that make communication and consensus difficult. Perlmutter and Sagafi-nejad found that the perceptual gaps (that is, gaps in perceptions of the other party's true intentions or true position) account for much of the difficulty in negotiating an international code.(4)

This theme of "mutual suspicion and mistrust" is also echoed by the Brandt Commission report of 1980. In addressing technology transfer issues, this report observed, "At present, relationships are too often prone to mutual suspicion and mistrust, which inhibit investment and cause conflicts and even political tensions over existing operations."(5)

Differences among recipient nations may be an even more serious obstacle to international regulations. If these nations could agree among themselves on a code, its implementation would have a significant effect on the technology market. But there are such great economic, political, and ideological differences among the recipient countries that it seems almost impossible to reach agreement. Each country enters into negotiations on an international code with a set of expectations and proposed regulations on the role of foreign technology. Differences exist among the countries in terms of the general restrictiveness of these attitudes and laws. The attitudes and laws of a given country depend on:

1. The bargaining strength of the country vis-a-vis the technology supplier. Countries with large, growing markets and stable political situations will be more attractive to suppliers; hence the nation can insist on more restrictions.
2. The perceived economic costs and benefits of foreign technology, which depend on the nation's need for the technology, the capabilities of its national enterprises, and the costs of foreign technology versus other alternatives.
3. The political importance of laws governing foreign technology. Some governments may use the control of foreign technology as a political issue.
4. The administrative capabilities of the government. Although the will to restrict foreign technology may exist, the capabilities to effectively control it may not.

Assuming that a country has arrived at an equilibrium level of restrictiveness in its laws on foreign technology, based on the above considerations, it will not wish to change its laws unless its perceptions change. Negotiations for an international code may serve an educational function and change some of its perceptions of its bargaining strength, the economic costs and benefits of foreign technology, or its administrative capabilities. Hence the country may make its rules more compatible with those of other countries; however, there is no assurance that this will occur.

Remaining differences among nations will influence each country's view of the desirability of various alternative forms of international regulation. A country with very restrictive laws, for example, probably would want a code with restrictive common standards, as this would make the country relatively more attractive to technology suppliers. A country with moderately restrictive laws would not wish to see a code that would force it to change to very restrictive rules, but it might want countries with looser standards to tighten their codes and, therefore, it might argue for common minimum standards. A country with loose rules probably would argue for flexible standards, which would allow it to achieve a better relative bargaining position by maintaining loose rules while other countries tightened theirs.

With such great differences among countries, it seems unlikely that a common set of rules could be derived and agreed upon; if a set of rules is nevertheless developed, it would be hard to ensure that they would be implemented effectively. It seems clear that unless countries are committed wholeheartedly to effective implementation, the technical difficulties will stymie efforts to make it work. So many problems of interpretation and regulation of a code exist that it would be easy for a country to adhere to the letter of a code while completely violating it in spirit. (Note that in the UNCTAD codes implementation is left pretty much completely to national authorities.) Implementation would require so much effort on the part of countries that they must have a strong incentive before they fulfill the obligations imposed by a code. So far the developing nations have been united in their efforts to write a code, but their efforts primarily have been in the form of demands on supplying countries. When implementation requires specific action on their part, the cohesiveness of the recipient countries is likely to disappear.

Disagreements also exist within the camp of the technology suppliers. Some industrialized countries have good relations with the Third World and hence are more willing to be conciliatory in negotiations on a technology transfer code. There are also differences among suppliers in their perceptions of the desirability of government intervention per se; some even have a much more dirigiste tradition than others. Finally, the industrial strengths of supplying countries vary greatly, and these differences color perceptions of what kinds of regulations are desirable. Countries with industries that are predominately high-technology in nature will have vastly different concerns than those whose strength lies in more mature industries. All differences among supplying countries inhibit the negotiation and effective implementation of an international technology transfer code.

Overall prospects. Although multilateral instruments for technology transfer are extremely hard to formulate, the alternatives are not better. Without a multilateral understanding, we may witness further proliferation of uncoordinated and even contradictory unilateral national laws and the ensuing confusion on the part of TNEs, and beggar-thy-neighbor policies and rivalries among the LICs. We also may witness direct conflicts with existing international agreements such as GATT, Paris Union, and ICSID, or with the national laws of individual countries. Indeed, the proponents of regulation argue that technology transfer is the only aspect of international trade that has not been brought under some form of order and control.

POLICY CONCLUSIONS

In light of the above discussion, we reach a number of conclusions regarding needed policies on the part of both technology-supplier countries and technology-recipient countries. These are outlined below.

Conclusions For Technology Supplier Countries

The technology suppliers must recognize that recipient nations have legitimate demands in seeking to regulate international technology transfer. The supplier countries regulate technology transfer themselves, and must expect similar actions by others. U.S. policy, for example, has been based on the fact that there are exceptions to the overall policy of promoting the perfection of the market systems. These have to do with basic strategic and national security considerations. As a corollary to the above, U.S. policy must acknowledge and tolerate unilateral measures other nations may take. Such controls (a) are aimed at promoting other nations' national development objectives, and (b) emanate from the basic sociocultural or philosophical premises of the society initiating them. A prime example is LDC policies and attitudes toward the use of the patent system as a control system. As pointed out by Timberg, the motives behind the adoption of patent systems in the less industrial countries today is different from what it was in advanced industrial countries when they were first promulgated. Whereas the initial intent in granting patents as a legal privilege was to encourage innovation and entrepreneurship, patents are used in the LICs today as a means of government control and state-led innovation.

With regard to controls on the outward flow of technology, technology suppliers must recognize the changes occurring in the international technology market. Trends toward less imperfection are manifested in the increased variety of the modes and channels of transfer, the increase in competition among suppliers from various national sources, and the increasing sophistication of technology buyers. National control systems, therefore, may end up making technology suppliers less competitive in the international market if these changes in the international technology market are not taken into account. In other words, excessive limits on the outflow of technology in the name of protecting jobs or other protectionist intentions is less likely to be effective as technology buyers become more educated and as their sources and choices expand.

Technology suppliers must show a genuine and demonstrable desire to incorporate into their technology transfer policies measures that could assist the LICs in their developmental objectives. This includes initiation, promotion, or support of interinstitutional linkages for information exchange and for the harmonization of policies and legislation on foreign investment, taxes, patents, and competition. It also includes providing assistance in upgrading R & D capabilities and training. It was pointed out by Driscoll and Wallender that the TNCs are becoming increasingly sensitive to the issue of strengthening the technological capabilities of host countries and are thus taking additional steps in this direction now. Such trends should be encouraged and promoted.

Finally, supplier countries must emphasize the imperatives of multilaterality and shared responsibility. Under ideal conditions, the market mechanism would be the most appropriate and efficient technology transfer control system, as Root, Dunning, Hawkins, Stewart and several others have pointed out. This line of argument would suggest that a laissez faire international technological order might be the "best strategy."

The fact remains, however, that many countries, including the United States, find it necessary, for one reason or another, to have some control systems for technology transfer. It was demonstrated repeatedly that unilateral measures at the individual national level are either ineffectual or have unintended and undesirable secondary consequences. It is therefore important that multinational measures prevail as a "second-best" strategy.

Conclusions for Technology Recipient Countries

It is important for recipient countries to recognize that overly restrictive unilateral measures will most likely have an adverse effect, discouraging the flow of technology and hampering the growth of their own technological capabilities. Thus as Stewart pointed out, certain tax measures aimed at reducing royalties may either discourage entry altogether or encourage a company to look for alternative forms of compensation. And as Root and Dunning pointed out, firms are seeking to maximize the total compensation package. They will succeed in doing so to the extent that the market is imperfect. Improving the overall competitiveness of the market, and tolerating multilateral controls where such controls are also aimed at removing market imperfections, appears to be the optimal strategy.

The protection of the proprietary rights of the private suppliers of technology is a sensitive issue with both the supplying and receiving parties. To diffuse this issue and to alleviate the concern of both TNCs and host countries, it must be pointed out that (1) technology transfer contracts are often placed on public record and therefore their contents (that is, their technology) are not amenable to absolute protection. And (2) much of the technology that the LICs are seeking is unpatented proprietary technology dealing with company know-how, and is transferred only if there is sufficient incentive for the company to do so. The consequences to the potential host country of not realizing these facts will be to receive less technology.

Areas for Further Research

Several fruitful issues and questions have been raised for which more answers are needed. Thus further research should focus on the following areas.

1. A comprehensive inventory of various control systems at all levels, assisted by parameters and dimensions discussed in this chapter, is highly desirable.

2. There is a need for the development of a methodology for assessing these instruments and systems, so as to check their performance vis-a-vis the objectives for which they have been devised. Case studies should then be undertaken with the aim of testing the methodologies for performance evaluation.

3. Added emphasis should be placed on both the theoretical and empirical fronts in measuring the technological absorptive capacity of recipient nations.

4. Deeper analysis of the natural causes of "perception gaps" would enhance our understanding of how best to promote international consensus and how to break out of deadlocks.

5. Finally, ongoing monitoring of the international market for technology, in light of the propositions put forth in this chapter regarding the use of the market as a control instrument, is clearly needed. Studying the extent to which market imperfections actuall exist would aid in developing policies to remedy these imperfections. On the other hand, if such monitoring shows a reduction in imperfections in a given technology market, such enhanced competition may render elaborate technology transfer control systems redundant.

NOTES

(1) For a comprehensive review of attempts at regulating the flow of technology and/or the activities of the transnational corporations, see Howard Perlmutter and Tagi Sagafi-nejad, International Technology Transfer: Codes, Guidelines and a Muffled Quadrilogue, Book #1 in the present Technology Transfer Trilogy (New York: Pergamon Press, 1981). The following four paragraphs draw on the above review.

(2) For the November 1979 text, see UNCTAD, "Draft International Code of Conduct on the Transfer of Technology as of 16 November 1979" (TC/CODE TOT/20. For the first compromise draft, produced in July of 1978, see TD/CODE TOT/1.

(3) A 1979 UNCTAD study contains seven options or variants at the international level alone, and provides legal precedent for each. See Richard Moxon and Tagi Sagafi-nejad, "Technology Transfer and the Transnational Enterprise: The Outlook for Regulation," Journal of Contemporary Business, Autumn 1977, p. 163, for a tabular summary.

(4) Perlmutter and Sagafi-nejad, International Technology Transfer.

(5) Independent Commission on International Development Issues (Brandt Commission), "Report to the UN Secretary-General," February 12, 1980, p. 6. Emphasis added. See also the full report, North-South: A Program for Survival (Cambridge: M.I.T. Press, 1980), chap. 12.

Author Index

Subject Index

About the Editors

TAGI SAGAFI-NEJAD is an international political economist specializing in political and economic development, the Middle East, technology transfer, and the TNCs. He has taught courses on international technology transfer, international business, the Third World, and the Middle East at the Wharton School of the University of Pennsylvania and the University of Washington. He was the coordinator and principle investigator of the International Conference on Technology Transfer Control Systems (Philadelphia, February 1979, Seattle, April 1979). Dr. Sagafi-nejad has published articles on the Middle East, the New International Economic Order, Technology Transfer, and Transnational Corporations. He has coordinated the Technology Transfer Trilogy, of which the present book is a part. Dr. Sagafi-nejad is currently a visiting faculty member at the School of Business and the Center for Middle Eastern Studies at the University of Texas at Austin.

RICHARD W. MOXON is Associate Professor of International Business at the University of Washington. He has done research in Latin America, Europe, and Asia and has published on foreign direct investment, technology transfer, offshore assembly, and industrial policy. Dr. Moxon's publications include Offshore Production in Less Developed Countries (1974) and "Effects of Offshore and Onshore Foreign Direct Investment in Electronics" (1980). During 1980-81, Dr. Moxon was a visiting professor at INSEAD, France.

HOWARD V. PERLMUTTER is Professor of Social Architecture and Director of the Worldwide Institutions Research Center, the Wharton School, University of Pennsylvania. Dr. Perlmutter has written numerous articles on the Transnational Corporations, International Business, Group Dynamics, and International Communication. He has been a visiting faculty member at the Stockholm School of Economics, Centre d'Etudes Industrielles, IMEDE, and Yale University, and has been advisor to various international organizations, TNCs, and governments. He is the author of The Theory and Practice of Social Architecture; co-author of Multinational Organization Development: A Social Architectural Perspective; and Mobilizing the Private Sector for Primary Health Care Delivery Systems in Developing Countries, with Dr. Foulie Psalidas-Perlmutter.

Contributors

Alvin Adams – U.S. Department of State, Washington, D.C.
Enrique Aguilar – United Nations Industrial Development Organization, Vienna
Michico Ariga – Chuo University and Research Center for Foreign Laws on Antitrust and Industrial Property, Tokyo
Igor E. Artemiev – Soviet Academy of Sciences, Moscow
Robert E. Driscoll – Fund for Multinational Management Education, New York
John H. Dunning – University of Reading, Reading, England
Eugene H. Eschbach – Battelle Memorial Research Center, Richland, Washington
A.A. Fatouros -- Indiana University Bloomington and Aristotelian University of Thessaloniki, Greece
Thomas N. Gladwin – New York University, New York
Denis Goulet – Overseas Development Council, Washington, and Notre Dame University, Notre Dame
Donald L. Guertin – Exxon Corporation, New York
Robert G. Hawkins – New York University, New York
Farouk I. Heiba – University of Utah, Salt Lake City, and Atlanta University, Atlanta
Thomas D. Hughes – University of Pennsylvania, Philadelphia
Hubert A. Janiszewski – United Nations Industrial Development Organization, Vienna
Endel-Jakob Kolde – University of Washington, Seattle
Herbert S. Levine – University of Pennsylvania, Philadelphia, and Stanford Research Institute, Washington, D.C.
R. Hal Mason – University of California, Los Angeles
Joseph McCann – Oregon State University, Corvalis
Bernard Mennis – Temple University, Philadelphia

George Modelski – University of Washington, Seattle
Richard W. Moxon – University of Washington, Seattle
Terutomo Ozawa – Colorado State University, Fort Collins
Howard V. Perlmutter – University of Pennsylvania, Philadelphia
Gustav Ranis – Yale University, New Haven
Thomas Roehl – University of Washington, Seattle
Franklin R. Root – University of Pennsylvania, Philadelphia
Tagi Sagafi-nejad – University of Washington, Seattle, and University of Texas, Austin
Louis B. Schwartz – University of Pennsylvania, Philadelphia
Stanislav J. Soltysinski – University of Poznan, Poland
Frances Stewart – Oxford University, Oxford, England
Sigmund Timberg – Lawyer, Washington, D.C.
Harvey W. Wallender III – Council of the Americas, New York
Louis T. Wells, Jr. – Harvard University, Boston
Gabriel Wilner – University of Georgia, Athens, and United Nations Conference on Trade and Development, Geneva
Miguel S. Wionczek – El Colegio de Mexico, Mexico City

*The views expressed by the contributors are their own and do not necessarily represent those of the organization with which they were affiliated at the time. The names of the organizations are for identification only.